REVIEW OF THE RESEARCH PROGRAM OF THE **FreedomCAR** AND **Fuel Partnership**

FIRST REPORT

Committee on Review of the FreedomCAR and
Fuel Research Program, Phase 1

Board on Energy and Environmental Systems
Division on Engineering and Physical Sciences
Transportation Research Board

NATIONAL RESEARCH COUNCIL
OF THE NATIONAL ACADEMIES

THE NATIONAL ACADEMIES PRESS
Washington, D.C.
www.nap.edu

THE NATIONAL ACADEMIES PRESS • 500 Fifth Street, N.W. • Washington, DC 20001

NOTICE: The project that is the subject of this report was approved by the Governing Board of the National Research Council, whose members are drawn from the councils of the National Academy of Sciences, the National Academy of Engineering, and the Institute of Medicine. The members of the committee responsible for the report were chosen for their special competences and with regard for appropriate balance.

This report and the study on which it is based were supported by Contract No. DE-AT01-04IG01153, Task Order No. 2, between the U.S. Department of Energy and the National Academy of Sciences. Any opinions, findings, conclusions, or recommendations expressed in this publication are those of the author(s) and do not necessarily reflect the views of the organizations or agencies that provided support for the project.

International Standard Book Number: 0-309-09730-4 (Book)
International Standard Book Number: 0-309-65277-4 (PDF)

Available in limited supply from:
Board on Energy and Environmental Systems
National Research Council
500 Fifth Street, N.W.
Keck W934
Washington, DC 20001
202-334-3344

Additional copies are available for sale from:
The National Academies Press
500 Fifth St., N.W.
Lockbox 285
Washington, DC 20055
800-624-6242 or 202-334-3313 (in the Washington metropolitan area)
http://www.nap.edu

Printed in the United States of America

THE NATIONAL ACADEMIES

Advisers to the Nation on Science, Engineering, and Medicine

The **National Academy of Sciences** is a private, nonprofit, self-perpetuating society of distinguished scholars engaged in scientific and engineering research, dedicated to the furtherance of science and technology and to their use for the general welfare. Upon the authority of the charter granted to it by the Congress in 1863, the Academy has a mandate that requires it to advise the federal government on scientific and technical matters. Dr. Ralph J. Cicerone is president of the National Academy of Sciences.

The **National Academy of Engineering** was established in 1964, under the charter of the National Academy of Sciences, as a parallel organization of outstanding engineers. It is autonomous in its administration and in the selection of its members, sharing with the National Academy of Sciences the responsibility for advising the federal government. The National Academy of Engineering also sponsors engineering programs aimed at meeting national needs, encourages education and research, and recognizes the superior achievements of engineers. Dr. William A. Wulf is president of the National Academy of Engineering.

The **Institute of Medicine** was established in 1970 by the National Academy of Sciences to secure the services of eminent members of appropriate professions in the examination of policy matters pertaining to the health of the public. The Institute acts under the responsibility given to the National Academy of Sciences by its congressional charter to be an adviser to the federal government and, upon its own initiative, to identify issues of medical care, research, and education. Dr. Harvey V. Fineberg is president of the Institute of Medicine.

The **National Research Council** was organized by the National Academy of Sciences in 1916 to associate the broad community of science and technology with the Academy's purposes of furthering knowledge and advising the federal government. Functioning in accordance with general policies determined by the Academy, the Council has become the principal operating agency of both the National Academy of Sciences and the National Academy of Engineering in providing services to the government, the public, and the scientific and engineering communities. The Council is administered jointly by both Academies and the Institute of Medicine. Dr. Ralph J. Cicerone and Dr. William A. Wulf are chairman and vice chairman, respectively, of the National Research Council.

www.national-academies.org

COMMITTEE ON REVIEW OF THE FREEDOMCAR AND FUEL RESEARCH PROGRAM, PHASE 1

CRAIG MARKS, *Chair*, NAE,[1] Altarum, Bloomfield Hills, Michigan
PETER BEARDMORE, NAE, Ford Motor Company (retired), West Bloomfield, Michigan
DAVID L. BODDE, Clemson University, South Carolina
GLENN A. EISMAN, Rensselaer Polytechnic Institute, Troy, New York
DAVID E. FOSTER, University of Wisconsin, Madison
JOHN B. HEYWOOD, NAE, Massachusetts Institute of Technology, Cambridge
JOHN G. KASSAKIAN, NAE, Massachusetts Institute of Technology, Cambridge
HAROLD H. KUNG, Northwestern University, Evanston, Illinois
JAMES J. MACKENZIE, World Resources Institute, Washington, D.C.
CHRISTOPHER L. MAGEE, NAE, Massachusetts Institute of Technology, Cambridge
MICHAEL P. RAMAGE, NAE, ExxonMobil Research and Engineering Company (retired), Moorestown, New Jersey
VERNON P. ROAN, University of Florida (professor emeritus), Gainesville
BERNARD ROBERTSON, NAE, DaimlerChrysler Corporation (retired), Bloomfield Hills, Michigan
R. RHOADS STEPHENSON, NASA Jet Propulsion Laboratory (retired), Consultant, La Canada, California
KATHLEEN C. TAYLOR, NAE, General Motors Corporation (retired), Birmingham, Michigan
BRIJESH VYAS, Bell Laboratories-Lucent Technologies, Murray Hill, New Jersey

Subgroup on Systems Analysis and Simulation

BERNARD ROBERTSON, *Chair*
PETER BEARDMORE
DAVID BODDE
DAVID FOSTER
JOHN HEYWOOD
CHRISTOPHER MAGEE
VERNON ROAN

[1]NAE, National Academy of Engineering.

Subgroup on Advanced Combustion Engines, Emissions Control, and Hydrocarbon Fuels

DAVID FOSTER, *Chair*
JOHN HEYWOOD
HAROLD H. KUNG
MICHAEL RAMAGE
BERNARD ROBERTSON
KATHLEEN TAYLOR

Subgroup on Electrochemical Energy Storage

BRIJESH VYAS, *Chair*
JOHN KASSAKIAN
CHRISTOPHER MAGEE

Subgroup on Fuel Cells

GLENN EISMAN, *Chair*
VERNON P. ROAN
KATHLEEN TAYLOR
BRIJESH VYAS

Subgroup on Electric Propulsion, Electrical Systems, and Power Electronics

JOHN KASSAKIAN, *Chair*
BERNARD ROBERTSON

Subgroup on Hydrogen Production, Delivery, and Storage

KATHLEEN TAYLOR, *Chair*
DAVID L. BODDE
JAMES MACKENZIE
CHRISTOPHER MAGEE
MICHAEL RAMAGE
RHOADS STEPHENSON

Subgroup on Safety

RHOADS STEPHENSON, *Chair*
JOHN HEYWOOD
HAROLD KUNG
CHRISTOPHER MAGEE

Subgroup on Materials

PETER BEARDMORE, *Chair*
GLENN EISMAN
CHRISTOPHER MAGEE
KATHLEEN TAYLOR

Project Staff

JAMES ZUCCHETTO, Director, Board on Energy and Environmental Systems (BEES)
JILL WILSON, Senior Program Officer, Transportation Research Board
PANOLA GOLSON, Program Associate (BEES)

BOARD ON ENERGY AND ENVIRONMENTAL SYSTEMS

DOUGLAS M. CHAPIN, *Chair*, NAE,[1] MPR Associates, Inc., Alexandria, Virginia
ROBERT W. FRI, *Vice Chair*, Resources for the Future (senior fellow emeritus), Washington, D.C.
ALLEN J. BARD, NAS,[2] University of Texas, Austin
DAVID L. BODDE, Clemson University, South Carolina
PHILIP R. CLARK, NAE, GPU Nuclear Corporation (retired), Boonton, New Jersey
E. LINN DRAPER, JR., NAE, American Electric Power, Inc. (emeritus), Austin, Texas
CHARLES GOODMAN, Southern Company, Birmingham, Alabama
DAVID G. HAWKINS, Natural Resources Defense Council, Washington, D.C.
MARTHA A. KREBS, California Energy Commission, Sacramento
GERALD L. KULCINSKI, NAE, University of Wisconsin, Madison
DAVID K. OWENS, Edison Electric Institute, Washington, D.C.
WILLIAM F. POWERS, NAE, Ford Motor Company (retired), Ann Arbor, Michigan
TONY PROPHET, Carrier Corporation, Farmington, Connecticut
MICHAEL P. RAMAGE, NAE, ExxonMobil Research & Engineering Company (retired), Moorestown, New Jersey
EDWARD S. RUBIN, Carnegie Mellon University, Pittsburgh, Pennsylvania
MAXINE SAVITZ, NAE, Honeywell, Inc. (retired), Los Angeles, California
PHILIP R. SHARP, Harvard University, Cambridge, Massachusetts
SCOTT W. TINKER, University of Texas, Austin

Staff

JAMES ZUCCHETTO, Director
ALAN CRANE, Senior Program Officer
MARTIN OFFUTT, Senior Program Officer
DANA CAINES, Financial Associate
PANOLA GOLSON, Project Assistant

[1]NAE, National Academy of Engineering.
[2]NAS, National Academy of Sciences.

Acknowledgments

The committee wishes to thank members of the FreedomCAR and Fuel Partnership, all of whom contributed a significant amount of their time and effort to this National Research Council (NRC) study by giving presentations at meetings, responding to requests for information, or providing valuable information. In addition, the chair wishes to recognize the committee members and the National Academies staff for their hard work in organizing and planning committee meetings and their individual efforts in gathering information and writing sections of the report.

This report has been reviewed in draft form by individuals chosen for their diverse perspectives and technical expertise, in accordance with procedures approved by the NRC's Report Review Committee. The purpose of this independent review is to provide candid and critical comments that will assist the institution in making its published report as sound as possible and to ensure that the report meets institutional standards for objectivity, evidence, and responsiveness to the study charge. The review comments and draft manuscript remain confidential to protect the integrity of the deliberative process. We wish to thank the following individuals for their review of this report:

William G. Agnew (NAE), General Motors Corporation (retired),
Paul Blumberg, Independent Consultant,
Andrew Brown, Jr. (NAE), Delphi Corporation,
Robert Epperly, Epperly Associates,
Norman Gjostein (NAE), Ford Motor Company (retired),
James G. Hansel, Air Products and Chemicals, Inc.,

Contents

Tables and Figures

TABLES

FIGURES

Executive Summary

THE FREEDOMCAR AND FUEL PARTNERSHIP

This is the first report of the Committee on Review of the FreedomCAR and Fuel Research Program, Phase I, formed in the fall of 2004 by the National Research Council (NRC). This FreedomCAR and Fuel Partnership is a collaboration among the U.S. government—in particular, the U.S. Department of Energy (DOE)—the U.S. Council for Automotive Research (USCAR), whose members are DaimlerChrysler Corporation, Ford Motor Company, and General Motors Corporation), and five major energy companies: BP America, Chevron Corporation, ConocoPhillips, ExxonMobil Corporation, and Shell Hydrogen (U.S.). At DOE, the program is managed through the Office of Energy Efficiency and Renewable Energy (EERE). This is a broad, very challenging research effort to assist in the development of high-risk technologies that will enable the vision of "a clean and sustainable transportation energy future" (DOE, 2004). To achieve that future, the program envisions a transition pathway involving more efficient internal combustion engines (ICEs), followed by increasing use of advanced ICE hybrid electric vehicles and then, by 2015, enablement of the private sector to make a decision about the commercialization of fuel-cell-powered personal transportation vehicles that run on economically competitive hydrogen produced from a variety of energy sources. Research goals have been established for 2010 and 2015 that, if attained, promise to overcome the multiple high-risk barriers to achieving that vision.

A major strength of the FreedomCAR and Fuel Partnership is that, like its predecessor, the Partnership for a New Generation of Vehicles (PNGV) program, it is organized around joint industry/government research teams. This structure brings the capabilities of the nation's federal laboratories and other research institutions to bear on overcoming the problems, identified by industry, that are critical to achieving the program vision. This kind of cooperation is a very effective way to develop technologies that will satisfy all of the requirements for the deployment of radically new systems in the marketplace on a large scale. However, unlike the PNGV program, which aimed at the development of concept and preproduction prototype automobiles, the FreedomCAR and Fuel Partnership addresses the development of advanced technologies for all light-duty passenger vehicles—for example, cars, sport utility vehicles, pickups, and minivans. Another strength of the new partnership is that it includes fuel production and infrastructure technologies and that it includes five energy companies, adding essential knowledge about fuels to the program.

The funding in FY05 for DOE programs falling under the purview of the FreedomCAR and Fuel Partnership is about $310 million and covers basic research, applied research, development, learning demonstrations, and deployment (including education that supports technology transfer and adoption). The complexity of the FreedomCAR and Fuel Partnership is evident from the broad scope of the technical areas addressed:

- Internal combustion engines (both petroleum- and hydrogen-fueled),
- Fuel cell power systems,
- Fuel cells,
- Hydrogen storage systems,
- Energy storage systems for hybrid vehicles,
- Electric propulsion systems,
- Hydrogen production and delivery systems, and
- Materials for lightweight vehicles.

There are 11 technical teams consisting of individuals from the national laboratories, the private sector, and the federal government:

- Advanced combustion and emissions control,
- Fuel cell systems,
- Onboard hydrogen storage,
- Electrochemical storage,
- Electrical and electronics,
- Materials,
- Hydrogen production,
- Hydrogen delivery,
- Fuel/vehicle pathway integration,
- Codes and standards, and
- Systems engineering and analysis.

DOE is the lead government agency in the Partnership, and a number of its offices are involved. EERE has primary responsibility for the program through its

FreedomCAR and Vehicle Technologies (FCVT) program and its Hydrogen, Fuel Cells, and Infrastructure Technologies (HFCIT) program. In addition, research and development (R&D) on hydrogen production from coal and nuclear energy is carried out in DOE's Office of Fossil Energy (FE) and its Office of Nuclear Energy, Science and Technology (NE). The Office of Science's Basic Energy Sciences (BES) program is focused on fundamental work in such areas as hydrogen production, hydrogen storage, and catalysts. The U.S. Department of Transportation also participates in safety-related work.

FOCUS OF THE COMMITTEE'S REPORT

An earlier NRC report, *The Hydrogen Economy*: *Opportunities, Costs, Barriers and R&D Need*s (NRC/NAE, 2004), addressed many of the R&D activities associated with the hydrogen parts of the program, such as hydrogen production, distribution, dispensing, and storage, as well as a transition strategy for making hydrogen more widely available. That report provides an excellent review of the challenges and potential benefits of using hydrogen as a transportation fuel and offers recommendations for the DOE R&D program. The current committee used the results of *The Hydrogen Economy* report and referred to its recommendations. The current report presents the committee's evaluation of DOE-sponsored research efforts directed at the goal of a hydrogen economy under the FreedomCAR and Fuel Partnership and offers comments and suggestions on the technical directions, strategies, funding, and management of the Partnership. Because *The Hydrogen Economy* report had just been published as the current committee was being constituted, with regard to the hydrogen technology parts of the Partnership, the committee reviewed just the plans of the three new hydrogen-fuel-related technical teams (hydrogen production; hydrogen delivery; fuel/vehicle pathway integration).

The primary charge to the committee was as follows:

- Review the challenging high-level technical goals and timetables for government and industry R&D efforts in the various technical areas being addressed by the Partnership.
- Review and evaluate progress and program directions since the inception of the Partnership towards meeting the Partnership's 2010 technical goals, and examine ongoing research activities and their relevance to meeting the goals of the Partnership.
- Examine and comment on the overall balance and adequacy of the FreedomCAR and Fuel research effort, and the rate of progress, in light of the technical objectives and schedules for each of the major technology areas.
- Examine and comment, as necessary, on the appropriate role for federal involvement in the various technical areas under development.

- Examine and comment on the Partnership's strategy for accomplishing its goals.

This Executive Summary presents only the main conclusions and recommendations of the committee's report. The body of the report contains additional observations, findings, and recommendations on specific aspects of the program. The rest of the Executive Summary presents the technical areas discussed in Chapters 3 and 4 and briefly addresses crosscutting issues.

AN EXTREMELY CHALLENGING PROGRAM

The FreedomCAR and Fuel Partnership is an extremely challenging program, whose ultimate vision involves a fundamental transformation of automotive technologies and the supporting fuel infrastructure. Many technical barriers exist and need to be overcome to achieve this vision, and fundamental invention is probably needed to meet the program's technical performance and cost targets. Even if the technical targets are met, transitioning from the current fuel infrastructure based on gasoline and diesel fuel to one based on hydrogen derived from a variety of sources will be a formidable social and economic challenge.

The committee believes that research in support of this vision is justified by the potentially enormous beneficial impact for the nation. At this early stage, no insurmountable barriers to achievement of this vision have been identified but several critical components of the program have been noted. Specific, quantitative 2010 and 2015 technology and cost goals have been established by the technical teams. These goals bear on each important element of the program, and the current status of the program relative to these goals is discussed in the body of this report. In view of the large number of unknowns and the need for breakthroughs, the committee does not feel that it is appropriate or useful at this time to speculate on the probability of this program achieving its long-term vision according to its current plan. Funding levels and the consequent research results during the next few years should allow future reviews to make a more firmly based assessment.

TECHNICAL AREAS

Advanced Combustion Engines and Emission Controls

Conclusion. The various types of ICEs will play a critical transitional role in achieving the FreedomCAR and Fuel Partnership's long-term goal. Even assuming the successful eventual transition to hydrogen as a primary transportation fuel, the ICE will be the automotive power plant that consumes most of the fuel in the fleet for several decades during the transition. Reducing the fuel consumption and emissions of ICEs is, therefore, critically important. Novel emission reduc-

comparable prospects for success. The committee suggests that high-energy batteries for pure battery electric vehicles might be such an alternative. The development of high-energy batteries would also increase the efficiency of advanced hybrid electric vehicles and fuel cell vehicles and accelerate the deployment of plug-in hybrid vehicles.

Recommendation. Searching for breakthrough technology in the area of high-energy batteries for electric vehicles should be a high priority of the program.

Electrical Systems and Electronics

Conclusion. The multiple systems in a fuel cell hybrid electric vehicle require both control and coordination. These functions will be provided by electronics, both for power and signal needs. The integrating role of a vehicle's electrical system makes it a critical-path technology, both functionally and economically. Thermal performance and cost are major challenges for both the propulsion and power electronics systems. Closer coordination of activities in this area is essential to meet the milestones of the program.

Recommendation. The electrical and electronics technical team should develop a process for coordinating the diverse activities it is overseeing. Integrating the electronics with the motor may well provide significant cost advantages. The team should consider such potential benefits and develop aggressive targets for an integrated system by 2010 and 2015. In addition, it should become aware of and leverage the high-temperature semiconductor, packaging, and thermal management work being funded by government agencies at universities, commercial organizations, and the national laboratories.

Hydrogen Fuel Production and Distribution

Conclusion. The committee compliments DOE on rapidly implementing most of the recommendations from *The Hydrogen Economy* and encourages program managers to ensure that sufficient efforts go into developing technologies and resolving issues for the transition period. Since the ultimate goal of a widespread hydrogen-fueled transportation system requires a massive infrastructure change, attention to the transition period and to how such change might be systematically achieved is critical. Systems analysis is an important tool for helping to understand and accelerate this transition.

Recommendation. DOE should pay special attention to the transition from the current ICE fuels infrastructure to a nascent hydrogen economy. As part of this attention, DOE should further focus the achievements of the fuel/vehicle pathway integration technical team by placing greater emphasis on the transition to hydrogen in its systems analysis work and should apply its systems capabilities to analyzing whether the cost goals for hydrogen production, established for a mature hydrogen economy, are appropriate for the transition. Specifically, this analysis should examine whether setting a hydrogen cost goal during the transition that is higher than the cost goal for a mature hydrogen economy would speed or impede the introduction of fuel-cell-powered vehicles.

Conclusion. Providing hydrogen for fuel-cell-powered vehicles during the transition period will initially require many refueling locations for a relatively small number of vehicles. This would probably be best accomplished by generating hydrogen at or near these locations rather than at large central hydrogen production facilities.

Recommendation. The committee believes that significant development efforts should be directed to distributed hydrogen production, including natural gas reforming and electrolysis as well as exploratory work on other distributed generation options.

Conclusion. Successfully dealing with the need for carbon sequestration is critically important to making coal and natural gas acceptable energy sources in a carbon-constrained world. Research in this area should be an integral part of the program.

Recommendation. DOE should create a carbon capture and storage (CCS) system subteam (under the hydrogen production team) in the FreedomCAR and Fuel Partnership and make it part of the overall Hydrogen Fuel Initiative.

Materials

Conclusion. Vehicle programs designed to achieve major fuel economy improvements must incorporate significant weight savings. The widespread application of lightweight materials and innovative manufacturing processes is necessary to attain this goal. FreedomCAR has set a vehicle weight reduction target of 50 percent, with the additional criterion "affordable cost." Affordability is the main barrier to meeting the 50 percent goal, and it is unlikely to be achieved within the time frame of this program. The alternative is to relax the weight reduction goal or allow a cost penalty or some combination of the two. The fundamental issue with carbon-fiber-reinforced polymers is the development of low-cost carbon fibers.

Recommendation. More extensive research on carbon-fiber-reinforced polymers and direct cooperation with the major fiber manufacturers appear necessary for any hope of success within the program time frame. Meanwhile, R&D on manufacturing of vehicle structures should continue.

Conclusion. Overall, although cost reduction is the most important need in many structural materials programs, the committee believes that research activities, with a few exceptions, will do little to achieve this goal.

Recommendation. DOE should review its expenditures on materials research to see if some of them should be applied instead to potentially more fruitful areas of research, such as hydrogen storage materials, batteries, fuel cells, and infrastructure.

CROSSCUTTING ISSUES

Safety

Conclusion. The transition to using hydrogen as a primary transportation fuel raises a multitude of safety questions that must be dealt with by many participants during each phase of the program. The committee believes that this is an extremely important subject that deserves continuing high-level attention and additional funding. Both real and perceived safety concerns exist, and all of them must be proactively and effectively dealt with to ensure the success of the program. The critical need to develop safety-related technology, codes, and standards (including vehicle standards) and inculcate widespread safety awareness before hydrogen vehicles can be widely introduced justifies a focused effort in this area. This effort should include the wide dissemination of DOE, National Highway Traffic Safety Administration (NHTSA) and USCAR reports and peer-reviewed papers on hydrogen safety issues.

Recommendation. DOE should form a new, crosscutting safety technical team with a mission that includes broad hydrogen-related safety issues, not only for the Office of Hydrogen, Fuel Cells and Infrastructure Technology (HFCIT), but for the other DOE offices as well. This new team should incorporate the existing codes and standards technical team as a subteam. Both DOE and NHTSA need enough resources to carry out their assigned safety roles.

Public Concerns

Conclusion. In addition to the very demanding technical challenges, some issues surrounding societal acceptance may be pivotal and ultimately determine the feasibility of creating a fleet of hydrogen-fueled vehicles and a supporting infrastructure. The present review of the FreedomCAR and Fuel Partnership focuses on the technical challenges. However, the committee considers it important to recognize that implementing the current program vision will require not only substantial technical breakthroughs but also successful efforts to address public concerns about the widespread use of hydrogen as a transportation fuel. The learning demonstrations program, just getting under way, should be a big step in this direction. The committee also notes that there is a need to understand the potential long-term ecological and environmental effects of the change to a hydrogen-fueled economy.

Recommendation. DOE, in collaboration with the Environmental Protection Agency, should systematically identify and examine possible long-term ecological and environmental effects of large-scale hydrogen use and production from various energy sources.

Importance of Systems Analysis

Conclusion. The FreedomCAR and Fuel Partnership has made an excellent start on developing significant systems analysis capability and has been particularly responsive to the relevant recommendations of the report *The Hydrogen Economy*.

Recommendation. The FreedomCAR and Fuel Partnership should use its systems analysis capability routinely in the program management process, establishing goals, evaluating trade-offs, setting priorities, and making go/no-go decisions.

Conclusion. To date the systems analysis technical team has focused on the development and refinement of systems analysis tools to predict vehicle and component characteristics. The committee believes that there is need for more than this. The complex nature of the program makes it critical to develop and use a robust, overall well-to-wheels systems analysis that will enable informed trade-off decisions throughout the program.

Recommendation. An ongoing, integrated well-to-wheels assessment should be made of the Partnership's progress toward its overall objectives of reducing the nation's oil dependence and introducing hydrogen as a transportation fuel, if appropriate. This assessment should examine possible trade-offs between the individual goals of the fuel program and the vehicle program, as well as between short-term goals and long-term goals, and between energy sources, to guide future research priorities and, ultimately, national transportation energy policy.

Program Balance and Funding

Short- and Longer-Term Goals

Conclusion. The Partnership involves both short-term goals related to hydrocarbon-fueled vehicles used during a transition period and much longer-term goals aimed at enabling "a clean and sustainable transportation energy future." The committee considers the current split of funding between the long-term and shorter-term goals to be appropriate. Hydrogen-related activities—for example, fuel cells, hydrogen production, distribution, and safety—absorb approximately 70 percent of the funds. The remaining funds support the development of transition technologies, where cost is often the most significant barrier.

Congressionally Directed Funding

Conclusion. During the last 2 years, congressionally directed funding to specific recipients and activities has diverted resources from efforts focused on critical program goals, particularly in the hydrogen portion of the program. The committee believes this earmarking increases the risk of missing critical program milestones and targets, places high demands on DOE management time, and signals to the industry partners somewhat less than full government support for the program goals. If this practice continues and appropriations are not increased to compensate for it, milestones for the program will most certainly have to slip. Congressional and administration leaders should be made aware of how congressionally directed funding affects program timing and leads to shortfalls in meeting its goals. In addition, DOE should ensure that these leaders understand the critical importance of the key parts of the FreedomCAR and Fuel Partnership to achieving its long-term, high-level goals.

Strategy for Accomplishing Goals

Program Management and Communications

Conclusion. Overall the committee is encouraged by the progress the FreedomCAR and Fuel Partnership has made in program management and communications across the many activities and interfaces of the DOE offices and contractors, USCAR, and the energy companies.

Setting Priorities

Conclusion. The committee believes that the setting of priorities needs more emphasis. It appears to the committee that several technical programs may not be contributing solutions to the most critical and important issues. There have not been any integrated assessments of overall progress toward the broad objectives of reducing petroleum demand and introducing hydrogen. Such assessments would be valuable in informing the Partnership's high-level program decision making.

Recommendation. The FreedomCAR and Fuel Partnership should perform an overall program evaluation, using go/no-go decisions and setting priorities that focus resources on programs that will contribute most to solving the problems critical to the success of the long-term program goals.

Value of Learning Demonstrations

Conclusion. The learning demonstrations program is very important to validate current component and systems concepts and to uncover previously unknown issues. Such demonstrations will establish many system and engineering parameters for a complete operating hydrogen supply and a fuel cell transportation system. These cooperative programs are well designed. Information will be collected from both vehicle and infrastructure components, pooled, and shared. It will guide the technical teams as well as the systems and modeling efforts and help to establish appropriate program priorities.

REFERENCES

DOE. 2004. *Partnership Plan. FreedomCAR & Fuel Partnership*. Washington, D.C.: U.S. Department of Energy, Energy Efficiency and Renewable Energy. Available on the Web at <http://www.eere.energy.gov/vehiclesandfuels/pdfs/program/fc_fuel_partnership_plan.pdf>.

NRC/NAE (National Research Council/National Academy of Engineering). 2004. *The Hydrogen Economy: Opportunities, Costs, Barriers, and R&D Needs*. Washington, D.C.: The National Academies Press.

1

Introduction

BACKGROUND

The U.S. Department of Energy (DOE) has been involved for almost 30 years in research and development (R&D) programs related to advanced vehicular technologies and alternative transportation fuels. In particular, in 1993, during the Clinton administration, the Partnership for a New Generation of Vehicles (PNGV) was formed between the federal government and the auto industry's U.S. Council for Automotive Research (USCAR).[1] The PNGV sought to significantly improve the nation's competitiveness in the manufacture of future generations of vehicles, to implement commercially viable innovations emanating from ongoing research on conventional vehicles, and to develop vehicles that achieve up to three times the fuel efficiency of comparable 1994 family sedans (NRC, 2001; PNGV, 1995; The White House, 1993).[2]

The election of President Bush in 2000 resulted in changes in direction and organization of a number of DOE R&D programs and the creation of new, multiyear program plans for vehicle and fuel R&D efforts (DOE, 2004a,b). In January 2002, the Secretary of Energy and executives of DaimlerChrysler, Ford,

---

[1]USCAR, which predated the formation of PNGV, was established by Chrysler Corporation, Ford Motor Company, and General Motors Corporation. Its purpose was to support intercompany, precompetitive cooperation that would reduce the cost of redundant R&D in the face of international competition. Chrysler Corporation merged with Daimler Benz in 1998 to form DaimlerChrysler. USCAR currently supports a number of consortia (Appendix A).

[2]The goal of PNGV that attracted the most attention from the news media was the development of a family sedan that would achieve a fuel economy of 80 miles per gallon (mpg) and cost the same as a comparable 1993 sedan. The media usually ignored the fact that the goal had been set as "up to 80 mpg," not "80 mpg."

and General Motors announced a new government-industry partnership between DOE and USCAR called FreedomCAR, with CAR standing for Cooperative Automotive Research. The new partnership supersedes and builds upon the PNGV program. In September 2003, FreedomCAR was expanded to include five large energy companies—BP America, Chevron Corporation, ConocoPhillips, ExxonMobil Corporation, and Shell Hydrogen (U.S.)—to address issues related to the supporting fuel infrastructure. The expanded scope of the partnership was acknowledged by changing the name to FreedomCAR and Fuel Partnership.[3] The long-term goal of the program is to "enable the full spectrum of light-duty passenger vehicle classes to operate completely free of petroleum and free of harmful emissions while sustaining the driving public's freedom of mobility and freedom of vehicle choice" (DOE, 2004a).

The FreedomCAR and Fuel Partnership differs in several ways from PNGV. PNGV focused on replacing the family sedan (that is, a midsize automobile such as the Concorde, Lumina, or Taurus) with a marketable, more fuel-efficient design. It included specific vehicle milestones—namely, a concept vehicle by 2000 and a preproduction prototype by 2004. The FreedomCAR and Fuel Partnership addresses the development of advanced technologies for all light-duty passenger vehicles: cars, sport utility vehicles (SUVs), pickups, and minivans. It also addresses technologies for hydrogen production, distribution, dispensing, and storage, which were not a part of the PNGV program. It is a partnership between USCAR and one government agency, DOE, which collaborates with other agencies as needed. In PNGV, many agencies were involved and the lead agency was the Department of Commerce.[4] No new government money was appropriated for PNGV. Each participating agency was expected to reprogram existing R&D funds to support PNGV goals. The FreedomCAR and Fuel Partnership started with a presidential commitment to request $1.7 billion over 5 years (FY04 to FY08), with FY05 appropriations of about $310 million and an FY06 presidential budget request of about $360 million (Garman, 2005).[5] Funding for research, development, and demonstration activities goes to universities, the national labo-

---

[3]In February 2003, before the announcement of the FreedomCAR and Fuel Partnership, the President announced the FreedomCAR and Hydrogen Fuel Initiative to develop technologies for (1) fuel-efficient motor vehicles and light trucks, (2) cleaner fuels, (3) improved energy efficiency, and (4) hydrogen production and nationwide distribution infrastructure needed for vehicle and stationary power plants, to fuel both hydrogen internal combustion engines (ICEs) and fuel cells (DOE, 2004a). The expansion of the FreedomCAR and Fuel Partnership to include the energy sector after the announcement of the initiative also supports the goal of the Hydrogen Fuel Initiative.

[4]The federal agencies involved in PNGV included the Department of Commerce, DOE, the Environmental Protection Agency (EPA), the Department of Defense (DOD), the National Science Foundation (NSF), the Department of Transportation (DOT), and the National Aeronautics and Space Administration (NASA).

[5]The FY05 appropriation breaks down as follows: hydrogen technology, $120 million; fuel cells, $75 million; vehicle technologies, $85 million; Office of Science, $29 million; DOT, $0.55 million (Chapter 5, Tables 5-1 and 5-2).

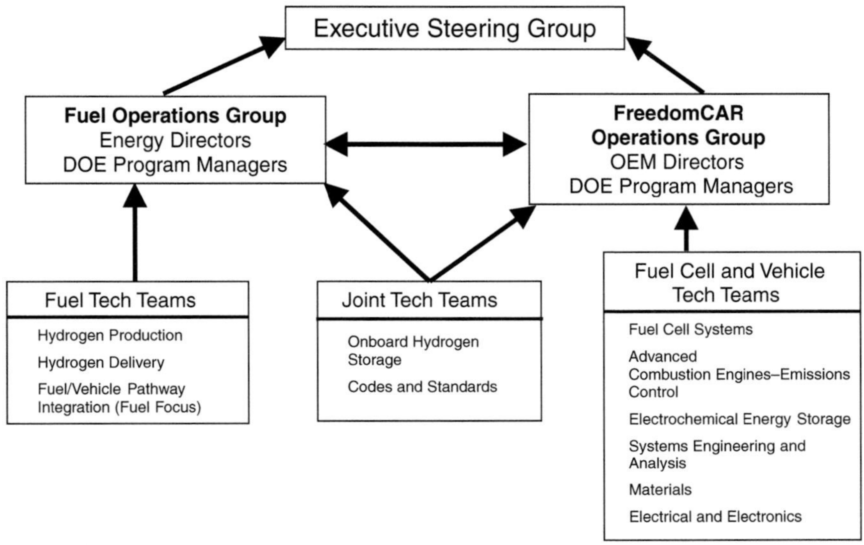

FIGURE 1-1 FreedomCAR and Fuel Partnership organizational structure. SOURCE: R.F. Moorer, "FreedomCAR and Fuel Partnership peer review," Presentation to the committee on November 17, 2004.

efficiency and renewable energy (EERE) and a vice-presidential- or presidential-level executive from each of the Partnership companies. It meets as needed. The FreedomCAR Operations Group, made up of DOE program managers and USCAR member company directors, is responsible for direction of the technical teams and prioritization of research issues. The Fuel Operations Group, made up of DOE program managers and energy company directors, is responsible for the direction of the fuel technical teams. Periodically, the FreedomCAR Operations Group and the Fuel Operations Group hold joint meetings to coordinate fuel and power-plant issues and identify strategic or policy issues that warrant attention by the Executive Steering Group (DOE, 2004c).

The Partnership has formed 11 industry-government technical teams responsible for R&D on the candidate subsystems (see Figure 1-1). Most of these technical teams focus on specific technical areas, but some, such as codes and standards and systems engineering and analysis, focus on crosscutting issues. A technical team consists of scientists and engineers with technology-specific expertise from the USCAR member companies, energy partner companies, and national laboratories, as well as DOE technology development managers. They may come from other federal agencies if approved by the appropriate operation group(s). A technical team is responsible for developing R&D plans and roadmaps, reviewing research results, and evaluating technical progress toward meeting established research goals (DOE, 2004c). Its discussions are restricted to nonproprietary topics. Fuel cell and vehicle technical team members come from the USCAR partners and DOE. They address the same topics as they did under the PNGV—namely, fuel cells, advanced combustion and emissions control, systems engineering and analysis, electrochemical energy storage, materials, and electrical systems and power electronics.

The three teams in the hydrogen fuel technical teams and the codes and standards technical team are new. In May 2003, the USCAR partners and DOE invited five energy companies to join the Partnership. In September 2003, the energy companies attended the joint meeting of the two operations groups and the meeting of the Executive Steering Group. Two Fuel Operations Group technical team reviews took place in 2004. Three teams—hydrogen production, hydrogen delivery, and fuel/vehicle pathway integration—each have members from the energy companies and DOE, and there are two joint technical teams connecting the fuel teams and the vehicle teams: an onboard hydrogen storage team and a codes and standards team.

At DOE, primary responsibility for the FreedomCAR and Fuel Partnership rests with EERE.[6] The two main program offices within EERE that are involved in the activities of the Partnership are the FreedomCAR and Vehicle Technologies (FCVT) program and the Hydrogen, Fuel Cells, and Infrastructure Technologies (HFCIT) program.

The FCVT program has the following specific goal: to support "R&D that will lead to new technologies that reduce our nation's dependence on imported oil, further decrease vehicle emissions, and serve as a bridge from today's conventional power trains and fuels to tomorrow's hydrogen-powered hybrid fuel cell vehicles" (DOE, 2004b). The FreedomCAR and Fuel Partnership, the focus of this report, and the 21st Century Truck Partnership are both within FCVT.[7]

The FreedomCAR and Fuel Partnership activities in the FCVT program are organized into a number of areas:

- Vehicle systems analysis and testing to provide an overarching vehicle systems perspective to the technology R&D subprograms and activities in the FCVT and HFCIT programs;
- Hybrid propulsion systems for light-duty vehicles, which includes activities on advanced internal combustion engine (ICE) power trains and hydrogen ICE power trains as well as testing on various fuel and propulsion system combinations;
- Energy storage technologies (batteries and ultracapacitors);
- Advanced power electronics and electric machines;

[6]EERE has a wide variety of technology R&D programs and activities related to renewable energy technologies, such as the production of electricity from solar energy or wind or the production of fuels from biomass, to the development of technology to enhance energy efficiency, whether for vehicles, appliances, buildings, or industrial processes. It also has programs on distributed energy systems (see Appendix B for an EERE organization chart).

[7]DOE supports several other programs related to the goal of reducing dependence on imported oil. The 21st Century Truck Program supports R&D on more efficient and lower emission commercial road vehicles.

- Advanced combustion engine R&D, which in concert with the work on light-duty hybrid propulsion systems focuses on enabling technologies for energy-efficient, clean vehicles powered by advanced ICEs using clean hydrocarbon-based and non-petroleum-based fuels and hydrogen;
- Materials technology for lightweight vehicle structures and for propulsion system materials, including power electronics and combustion engines; and
- Fuels technologies to allow current and emerging advanced ICEs and emission control systems to be as efficient as possible while meeting future emission standards and to reduce reliance on petroleum-based fuels.

The HFCIT program directs activities in hydrogen production, storage, and delivery and integrates them with transportation and fuel cell development activities. The proton exchange membrane (PEM) fuel cell R&D is undertaken in the HFCIT program. The program is focused on

- Overcoming technical barriers through R&D on hydrogen production, delivery, and storage technologies, as well as fuel cell technologies for transportation, distributed stationary power, and portable power applications;
- Addressing safety concerns and developing model codes and standards;
- Validating and demonstrating hydrogen fuel cells in real-world conditions; and
- Educating key stakeholders whose acceptance of these technologies is critical to their success in the marketplace (DOE, 2004a,b).

The manager of HFCIT is the overall DOE hydrogen technology program manager.

Some activities related to the HFCIT program focus are not within EERE. The Office of Fossil Energy (FE) supports the development of technologies to produce hydrogen from coal, as well as carbon capture and sequestration programs. The Office of Nuclear Energy, Science and Technology (NE) supports research into the potential use of high-temperature nuclear reactors to produce hydrogen, while the Office of Science (SC) supports fundamental work on new materials to store hydrogen; catalysts; fundamental biological or molecular processes for hydrogen production; fuel cell membranes; and other basic science areas (DOE, 2004d,e). An overall evaluation and strategic review of these hydrogen technology R&D activities was undertaken by the NRC's Committee on Alternatives and Strategies for Future Hydrogen Production and Use, and the current study references the results of that study and its recommendations, which are contained in the report *The Hydrogen Economy: Opportunities, Costs, Barriers, and R&D Needs* (NRC/NAE, 2004), referred to as *The Hydrogen Economy* in the remainder of the present report.

INTERRELATIONSHIP OF VEHICLE AND FUEL TECHNOLOGIES

Historical examples illustrate the importance of linking vehicles and fuels, not only for setting technical targets but also for addressing infrastructure barriers to widespread use. As noted in the preceding sections, the FreedomCAR and Fuel Partnership addresses R&D for both vehicles and fuels. The critical interrelationship between vehicles and fuels has been recognized for many years. This recognition led, for example, to the formation of the Coordinating Research Council (CRC) by the automobile companies and the major oil companies in 1942. CRC worked on technical issues at the vehicle/fuel interface and, during the two decades that followed, enabled the introduction of high-octane gasoline and higher compression engines with increased specific power and efficiency. NRC reviews of the PNGV program during the 1990s also called for cooperation between the PNGV program and the fuel industry.

Examples of important advances brought about by technological change at this vehicle/fuel interface include the introduction of unleaded gasoline in 1971 and reformulated gasoline (RFG) in the early 1990s. Unleaded gasoline provided an immediate reduction in vehicle exhaust emissions and, more important, enabled the introduction of first-generation catalytic converters. Since these were essential to comply with the 1970 Clean Air Act, the phase-in of unleaded fuel was mandated by EPA.

The Auto-Oil program, launched as a collaborative effort between the automotive and oil industries in the mid-1980s, led to RFG, adopted by EPA and mandated in the 1990s. Similarly, reduced sulfur in diesel fuel has been shown to reduce exhaust emissions significantly and also is essential for facilitating the introduction of advanced exhaust aftertreatment devices. Consequently, EPA has mandated the phase-in of low-sulfur diesel fuel commencing in 2006.

Because infrastructure and availability are just as important as the technical specifications of the fuel and its compatibility with the power plant, EPA mandates the availability of unleaded gasoline and low-sulfur diesel fuel. Efforts to introduce on a wide scale alternative fuels such as methanol, ethanol, and compressed natural gas have all foundered, in part owing to unavailability. Apart from the incentive created by the Energy Policy Act of 1992 for fleets to use alternative fuels, there is no compelling reason for consumers to demand them. Indeed, there are disincentives, such as economics and/or inconvenience, that discourage their use. It has been repeatedly demonstrated that unless a fuel is widely available, easy to refuel, and competitively priced it will not enjoy widespread use as a replacement for gasoline.

Fuels composed mainly of alcohol, such as 85 percent methanol (M85) and 85 percent ethanol (E85), work well in vehicles designed to accept them, and there are now well over 4 million vehicles in the United States equipped to operate on M85 or E85. However, despite their ease of use, there are fewer than 200 predominately alcohol filling stations nationwide, compared with about 168,000 retail gasoline stations (National Petroleum News, 2004). Since alcohol fuels cost more than gasoline (on an equivalent-energy basis) and limit vehicle range, there is no incentive for customers to use them and hence no business case for producing the fuel or building refueling stations.

Compressed natural gas (CNG) also performs well in CNG-compatible vehicles, especially dedicated CNG vehicles (more so than dual-fuel vehicles), but it also has enjoyed only limited success. CNG is mainly used in niche markets such as city buses, airport vans, some taxis, and other fleets. Again, limited availability and reduced vehicle range, combined with inconvenient refueling and lack of a secondary vehicle market, resulted in the absence of a positive business case for operators or energy producers.

To overcome the challenges, whether they be technical, economic, or public policy, of vehicle/fuel compatibility and fuel availability, any attempt to introduce a radically different vehicle fuel, such as hydrogen in any form, must involve a dedicated effort by all interested parties, including vehicle manufacturers, the energy industry, and the government. While there is no guarantee that such collaboration will ultimately be successful, its absence will guarantee failure. One of the major strengths of the FreedomCAR and Fuel Partnership is that its membership includes three of the essential stakeholders to identify, define, and oversee the needed research, as described in the previous sections in this chapter.

ADVANCED INTERNAL COMBUSTION ENGINES AND FUELS

Even in the most optimistic scenario postulated in the National Academies report *The Hydrogen Economy*, only 10 percent of new vehicles and 6 percent of the total miles traveled in 2024 are projected to be hydrogen-fueled fuel cell vehicles (NRC/NAE, 2004). The remaining 90 percent of new vehicles are projected to be conventionally powered vehicles, either hybrid or nonhybrid. Consequently, by far the greatest contribution to reduced energy use and emissions by and from the U.S. vehicle fleet over the next 20 years and beyond will come from continued improvement in ICEs, hybrid electric vehicles, and their fuels.

Despite increasingly stringent emissions requirements and the seemingly insatiable demand by vehicle customers for increased performance, the fuel efficiency of domestic cars and light trucks (pickups, SUVs, vans) has been increasing steadily at 1.5 percent per year for at least 20 years. Figure 1-2 shows fuel efficiency as ton-miles per gallon (mpg) for cars and trucks, respectively.

This steady increase in fuel efficiency has been masked by increasing vehicle content, hence weight, acceleration performance, and a shift in the fleet mix from cars to light trucks. Consequently, overall U.S. fleet fuel economy (as indicated by the Corporate Average Fuel Economy [CAFE] measure) has remained relatively stable, or even declined, in recent years (Figure 1-3).

To reduce transportation fuel use, current industry-wide efforts to improve ICE efficiency and further develop relevant fuels must continue or accelerate. This is true regardless of the degree to which hybrid electric vehicle power trains

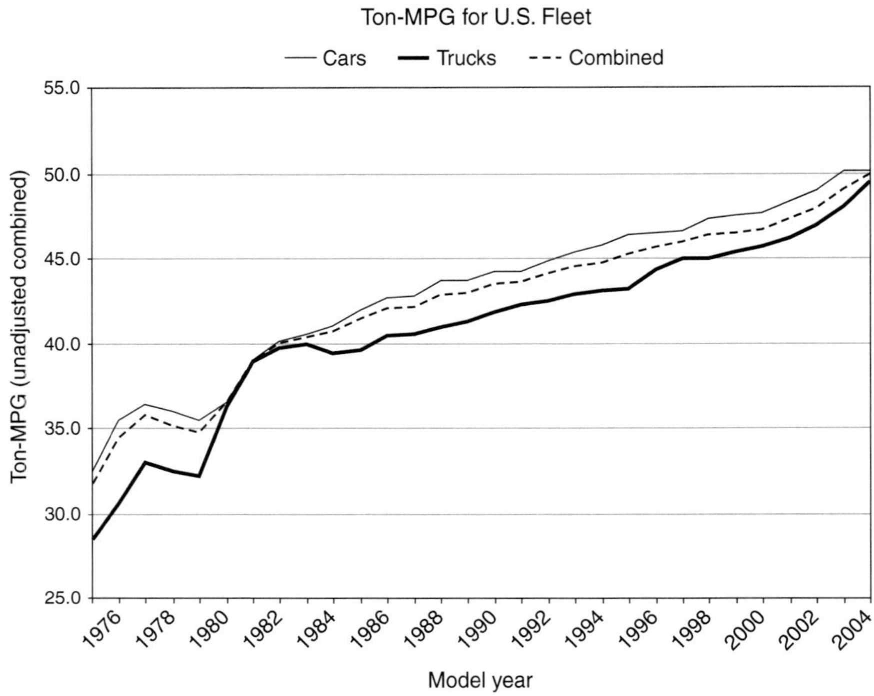

FIGURE 1-2 Fuel efficiency for U.S. fleet by model year for cars and trucks, expressed as ton-mpg. The metric ton-mpg provides an indication of a vehicle's ability to move weight. It is a measure of power train/driveline efficiency. SOURCE: EPA, 2004.

proliferate or regardless of whether advanced diesel engines achieve customer acceptance and meet emission standards. The urgency of this task is amplified by the reality that with approximately 16 million new vehicles sold in the United States every year, it takes almost 15 years to turn over the national fleet of roughly 225 million vehicles.

While much of the FreedomCAR and Fuel Partnership activity is devoted to fuel cell vehicles and hydrogen fuel, further improvement in conventional ICEs and hybrid electric vehicles can contribute significantly to the goals of energy independence and reduced carbon emissions and should benefit from continued collaboration between industry engineers and the national laboratories in this area.

The four-stroke, direct-injection engine technical team accomplished a great deal in the PNGV program, especially in the diesel (compression ignition direct injection [CIDI] engines) and four-stroke gasoline direct injection (4SDI) areas, and it has continued many of the most promising concepts under the FreedomCAR and Fuel Partnership umbrella. In particular, nitrogen oxides (NO_x) and particulate emissions objectives in PNGV were far more stringent than those anywhere

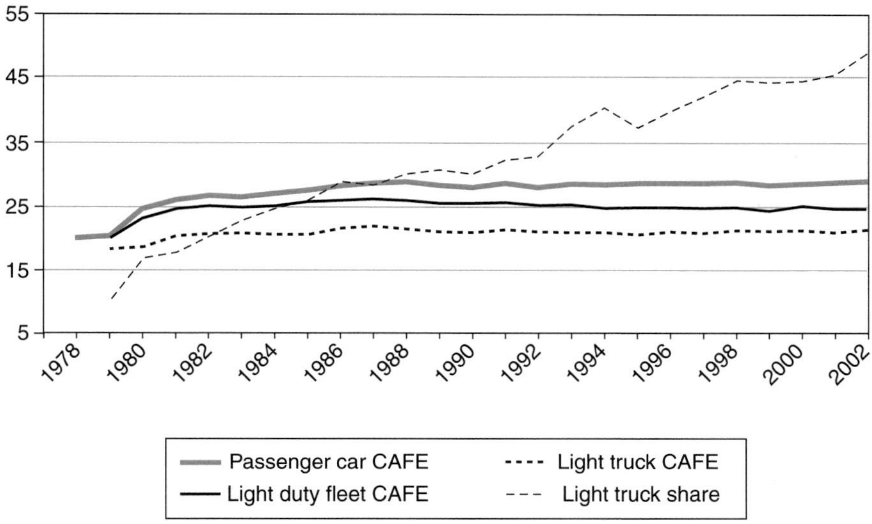

FIGURE 1-3 Fuel economy of the U.S. automotive and light truck fleet for model years 1978 to 2002. The fuel economy curves are in miles per gallon (mpg) and the light truck share is in percent. SOURCE: NHTSA, 2003.

else in the world, in anticipation of the upcoming EPA Tier 2 requirements. The PNGV partners drew on all of their global resources to develop candidate exhaust aftertreatment systems to address these goals, and the results of this development, which were shared with the entire OEM and supplier industry, are beginning to appear on production vehicles in Europe, Japan, and the United States, as increasingly stringent regulations require, and improved, low-sulfur fuels permit.

Given the limits of aftertreatment, renewed emphasis is also being applied to reducing engine-out emissions, again building on work commenced within PNGV; in particular, the so-called homogeneous charge compression ignition (HCCI) and low-temperature combustion (LTC) concepts are of great interest. The status of FreedomCAR and Fuel Partnership efforts on ICE and emissions control development is discussed further in Chapter 3.

ADVANCED FUEL CELL VEHICLES AND FUELS

The basic concept of the fuel cell was invented in 1839, and attempts to apply it as a vehicle prime mover date back to the 1960s. However, serious development of the PEM fuel cell began worldwide only in the 1990s, with the growing awareness that a hydrogen-fueled fuel cell is one of the very few candidates capable of achieving the holy grail of zero vehicle emissions, high efficiency, reduced dependence on petroleum, and—potentially—zero (systemwide) carbon dioxide (CO_2) emissions.

Successive generations of fuel cell vehicles have demonstrated increasingly compact systems and improved functionality. Various methods of providing on-board hydrogen, from the on-board reformation of hydrocarbon fuels (e.g., gasoline or methanol) to the storage of compressed or liquefied hydrogen gas, have been demonstrated, with most current demonstrations using compressed storage. The remaining vehicle issues are primarily on-board storage of sufficient hydrogen, system functionality, durability and reliability, and total system cost. Technical team activity in these areas is covered in detail in Chapter 3.

While a major obstacle to the deployment of fuel cell vehicles is clearly the absence of a hydrogen fuel production and distribution infrastructure, work to address this has only recently begun. The primary issues here are how to produce hydrogen economically without exacerbating carbon dioxide emissions and how to deliver it safely and cost-effectively to the point of vehicle refueling. The enablers of hydrogen production and distribution were fully enunciated in *The Hydrogen Economy* (NRC/NAE, 2004). Chapter 4 describes progress in this area.

The issue of safety pervades virtually every aspect of the pursuit of a hydrogen economy. The propensity of hydrogen to find even the most infinitesimal leak path, its low ignition energy, its flammability over a very wide range of concentrations, and the invisibility of its flame are well known. Less well understood are its behavior in the event of vehicle impact and the adequacy of emerging codes and standards. This important subject is discussed in Chapter 2.

COMMITTEE APPROACH AND ORGANIZATION OF THIS REPORT

The Hydrogen Economy discusses many of the R&D activities associated with the hydrogen technology parts of the FreedomCAR and Fuel Partnership, such as hydrogen production, distribution, dispensing, and storage, as well as the transition strategy for making hydrogen more widely available (NRC/NAE, 2004). It provides an excellent review of the challenges and potential benefits of using hydrogen as a transportation fuel and offers recommendations for the DOE R&D program, and the current committee has used its results and referred to its recommendations. The current report presents the committee's evaluation of the DOE-sponsored research efforts to achieve the hydrogen economy through the FreedomCAR and Fuel Partnership and offers comments and suggestions on the technical directions, strategies, and funding and management of this very important program. Because *The Hydrogen Economy* had just been published as the current committee was being constituted, with regard to the hydrogen technology parts of the Partnership, the committee reviewed just the plans of the three new fuel-related technical teams.

The statement of task for this committee is as follows:

> The National Academies' National Research Council (NRC), through its Board on Energy & Environmental Systems and Transportation Research Board (TRB) established a committee to conduct an independent, credible and unbiased review of the research pro-

REFERENCES

DOE. 2004a. *Hydrogen, Fuel Cells and Infrastructure: Multi-Year Research, Development and Demonstration Plan.* DOE/GO-102003-1741. Washington, D.C.: U.S. Department of Energy, Energy Efficiency and Renewable Energy. Available on the Web at <http://www.eere.energy.gov/hydrogenandfuelcells/mypp/>.

DOE. 2004b. *FreedomCAR and Vehicle Technologies Multi-Year Program Plan*). Washington, D.C.: U.S. Department of Energy, Energy Efficiency and Renewable Energy. Available on the Web at <http://www.eere.energy.gov/vehiclesandfuels/resources/fcvt_mypp.shtml>.

DOE. 2004c. *Partnership Plan. FreedomCAR & Fuel Partnership.* Washington, D.C.: U.S. Department of Energy, Energy Efficiency and Renewable Energy. Available on the Web at <http://www.eere.energy.gov/vehiclesandfuels/pdfs/program/fc_fuel_partnership_plan.pdf>.

DOE. 2004d. *Basic Research Needs for the Hydrogen Economy: Report of the Basic Energy Sciences Workshop on Hydrogen Production, Storage, and Use,* May 13-15, 2003. Washington, D.C.: U.S. Department of Energy, Office of Science. Available on the Web at <http://www.er.doe.gov/production/bes/hydrogen.pdf>.

DOE. 2004e. *Hydrogen Posture Plan: An Integrated Research, Development and Demonstration Plan.* Washington, D.C.: U.S. Department of Energy. Available on the Web at <http://www.eere.energy.gov/hydrogenandfuelcells/pdfs/hydrogen_posture_plan.pdf>.

EPA (Environmental Protection Agency). 2004. *Light Duty Automotive Technology and Fuel Economy Trends, 1975-2004.* EPA420-R-04-001 (Appendix M). Washington, D.C. Available on the Web at <http://www.epa.gov/otaq/cert/mpg/fetrends/420r04001.pdf>.

Garman, D. 2005. *The FY2006 Budget Request.* Washington, D.C.: U.S. Department of Energy, Office of Energy Efficiency and Renewable Energy. Available on the Web at <http://www.eere.energy.gov/office_eere/pdfs/fy06_budget_request.pdf>.

NHTSA (National Highway Traffic Safety Administration). 2003. *Automotive Fuel Economy Program, Annual Update Calendar Year 2002.* DOT HS 809 512. Washington, D.C.: U.S. Department of Transportation. Available on the Web at <http://www.nhtsa.dot.gov/cars/rules/cafe/FuelEconUpdates/2002/index.htm>.

National Petroleum News. 2004. *Market Facts 2004.* Available on the Web at <www.npn-net.com>.

NRC (National Research Council). 2001. *Review of the Research Program of the Partnership for a New Generation of Vehicles, Seventh Report.* Washington, D.C.: National Academy Press.

NRC/NAE (National Research Council/National Academy of Engineering). 2004. *The Hydrogen Economy: Opportunities, Costs, Barriers, and R&D Needs.* Washington, D.C.: The National Academies Press.

PNGV (Partnership for a New Generation of Vehicles). 1995. *Partnership for a New Generation of Vehicles Program Plan* (draft). Washington, D.C.: U.S. Department of Commerce, PNGV Secretariat.

The White House. 1993. "Historic partnership forged with automakers aims for threefold increase in fuel efficiency in as soon as ten years." Washington, D.C.: The White House.

2

Major Crosscutting Issues

This chapter addresses the main crosscutting issues that the committee identified in its review of the FreedomCAR and Fuel Partnership. Given that much of the Partnership is relatively recent, the committee deems it important to address some issues early: program decision making, safety, learning demonstrations, and program structure. Some of these issues touch on the broader context on which the successful adoption into the marketplace of the technologies under development depends. The committee believes that proper attention to its recommendations will help move the program's progress and increase its chances of success. It focuses on specific technical areas in Chapters 3 and 4.

PROGRAM DECISION MAKING

Program Management and Organization

As described in Chapter 1, most of the government programs on more efficient vehicle and engine technology, fuel cells, and hydrogen are in DOE, but some programs are in other agencies, such as the Department of Defense (DOD), the Department of Transportation (DOT), and the Environmental Protection Agency (EPA). The FreedomCAR and Fuel Partnership under review by the committee is a substantial piece of the larger whole. Coordination of all advanced vehicle and hydrogen programs is essential if they are to have an optimal impact. The vehicle systems programs form a fairly coherent whole, while the hydrogen production delivery and storage programs are diverse and explore many different pathways to the use of hydrogen as a transportation fuel.

To integrate the planning, budgeting, and management of the many DOE programs constituting the FreedomCAR and Fuel Partnership, DOE's Office of Energy Efficiency and Renewable Energy (EERE) issued the Hydrogen Posture Plan in February 2004 (DOE, 2004a). The plan describes DOE's intended role in hydrogen energy R&D and its pursuit of an accelerated path to the deployment of hydrogen fuel cells, and the associated infrastructure. Within DOE, a working group was established with representatives from the Offices of Energy Efficiency and Renewable Energy; Fossil Energy; Nuclear Energy, Science, and Technology; Science; Management, Budget, and Evaluation/chief financial officer; and Policy and International Affairs (in an oversight capacity). While the broader management of many of these DOE program activities lies beyond the statement of task for the committee, it is important to note that the planning, budgeting, execution, evaluation, and reporting of the government's hydrogen-related programs be well coordinated and integrated.

The committee finds that, while there has been commendable progress in managing the various transportation-related hydrogen activities across DOE, further improvements are needed. The committee identified two areas that need special attention: (1) carbon capture and sequestration and (2) basic energy research. The cost-effective, large-scale production of hydrogen may require that coal be the primary energy source. Carbon capture and sequestration would then be essential to reduce the emission of greenhouse gases. The potential of carbon sequestration to enable hydrogen production appears to play only a minor role in the current work on capture and sequestration (see Chapter 4).

DOE's various presentations to the committee indicated a strong focus on technology development and technology demonstration. It was also apparent that new technologies are likely to be required in hydrogen storage, fuel cell membranes, and electrodes. The committee encourages DOE to ensure that its Basic Energy Sciences Division in the Office of Science is appropriately involved in fundamental research critical to the FreedomCAR and Fuel Partnership.

Congressionally Directed Funding

The committee's review of the FreedomCAR and Fuel Partnership found that in certain program areas, congressionally directed activities (earmarking) of funds had a serious negative impact on the program. Of concern to the committee is the allocation by Congress of significant funds to specific organizations for activities that will contribute little to achieving the Partnership's objectives. Although DOE has some discretion over the allocation of funds not earmarked, over the past 2 years, earmarking has effectively removed about $80 million from the funding for planned programs. This has negatively impacted projects in safety, the production of hydrogen from fossil fuel and renewable energy sources, and hydrogen storage. One possible result is that not enough knowledge and technology will be available by 2015, when commercial feasibility will be assessed, making a positive assessment less likely. In addition to increasing the chances that critical program milestones will be missed, earmarking has forced DOE managers to spend a lot of time trying to adapt programs so they contribute as much as possible to priority goals. It also signals to industry partners somewhat less than full government support for the program goals. This committee feels strongly that all of the funds appropriated for this FreedomCAR and Fuel Partnership should contribute directly to achieving the Partnership's objectives.

Congressional and administration leaders should be made aware that congressionally directed funding affects program timing and leads to shortfalls in meeting goals. DOE should ensure that these leaders understand how critical key parts of the FreedomCAR and Fuel Partnership are to achieving the program's long-term, high-level goals.

Determining Priorities, Milestones, and Go/No-Go Decisions

Priorities, key milestones, and critical decisions within the FreedomCAR and Fuel Partnership must be determined in the appropriate systems context. For example, engine improvements must be evaluated in the context of the vehicle. It is the impact of vehicle improvements in context of the total vehicle fleet that really matters, and this impact depends on production volumes and the fraction of total fleet mileage accounted for by these improved vehicles. Several questions are raised by the need for a technology-development program focused on component technologies to conduct integrated evaluations in the appropriate system context.

For example, each technical team has a detailed set of milestones and decision-making points in its strategic plan. These milestones allow assessing the progress of individual technology development projects, but it is not clear how individual team plans can be integrated into a broader assessment of progress and consequent decision making. Ultimately, the Partnership succeeds when car companies incorporate these new technologies into specific vehicle designs that are attractive to the public and subsequently become part of the vehicle fleet.

The FreedomCAR and Fuel Partnership needs to identify and define appropriate priorities, milestones, and decision criteria in a context that goes beyond the individual technology areas. For example, if substantial demand for electricity for hydrogen generation is anticipated, multisector energy perspectives and inputs will be necessary to assess the prospects for carbon capture and sequestration as well as the electricity generating requirements and the impact of hydrogen generation on natural gas demand.

The committee feels that the FreedomCAR and Fuel Partnership management teams have not yet resolved how best to address these broader program assessment issues. Nor do they appear to have developed plans to carry out analysis of these issues that would support the decision-making process. The management structure outlined in Chapter 1 appears to the committee to be appropriate for the management tasks, but it is not clear how the structure allows for decisions about the Partnership's progress toward commercialization of fuel cell, hydrogen infrastructure, and transition propulsion systems and vehicle technologies. DOE should consider identifying a working group to support overall program management. The working group should perform technology assessments in an appropriate systems framework; check targets and revise them if necessary; evaluate the broader impacts of the technologies being developed in the program on major problems such as overall fleet petroleum consumption and greenhouse gas emissions; and assess progress towards commercialization. While DOE has established a systems analysis capability to do much of this analysis, it is not clear whether or how it is planning to do so, nor does it appear to be adequately staffed to do so. As noted in the next section, "Systems Analysis and Simulation," the systems analysis focus appears to be on developing the systems analysis tools rather than using them for more broadly based and integrated assessments.

An important aspect of the Partnership's decision making is that because the context for an assessment changes with time, so do the targets set for each component technology. One example of this is that the cost targets depend on the price of crude oil and natural gas. Also, because it is the total integrated cost of all the components that is critical to marketability, program management must have processes for reevaluating technology and cost targets as the program progresses.

Because of the importance and challenge of making Partnership decisions in these broader contexts, the committee makes the following recommendation:

Recommendation. An ongoing, integrated, well-to-wheels assessment should be made of the Partnership's progress toward its overall objectives of reducing the nation's dependence on oil and introducing hydrogen as a transportation fuel, if appropriate. This assessment should examine possible trade-offs between individual goals of the fuel program and the vehicle program, between short-term and long-term goals, and between energy sources, to guide future research priorities and, ultimately, national transportation energy policy.

Systems Analysis and Simulation

The previous discussion makes it evident that extensive systems analyses and simulation models are required to make informed decisions and manage the FreedomCAR and Fuel Partnership intelligently. Over the past several years, DOE has successfully developed many models to predict vehicle and component characteristics. Similarly, the production and distribution of fuels can be modeled to permit the comparison of production methods and of candidate fuels.

One important model, the so-called "well-to-wheels" analysis, enables comparisons of alternative fuels and vehicle system architectures by predicting overall energy efficiency and emissions performance of the entire system, taking into account the widely varying energy efficiency of alternative fuel production processes and the behavior of corresponding vehicle systems.

The critical importance of developing systems analysis and simulation was a recurring theme throughout the Partnership for a New Generation of Vehicles (PNGV) program. The seventh and final report on the PNGV program (NRC, 2001) recognized considerable progress in this area, but it encouraged increased effort in system modeling. The report also criticized the lack of progress in cost modeling.

The Hydrogen Economy (NRC/NAE, 2004) devoted one of its main recommendations to the urgent need for systems analysis, offering guidance in 10 areas. In response to these recommendations, DOE has been developing a systems analysis plan, a preliminary draft of which was presented to the committee. The key resources are in place, and individuals have been identified to take the lead on key analysis elements. Figure 2-1 depicts the main analysis domains—technical, cost, and market/benefits—and gives examples of applicable models.

The overall organization structure, as discussed in Chapter 1, is depicted in Figure 2-2, with the systems analysis functions added. The overall systems analysis budget for FY05 is $17.592 million, over half of which is devoted to Vehicle Systems Analysis and Hydrogen Infrastructure Analysis, with the remaining budget devoted to analysis of individual technologies.[1]

Progress to date is shown in Table 2-1. Six areas register significant progress, and two register partial progress; plans are in place for the remaining two.

In summary, the committee commends DOE on progress to date in addressing the systems analysis issue and especially on its response to the recommendations in *The Hydrogen Economy*. The plans and the personnel in place appear sound, and the overall approach is robust. If there are weaknesses, they probably lie in modeling (1) cost to the consumer and (2) consumer behavior in the face of a market transition to a radically different vehicle. Modeling cost is particularly difficult since it must rely on assumptions about future processes and products, and the necessary input data are highly proprietary.

Recommendation. The FreedomCAR and Fuel Partnership should develop and refine its models for consumer behavior during a market transition to radically different vehicles and should also explore ways to enhance the effectiveness of its cost models.

Finally, it appears that the systems analysis efforts are focused on developing and refining the systems analysis tools. While important, there is an additional responsibility to use the tools that are developed to provide overall program management or at least to perform technology assessments, goal checking, evalu-

[1]S. Chalk, "Systems Analysis Introduction," Presentation to the committee on January 24, 2005.

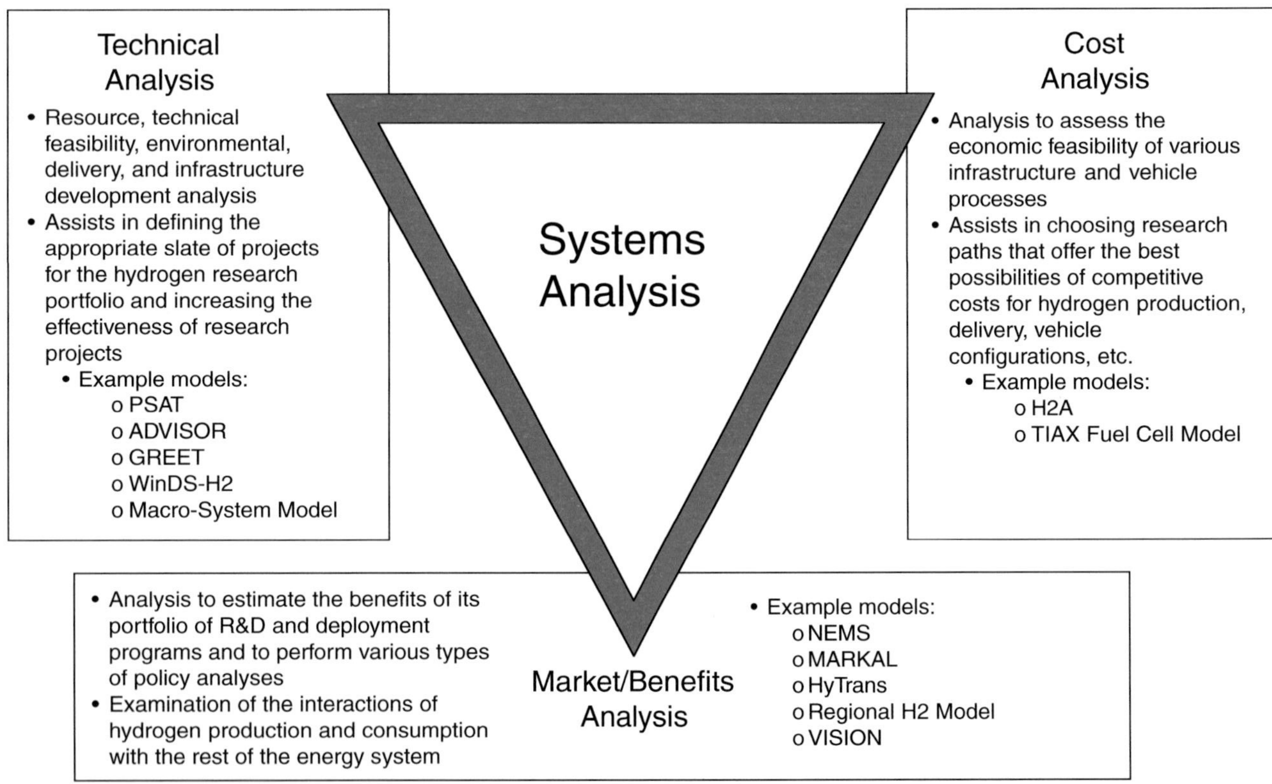

FIGURE 2-1 Analysis domains for systems analysis efforts by the FreedomCAR and Fuel Partnership. SOURCE: S. Chalk, "Systems analysis introduction," Presentation to the committee on January 24, 2005.

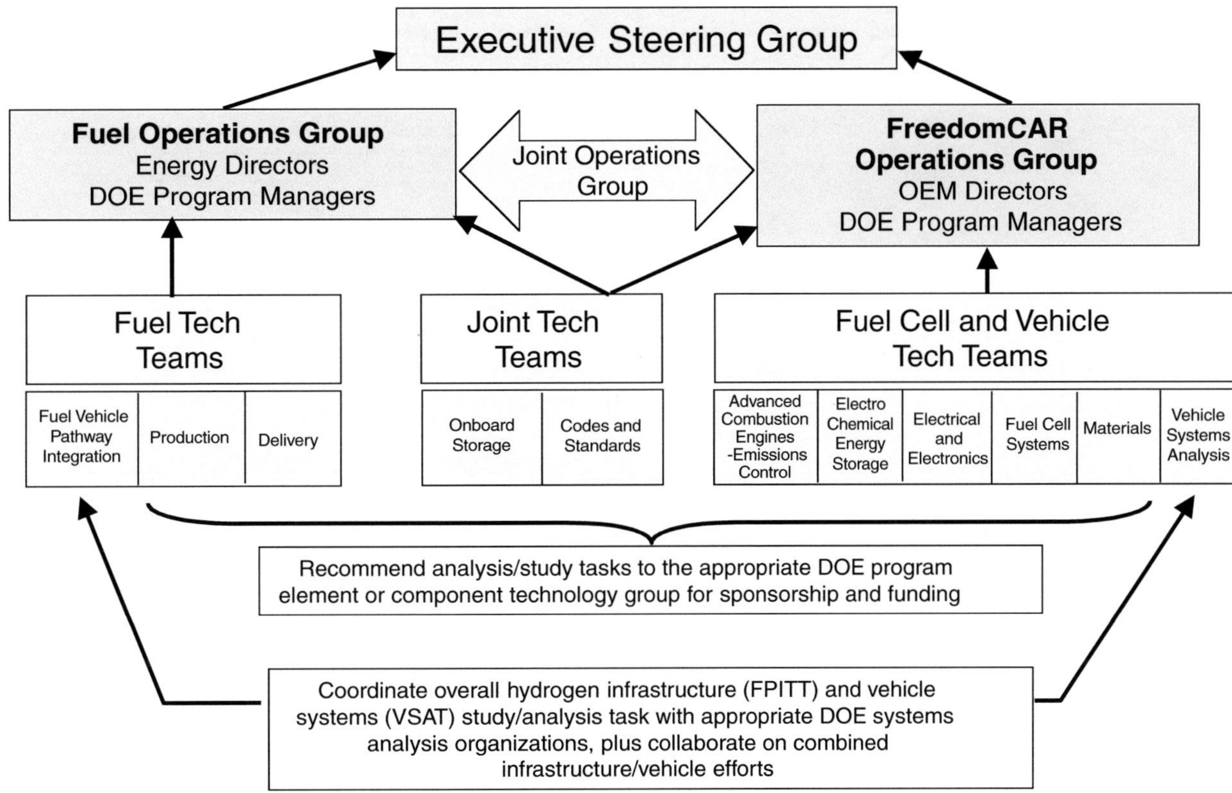

FIGURE 2-2 Overview of the systems analysis activity. SOURCE: S. Chalk, "Systems analysis introduction," Presentation to the committee on January 24, 2005.

ation of the broader impacts of competing technologies, and commercialization assessment. Responsibility for performing these tasks needs to be assigned and a work plan drawn up to address this expectation. Additional resources may well be required. It is in these areas that the committee recommends emphasis as the plan for systems analysis is solidified.

Recommendation. The FreedomCAR and Fuel Partnership should assign responsibility for overall program management and for the complex analyses to support program management, such as technology assessments, goal checking, evaluating the broader impacts of the technologies on the major problems, commercialization assessment, and decision making, among others.

SAFETY

Safety is a topic that pervades virtually every aspect of the pursuit of a hydrogen economy, and it is one of the major hurdles to the use of hydrogen as a transportation fuel. Large quantities of hydrogen are manufactured and used today throughout the world without undue safety hazards. Safety becomes an issue, however, when it is in consumer hands and on board a vehicle. *The Hydrogen Economy* (NRC/NAE, 2004) emphasized safety from both technical and societal perspectives. Some of the issues are well known—for example, hydrogen's propensity to find even the most infinitesimal leak path, its low ignition energy, its flammability over a wide range of concentrations, and its lack of a visible flame. Other issues, such as the potential consequences when a hydrogen powered vehicle crashes, and the adequacy of emerging codes and standards, are less well understood. In addition, an excellent safety record is essential to the public acceptance of hydrogen vehicles. Addressing public concerns about the widespread use of hydrogen as a transportation fuel could be critical in determining the feasibility of creating a fleet of hydrogen-fueled vehicles and their supporting infrastructure.

Hydrogen Safety Program

The purpose of the DOE Safety, Codes, and Standards program is to ensure that DOE's R&D is conducted in a safe, exemplary manner and that deployed elements of the hydrogen-fueled transportation system have an acceptable level of risk. The elements of the program include producing and maintaining a DOE Safety Plan, a best engineering practices document for hydrogen systems, and extensive support for the development of national and international codes and standards that will allow the deployment of the hydrogen infrastructure and hydrogen-fueled vehicles.

The Hydrogen Safety Review Panel (HSRP) is an important part of DOE plans to ensure that its hydrogen research is conducted in a safe manner. The panel has updated a safety guidance document specifying requirements for DOE contractors (DOE, 2004b). It has made safety site visits to approximately 10 contractors. It is also involved with the review of safety plans that have recently been required on new DOE procurements. The panel reports to the manager of the Office of Hydrogen, Fuel Cells, and Infrastructure Technology (HFCIT). The manager of HFCIT also has overall DOE responsibility for the hydrogen technology program, which, as noted in Chapter 1, encompasses activities in other offices within DOE, including the FreedomCAR and Vehicle Technologies Office (FCVT); the Office of Fossil Energy (FE); the Office of Nuclear Energy, Science and Technology (NE); and the Office of Science (SC). These other offices do not have hydrogen safety offices and intend to rely on HFCIT for their safety support.

The budget of the Safety, Codes, and Standards program was about $6 million in FY04 and FY05. It was appropriated $16 million in FY05, but this money had to be reallocated to balance other parts of the program that were severely impacted by congressionally directed activities (earmarks). The Safety, Codes, and Standards program was the most severely impacted by congressional earmarks, making it, in effect, level-funded in FY05. The President's FY06 budget request recommends $13.1 million for this program.

The top-level allocation of FY05 funds is shown in Table 2-2. Four national laboratories (Sandia National Laboratories at Livermore, California, Pacific Northwest National Laboratory, Los Alamos National Laboratory, and the National Renewable Energy Laboratory) will perform the major portions of the work. There are just two federal employees working in the DOE headquarters Safety, Codes and Standards program.

DOE and USCAR set up a codes and standards technical team about a year ago. The team has representatives from the automotive and energy companies, DOE, and National Highway Traffic Safety Administration (NHTSA). It has spent the first year preparing a roadmap, which is now in first draft form. Detailed descriptions of deliverables for the team do not exist yet. The near-term need is for interim codes and standards for the transition period.

Priority

The transition to hydrogen as a primary transportation fuel raises a multitude of safety questions that must be dealt with by many participants during each phase of the program. It will require a set of codes and standards, including those for on-site hydrogen production and dispensing at fueling stations. The committee observed that the DOE Safety, Codes, and Standards program addresses one of the highest priority areas of the entire FreedomCAR and Fuel Partnership but is also one of the least mature efforts at this early stage. The following discussion and recommendations elaborate on the importance of the Safety, Codes and Standards program.

TABLE 2-2 FY05 Budget for Hydrogen-Related Safety Codes and Standards Activities

Project	FY05 Funding ($)
Hydrogen release R&D	500,000
Materials R&D and handbook	412,000
Risk assessment and analysis	450,000
Hydrogen R&D subcontract support (modeling, SRI facilities, and International Energy Agency Safety Annex)	360,000
Safety Panel	350,000
Codes and Standards National Template (National Renewable Energy Laboratory execution of National Codes and Standards Template)	850,000
National template for contracts and subcontracts (support of codes and standards organization activities, i.e., NFPA, ICC, ASME, SAE, ANSI, CSA-America, NHA)	975,000
International codes and standards development	400,000
Office safety plan, lessons learned, best practices	329,400
Funding of DOT activities (NHTSA, RSPA)	790,000
HAMMER—safety training of officials and first responders	300,000
Taxes (7.8%) (SBIR, overarching analysis, international activities)	483,600
Total	6,200,000

NOTE: ANSI, American National Standards Institute; ASME, American Society of Mechanical Engineers; HAMMER, Hazardous Materials Management and Emergency Response facility at Pacific Northwest National Laboratory, Richland, Washington; ICC, International Codes Council; NFPA, National Fire Protection Association; NHA, National Hydrogen Association; NHTSA, National Highway Traffic Safety Administration; RSPA, Research and Special Projects Administration; SAE, Society of Automotive Engineers; SRI, Stanford Research Institute; SBIR, Small Business Innovation Research program.
SOURCE: Information supplied by DOE in response to questions from the committee.

Technical Teams

The current technical team is focused just on codes and standards, not on overall safety issues. Codes and standards are necessary to allow the deployment of the hydrogen infrastructure, vehicles, and other equipment into the public domain. They are necessary but are not by themselves sufficient to ensure the safety of the overall activity. There does not appear to be any top-down process in place to allocate risk to the various parts of the end-to-end system. Safety is inherently a systems issue. To enhance the safety mission, the current codes and standards technical team should be made part of a safety team with a broader mission that would interact closely with the other technical teams involved in hydrogen use learning demonstrations (these will involve the public; see section "Learning Demonstrations"). This new safety technical team should identify needed information related to safety issues at different levels, perform system-based safety analysis and definitions, develop draft codes and standards, set up a timetable and milestones to examine available models to predict safety, collect and generate information related to safety when lacking, and coordinate with other technical teams to set priorities and provide timely guidelines for setting safety performance targets.

The responsibility of a safety technical team should start where the hydrogen is produced and end where it is consumed ("follow the hydrogen"). It should cover all of the equipment in the entire system and each component from its manufacture to its ultimate recycling and disposal. Special emphasis should be given to the interface between the consumer and the hydrogen supply system at the filling station.

Safety should not be limited to just hydrogen hazards—it should also include crashworthiness, high-pressure and -temperature gas, toxicity, asphyxiation, high voltage, energy storage in batteries, and other potential hazards.

A safety technical team should also address the presence of hydrogen-fueled vehicles in buildings, tunnels, and other confined spaces. A recent study by Parsons-Brinkerhoff (2004) sponsored by the California Fuel Cell Partnership studied hydrogen cars in four types of parking structures. For the assumptions made, it was concluded that no modifications were necessary to the buildings. This result needs to be independently validated and a wider range of hydrogen release scenarios considered. Clearly, if *any* modifications to existing buildings are required (such as hydrogen sensors and/or increased ventilation), this will constitute another major infrastructural barrier to the widespread deployment of hydrogen cars.

It would be desirable for all the other technical teams involved with hydrogen to designate a safety person as a liaison to the safety technical team, which should provide safety guidance to the other technical teams. It would also be appropriate for the non-HFCIT DOE offices to have representatives on the safety technical team. Currently, the HFCIT Safety, Codes, and Standards program element is the focal point for safety for the other parts of DOE that are engaged in hydrogen-related work. However, there is little evidence that those offices are actively making use of the program.

The safety technical team, DOE, NHTSA, and the various codes and standards organizations should attempt to have draft codes and standards published as soon as possible. Some currently exist, and some may be issued soon enough to affect the learning demonstrations that are under way. Those released later will help with the transition from 2009 to 2015 and beyond.

Recommendation. DOE should form a new crosscutting safety technical team with a mission that includes broad hydrogen-related safety issues not only for HFCIT but for the other DOE offices as well. The new team should incorporate the existing codes and standards technical team as a subteam. The other offices should assign a person to be responsible for safety and to interface with the safety technical team. The safety, codes and standards effort needs adequate resources so that it can accomplish the goals identified in its roadmap.

Vehicle Standards and NHTSA

NHTSA is responsible for vehicle safety standards. It recently published a 4-year R&D plan for hydrogen-fueled vehicles (NHTSA, 2004). A NHTSA representative is appropriately included on the codes and standards technical team, and the codes and standards roadmap includes a number of vehicle safety deliverables; it does not, however, include all the important milestones from the NHTSA R&D plan. The codes and standards team should examine the NHTSA plans and integrate into its roadmap any milestones that are appropriate. Also, NHTSA can benefit from the various hydrogen release experiments being conducted by Sandia National Laboratories at Livermore.

NHTSA has been asked by the automotive industry and by DOE to have draft standards for hydrogen-fueled vehicles by 2010. It would be desirable to have interim standards in place for the learning demonstrations that began recently. Unfortunately, however, it will most likely take several years to develop such standards, which should be performance-based and not prematurely lock in any particular technology.

Although NHTSA has been planning hydrogen R&D activities for several years, a lack of funding has kept it from pursuing them. NHTSA estimates that it will need about $4 million to $5 million per year to develop these standards (a total of about $20 million). It is important that when such standards are developed they be harmonized internationally.

DOE is planning to transfer a small amount of money in FY05 to NHTSA to get its program started. The President's proposed budget for FY06 includes $2.3 million for DOT for its hydrogen activities. A portion of that funding will go to NHTSA.

The existing NHTSA plan relies heavily on crash testing prototype hydrogen vehicles. The committee thinks this activity is premature because the technology of the hydrogen vehicles will change dramatically over the next 5 to 10 years. The committee believes that NHTSA should instead focus more on analysis and research to establish the technical basis for its standards. Using finite element models to investigate crashworthiness issues and computational fluid dynamics models to examine fire risks from and to the hydrogen-containing components of a vehicle are examples of appropriate near-term activities. NHTSA should delay its crash testing of prototype vehicles until an appropriate generation of vehicles is available.

A separate but related DOT issue is pipelines. DOT's Pipeline and Hazardous Materials Safety Administration (PHMSA)[2] is responsible for pipeline safety.

---

[2]Prior to the creation of PHMSA in February 2005, pipeline safety was the responsibility of the Research and Special Projects Administration (RSPA).

As the FreedomCAR and Fuel Partnership develops a clearer view of the likely form and timing of a hydrogen delivery system, DOT may need to develop codes and standards for carbon dioxide (CO_2) pipelines as well, which may be needed for CO_2 delivery to carbon sequestration sites.

Recommendation. NHTSA should begin its hydrogen R&D program in FY05 by focusing on the effects of hydrogen releases and other potential hazards with hydrogen-fueled vehicles, as well as analyses and research to determine the right mix of system-level and component-level standards. NHTSA should also work with other U.S. and international safety groups to establish global standards for hydrogen-fueled vehicles.

Publication, Openness, and Safety Documents

Safety is an overriding consideration for the successful transition to a hydrogen economy based on hydrogen-fueled vehicles. An excellent safety record is essential for public acceptance of hydrogen vehicles. For this reason, everyone in the field should openly share information related to safety. In the NHTSA Docket (NHTSA, 2004, Document 19), the auto company Bayrische Motor-Werken" (BMW) states as follows: "The issue of safety in the use of hydrogen should always be treated in the same way along commonly agreed lines, not as a competitive feature distinguishing one company from another."

Accordingly, safety research results should be presented at professional societies and in peer-reviewed journals. DOE and others have been working on hydrogen fuel cells, tanks, and other components for many years. This work goes at least as far back as the Spark M. Matsunaga Hydrogen Research, Development and Demonstration Act of 1990 and was also part of the PNGV program, yet very little research related to safety has been published.

As the amount of hydrogen research increases and the number of hydrogen-fueled vehicles increases, incidents will occur in which hydrogen is unintentionally released, sometimes causing fire and/or explosion. It is important that accidents of this type be investigated and that the lessons learned be made available to the public. Any attempts to cover up such incidents will backfire and cause public skepticism. Transparency is the best long-term policy.

The National Transportation Safety Board's (NTSB's) methods for transportation accident investigation may be a useful model. In the early days NHTSA may also want to investigate high-delta-V (large changes in velocity) crashes of hydrogen vehicles that do not result in a hydrogen release. While there are significant differences between hydrogen-fueled and natural-gas-fueled vehicles, there are also some similarities, especially for vehicles that store their fuel as high-pressure gas. Since there are over 100,000 compressed natural gas vehicles on the road, there is a much larger accident pool to investigate.

Incidents should be written up and made available to the public and to others in the field. Learning from past mistakes will take place, and the engineering and the codes and standards will gradually improve. For example, a recent hydrogen leak during fueling at Ballard Systems in Canada was written up in a noncontroversial way and published in the newsletter of a fuel cell conference (Kinzey, 2004). This is a positive example of how an incident should be reported.

It is important to have factual information, scientific and engineering data, and lessons learned readily available. The Web sites of DOE, NHTSA, and USCAR could be improved (and interlinked) to make it easier to find hydrogen safety information.

The Safety, Codes, and Standards plan identifies two safety documents—a safety plan and a best engineering practices document. The former is due this year and the latter in 2007. Both are important, and the latter should be expedited if possible. HSRP membership, meeting minutes, site visit reports, and other products should be made available on the DOE Web site.

Recommendation. DOE, USCAR, and NHTSA should prepare and maintain a bibliography of hydrogen-safety-related reports and papers and make that information available on their Web sites in a user-friendly manner. NHTSA and DOE should develop investigation protocols and have investigation teams ready to visit serious incidents anywhere.

Budget and Schedule

The amount appropriated for the Safety, Codes and Standards program is much less than was requested. There are many milestones on the Codes and Standards Roadmap coming up in the next 2 to 3 years. The committee is concerned that many milesones will slip as a result of the severe cuts in this budget in FY05, jeopardizing the goal of having all of the codes and standards in place by 2010.

Recommendation. DOE should examine the budget and schedule estimates for each of the codes and standards deliverables and also for the other safety activities of the Safety, Codes and Standards program. To the extent that the budget and schedule are incompatible, changes should be reflected in the next update of the roadmap.

LEARNING DEMONSTRATION: NATIONAL HYDROGEN VEHICLE/INFRASTRUCTURE PROGRAM

The FreedomCAR and Fuel Partnership includes a variety of R&D and demonstration activities for fuel cell vehicles and hydrogen fuel systems. Approximately 13 percent of the FY05 budget for the program was focused on demonstration activities, and what DOE calls learning demonstrations that will operate during the next few years were announced. Any advanced technology, no matter how well tested by its developers, will show unanticipated characteristics when placed in the hands of the users. Some technologies—software, for example—require alpha and beta versions before a truly commercial product can be claimed. This kind of feedback will be especially important for the FreedomCAR and Fuel Partnership because of the long-term, high-risk research agenda and because public safety must be ensured in the face of highly energetic materials—for example, hydrogen and high-voltage batteries. To be sure, the private participants (e.g., the automotive and fuel companies) provide some feedback to DOE, but this would remain incomplete without feedback from the actual users.

Thus the learning demonstration program should be considered an essential component of the FreedomCAR and Fuel Partnership. Rather than attempting to demonstrate that these technologies are commercially ready, the program will collect and analyze the experience of the early adopters of hydrogen vehicles and fuels infrastructure technologies in order to inform the research programs. Further, the use of private companies as partners will help disseminate the learning beyond DOE. Recognizing that the learning demonstration program is in its early stages, the committee recommends several points for DOE to consider as this program unfolds.

Recommendation. The FreedomCAR and Fuel Partnership should continue to develop prompt and effective channels of communication among its members to disseminate the learning from the demonstrations. The results should also be disseminated to supporting organizations outside the Partnership in order to promote widespread innovation and competition. But once the learning demonstration for a project has been carried out, the project should be reassessed to see whether further operation is warranted.

Recommendation. DOE management should keep the demonstration projects focused on their primary purpose—the accumulation, analysis, and dissemination of experience from the field.

Safety should be stressed throughout the learning demonstration program, because an accident early on could attract publicity out of proportion to its true consequences.

Recommendation. Among the high priorities for feedback, DOE should identify precursor incidents that point to incipient safety problems and should develop appropriate methods for training first responders to deal with hydrogen-related emergencies.

Recommendation. The FreedomCAR and Fuel Partnership should develop effective channels of communication among its members to disseminate lessons learned and communicate to appropriate organizations outside the Partnership to

The Executive Steering Group (see Figure 2-2), with high-level representation from the three automotive companies, the energy companies, and the government, is responsible for the overall direction of the program. One of its most important responsibilities is to ensure that the strategic direction of the program adjusts as technology advances, the marketplace changes, and new information becomes available. Developing an entirely new, radically different transportation system that meets the goals of this program and our nation is a formidable task. Finding the best way forward demands the best thinking and close cooperation of people with backgrounds and knowledge in many areas. The Executive Steering Group may want to consider chartering an ongoing strategic planning activity to carry out this important task.

The committee has observed that there are many pathways by which to achieve the ultimate goals of this program. Some could involve energy sources other than petroleum or natural gas—perhaps coal or nuclear power. A comprehensive strategic plan should certainly consider all of these options and suggest research in those areas that might contribute to achieving the goals of the program.

Roles of the Federal Government and Industry

The FreedomCAR and Fuel Partnership, like its predecessor, the PNGV program, is based on the sponsorship of research projects which, if successful, will enable the production of new vehicles that enjoy widespread customer acceptance and help to achieve some important societal goals. The research projects are chosen jointly by representatives of government and industry with the intent of developing technologies able to achieve the desired results and capable of being widely deployed in mass-produced vehicles. This is the essence of the Partnership. The government establishes the societal goals to be addressed and funds and manages the program. Industry identifies the technologies needed, sets critical performance and cost parameters, and, in some instances, participates in the research.

Figure 2-3 illustrates this relationship. It shows the FreedomCAR and Fuel Partnership as a framework for directed, focused communication of marketplace performance requirements, public sector needs, and research results. The framework embodies three premises:

- That high-risk, precompetitive research can expand the technical options available to the participating automotive and energy companies;
- That the effectiveness of this precompetitive research will be increased by a more thorough understanding of the realities of the marketplace; and
- That focused channels of communication among these companies as well as between them and the DOE research programs can accelerate the application of this new technology to meet public goals in a competitive marketplace.

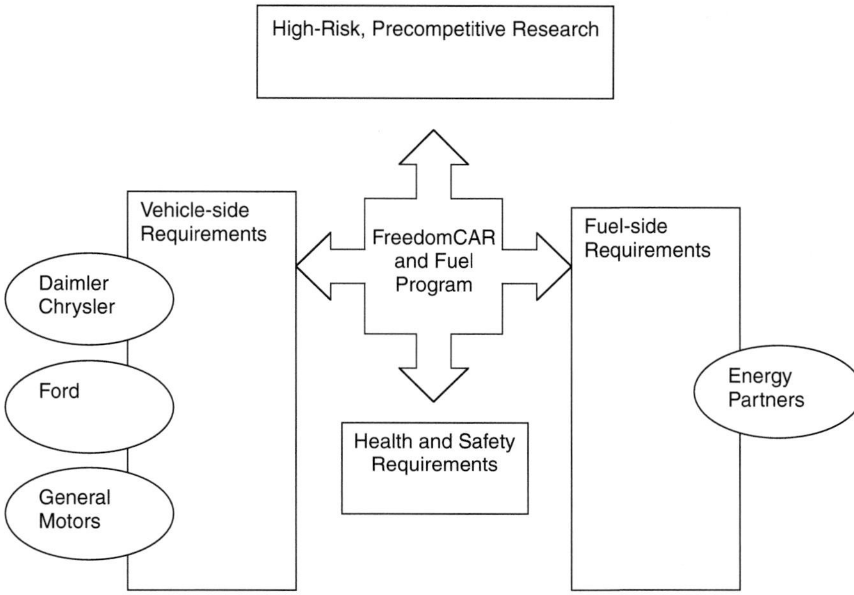

FIGURE 2-3 FreedomCAR and Fuel Partnership as a framework for communication.

For this relationship to work, the program's precompetitive research must be adopted by the industrial partners, developed and made a part of their product and process development programs and, ultimately, successfully introduced in commercially acceptable products. Achieving this result can be a major challenge. Many large companies have experienced difficulty in transitioning even their internal research results to commercial products. It is too early in the FreedomCAR and Fuel Partnership to expect to find significant examples of this part of the process, and the process itself is not clearly defined. But it is not too early for the partners to examine the organizational and other barriers to successful commercialization and find ways to overcome them. This should become an ongoing, shared activity as the partnership matures.

Government and industry leaders in the Partnership appear to recognize that it is a high-priority management task to develop better ways to ensure that this transition of research results to commercial products occurs smoothly. While the target date for deciding on commercialization of a hydrogen-fueled transportation system is 2015, 10 years away, the commercial potential of each technology must be assessed on an ongoing basis. To stress the importance of this step, the committee makes the following recommendation:

Recommendation. The FreedomCAR and Fuel Partnership and USCAR leadership should examine the effectiveness of the current process for transferring technology from DOE projects to within-the-industry activities and develop and implement procedures that will make such transfer as effective as possible.

FreedomCAR in the Policy Context

Two broad public objectives drive federal support for the FreedomCAR and Fuel Partnership:

- To enhance the nation's energy security by reducing and eventually eliminating the use of petroleum in light-duty vehicles.
- To improve the global environment by reducing and eventually eliminating greenhouse gas emissions and criteria pollutants from light-duty vehicles.

If these public objectives are to be achieved in a society that allocates resources chiefly through the marketplace, then consumers must find the new generation of vehicles more attractive than current vehicles. Vehicles with advanced petroleum-saving technologies—improved internal combustion engine (ICE) vehicles, hybrid electric vehicles (HEVs), and hydrogen fuel cell vehicles (HFCVs)—must compete successfully with the market incumbent, the conventional ICE vehicle with a mechanical drive train.

The FreedomCAR and Fuel Partnership has interpreted this competition strictly as a technological challenge on the supply side of the market—vehicles and fuels. Thus the program goals require the new technologies to achieve the road performance, carrying capacity, variety, cost, and safety of the conventional vehicle fleet. The HFCVs face an additional challenge—fueling stations must be everywhere when new HFCVs are introduced on a broad scale.

This section contains some committee observations regarding two policy issues that could influence the pace and ultimate success of the FreedomCAR and Fuel Partnership:

- Demand-side policies that could add market pull to advance the fuel-efficient technologies offered by the program and
- The societal implications of alternative pathways to a hydrogen economy.

Beyond Technology: Policies for Greater Demand Pull

To understand how policies can influence the demand side of the market, consider the basic economics of the consumer purchase decision. As long as fuel remains readily available for about $2.00 per gallon, new car buyers have reason to be ambivalent about paying more for a vehicle with improved fuel economy.[3]

[3]D.L. Greene, "Improving the nation's energy security: Can cars and trucks be made more fuel efficient?," Testimony to the U.S. House of Representatives' Science Committee on February 9, 2005.

This implies that some policy intervention beyond the availability of new technology might accelerate a transition to HEVs and, ultimately, fuel cell vehicles.

The FreedomCAR and Fuel Partnership is currently configured to enhance the capacity of automakers to offer fuel efficiency improvements and other desirable attributes at a cost that attracts buyers, even at current fuel prices. However, these technical programs operate only on the supply side of the market. If the hydrogen economy is to emerge as a reality in the 2015-2020 time period envisioned by Presidential policy statements, then it would be appropriate to consider policies that could stimulate the demand for vehicles offering greater fuel economy and thereby reduce the risk of delaying the hydrogen economy.

The technology goals for the FreedomCAR and Fuel Partnership assume that federal policy will not intervene decisively in the marketplace to tip the competitive balance in favor of fuel cell or hybrid vehicles. Thus, the technology goals are set to match the vehicle performance that U.S. consumers have become accustomed to, and the cost goals are meant to achieve vehicle cost parity. If, especially during the transition period, fuel cell vehicles cost more than vehicles using the competing technologies, policy interventions might be used to facilitate the transition, moderating its targets and speeding the introduction of the new technologies into the marketplace.

Four broad classes of market intervention promise to complement the technology development of the FreedomCAR and Fuel Partnership. The first two, cap-and-trade programs and motor fuel taxes, influence the demand side of the market—that is, they motivate customers to prefer more fuel-efficient vehicles—by raising the price of transportation fuel. The third, Corporate Average Fuel Economy (CAFE) standards, operates chiefly on the supply side by requiring auto and light truck manufacturers to increase vehicle efficiencies or modify the composition of their fleets. And the fourth class, subsidies, can operate on either the supply side or the demand side of the market. To the extent that these or similar policies are implemented, the market penetration of improved ICE drive trains, HEVs, and HFCVs would accelerate. Each will be discussed in turn.

Cap-and-Trade Programs Trading programs have long been used by the federal government to achieve environmental goals—reducing sulfur dioxide emissions from stationary plants and reducing the use of ozone-depleting chemicals, for example. As envisioned in recent literature, the government would set a cap on carbon emissions from all sources, including the production of motor fuels, and issue allowances to burn, produce, and import only the amounts of fuel that correspond to that cap (Pizer and Kopp, 2003). After the initial allocation, firms would be allowed to trade allowances, thus creating a market value for them. The net effect would be an increase in the cost of conventional motor fuels and a greater incentive for customers to prefer HEVs or HFCVs. (The incentive for vehicle owners to modify their driving behavior with the current fleet would provide an ancillary benefit.)

Motor Fuel Taxes In addition to the federal government, states and localities tax motor fuels. These taxes averaged $0.41 per gallon of gasoline in 2002 according to the Congressional Budget Office (CBO). Such taxes directly create an incentive to purchase more fuel-efficient vehicles and influence driving patterns for all vehicles. However, recent analyses suggest that the per-gallon tax would have to be very substantial to make consumers willing to pay a premium for a more efficient vehicle. For example, the CBO estimates that an additional $0.46 per gallon fuel tax would cause a reduction of only 10 percent in gasoline consumption over a 14-year period (CBO, 2004). In addition, the strength of the incentive depends on a consumer perception that the tax-induced increase in the price of gasoline is permanent and structural, not to be offset later by declining fuel costs.

CAFE Standards Currently, each automaker's annual production is divided into three "fleets": imported passenger cars, domestically produced passenger cars, and light trucks (pickups, minivans, and SUVs). If the average fuel economy of each fleet does not meet or exceed the standard set for it, the automaker must pay a penalty. The current standards are set at 27.5 miles per gallon (mpg) for domestic and imported cars, and, until recently, 20.7 mpg for light trucks. In 2003 NHTSA issued a new rule for light trucks that increased the standard to 21.0 mpg for model year (MY) 2005, 21.6 mpg for MY 2006, and 22.2 mpg for MY 2007. Vehicles weighing over 8,500 pounds are exempt. However, fuel economy standards offer no incentive for further innovation once the standards have been achieved. And because they operate only on the supply side of the market, they have no influence on driver behavior, miles traveled, or other important aspects of fuel consumption (NRC, 2002).

If new CAFE standards were proposed, the policy might consider the economic efficiency improvements that could come from a trading approach suggested by the NRC committee that wrote the aforementioned report. Under this approach, automakers facing high costs of improving the fuel economy of their fleets would be allowed to purchase fuel economy credits from automakers able to exceed the standards. This would reduce the disruptive effects of the regulation and improve its economic efficiency (NRC, 2002).

Subsidies A simple buy-down of the costs of the transition could also tip purchase decisions toward a vehicle with improved fuel economy. Though direct payments might be possible, so might be tax incentives. Typically these take the form of tax credits for the purchase of specified vehicle types, such as the credit currently offered for purchase of an HEV. However, the effectiveness of a tax credit incentive varies with the tax status of the individual.

An alternative incentive scheme at time of vehicle purchase is "feebates." Purchasers of low fuel economy vehicles pay an additional fee; those who buy a fuel-efficient vehicle receive a rebate. This fee and rebate system can be revenue neutral. Feebate policies typically charge maximum fees of about $1,500 per vehicle and give maximum rebates of about $500. Such a scheme could provide market incentives that support future CAFE requirements.

The challenges posed by a transition to a hydrogen economy within the time implied by the President's 2003 State of the Union address are substantial, and the likelihood of meeting them might be increased by market intervention. Such intervention would provide a demand-pull to complement the technology-push during the early phases of the transition. Such interventions, of course, lie beyond the scope of the FreedomCAR and Fuel Partnership. Nevertheless, because they could certainly influence the success of the Partnership, they should be included in general discussions of energy policy.[4]

Recommendation. DOE should analyze the implications of alternative market interventions for the technical goals of the FreedomCAR and Fuel Partnership. These implications then could be included in DOE's policy deliberations.

ENVIRONMENTAL IMPACTS OF ALTERNATIVE PATHWAYS

In the long-range portion of the FreedomCAR and Fuel Partnership, DOE has focused on the production of hydrogen and fuel cells to achieve a variety of national goals. In this connection, some observations relating to the environment are appropriate. It is clear that the hydrogen needed to power the vehicles must be based on fossil-free feedstocks, or, if it is based on coal, oil, or methane (natural gas), a huge carbon sequestration program will be required. It is also clear that the one incentive to use hydrogen to carry energy to vehicles is the elimination of their carbon dioxide emissions. Liquid hydrocarbon fuels can be made readily from natural gas or coal and would avoid the difficult production, distribution, and storage problems of hydrogen.

Using hydrogen to reduce greenhouse gases while shifting to a greater use of domestic sources of energy significantly constrains the energy sources that can be used. Specifically, the energy must come from wind machines, photovoltaic cells (PVs), fossil fuels (with sequestration), sustainably grown biomass, nuclear power plants, or hydroelectric dams.[5]

Although hydrogen itself poses little environmental threat when it is oxidized to water in a fuel cell, the same cannot be said for most of the primary energy sources that would be used in its production. A few impacts include the setting aside of large amounts of land to make electric power for electrolysis using PVs, wind turbines, and biomass; the disturbance of wildlife and birds of prey by the large-scale deployment of PV and wind machines; the creation of

---

[4]For a recent discussion of energy policy, see the report by the National Commission on Energy Policy (NCEP, 2004).

[5]Biomass can also serve as a limited gasification feedstock for hydrogen production.

large amounts of low- and high-level radioactive wastes from nuclear power plants; the sequestration of large amounts of CO_2 from fossil sources of hydrogen; and the consumption of large amounts of fresh water if electrolysis is employed to make the hydrogen.

In June 2003, a paper appearing in *Science* reviewed the "expected" impacts on the stratosphere of the hydrogen gas emissions produced for a hydrogen economy (Tromp et al., 2003; Ananthaswamy, 2003). These impacts include a cooling of the stratosphere and a loss of some of the ozone that protects humans and other components of the biosphere from ultraviolet radiation.

Hydrogen has also been associated with increased global warming through atmospheric reactions of trace amounts of hydrogen with ozone and methane, both themselves greenhouse gases (Derwent, 2003, 2004). This effect may require setting a cap on the overall well-to-wheels leakage into the atmosphere.

Although the production of enough hydrogen to meet the entire transportation needs of the country could have significant deleterious environmental impacts, there appears to be no systematic DOE program to identify, study, and model them. The impacts need to be carefully evaluated to ensure that long-term national energy planning succeeds. Such a study would allow comparing potential hydrogen sources, making trade-offs among them, and developing mitigation measures.

Recommendation. DOE, in collaboration with the Environmental Protection Agency, should systematically identify and examine possible long-term ecological and environmental effects of the large-scale use and production of hydrogen from various energy sources.

REFERENCES

Ananthaswamy, A. 2003. "Reality bites for the dream of a hydrogen economy." *New Scientist*, November 15, 2003.

CBO (U.S. Congressional Budget Office). 2004. *Fuel Economy Standards Versus a Gasoline Tax*, Economic and Budget Issue Brief. Available on the Web at <http://www.cbo.gov/showdoc.cfm?index=5159&sequence=0>.

Derwent, R. 2004. "Global warming consequences of a future hydrogen economy." *Issues in Environmental Science and Technology*. Vol. 20: *Transport and the Environment*. London, England: Royal Society of Chemistry.

Derwent, D. 2003. "Climate implications of a hydrogen economy." Berkshire, England: Climate Research, The Meteorology Office. Available on the Web at <http:www.cambrensis.org/hydropresentation.htm>.

DOE (U.S. Department of Energy). 2004a. *Hydrogen Posture Plan: An Integrated Research, Development and Demonstration Plan*. Washington, D.C.: U.S. Department of Energy. Available on the Web at <http://www.eere.energy.gov/hydrogenandfuelcells/pdfs/hydrogen_posture_plan.pdf>.

DOE. 2004b. *Guidance for Safety Aspects of Proposed Hydrogen Projects*. Washington, D.C.: Hydrogen, Fuel Cells and Infrastructure Technologies Program, Office of Energy Efficiency and Renewable Energy. Available on the Web at <http://www.eere.energy.gov/hydrogenandfuelcells/pdfs/safety_guidance.pdf>.

DOE. 2004c. *Partnership Plan, FreedomCAR and Fuel Partnership*. 2004. Washington, D.C.: Office of Energy Efficiency and Renewable Energy, U.S. Department of Energy.

Kinzey, B. 2004. "Hydrogen fueling incident in Burnaby, Canada." *Fuel Cell Summit Newsletter* 5(3). Available on the Pacific Northwest National Laboratory (PNNL) Web site at <http://www.pnl.gov/fuelcells/>.

NCEP (National Commission on Energy Policy). 2004. *Ending the Energy Stalemate: A Bipartisan Strategy to Meet America's Energy Challenges*. Washington, D.C.: National Commission on Energy Policy. Available on the Web at <www.energycommission.org>.

NHTSA (National Highway Traffic Safety Administration). 2004. *NHTSA's Four-Year Plan for Hydrogen, Fuel Cell and Alternative Fuel Vehicle Safety Research*. Available on the Web at <http://dms.dot.gov/>. Search on docket number 18039 and download document 3.

NRC (National Research Council). 2001. *Review of the Research Program of the Partnership for a New Generation of Vehicles*, Seventh Report. Washington, D.C.: National Academy Press.

NRC. 2002. *Effectiveness and Impact of Corporate Average Fuel Economy (CAFE) Standards*. Washington, D.C.: National Academy Press.

NRC/NAE (NRC/National Academy of Engineering). 2004. *The Hydrogen Economy: Opportunities, Costs, Barriers, and R&D Needs*. Washington, D.C.: The National Academies Press.

Parsons Brinckerhoff. 2004. *Support Facilities for Hydrogen-Fueled Vehicles: Conceptual Design and Cost Analysis Study*. Available on the Web at <www.cafcp.org>.

Pizer, W.A., and R.J. Kopp. 2003. *Summary and Analysis of McCain-Lieberman Climate Stewardship Act of 2003*, S.139. Washington, D.C.: Resources for the Future.

Tromp, T.K., R.L. Shia, M. Allen, J.M. Eiler, and Y.L. Yung. 2003. "Potential environmental impact of a hydrogen economy on the stratosphere." *Science* 300 (June 13):1740-1742.

White, J.B. 2005. "An auto-industry dilemma." *The Wall Street Journal Online*, March 21.

this transition to take place, the industry will require enhanced understanding in many areas so that it can develop new vehicle subsystems and vastly improved vehicles. The DOE-sponsored activities described in this section are intended to provide such understanding.

Near-term reductions in fuel consumption and emissions can be accomplished by improving ICEs. Specifically, better understanding of the combustion process and how emissions are produced could both increase efficiency and decrease engine-out emissions. Higher thermal efficiency means reduced fuel consumption and lower engine-out emissions means less extensive, and probably less expensive, exhaust aftertreatment systems. Improved ICEs, which could come in the near term, would benefit both conventional vehicles and HEVs.

The fuel cell subsystem is an energy converter that has the potential to be more efficient than an ICE. However, fuel cell systems of the type deemed appropriate for transportation systems use only hydrogen as fuel. The hydrogen can be stored onboard the vehicle in pure form or it can be extracted from hydrogen-bearing hydrocarbon fuels and water using onboard fuel processors. However, DOE effectively eliminated the latter alternative from its R&D portfolio after years of R&D offered little prospect of meeting essential cost and performance targets within the program time frames. Without this option, sufficient pure hydrogen must be carried onboard the vehicle to meet range requirements. Further, since it is extremely difficult with typical light-duty vehicles to carry hydrogen quantities with an energy content equivalent to that of a typical fuel tank filled with gasoline, it is imperative to minimize fuel consumption. This implies reducing the mass of the vehicle and maximizing the efficiency of the energy converter.

Current experimental hydrogen-fueled fuel cell systems demonstrate efficiencies approaching 50 percent over a fairly wide range of operation. Further, such systems produce zero criteria emissions (occasional discharges of small quantities of hydrogen may occur). However, there are performance, durability, and cost issues to be resolved if fuel cells are to become viable options for personal transportation vehicles.

Hybrid electric vehicles require compact, efficient, and low-cost power electronics and energy storage systems as well as other advanced electrical components to make vehicle costs and weights competitive with conventional vehicles. Many of the same technologies also are applicable to fuel cell vehicles since fuel cell vehicles will be basically electric vehicles with various degrees of hybridization. Consequently, advances in the power electronics and electrical subsystems are critical for improved viability of both mid-term HEVs as well as longer-term fuel cell vehicles.

One important means of minimizing fuel consumption for mid-term HEVs and longer-range fuel cell vehicles is the partial recovery of vehicle kinetic energy during deceleration and stopping. Thus, these vehicles will need some form of energy storage capable of accepting some of this energy (regenerative braking) and providing it back to the drive train for propulsive power. The mostly likely form of such energy storage is electrochemical (batteries), but ultracapacitors are also being investigated. For such relatively small-scale energy storage, the most important parameters are cost per kilowatt, specific power (kW/kg), and cycle and calendar life.

Even though hybrid electric vehicles are currently on the market, the projected cost savings due to higher fuel mileage will probably not offset the higher initial cost of the vehicle at foreseeable fuel prices. This implies that further cost reductions may be necessary for the hybrid vehicles to gain widespread acceptance and have a significant impact on fleet fuel mileage. Such cost reductions require additional understanding in the areas discussed.

Beyond the need for small-scale energy storage required to handle energy from regenerative braking is the need for sufficient on-board energy storage to propel the vehicle for a reasonable range without use of the power plant (e.g., fuel cell or engine). Moving in this direction could add design flexibility to HEVs and reduce some of the performance requirements for the fuel cells (e.g., start-up time and power ramp-up rate) in a fuel cell vehicle. Further increases in on-board energy storage capacity could enable plug-in hybrid vehicles (a vehicle whose battery could be recharged by plugging into a source of electricity while it is parked) or even all-electric vehicles. Both plug-in hybrids and all-electric vehicles could provide the immediate benefit of shifting some transportation energy demand from onboard petroleum-based fuels to the electric grid, which is mostly non-petroleum-based but, of course, not emission-free. The most important parameters for these energy storage systems will be cost per kilowatt-hour, specific energy (kWh/kg), cycle life, and calendar life. These storage systems would also have to maintain adequate specific power (kW/kg), even at low states of charge and low ambient temperatures.

Irrespective of the propulsion technology, reducing the mass of a vehicle for a given mission will have the effect of reducing fuel consumption. However, to conform to FreedomCAR goals, any such mass reduction would have to be accomplished without compromising safety or overall vehicle utility. To accomplish significant weight reductions, several materials, including aluminum, high-strength steel (HSS), and carbon-fiber-reinforced polymer composites could replace a large part of the (mostly) mild steel currently used. Other material substitutions, such as cast magnesium, in other vehicle components could further decrease vehicle weight. Unfortunately, thus far all of these potential material substitutions would result in large cost penalties. Therefore, research in materials production and manufacturing techniques is essential if the mass-reduction benefits of these materials are to be realized.

The following sections discuss in more detail the issues associated with the alternative technologies for vehicle components.

ADVANCED COMBUSTION ENGINES, EMISSION CONTROLS, AND HYDROCARBON FUELS

Introduction

The ICE plays a critical transitional role in achieving the FreedomCAR and Fuel Partnership's long-term goal. If the Partnership meets its objective—namely, of enabling the private sector to make a commercialization decision on fuel cell vehicles by 2015—it would still be decades after that before these vehicles penetrate the market sufficiently to have a measurable impact on total fleet fuel consumption. If commercialization is delayed beyond 2015, the impact will be pushed even further into the future. In contrast, improvements in engine and aftertreatment technologies could be incorporated into a large spectrum of new vehicles quite rapidly. With approximately 16 million new vehicles sold in the United States every year, improving the energy efficiency of vehicles sold now will have near-term and growing impact on the petroleum consumption of the entire vehicle fleet.

FreedomCAR's transition strategy to hydrogen-fueled vehicles envisions a sequence of improved ICEs, increasing use of advanced ICE hybrid vehicles and hydrogen-fueled ICE hybrid vehicles, and—ultimately—a transition to hydrogen-fueled fuel cell vehicles (DOE, 2004a). The focus of the advanced combustion engines and emission controls (ACEC) activity of the FreedomCAR and Fuel Partnership is to improve the efficiency of the engines of these transitional vehicles and reduce their emissions.

To this end ACEC has established a sequence of technical targets during the transition (Table 3-1). The benefits of improved ICEs could begin in the very near term. However, the total impact would be limited by the slow rate of market penetration and the large number—roughly 225 million—of light-duty vehicles in the current fleet. As part of the Government Performance and Results Act, EERE estimated the potential fleet fuel savings from introducing these new technologies to the market. In performing this analysis, it was assumed that the technical targets of Table 3-1 were met, and because these new technologies would add to the cost of the vehicle, the analysis was performed on a cost-competitive basis, assuming that the incremental cost of the technology is paid back by fuel savings in 3 years. Vehicle price and fuel economy were the two most important attributes characterized (DOE, 2004b). The results of the analysis indicated that for light-duty vehicles, oil savings, in millions of barrels per day (mbpd), from diesels and diesel hybrid vehicles would be approximately 0.05 mbpd in 2015, 0.22 mbpd in 2020, and 0.57 mbpd in 2025. These are small reductions considering the light-duty vehicle petroleum consumption in 2004 was approximately 8 mbpd. However, a different rate of market penetration would change these projections. These new technologies would be incorporated into the market more quickly, as enabling technologies, if the market drivers were

TABLE 3-1 Goals and Status of the Advanced Combustion Engines and Emission Controls Activity

		2004 Status		Goals by Fiscal Year			
Goals	Unit	PFI	DI	FY07	FY10	FY13	FY15
For hydrocarbon fuel							
ICE peak brake thermal efficiency	%	30	41	43	45	46	
ICE powertrain cost[a]	$/kW	20	30	35	30	30	
Projected vehicle emissions	Tier 2	<Bin 10	Bin 10	Bin 5	Bin 5	Bin 5	
Emission control fuel economy penalty[b]	%			<5	<4	<3	
Emissions durability	1,000 miles	120	120	120	120	120	120
For hydrogen fuel							
H_2 ICE peak brake thermal efficiency	%	38			45		45
H_2 ICE powertrain cost[a]	$/kW				45		30
Projected vehicle emissions	Tier 2	<Bin 5			Bin 5		Bin 5

NOTE: PFI, port fuel injection; DI, direct injection. The emission standards are based on EPA Tier 2 emission regulations. The description of the test procedures and the regulated levels for each Bin may be found at <http://www.dieselnet.com/standards/us/light.html>.

[a]High-volume production of 500,000 per year.

[b]Fuel economy penalty over combined federal test procedures due to emission control relative to diesel vehicle with 2003 emissions.

SOURCE: K. Howden and R. Peterson, "Advanced combustion and emission controls (ACEC) activities," Presentation to the committee on November 17, 2004.

to change—for example, if the price of fuel were to increase or by any of the policy alternatives discussed in Chapter 2.

Program Technologies

Because a primary goal of the program is to reduce fuel consumption, the most fuel-efficient power plant available is being considered as the basis for the research effort. For light-duty vehicle applications, the compression ignition direct injection (CIDI, diesel) engine is the most fuel-efficient engine currently in production. It is well known, however, that current diesel engines will not meet future emission standards. Therefore, to reduce fuel consumption through the more widespread introduction of diesel engines into the market, advances must be made in emission reduction technologies. Here, the most significant barriers are cost and insufficient fundamental understanding of engine combustion phe-

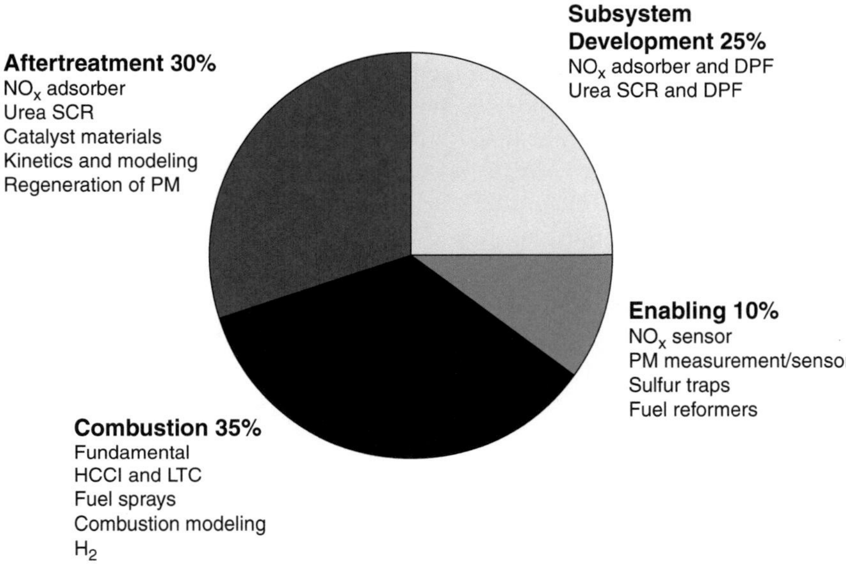

FIGURE 3-1 Technical areas and relative funding for the ACEC activity, FY04. DPF, diesel particulate filter; HCCI, homogeneous charge compression ignition; LTC, low-temperature combustion; PM, particulate matter; SCR, selective catalytic reduction. SOURCE: Response to questions from the committee to DOE, received January 19, 2005.

nomena, exhaust emission control technologies, and engine controls. It is important to realize that the phenomenon of diesel combustion and its emission reduction is fundamentally different from that of the conventional spark ignition engine. The technologies are not transferable. New technologies are required. The research directly addresses all of the barriers except cost. The operating paradigm of the program is to expand the fundamental understanding of combustion, aftertreatment, and controls phenomena in a precompetitive research environment and then let industry address cost as it works to incorporate the new technologies into vehicle power plants.

The individual project topic areas within each research focus are shown in Figure 3-1. Details on the specific research projects within those topics are available in the DOE annual report (DOE, 2003). The vehicle manufacturers all have in-house programs that could be grouped under the topic headings given in Figure 3-1. Government-supported research efforts in these areas differ from industry efforts in the nature of the understanding being sought. Industry is focused on trying to find workable engineering embodiments of the various technologies—for example, establishing the operating parameters of an engine that facilitate low-temperature combustion (LTC); a classification of combustion processes that includes homogeneous charge compression ignition (HCCI) combustion;

and devising strategies to switch from low-temperature combustion to conventional diesel combustion when loads outside the LTC regime are required. The activities supported by federal money at the national laboratories and universities are pursuing a fundamental understanding of the processes that will enable LTC technology to be extended or optimized. For example, federal programs are working to understand the thermochemical interactions that constrain LTC to its current regime. If these are better understood, a wider range of LTC will be possible, with a corresponding improvement in efficiency and reduction in emissions.

As seen from the budget distribution (Figure 3-1), the research effort tilts strongly toward the fundamentals of combustion and aftertreatment, with smaller efforts addressing component subsystems and sensors. The subsystem and sensor research programs are aimed at the question of controls. Control algorithms for the power train system will need inputs from sensors that are monitoring component performance; commands would then be issued, adjusting the engine and aftertreatment system operation.

During this last year the ACEC technical team has shifted the emphasis of its research programs. If the current diesel engine combustion process were left unaltered, the conversion efficiency for the nitrogen oxides (NO_x) and particulate matter (PM) aftertreatment systems would need to be maintained at levels in excess of 90 percent for the lifetime of the vehicle—a huge challenge. The emphasis has therefore shifted from controlling emissions with aftertreatment technologies to reducing the in-cylinder formation of emissions, thereby reducing the burden on exhaust gas aftertreatment. Research has demonstrated that LTC, of which HCCI combustion can be viewed as a subset, has the potential to generate very low levels of NO_x and PM (Akihama et al., 2001; Siebers and Pickett, 2004). The challenge is that to date, LTC has been limited to low-load operation, and the parameter space for controlling it is not well understood.

The research effort on NO_x and particulate matter aftertreatment is very closely aligned with the effort of industrial partners. The formulation and development of new or improved catalysts is conducted primarily in industry, where catalyst suppliers team with vehicle or engine manufacturers. Catalyst mechanisms such as sulfur poisoning, desulfation of lean NO_x traps, thermal aging, and soot filter regeneration are being investigated at the national laboratories using diagnostic microscopy and spectroscopy techniques not readily found in industry. These investigations use both model catalysts and real formulations. The data are being made available for the development of computer simulations of emission control devices, being done at the national laboratories and universities involved in the Crosscut Lean Exhaust Emission Reduction Simulation (CLEERS) activity.

The direct fueling of ICEs with hydrogen is also under investigation. This is an area where partner companies have in-house programs, so the DOE-supported effort is minimal. Using hydrogen in an ICE would of course provide some of the emission benefits at much lower capital cost than changing to fuel cells and electric propulsion. The focus of DOE-funded research in this area is direct injection (DI) hydrogen engines. DI hydrogen engines offer higher power density than engines in which the hydrogen is introduced in the intake manifold.

Fueling of ICEs with hydrogen through an intake manifold is also under investigation. Using hydrogen as an ICE fuel is not new. Because hydrogen has such wide flammability limits and high flame speeds, it may be possible to extend the lean operating limits of the engine, which would reduce fuel consumption and emissions. The issues associated with implementing this are pragmatic and not fundamental. Consequently much of the research on direct fueling on ICEs with hydrogen is being done by industry, with little or no DOE involvement.

One challenge with direct fueling an engine with hydrogen is the loss of volumetric efficiency because the fuel is gaseous. In addition, lean burn mixtures—combined with loss of volumetric efficiency—causes a large power reduction (for the same displacement engine), necessitating a supercharger or turbocharger to bring power levels back up. If a stoichiometric mixture of hydrogen and air is introduced into the engine in the intake manifold, hydrogen will comprise approximately 30 percent of the mixture by volume. This disadvantage can be overcome if hydrogen is directly injected into the cylinder. In this connection some fundamental issues need investigation. The penetration and mixing phenomenon surrounding low-density, high-velocity gas inside a cylinder during direct injection is the subject of DOE-supported investigations at the Sandia Combustion Research Facility. These activities are aimed at increasing the power density of a hydrogen-fueled engine.

Budget and Organization

The FreedomCAR and Fuel Partnership is focused on light-duty passenger vehicles. However, the fundamental knowledge being pursued to enable fuel-efficient technologies is not exclusive to light-duty passenger vehicles. It also applies to engines used in the commercial sector—for example, heavy-duty trucks. To capitalize on these synergies in the combustion engines and emissions technical area, the FreedomCAR and Fuel Partnership is collaborating with the 21st Century Truck Partnership, a partnership of DOE, DOD, EPA, and DOT and 15 industrial partners. The combined budget for advanced combustion for both the FreedomCAR and Fuel Partnership and the 21st Century Truck Partnership for FY04 was $54.4 million. Of this total, $19.5 million was under the direct control of the FreedomCAR and Fuel Partnership. The distribution of these directly controlled funds to the research topics is shown in Figure 3-1.

The vision of the Partnership is that hybrid vehicles will be an important part of the transition. However, because hybrid vehicles are already on the market and are being further developed by the individual automotive companies that are part of the partnership, government-funded efforts for the ACEC activity aim to get a better fundamental understanding of ICEs and aftertreatment systems. This should lead to better engines, which would become part of better hybrid power trains. The technical goals in Table 3-1 are related to engine and aftertreatment performance; hybrid vehicles as such do not appear.

The technical teams, made up of researchers at government laboratories, including the DOE national laboratories, and industry and university laboratories, have established a good process for interaction, feedback, and review. There is interaction between the technical teams of FreedomCAR and 21st Century Truck in the form of crosscut teams, workshops, biannual program reviews, and discussions facilitated by memoranda of understanding. The industrial partners provide input for the setting of research priorities through the workshops and technical reviews, whose outcomes are reflected in DOE solicitations for proposals. As technologies are considered for commercialization the developmental research is performed by vertically integrated teams of industry partners.

Achievements

Quantifying the achievements of the ACEC activity is challenging in that the primary outcome of the government-supported research is new knowledge. In this sense, progress is good. The advanced combustion and emissions control technical team has demonstrated new understanding of the LTC process, including HCCI, and has achieved low-temperature operation in running engines. New understanding of phenomena occurring at the spray nozzle tip, where fuel atomization and air entrainment begin, has been obtained through x-ray imaging, and the boundaries of clean, injection-driven combustion are being expanded. Operational windows of lean catalyst and mechanisms of catalyst poisoning are being studied, and a real-time exhaust stream particulate sensor is being tested. These are important accomplishments; however, it is not known at this time to what extent these advancements in knowledge will be integrated into light-duty vehicle power plants in the near term.

Comments and Recommendations

The various types of ICEs will play a critical transitional role in achieving the FreedomCAR and Fuel Partnership's long-term goal. Even assuming the eventual success of hydrogen as a primary transportation fuel, for several decades ICE will be the automotive power plant that consumes most of the fuel in the fleet. Reducing its fuel consumption and emissions is therefore critically important. Novel emission technologies are needed, and the cooperation of energy companies in such research will increase the likelihood of finding solutions.

The energy companies joined the Partnership in September 2003, adding a new dimension to the program: Now, the impact of fuel modification or substitution on the combustion, emission, and aftertreatment performance can be examined. The variety of fuels that could be investigated is huge, as is the number of

look for solutions to potential emissions problems and solutions for emerging fuels, the fuel infrastructure, and propulsion systems. For example, the emissions problem associated with a distributed hydrogen production system could be quite different and costly, and treatment of emissions from low-temperature combustion could pose new challenges.

Recommendation. Increased emphasis should be placed on novel emission control technologies, and the advanced combustion and emission controls technical team should plan for, analyze, and seek solutions for emission problems associated with emerging fuels, fuel infrastructure, and propulsion systems.

FUEL CELLS

If hydrogen is to account for a significant share of the fuels used for transportation, the transition will be greatly facilitated by fuel cell power systems whose performance and cost are compatible with automotive requirements. This is especially true if costs and onboard storage continue to be problem areas for hydrogen. Fuel cells promise higher conversion efficiencies for hydrogen than ICEs, thus reducing fuel consumption and onboard storage requirements by increasing equivalent fuel economy.

Fuel cell systems are operating successfully on hydrogen in dozens of experimental vehicles in the United States and several other countries. These systems are not compatible, however, with the requirements for mass-manufactured automobiles. They are too expensive, too large, and too heavy, and they have performance problems such as slow start-up and slow power transients, and poor durability, such as degraded performance and limited component life. The status in 2004 of these and other characteristics as well as the targets for 2005 and 2015 are shown in Table 3-2. As can be noted, the 2005 targets (established in 2003) were essentially being met in 2004 in some areas but still had a long way to go in others, such as durability, survivability, and start-up time. Indeed, a review of these parameters not only through the early stages of the FreedomCAR program but also through the entire PNGV program preceding FreedomCAR would show impressive and continuing progress in every area. However, comparing the 2005 targets with the 2015 targets shows clearly that much additional progress is needed.

As delineated in Table 3-3, the fuel cell program is focused on R&D to improve fuel cell technologies for both transportation and stationary applications. The fuel cell program is being implemented by DOE's Office of Hydrogen, Fuel Cells, and Infrastructure Technology Program (HFCIT), which is identifying and developing the critical technology and knowledge needed. The fuel cell part of the FreedomCAR and Fuel Partnership is organized to facilitate the engagement of automobile developers, component suppliers, and related participants so as to meet the 2015 objectives. It is a multidimensional, complex effort spanning many

TABLE 3-2 Technical Targets for an 80-kWe (net) Integrated Transportation Fuel Cell Power System Operating on Direct Hydrogen[a]

			Goals		
Characteristic	Unit	2004 Status	2005	2010	2015
Energy efficiency[b] at 25% rated power	%	59	60	60	60
Energy efficiency at rated power	%	50	50	50	50
Power density	W/L	450	500	650	650
Specific power	W/kg	420	500	650	650
Cost[c]	$/kWe	125	125	45	30
Transient response time	s	<3	2	1	1
Cold start-up time to max power					
at –20°C ambient	s	120	60	30	30
at +20°C ambient	s	60	30	15	15
Emissions		zero	zero	zero	zero
Durability[d]	hr	1,000	2,000[e]	5,000[f]	5,000
Survivability[g]	°C	–20	–30	–40	–40

[a] Targets exclude hydrogen storage and are based on an aerodynamic 2500-lb vehicle.
[b] Ratio of DC output energy to the lower heating value of the input fuel (hydrogen). Peak efficiency occurs at about 25% rated power.
[c] Includes projected cost advantage of high-volume production (500,000 units per year).
[d] Performance targets must be achieved at the end of the durability time period.
[e] Includes thermal cycling.
[f] Includes thermal cycling and realistic drive cycles.
[g] Achieves performance targets after 8-hour cold-soak at temperature.
SOURCE: DOE, 2005.

TABLE 3-3 Funding for Fuel Cell Technology Programs (thousands of dollars)

Technology Component (Interior Appropriations)	FY04	FY05	FY06 Request
Transportation systems	7,317	7,495	7,600
Distributed energy systems	7,249	6,902	7,500
Fuel processor R&D	14,442	9,721	9,900
Stack component R&D	24,551	32,541	34,000
Technology validation	9,828	17,750	24,000
Technical program mgmt support	395	535	600
Total	63,782	74,944	83,600

SOURCE: Provided by DOE in response to a request by the committee. Interior Appropriations refer to the Congressional Subcommittee on the Interior and Related Agencies, which funded these activities.

technologies across a number of organizations, including government agencies, national laboratories, automotive fuel cell developers, original equipment manufacturers (OEMs), and potential suppliers. Owing to the interdependency of the fuel cell and hydrogen, there is also a growing and evolving relationship with the fuel component of the Partnership. To maximize the chances for a successful outcome, plans and schedules must be adhered to, yet the fuel cell program must be flexible enough to allow evolutionary changes to it.

A successful fuel cell development effort will, by itself, not guarantee that the key objectives of the FreedomCAR and Fuel Partnership are met. Success also depends on achieving on-board hydrogen storage, related systems integration, and the broader requirement of a hydrogen fueling infrastructure. The efforts must be fully interactive. For example, fuel quality standards are partly driven by fuel cell developers.

The fuel cell program is focused on the high-risk, precompetitive, industry-wide issues that are hindering the technological and commercial success of the power generation module (stack) and related ancillary processes. The program is broken down into critical and enabling components, with the funding allocated to diverse teams in industry and academia. Selection of the critical components was based on the outcome of workshops organized and facilitated by DOE. Specific technology and cost targets were set by the industry participants and then formulated into a technology milestone plan (DOE, 2005). The primary themes are stack subcomponent (e.g., catalyst, plate hardware, membranes) development, operation, durability, efficiency, and cost; hydrogen fuel as it relates to stack performance; and transportation power systems (cost and performance analyses). The funding for each technology component is presented in Table 3-3. The results of the development efforts are communicated among the technical teams and may ultimately be integrated into the program of an individual developer of automotive fuel cells.

Technology Issues

The PEM fuel cell is based on compartmentalized hardware (cells), in which the reactant gases (H_2 and air) are separated by a membrane. The catalyst, which is in intimate contact with the membrane, initiates the chemical reactions that generate power. Coolants, cell separators, and sensors are other parts of the package. The properties of the membrane and catalyst layers, which impact performance and reliability, are highly dependent on water content. If water is not properly managed in the cell, failure modes are enabled, reducing lifetime, reliability, and durability (the focus of many FreedomCAR projects). The key goals are for a fuel cell vehicle to achieve characteristics, including power and drivability, manufacturability, and cost, that do not compromise consumer expectations. The technical targets necessary to meet these key goals (see Table 3-2) do not take into account the benefits of overall fuel efficiency and protection of the environment. Some of the more important 2015 technical targets are these: 60 percent peak efficiency; 300-mile driving range (vehicle target); 5,000-hr lifetime; \$30/kW cost for the fuel cell system; low Pt catalyst loading (<0.2 g/kW); and 30 s at 20°C cold start-up capability.

Although the near-term FreedomCAR objectives for the development and eventual commercialization of membrane fuel cell technology are currently on target, deployment in the automotive sector in the long term is still at considerable risk. Performance, reliability, durability (current stack lifetimes of 1,000 hr vs. the project goal of 5,000 hr), and cost (estimated at ~\$125/kW in 2004 vs. the goal of \$30/kW, not including the hydrogen storage system) remain major obstacles (DOE, 2005). (The estimated cost of a fuel cell with a hydrogen storage system is about \$175/kW [TIAX, 2004]. However, the validity of these cost estimates is questionable and is discussed further in the section "Cost Issues.")

Some of the critical technical barriers will need breakthrough invention, not just incremental improvements to existing technology. For example, to meet the 2015 targets, it is expected that new materials for membranes, catalysts, catalyst supports, and plate compositions will have to emerge. This adds a high degree of risk to the overall development effort, and the large improvements that are needed make it likely that additional unforeseen roadblocks will arise as the program proceeds.

Because many elements of the program are carryovers from earlier DOE-funded initiatives—for example, the PNGV program (NRC, 2001)—selected activities and development plans are already in place to address the known critical issues. For example, the technical roadmap (DOE, 2005) contains detailed plans and schedules for new membrane development, lower catalyst loadings and increased electrode durability. However, in order to meet the overall targets for fuel cells, it is likely that the various technical scenarios will probably have to be revised as new technologies and issues emerge. The roadmap calls for several specific development activities to take place in parallel. However, the most critical issues may have to be resolved before secondary, less important issues can be addressed. DOE leadership, along with the technical teams, must be alert and make needed changes to the roadmap as they are required.

A number of critical technical issues that are shared by the entire fuel cell industry must eventually be resolved if the overall program goals are to be achieved. Because many of these issues relate to the subcomponents of the fuel cell stack, they involve vendors, suppliers, and OEMs. DOE understands such issues and dynamics and has provided substantial project funding for these areas. The primary technical issues that are being addressed are short-lived membranes, degraded catalysts, suboptimal stack design, and complicated operating strategies. In addition, the high costs of the membrane materials (>\$200/m^2 vs. the target of \$50/m^2 at volume) are problematic. Since most membrane properties are affected by the water concentration, new materials with a lower hydration dependency are required. Current and planned efforts focus on these challenges.

The quantities of platinum catalyst required to meet performance and reliability targets are currently excessive (~1 mg/cm^2 vs. a target of 0.2 mg/cm^2) yet, even with this high loading, cells still exhibit gradual degradation in performance. Existing and new projects will have to come up with novel advances in both catalysts and catalyst layer architecture to resolve this problem. Since catalyst performance is related to operating conditions, gas distribution, and membrane interactions, the technical teams must continue to coordinate and enhance their activities. The 2004 Annual DOE Merit Review, held in Philadelphia in May 2004, reports initial progress in these key areas of more durable membranes and lower catalyst loadings.[1]

The inadequate lifetime and performance characteristics of today's vehicular fuel cell systems can be contrasted with those of stationary applications that use nearly the same technology. Stationary fuel cell systems have demonstrated lifetimes of 8,000 hr compared to the FreedomCAR goal of 5,000 hr (DOD, 2003).[2] DOE is taking a proactive role in learning the causes of such differences by funding selected stationary projects to obtain additional understanding. These projects could be extremely valuable to the project teams by providing an understanding of the importance of differences in operating modes, water management, and impact of environmental conditions. It is difficult to assess the results of such programs at this time but it is expected that recent and future workshops facilitating direct interactions between vehicle and stationary developers will accelerate successful solutions.

There are significant schedule risks attached to current development efforts since there is no clear path to achieving acceptable reliability, performance, and cost solutions. Therefore, the fuel cell program should be carefully monitored and frequently assessed. The present limitations of membranes and electrodes and the status of related stack development should be given the highest priority within HFCIT and in efforts funded in other DOE programs (such as Basic Energy Sciences [BES], in the Office of Science). Longer term, next-generation, outside-the-box concepts of the kind that are typically funded by BES are highly encouraged by the committee. It is not apparent that there are any significant *radical* technical R&D initiatives in the current fuel cell program, and even if a decision to commercialize membranes and electrodes can be made in 2015, DOE should continue to support the development of next-generation materials.

While it is possible that viable solutions may emerge during the remaining decade, there is no guarantee that any of the solutions will meet the stringent requirements of the fuel cell program. Understanding the failure mechanisms of current materials is often the way to develop new concepts and solutions. Such

---

[1]Annual DOE Program Review Proceedings available on the Web at <http://www.eere.energy.gov/hydrogenandfuelcells/2004_annual_review.html>.

[2]Personal communication between committee member Glenn Eisman and Alan Feitelberg, Plug Power. Also see Roger Saillant, "Stationary fuel cells," Plenary address at the 2004 Fuel Cell Seminar, San Antonio, Texas.

mechanisms must be painstakingly studied and delineated and, therefore, it is important that some of our strongest and most talented scientists and engineers, particularly at the national laboratories, focus on them in a precompetitive, open environment.

There are also significant government activities outside the Department of Energy—for example, in the Defense Advanced Research Projects Agency (DARPA), DOD, the Department of Commerce's National Institute of Science and Technology (NIST), and the National Science Foundation—that are contributing to related fuel cell technical issues. FreedomCAR can learn from the knowledge and technical direction of such activities and incorporate them into its own effort. In some cases DOE is already funding such activities. A case in point is the development of a new process to assist in the understanding of fuel cell performance related to water dynamics. The technique, based on the imaging of fuel cells, is under development at the Neutron Research Center at NIST. The NIST effort is important in that it has been able to develop a CAT-scan-like technique to "see" the water within a working cell, enabling the development of concepts that might improve performance. This is a considerable achievement and one of the most significant analytical advances in the membrane fuel cell realm in decades. The NIST facility offers the entire fuel cell community unique research opportunities that previously eluded them.

Cost Issues

The program is in the process of developing a comprehensive cost model for fuel cell systems (TIAX, 2004). Because the program is so complex, the model is a work in progress, and the technology is evolving, it is too early for the committee to assess the viability and accuracy of specific findings at this time. In its present form, the model should be useful for tracking cost changes with design modifications, for establishing goals for component cost, and for prioritizing cost reduction targets. However, the committee found that TIAX predictions had not yet been validated against the cost of existing entire fuel cell systems, making such predictions speculative. Furthermore, details of the economies expected with manufacturing improvements and the large-volume production of components must be carefully documented and validated for reasonableness before a realistic systems cost estimate can be established. The program is currently reporting a large-volume cost for the fuel cell system of $125/kW and $175/kW with hydrogen storage (TIAX, 2004). These estimates are well below the published costs for many experimental systems (as they should be), but it was not possible for the committee to verify their validity.

For viable fuel cell systems costs to reach the FreedomCAR goal of ~$30/kW, low-cost materials, new, high-volume manufacturing technologies, and better performance and reliability must converge. Currently, membranes and catalysts are both costly. Volume manufacturing may alleviate some of the cost burden, but the catalyst pricing is dictated by market dynamics (the price of platinum in 2004 was $900 per ounce, nearly twice the average price during the 1990s). The unpredictable cost of platinum provides further incentive to reduce catalyst loadings and/or develop a nonprecious metal system. While the balance-of-plant costs are not insignificant, they are based on more conventional engineering processes.

The analysis of costs associated with the fuel cell (TIAX, 2004) points to and supports the conclusion that significant advancements must be made in order for the FreedomCAR cost targets to be met. Such analyses are extremely valuable, but because they are based on numerous assumptions, it is too early to attach any great significance to them. This is especially true for elements related to the power generation module, because such technology is still evolving.

Findings and Recommendations

Overall, the committee finds that the DOE fuel cell program is well organized, has a well developed and comprehensive technology roadmap, has focused on the appropriate priorities and initiatives, and has effectively budgeted and applied program and project management processes. The facilitation of the relationships between the various groups and the ability to set targets that are appropriate yet not without risk are examples of successful program implementation by EERE, with execution by the supporting teams.

The results of this review are indicative of a program that is in its early stages. The review finds that a foundation has been laid that will address the critical issues. The reallocation of project funds to more basic and applied research addressing more fundamental issues is an example of recent proactive changes in the program. It should be pointed out that although the program is sound on numerous fronts, its success is highly dependent on communication and cooperation among the members of the fuel cell technical teams. Fortunately, communication between different parts of EERE, as well as among EERE managers at the various technical teams, is well established, though in some cases not formalized.

Procedures are in place for effective communication and project coordination, leading to the aforementioned conclusion that the fuel cell program is being efficiently managed. The annual merit program review is one such procedure. Because the program is so complex and the detailed objectives are spread over so many technical teams and academic and industrial projects, an additional level of scrutiny for specific efforts would be beneficial. It is also expected that the go/no-go decision making process will continue and that additional evaluation mechanisms will be incorporated into the effort.

Recommendation. DOE should broaden its collaboration with industry, academia, and other government agencies on precompetitive, industry-wide technical issues and solutions. Stationary fuel cell developers should be included as well. For example, DOE could sponsor one or more conferences, workshops, debates, or forums to facilitate in-depth interactions or it could set aside some discretionary funds that would allow program managers to accelerate progress on promising new ideas.

Recommendation. To promote new fuel cell water and hardware imaging techniques that could address technical barriers, DOE should enhance its existing collaboration with the NIST Neutron Research Center. DOE should also determine whether similar capabilities exist at the national laboratories and related academic centers so it could capitalize on this significant analytical advancement.

Recommendation. DOE should expand activity and place a higher priority on membrane R&D, new catalyst systems, and electrode design (with the BES program). In particular, the national laboratories and other appropriate scientific centers should be focused on the fundamental failure mechanisms, including a better understanding of the chemistry, physics, and materials involved.

HYDROGEN STORAGE

Hydrogen storage activities are organized within the DOE Hydrogen Fuel Initiative (HFI), with oversight by the hydrogen storage technical team of the FreedomCAR and Fuel Partnership. As the technical team noted in its presentation to the committee's November 2004 meeting, "Hydrogen storage is critical to the success of the hydrogen economy! No current technology meets the needs for hydrogen storage." The goal of the hydrogen storage technical team is to drive the development and demonstration of commercially viable hydrogen storage that meets FreedomCAR goals.

Hydrogen storage is a key enabling technology for the advancement of fuel-cell-powered technologies for all applications—transportation, stationary, and portable. The goals for hydrogen storage on board the vehicle are shown in Table 3-4. These goals are chosen to drive the development of technologies that will compete with current vehicles in terms of cost, performance, and durability. The mass goal is based on providing a vehicle with a driving range of more than 300 miles. The volume required must leave enough space to satisfy other functional needs. Gravimetric, volumetric, and cost targets have been developed for 2010 and 2015. These goals represent a consensus within the DOE hydrogen technology program. They are used by the technical teams to select and evaluate prospective hydrogen storage materials and related technologies. The 2015 energy density goal of 2.7 kWh/L and specific energy goal of 3.0 kWh/kg are half of what is provided by gasoline (6 kWh/L). A bulky hydrogen storage technology that does not meet these goals would increase fuel consumption and decrease the useable space available on-board the vehicle. Other targets include system fill time, temperature, pressure, flow rates, cycle life, and transient response time.

No current hydrogen storage technology meets the 2015 target. Figure 3-2 shows estimates provided by DOE to the committee for current storage technolo-

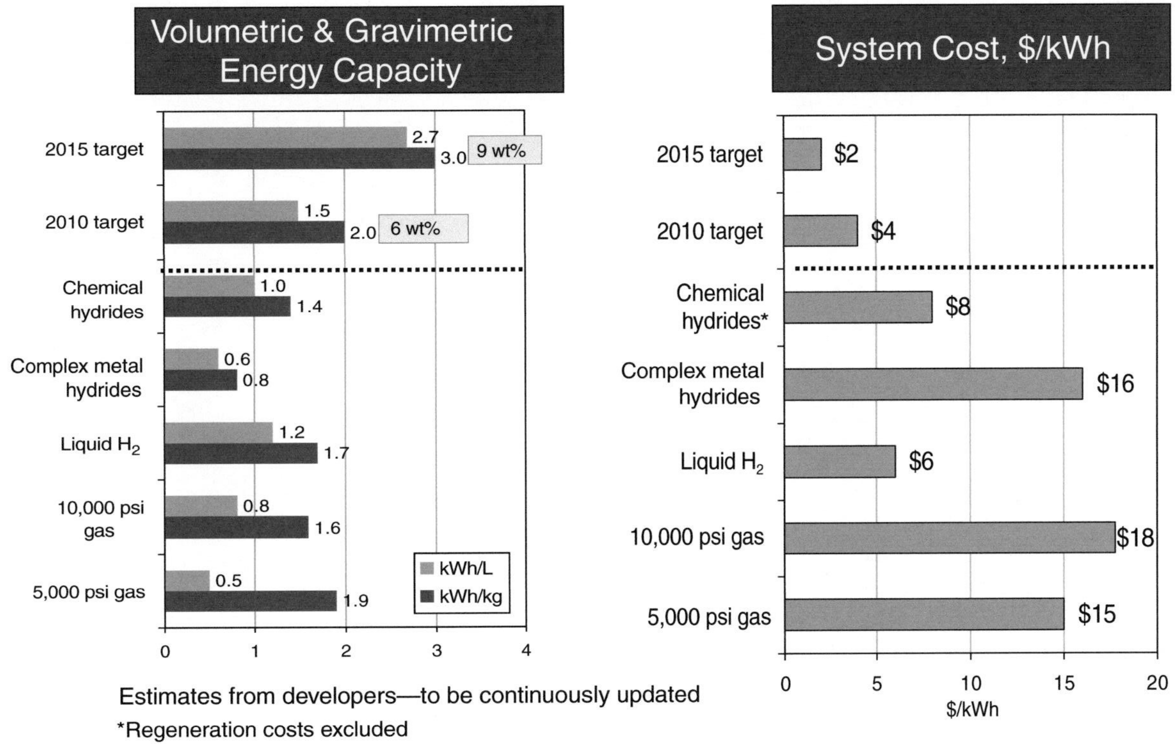

FIGURE 3-2 Status of hydrogen storage technologies relative to targets. No current technology meets these targets. SOURCE: S. Satyapal, S. Jorgensen, and F. Bavarian, "Hydrogen storage joint technical team," Presentation to the committee on November 18, 2004.

In all, 71 hydrogen storage projects were funded in FY05, for a total of $23,654,000. Three centers of excellence were established under the leadership of the national laboratories. Other areas receiving funding are the new, independent, so-called Grand Challenge projects being carried out at universities, industry, and national laboratories. New materials and concepts, physical storage off-board, and analysis are also being studied. Physical storage off-board is important because off-board storage will be required at all hydrogen fueling stations.

Tanks for the on-board storage of compressed and liquid hydrogen are being developed by the private sector and are now only a small effort within the program. The focus of the work on tanks within the program is on cost reduction and advanced concepts such as cryocompressed hydrogen and conformable tanks in various shapes. The primary cost driver is tank materials, with carbon fiber representing 40-80 percent of the material cost. Future efforts in compressed and liquid hydrogen storage tanks will involve off-board storage. Metal hydride hydrogen storage materials are not expected to meet the targets, but work in this area is providing an understanding of storage capacity and reversibility. Carbon-based materials for hydrogen storage will focus on reproducibility of results, on-site peer review of projects, and verification of results using standard materials. A go/no-go decision point on carbon nanotubes is scheduled for the fourth quarter of FY06. In chemical hydrogen storage, the storage material is recharged with hydrogen off-board the vehicle and, following use, is recovered and recycled. A chemical composition with a hydrogen capacity of 5.5 percent by weight has been demonstrated. The main technical issue that needs to be addressed for this approach to be viable is off-board regeneration efficiency. Reversible chemical hydrogen is the focus of many exploratory projects. Reversible storage would permit charging the system with hydrogen on-board the vehicle and eliminate recovery and reprocessing of the spent fuel material.

Findings and Recommendations

The hydrogen storage technical team is newly formed and received significant funding only in FY05. Further, DOE's Office of Science basic research on hydrogen storage was first funded in FY05 as well. Although many approaches are worthy of consideration, most will not be able to demonstrate adequate capacity and/or suitable storage release characteristics.

Recommendation. In view of the exploratory nature of the work and the need to take technical risk and thereby foster discovery, DOE should check progress at appropriate times with go/no-go decisions. In this way, new ideas are able to emerge and the most promising approaches are adequately supported.

A center-of-excellence approach involving collaboration among the national laboratories, universities, and industry is being taken by the hydrogen storage program. Three centers of excellence have been selected based on a program solicitation: one on metal hydrides, one on chemical hydrides, and one on solid state, carbon-based materials. The center of excellence approach should enable focused interdisciplinary research by teams of researchers that provide for sharing of facilities and rapid dissemination of findings. The FY04 and FY05 hydrogen storage project participants at the time of the committee review are shown in Appendix E.

Recommendation. The center-of-excellence research model should be carefully evaluated in parallel with peer review of the research. The committee believes centers of excellence are a good concept, but DOE should wait for an evaluation of the three centers' performance before expanding the concept to other areas of research.

Recommendation. In view of the risk posed to the entire hydrogen program by the need for a viable hydrogen storage system, the hydrogen storage technical team and the FreedomCAR and Fuel Partnership leadership team should report annually to all program participants, DOE, and Congress on the state of hydrogen storage technology worldwide relative to the goals and targets of the program.

ELECTRICAL ENERGY STORAGE

The FreedomCAR and Vehicle Technologies (FCVT) program is responsible for advancing the development of energy storage systems. The primary focus of the effort is the development of advanced batteries and includes work on ultracapacitors. Advanced battery work in support of all light- and heavy-duty vehicles is conducted through this program. Energy storage technologies are critical enablers for the development of fuel-efficient ICE hybrid electric vehicles (HEVs) as well as fuel cell hybrid electric vehicles (FCHEVs). Advanced batteries can also be the primary source of power for plug-in hybrids and electric vehicles that provide alternatives to reduced petroleum consumption.

DOE has supported battery research for a long time. With the formation of the United States Advanced Battery Consortium (USABC) in 1991, followed by the establishment of the PNGV program in 1993, the battery effort was directed more toward large battery modules. The PNGV program advanced technologies for nickel metal hydride (NiMH), lithium ion (Li-ion), and other rechargeable batteries for hybrid automotive applications. At that time it was realized that most battery systems under consideration had one or more problems—such as abuse tolerance, cost, and calendar life—limiting their potential use. Thus in 1997 the DOE initiated applied battery research, mainly based at five DOE national laboratories. In 2000, a long-term exploratory research activity program was organized to understand fundamental impediments to the development of advanced batteries.

Today the technology effort for storing electrical energy is organized into three subactivities:

- Battery development, the primary activity, consists of full battery system module development, technology assessment, and benchmark testing.
- Applied battery research focuses primarily on gaining an understanding of failure modes and limiting parameters of the Li-ion system that currently is closest to meeting the technical goals of HEVs.
- Long-term exploratory research focuses on specific electrochemical systems to solve fundamental problems limiting the performance of advanced battery systems.

DOE undertakes these efforts in collaboration with USABC, and the work is conducted at battery developers, DOE national laboratories, and universities and through Small Business Innovation Research (SBIR) grants. The total budget for all energy storage technologies in FY04 and FY05 was $22.3 and $23 million, respectively. Of this total, most of the funds, over 75 percent, are used to fund the battery development effort. In FY05, $17.4 million is allocated to battery development, $1.4 million to applied battery research, and $4.2 million to long-term exploratory research.

The plan of FCVT is to develop affordable advanced batteries covering the full range of applications, including start/stop 42-V systems, power assist for HEVs, FCHEVs, and battery electric vehicles (EVs). The technical targets for the various applications are listed in Table 3-5. As noted in FCVT's multiyear plan, a primary goal of the energy storage program is to develop by 2010 an electric drive train that includes a battery with a 15-year life at 300 Wh of available energy, discharge power of 25 kW for 18 s, and a $20/kW cost (DOE, 2004a).

Program Status and Assessment

The main achievement of the battery effort is the demonstration of high-power Li-ion batteries that will be able to meet or exceed most of the performance targets for an HEV, including specific power, power density, specific energy density, and cycle life. Battery calendar life is now estimated to be greater than 10 years vs. the target of 15 years. These batteries have two major deficiencies: abuse tolerance and cost. Li-ion high-power batteries are not intrinsically tolerant of abuse such as short circuits, overcharge or overdischarge, vibration, and fire. Some of these issues can be addressed by external electronic control; however, it is imperative to continue to look for battery chemistries that are resistant to voltage or thermal abuse.

Partly to address abuse tolerance, the battery effort has programs to develop Li-ion/gel polymer and lithium sulfur (Li/S) batteries. The gel electrolyte in a Li-ion battery is expected on the one hand to reduce the rate at which the electrolyte reacts with the electrodes during abuse conditions. On the other hand, it may reduce the power capability of the battery, particularly at low temperatures. The Li/S battery is expected to be more abuse tolerant due to a sulfur shuttle reaction. In addition, it has the theoretical potential to meet all the EV performance targets.

However, several technical barriers have to be overcome to make the Li/S chemistry work efficiently, particularly dendritic growth of metallic lithium during cycling. The Li-ion/gel polymer battery and the Li/S battery are being tested to determine how they will meet the various performance targets in Table 3-5. A go/no-go decision based on the performance data will be made in FY06.

Further understanding and control of thermal abuse is being done through battery thermal management studies at the National Renewable Energy Laboratory (NREL) using thermal modeling, characterization, and control. This work is critical, not only for managing thermal abuse conditions but also for achieving more uniform temperature control over the battery module, which should increase battery life.

Battery development teams are working on several fronts to reduce the cost of the Li-ion battery. They are assessing new cathode materials, such as a spinel-based lithium manganese oxide and lithium iron phosphate as an alternative cathode material for Li-ion batteries. These materials are a lower cost alternative to the lithium cobalt oxides used today and may have greater stability under abuse conditions. Other cost reduction efforts are based on earlier studies showing that the cost of nonactive material in a high-power battery, particularly the separator material, can exceed the cost of the active material. Thus, support is being provided to develop low-cost polypropylene-based materials with a cost goal for the separator of $\$1/m^2$. The goal is to have a direct replacement for the current separator material that is cheaper and more stable at high temperature.

The main barriers for high-power batteries—abuse tolerance, cost, and calendar life—are also being pursued by the applied battery research and the long-term exploratory research groups. Both groups are looking at newer materials and electrochemical couples to obtain a basic understanding of the failure mechanisms and factors that limit performance of the systems. The applied battery research activity is conducted primarily at five national laboratories—Argonne National Laboratory, Brookhaven National Laboratory, Idaho National Engineering and Environmental Laboratory (INEEL), Lawrence Berkeley National Laboratory (LBNL), and Sandia National Laboratories (SNL). The focus here is to work on a second-generation Li-ion system, determine performance and failure mechanisms, and relate them to the individual components of the system. The long-term battery research is carried out primarily at LBNL and looks at the performance of a large number of promising electrode materials and electrolytes and determines their limitations by advanced material diagnostics and sophisticated modeling studies.

The committee commends the FCVT program for expanding its primary focus on battery development to include both the basic and applied research necessary to enhance the performance of the battery. The committee believes strongly that this expanded process should continue and that funding and effort on long-term exploratory research and applied battery research should be accelerated. It is clear that solutions to the main barriers of abuse tolerance, cost, and calendar life for high-power batteries will come only with the introduction of new materials and electrochemical couples and from a better understanding of the factors that limit battery performance. Thus the efforts being conducted in the

TABLE 3-5 Technical Targets for Electrochemical Storage

		42-volt			HEV (Power-Assist)		Fuel Cell Vehicle	Battery EV	
Target	Unit	Stop Start	M-HEV	P-HEV	Low Power	High Power	Low Power	Commercial-ization	Long Term
Discharge power	kW	6 for 2 s	13 for 2 s	18 for 10 s	25 for 10 s	40 for 10 s	25 for 18 s		
Specific power-discharge, 80% DOD/30 s	W/kg							300	400
Regenerative pulse	kW	N/A	8 for 2 s	18 for 2 s	20 for 10 s	35 for 10 s	25 for 5 s		
Specific power-regeneration, 20% DOD/10 s	W/kg							150	200
Engine-off accessory load	kW	←	3 for 5 min	→					
Recharge rate	kW	2.4	2.6	4.5					
Power density	W/L							460	600
Available energy at 3 kW	Wh	250	300	700	300	500	250		
Specific energy—C/3 discharge rate	Wh/kg							150	200
Energy density—C/3 discharge rate	Wh/L							230	300
Specific power/specific energy ratio	h^{-1}							2:1	2:1
Total energy	kWh							40	40

Energy efficiency on load profile	%	90	90	90	90	90	90		
Cycle life profiles (engine starts)	cycle	←	450,000	→	300,000	300,000	TBD	1,000 at 80% DOD	
Calendar life	year	15	15	15	15	15	15	10	10
Cold cranking power at –30°C	kW	←	8 at 21V minimum for 2 s	→	5 for 2 s	7 for 2 s	5 for TBD min		
Maximum system weight	kg	10	25	35	40	60	32		
Maximum system volume	liter	9	20	28	32	45	26		
Price at 100,000 units/year	$	150	260	360	500	800	400		
Price for 25,000 units (40 kWh)	$/kWh							<150	100
Maximum operating voltage	Vdc	48	48	48	400	400	440		
Minimum operating voltage	Vdc	27	27	27	←	>0.55 x V_{max}	→		
Maximum self-discharge	Wh/d	<20	<20	<20	50	50	50		
Operating temperature range	°C	←	–30 to +52	→	←	–30 to +52	→	–40 to +50	–40 to +85
Survival temperature range	°C	←	–46 to +66	→	←	–46 to +66	→		

NOTE: M-HEV, mild hybrid electric vehicle; P-HEV, power assist hybrid electric vehicle; DOD, depth of discharge.
SOURCE: T. Duong and A. Habb, "Electrochemical energy storage," Presentation to the committee on November 17, 2004.

applied and long-term research groups are crucial to meeting all the target goals of a high-power battery. At present less than 25 percent of the total energy storage budget is allocated to the applied and long-term research. This share should be increased significantly.

Most of the work sponsored by the energy storage technology team is directed to the development of high-power batteries. Specifically, in FY04, 80 percent of the funds were spent on high-power batteries for HEVs. Some funds were used to develop high-energy batteries to meet the target for an EV. The applied and long-term exploratory research on new materials and electrochemical couples is also primarily for high-energy batteries. The committee recognizes that the distinction between a high-power and high-energy battery is somewhat arbitrary, particularly when attempting to gain a basic understanding of the factors limiting the performance of the battery. The fundamental property of a battery is its specific energy, and generally the specific power is determined by design optimization for a given application. Thus, the Li-ion battery is a high-energy battery that has been optimized to meet the power and cycle life requirements of a high-power battery for HEV applications, and increased effort on high-energy batteries will increase the likelihood of meeting the high-power battery goals for hybrid application.

The target requirements for the HEV listed in Table 3-5 show that as one moves from a 42-V application to a more demanding power-assist HEV and then to a FCHEV, the power requirement increases from 13 kW to 25 kW. Not only is the power increased for the more demanding hybrid applications, but it is also required for a longer time, increasing from 2 s in the 42-V application to 18 s for a FCHEV (DOE, 2004a). Thus one would require a battery with not only a higher power rating but also significantly higher specific energy. It is clear that various hybrid designs will have different power requirements and there is continuum of energy requirements. Higher energy can be utilized to gain a better mix of power and energy density in battery design for a given application or it can also be used to optimize the size of the ICE engine.

The target requirements for an EV are listed in Table 3-5. The main requirement is a high-energy battery with specific energy of 200 Wh/kg and 2:1 power to energy ratio. An EV represents an alternative route to achieving the primary goal of the FreedomCar and Fuel Partnership: energy independence and an environmentally friendly transportation system. The development of high-energy batteries is consistent with DOE's goal of investing in high-risk technologies. The challenge of such an effort is probably no greater than the challenges of hydrogen storage and the hydrogen infrastructure requirements for a hydrogen fuel cell vehicle. The committee feels that the effort for high-energy batteries should be significantly increased.

Ultracapacitors

The double-layer capacitor (DLC), commonly referred to as an ultracapacitor, is an energy storage device having a state-of-the-art energy density about 1/10

a fuel cell (FC) alone may be compatible with the dynamic requirements of electrical traction, regenerative braking requires that the FC be augmented by an energy storage device—e.g., a battery or an ultracapacitor. This electrical energy storage may also be used to enhance the dynamic performance of the propulsion system by, for instance, providing fast transient power for acceleration. The battery may also be used for drive-away when fuel cell start-up time is excessive. The charge/discharge cycles of this device will require power electronics. Control of the FC itself will also require electrical system controls. The integrating role of the vehicle electrical system makes it an important technology, both functionally and economically. However, the FreedomCAR goal for the electrical and electronics (EE) technical team, as stated, is limited to the propulsion system—that is, the electric machine and the power electronics to drive it.[3] Although the multiyear plan of the FCVT program also states this goal, the surrounding discussion does refer to many of the other EE functions necessary for the complete system (DOE, 2004a; FCVT, 2004).

Program Status and Progress

Within the restricted goals established for the EE systems, the EE technical team has built on the results of the motor and electronics development in the PNGV program. The power electronics technology for FreedomCAR is evolving from the PNGV-funded automotive integrated power module (AIPM) developments. While several companies (one is Rockwell Automation) continue to develop AIPMs that may be applicable to FreedomCAR, the FCVT program is currently funding work at Semikron.

Table 3-6 shows the power electronics and traction motor status in 2003, the 2010 targets, and the gap between status and target. The reported status of the power electronics in November 2004 had not changed from that in 2003. The principal challenges are thermal performance, lifetime, and cost. Although the 2003 status for power electronics shows a lifetime of 15 years, this is with a coolant temperature of 70°C. The lifetime would be considerably less at the 2010 target temperature of 105°C. Although the specific power and volumetric power density of the motor are still shy of their 2010 targets, it is the motor cost and thermal performance that remain the most significant challenges. The committee is not convinced that the motor cost goal of $7/kW is achievable, since the cost is principally a matter of commodity prices (e.g., copper and iron prices), and it seems unlikely that cost can be reduced much through research.

Thermal performance of the power electronics has been a challenge using conventional component technology, and the EE technical team is exploring

[3]S. Rogers and V. Garg, "Electrical and electronic tech team—NAS review," Presentation to the committee on November 18, 2004.

TABLE 3-6 Technical Targets for Power Electronics and Electric Motors

	2003 Status	2010	Gap
Power electronics (inverter/controller)[a]			
Specific power at peak load (kW/kg)	11	>12	1
Volumetric power density (kW/L)	11.5	>12	0.5
Cost[a] ($/kW peak)	6	<5	1
Efficiency (%)	97	97	0
Coolant inlet temperature (°C)	70	105	35
Lifetime (yr)	15	15	0
Electric motors (traction)[a, b]			
Specific power at peak load (kW/kg)	1.0	>1.3	0.3
Volumetric power density (kW/L)	3.5	>5	1.5
Cost ($/kW peak)	15	<7	8
Efficiency (%)	>90 at 35% to 100% maximum speed	>93 at 10% to 100% maximum speed	

[a]The targets are based on a series power train with 30-kW continuous power and 55-kW peak power. Entries for 2003 are taken from AIPM and automotive electric motor drive (AEMD) specifications.
[b]Technical targets include the gearbox and connectors.
SOURCE: FCVT, 2004.

alternatives through an advanced R&D program whose salient elements are the development of silicon carbide (SiC)-based converters and high-temperature capacitors. The Semikron inverter is being retrofitted with SiC diodes, and a number of research programs at different organizations are directed toward higher temperature capacitors. New thermal management techniques applicable to both the electronics and the motor are being explored.

The EE technical team is pursuing the development of an integrated motor controller chip. The preliminary design was completed in October 2004 and testing was to have occurred in December 2004. The vision is that such integration will reduce system costs.

The FreedomCAR and Fuel Partnership and INEEL have benchmarked the commercially available DLCs and identified their potential for HEVs (see discussion of DLCs in the section on electrical energy storage). Oak Ridge National Laboratory (ORNL) has done a detailed analysis of the Toyota Prius second-generation hybrid motor to evaluate its design and performance and the processes used in its manufacture. A very preliminary analysis of the Prius drive electronics has also been performed, with a more detailed analysis of performance and construction planned for the near future. To the committee's knowledge, the results of these analyses have not yet been used for guidance in the specification or design of the FreedomCAR vehicle EE components.

Assessment of the Program

The EE component of the FreedomCAR program is addressing a diversity of challenges. Development programs are spread among national laboratories, universities, commercial contractors, and the three automotive companies (OEMs). The particular problems being addressed by these organizations appear to be well defined and relevant to the FreedomCAR and Fuel Partnership goals. The quality of these activities and their results to date are also good, though several of them are in the early stages.

While all these activities are addressing important EE issues, a process for coordinating their output to address systemic Partnership goals was not apparent in either the presentations or written material provided to the committee. This will be no small task given the diversity and large number of activities.

Of particular interest will be the benchmarking of FreedomCAR EE developments against both the components and the integrated systems of the Toyota Prius. For example, since the Prius's physical and performance characteristics are similar to those of the FCHEV, many of the metrics for the latter should be similar to those for the Prius—for example, thermal performance, cost, efficiency, and energy/power densities. Toyota has invested considerable resources in the continuing redesign of the Prius motors and power electronics, and the FreedomCAR program should exploit this investment to its advantage. The results of the ORNL benchmarking exercise should be used to help establish the starting point for the EE technical team's more aggressive research agenda and goals.

Since electronic controls serve as the interface for all the subsystems in the FCHEV, coordination among the technical teams to assure that the interfaces are correctly defined is crucial for system integration. To date the interaction among the technical teams has been informal and infrequent. The quarterly meetings among technical team chairs and the USCAR FreedomCAR directors address operational issues, not the tactical issues necessary for specifying interface tasks. There is an all-technical-team meeting every other year, but this relaxed schedule cannot achieve the coordination needed. Given its central role at the interfaces, the EE technical team should probably serve as the catalyst for the coordination process. Furthermore, it is not clear to the committee that progress in one area that has implications for the specifications or parameters in other areas is communicated to or recognized by those other areas.

Recommendations

The interfaces among the many subsystems in the FCHEV are not only critical to safe and proper vehicle operation but also may contribute significantly to vehicle cost, and the committee is concerned that this issue is not receiving adequate attention.

Recommendation. The EE technical team should play a leading role in coordinating the specifications for the interfaces among the many vehicle subsystems, using established standards where they exist and accelerating the development of new ones where they are needed.

Recommendation. The EE technical team should identify the R&D path leading to the motor cost goal, or it should reassess that goal.

Recommendation. The EE technical team should use its evaluation of the state of the art of HEV technology to update and establish the team's future research agenda and goals.

Recommendation. The EE technical team should develop a process for coordinating the diverse activities it is overseeing.

Recommendation. Integrating the electronics with the motor may well provide significant cost advantages. The EE technical team should consider these potential advantages and extend Table 3-6 to include aggressive targets for an integrated system in 2010 and 2015.

Recommendation. High-temperature power electronics and advanced thermal management systems will significantly impact the size, weight, cost, and reliability of the EE subsystems. FreedomCAR work in this area appears to be limited to the application of SiC devices to the Semikron inverter. The EE technical team should be aware of and leverage the work on high-temperature semiconductors, packaging, and thermal management being funded by government agencies at universities, commercial organizations, and the national laboratories.

STRUCTURAL MATERIALS

Vehicle programs designed to achieve major fuel economy improvements must incorporate significant weight savings. The widespread application of lightweight materials and innovative manufacturing processes are necessary to attain this goal. The FreedomCAR and Fuel Partnership has set a vehicle weight reduction target of 50 percent, adding the criterion "affordable cost." These objectives in weight and cost are not dissimilar to those in the predecessor PNGV program and thus allow continuation of the materials programs already in place. Perhaps more important, the same materials technical team is in place to continue these efforts, which have been under way for a number of years, without major perturbations in objectives or content. These programs were reviewed extensively by the NRC Standing Committee to Review the Research Program of the PNGV Program in its seventh report (NRC, 2001). The current review of the FreedomCAR and Fuel Partnership will concentrate on the relevance and adequacy of the overall materials programs rather than the specific details of individual programs, which were covered in the previous PNGV report.

Material Thrusts

Virtually all of the important materials programs in place when the FreedomCAR program was initiated have been continued. In summary, the programs consist of R&D on materials known to be capable of producing very significant weight savings when applied extensively throughout the vehicle structure—namely, high-strength steels, aluminum alloys, and carbon-fiber-reinforced polymer (CFRP) composites. In addition, work on the selective application of cast aluminum alloys, cast magnesium alloys, aluminum metal matrix composites, and titanium alloys is under way. Owing to the lower densities (except HSS), all these materials achieve the weight savings, but usually at a very significant cost penalty compared with current materials. Thus, while there are some major technical obstacles to the extensive application of these lightweight materials, the paramount challenge to the program is achieving cost parity, or "affordability." The difficulty of achieving affordability cannot be overemphasized, and all the large research programs should include a roadmap showing how to reach the cost target.

HSS structures are probably the closest to approaching affordability but are only likely to achieve about half of the targeted weight reduction. As documented in the last NRC review of the PNGV program, there is an extensive program under way in the steel industry to maximize weight savings in body structures through optimal use of HSS and innovative manufacturing practices (NRC, 2001). This program, known as the Ultralight Steel Auto Body-Advanced Vehicle Concept (ULSAB-AVC), adequately covers HSS development and capability. Additional research on HSS within the FreedomCAR and Fuel Partnership does not appear to the committee to be necessary. The NRC report recommended that the materials technical team closely monitor the ULSAB-AVC program, not only for the potential use of HSS but for the possible transfer of innovative (weight-saving) manufacturing processes to other low-density materials, in particular aluminum alloys (NRC, 2001). The current committee enthusiastically endorses this recommendation.

The only current competitors to HSS for extensive vehicle applications are aluminum alloys and CFRP composites. These material families are capable of meeting the overall weight-saving target, but their cost penalty is large. While not minimizing the manufacturing difficulties associated with both classes of materials, it is the cost of the feedstock material that will most seriously prevent widespread application. In the case of aluminum alloys, previous projects for demonstrating lower cost feedstock—for example, continuous cast sheet for body structures—have not resulted in the commercial development of any such material. Similarly, CFRP composites await the arrival of low-cost carbon fibers, a holy grail that has eluded the fiber industry for decades and is unlikely to be achieved without the enthusiastic participation of large carbon fiber producers. While the current committee supports some research activities in both aluminum alloys and CFRP, greater

TABLE 3-7 Weight Savings for Lightweight Materials

Lightweight Material	Material Replaced	Mass Reduction (%)	Relative Cost (per part)[a]
HSS	Mild steel	10-24[b]	1
Aluminum	Steel, cast iron	40-60	1.3-2
Magnesium	Steel or cast iron	60-75	1.5-2.5
Magnesium	Aluminum	25-35	1-1.5
Glass FRP[c]	Mild steel	25-35	1-1.5
Carbon FRP[c]	Mild steel	50-65	2-10+
Aluminum MMC[d]	Steel or cast iron	50-65	1.5-3
Titanium	Alloy steel	40-55	1.5-10+
Stainless steel	Mild steel	25-40	1.2-1.7

[a]Includes both materials and manufacturing costs; the lower bound of unity is a future projection.
[b]The lower bound is taken from Powers (2000) and the upper bound from NRC (2000).
[c]FRP, fiber-reinforced polymer.
[d]MMC, metal matrix composite.
SOURCE: Powers, 2000.

efforts to gain the cooperation of the major material manufacturers would clearly be critical to any future long-term use of these materials.

In the R&D areas for more selective applications, the materials technical team reported an increasing interest in magnesium alloys based on their potential to offer major weight savings in cast applications. Programs in this arena seem very appropriate because of the low density of the materials and the significant opportunities in materials development that are necessary for successful applications. It is likely that magnesium materials will be useful only in cast applications, and these should be the focus of the FreedomCAR programs, including significant basic materials research. While the technical team expressed guarded optimism that magnesium sheet structures might have some potential, this topic would not appear to be fruitful without some major substantial participation from the raw materials industry.

For reference purposes, Table 3-7, adapted from the PNGV report (NRC, 2001; Powers, 2000), illustrates the relative costs of low-density materials and associated manufacturing costs. The relative cost column indicates the potential cost penalties resulting from application of the various material technologies under consideration.

Recommendations

The materials technical team has the benefit of several years' experience and obviously operates very cooperatively. It has clearly benefited from the earlier NRC reviews of the PNGV program and encompasses in its R&D portfolio all the opportunities for weight reduction afforded by current and future materials.

Recommendation. The only FreedomCAR effort on HSS should be careful monitoring of outside programs with the objective of adopting novel manufacturing and assembly methods to aluminum structures. This recommendation mirrors the previous NRC recommendation on the PNGV program.

The cost of a high-volume material such as aluminum is not likely to be reduced by federally sponsored research. Such cost reduction will be achieved only with increased application and production.

Recommendation. The most important aspect of the stamped aluminum program is cost reduction, particularly for the feedstock material. Efforts in manufacturing should be limited until progress in the cost area has been achieved.

The fundamental issue with CFRP composites is the development of low-cost carbon fibers. The award of a single research grant to a national laboratory does not reflect the importance of this problem. Low-cost carbon fibers would also reduce the costs of several of the hydrogen storage options.

Recommendation. More extensive research programs on CFRPs, combined with the direct cooperation of the large fiber manufacturers, appear mandatory for any hope of success within the program time frame. Meanwhile, R&D for manufacturing of structures should continue.

Recommendation. Longer-term research programs in magnesium alloys should be funded because of the weight savings these materials could offer. Cast materials should be the primary emphasis, with limited exploratory work on wrought materials. Increased activity in this area is highly recommended.

Recommendation. The materials technical team should provide technical materials input to other technical teams—for example, electronics, the hydrogen onboard supply system, magnets, motors, fuel cell structural issues—where such input would be useful. The team has never been asked to do this, but it could be extremely useful to the overall program.

Recommendation. The materials technical team should provide models of weight reduction/cost trade-offs to the systems analysis and engineering team. This would help define the singular objectives for individual systems and allow some flexibility in the focus of cost reduction efforts.

Recommendation. Overall, since cost reduction is the main need in many of the materials programs, the committee suspects that research activities are of somewhat limited benefit. Thus, much of this research funding might better be expended on other more challenging research areas, such as hydrogen storage materials, batteries, fuel cells, and the infrastructure.

Comment on Vehicle Weight Reduction Target

In the opinion of this review committee, it is extremely unlikely that a 50 percent reduction in vehicle weight (at anywhere close to cost parity) can be achieved in the 2010 to 2012 time frame without increasing vehicle costs substantially. It might be prudent for FreedomCAR to reconsider this probably unattainable target and adopt a more realistic goal–for example, a 30 percent overall weight reduction with minimal (say, <5 percent) cost penalty. The overall system objectives could either directly reflect this change, or targets in other systems areas might be adjusted in compensation. The only other alternative is to maintain the current 50 percent weight saving goal but allow for a significant cost penalty—probably unacceptable to both the automotive industry and the automotive consumer.

REFERENCES

Akihama, K., Y. Takatori, K. Inagaki, S. Sasaki, and A.M. Dean. 2001. "Mechanism of the smokeless rich diesel combustion by reducing temperature." SAE paper 2001-01-0655.

DOD. 2003. *Residential Fuel Cell Demonstration Site Data.* Available on the Web at <http://dodfuelcell.com/res/site_performance.php4>.

DOE (U.S. Department of Energy). 2003. *Advanced Combustion Engine R&D. 2003 Annual Progress Report.* Office of Energy Efficiency and Renewable Energy. Available on the Web at <www.eere.energy.gov/vehiclesandfuels/pdfs/program/2003_pr_adv_cidi.pdf>.

DOE. 2004a. *FreedomCAR and Vehicle Technologies Multi-Year Program Plan.* Washington, D.C.: U.S. Department of Energy, Office of Energy Efficiency and Renewable Energy. Available on the Web at: <http://www.eere.energy.gov/vehiclesandfuels/resources/fcvt_mypp.shtml>.

DOE. 2004b. *Projected Benefits of Federal Energy Efficiency and Renewable Energy Programs (FY2005-2050), Appendix J.* Available on the Web at <http://www.eere.energy.gov/office_eere/pdfs/gpra_fy05/appendix_j.pdf>.

DOE. 2005. *Hydrogen, Fuel Cells and Infrastructure Technologies Program, Multi-Year Research, Development and Demonstration Plan.* Office of Energy Efficiency and Renewable Energy.

FCVT (FreedomCAR and Vehicle Technologies). 2004. The Advanced Power Electronics and Electric Machines Roadmap (April 14, 2004), EE Technical Team, FreedomCAR and Fuel Partnership. Washington, D.C.: Office of Energy Efficiency and Renewable Energy.

NRC (National Research Council). 2000. *Review of the Research Program of the Partnership for a New Generation of Vehicles, Sixth Report.* Washington, D.C.: National Academy Press.

NRC. 2001. *Review of the Research Program of the Partnership for a New Generation of Vehicles, Seventh Report.* Washington, D.C.: National Academy Press.

Powers, W.F. 2000. "Automotive materials in the 21st century." In *Advanced Material Processes.* Materials Park, Ohio: ASM International.

Siebers, D.L., and L.M. Pickett. 2004. *Aspects of Soot Formation in Diesel Fuel Jets.* THIESEL 2004 Conference on Thermo- and Fluid Dynamic Processes in Diesel Engines.

TIAX LLC. 2004. *Cost Analysis of Fuel Cell Systems for Transportation*, October 20, Reference D0006, SFAA No. DE-SC02-98EE50526, Topic 1 Subtopic 1C. Cambridge, Mass.: TIAX LLC.

Office of Fossil Energy (FE); the Office of Nuclear Energy, Science and Technology (NE); and the Office of Science's Basic Energy Sciences (BES) Program. Overall responsibility for HFI rests with the Hydrogen, Fuel Cells, and Infrastructure Program Manager in EERE. Important elements of the program are hydrogen production, hydrogen delivery and dispensing, hydrogen storage, safety codes and standards, infrastructure validation, and education. For FY05, funding is $169 million for the entire HFCIT program, which includes about $75 million for fuel cells and $38 million for projects in hydrogen production, delivery, and storage (see Chapter 5, Table 5-1). For FY05, $37 million of the HFI program funds are congressionally directed (earmarked). (See Chapter 3, "Vehicle Technologies," for discussion of onboard hydrogen storage for the vehicle and Chapter 2, "Major Crosscutting Issues," for discussion of safety, codes, and standards.)

NRC Report *The Hydrogen Economy*

The NRC/NAE report *The Hydrogen Economy: Opportunities, Costs, Barriers, and R&D Needs* (NRC/NAE, 2004) noted the central importance of the HFCIT program to improving U.S. energy security and environmental protection. It presented recommendations on program plans and operations. In particular, the report emphasized the development of technologies both to facilitate the early transition to the hydrogen economy and to ensure its long-term viability. That report also recommended that the program shift its emphasis in several key areas. The program's management has responded rapidly to these recommendations, most of which have been incorporated into the program during the past several months.

The principal recommendations of the 2004 report on *The Hydrogen Economy* may be summarized as follows:

- DOE should take a systems approach to understand the complex interactions across the well-to-wheels hydrogen system.
- Increased emphasis should be placed on breakthrough research in on-vehicle hydrogen storage systems, fuel cell cost and performance, and photoelectrochemical hydrogen processes. In addition, efforts on distributed—at the filling station—hydrogen generation technologies should be increased to support the early introduction of hydrogen fuel cell vehicles into the market. Further, given the potential importance that coal may play in a future hydrogen system, there should be closer coupling among DOE's hydrogen, fuel cell, and carbon capture and sequestration efforts.
- Increased emphasis should be placed on developing technologies for hydrogen generation and on developing solutions to nontechnical issues for the transition period to the fully functional hydrogen economy.

The committee compliments DOE on rapidly implementing most of the recommendations in *The Hydrogen Economy* and encourages program management to ensure that sufficient effort is devoted to developing technologies and resolving issues for the transition period.

Recommendation. Recognizing that changes in large, complex programs necessarily occur at a measured pace, the committee nevertheless recommends special attention to three areas: the transition from the current ICE/fuels infrastructure to a nascent hydrogen economy; on-vehicle hydrogen storage; and carbon capture and sequestration.

Specific recommendations on hydrogen production, delivery, and dispensing activities in the FreedomCAR and Fuel Program are offered in other sections of this chapter.

Earmarking

The interim milestones of the hydrogen production, delivery, and storage component of the program have been delayed owing to significant congressionally mandated activities (earmarking)—approximately $37 million—in both the FY04 and the FY05 hydrogen program budgets (DOE, 2005). Although DOE has some flexibility to allocate funds not earmarked, budgets for the hydrogen production, delivery, infrastructure, and safety parts of the HFI were reduced by 50 percent. DOE continued to fund 80 percent of the hydrogen storage program because of its critical importance to the success of the FreedomCAR and Fuel Partnership. In contrast, the vehicle and fuel cell efforts in the overall FreedomCAR and Fuel Partnership were not earmarked, creating a disconnect in the ability to reach milestones between the two parts of the program. In the opinion of the committee, the earmarked projects will not help the program meet its goals, and the lower funding on critical projects will reduce its chances of success. The earmarked projects upset the balance of the program because they prevent some work from being done. The earmarked projects do not benefit from technical team input and oversight and are not selected by peer review, nor were they subject to review by the committee.

Recommendation. The committee strongly recommends that the Hydrogen Technology R&D be fully funded at the $99 million level for the areas indicated in the FY06 Presidential budget request to Congress.

FreedomCAR and Fuel Partnership

Within the FreedomCAR and Fuel Partnership, part of the HFI is managed by three new technical teams: (1) fuel/vehicle pathway integration, (2) hydrogen production, and (3) hydrogen delivery. These technical teams were established in 2004, when the energy companies joined the Partnership. While HFI has been progressing for several years, the energy companies have only been part of the

FreedomCAR and Fuel Partnership for the past year. The accomplishments of the new teams so far have been primarily defining and scoping the work to be done.

The committee is impressed by the rapid start-up of the fuel technical teams and impressed by how well the members and DOE are working together. The teams have set aggressive completion targets consistent with the overall hydrogen program goals. The committee encourages the teams to keep moving forward and frequently check their progress with go/no-go decisions and adjust their efforts accordingly. This is particularly important since the Partnership is in its early stages and its direction may shift as new knowledge is acquired.

HYDROGEN FUEL PATHWAYS

The hydrogen fuel/vehicle pathway integration effort is new to the Partnership. This is a very significant addition as the team is charged with looking across the full hydrogen supply chain—well to tank. Specifically, the goal of the integration effort is to (1) analyze issues associated with complete hydrogen production, distribution, and dispensing pathways, (2) provide input to the Partnership on setting targets for individual components, (3) provide input to the Partnership on needs and gaps in the hydrogen analysis program, and (4) work toward full transparency of all analysis activities.[1]

This technical team has an important role to play in providing input and guidance to the new systems analysis efforts in DOE. It also must work with the vehicle systems and engineering analysis team to integrate the entire hydrogen program on a well-to-wheels basis. Targets and milestones for individual components can be analyzed and reset, improving program prioritization and management.

The team has made a lot of progress. It established a set of principles to shape its effort. One of the most important principles is this: "Targets for the cost of hydrogen from energy source to vehicle should be pathway independent." In addition, the team has developed a framework to evolve the hydrogen program technical targets and addressed difficulties in using current DOE technical targets to assess complete hydrogen fuel pathways.

The committee is encouraged by the approach the team has adopted in its "Framework for Pathway Analysis," which will encompass well-to-tank costs, energy use, and CO_2 emissions. There are some significant challenges ahead. Providing input and guidance to layered models created by others, such as the Macro Systems model, the Transition model, and the Systems Analysis Plan, poses a serious coordination challenge.

Systems analysis is an especially important tool to help understand and prepare for the transition to a hydrogen economy. Many technologies may emerge

[1]D.J. Gardner, Jr., and D. Joseck, "Fuel pathway integration tech team-NAS review," Presentation to the committee on November 17, 2004.

but not become widespread. During the transition, a highly dynamic and uncertain phase, divergent vehicle and dispensing designs may appear in niche markets, offered by those seeking to demonstrate the technology and purchased by the early adopters. It will be important to try to understand the conditions under which such technologies might remain in these niche applications or, perhaps, become more widespread. As an example, high-technology all-electric vehicles recently failed to gain market share in competition with the ICE for mainstream auto markets; yet, scaled-down versions of these electric vehicles compete quite well in certain personal transportation applications, such as are found in retirement or private residential communities.

Consequently, although it is not possible to predict which technologies will emerge from their niches to capture mainstream markets, it will be important to understand the technology adoption mechanisms. Therefore, the immediate contribution of these or other niche concepts are the lessons that they provide for other, possibly superior technologies that will eventually prevail in the mainstream marketplace. An appropriate set of systems analyses that model technology adoption could help in understanding and accelerating the transition. For example,

- DOE could learn from and aggregate the experience of niche demonstrations around the country to ensure that others benefit from them;
- The models could guide DOE's technology programs so that they provide the precompetitive technology base that would best support a rapid and effective transition to the mature hydrogen economy; and
- They could address whether DOE's current goals for the cost of delivered hydrogen match the needs of the transition (as opposed to the mature) marketplace.

Recommendation. The committee recommends as follows:

- That DOE further focus the achievements of the fuel/vehicle pathway integration team by placing greater emphasis on the hydrogen transition in its systems analysis work;
- That the results of this systems analysis work be used to assist in identifying needs for the development of codes and standards and for the training of local zoning officials and emergency responders; and
- That DOE apply its systems capabilities to analyze whether the cost goals for hydrogen production, established for a mature hydrogen economy, are appropriate for the transition.

HYDROGEN PRODUCTION

The hydrogen production goals assume that U.S. energy security will best be enhanced by producing hydrogen from a diverse set of feedstocks and that no single candidate feedstock will probably be capable of providing all energy needs over the long term. The mission of the hydrogen production technical team is to "drive the development of commercially viable centralized and distributed hydrogen production technologies that meet the FreedomCAR and Fuel Partnership goals."[2]

The program encompasses the following (energy) sources for the generation of hydrogen: natural gas, coal, nuclear heat, biological systems, wind, and the sun. The overarching technical challenges in all areas are cost reduction, improved energy efficiency, and technological feasibility. The program considers distributed hydrogen generation (where hydrogen is produced at the filling station) as the most viable approach for hydrogen production and hydrogen delivery for the transition period. *The Hydrogen Economy* (NRC/NAE, 2004) estimated that in the most optimistic plausible case, significant hydrogen-fueled vehicle penetration (>50 percent) would not occur before 2035). Initially, then, hydrogen will be produced locally and a hydrogen delivery infrastructure is not required. Furthermore, it is unlikely that there would be investors willing to put significant capital at risk to distribute hydrogen, given all the uncertainties. The first fueling stations will need to be in areas that serve a small local market of vehicles. The volume of hydrogen demand will not be great enough to support central production, and distributed production might not achieve the fuel savings and carbon capture of the ultimate solutions. The reason to start with distributed stations is to get the number of hydrogen-fueled fuel cell vehicles large enough to justify centralized production. In other words, it could solve the significant chicken-and-egg barrier to widespread penetration of hydrogen-fueled fuel cell vehicles into the market; indeed, it might be the only solution.

The technologies for near-term distributed generation are the reforming of natural gas and small-scale water electrolysis. For the longer term, the vision is centralized production of hydrogen that will take advantage of economies of scale and use a more diverse set of feedstocks. However, the centralized approach requires the development of a massive hydrogen distribution infrastructure for hydrogen delivery and dispensing. In addition, since coal would probably play a significant role in a hydrogen economy, carbon capture and storage (CCS)—or sequestration—technologies and systems will have to be developed (NRC/NAE, 2004). From a societal standpoint, these infrastructure issues are some of the most difficult barriers to the program's realization.

In summary, the most cost-efficient means of providing hydrogen in the long term is centralized plants and a network of distribution pipelines. However, distributed hydrogen production will be the means of hydrogen production and delivery during the transition period, which could be long, and, as stated in *The*

[2]P. Devlin and S. Schlasner, "Hydrogen production tech team—NAS peer review," Presentation to the committee on November 17, 2004.

Hydrogen Economy, resources must be applied to distributed technologies in order to meet the need from 2010 to 2030 (NRC/NAE, 2004).

Technical targets for hydrogen production, delivery, and dispensing have been set for 2010 and 2015. The targets are R&D milestones and are different for each feedstock based on factors like the feedstock characteristics and cost, the state of development of the technology, and the expected production unit size. A total of 54 projects are being funded during 2005 in the hydrogen production and delivery area. The total funding is $14,218,000, and projects range in size from $100,000 to $800,000.

The committee considers the interrelationships among the elements of the program to be an essential feature that is being very appropriately addressed both through the working relationships of the DOE program managers and through the coordination of program goals and objectives. However, the overall program would be improved by expanding the scope and frequency of the coordination efforts.

Recommendation. Even closer coordination with other DOE programs would be beneficial, including programs in the Office of Fossil Energy (FE) and the Office of Nuclear Energy, Science and Technology (NE). Representatives from FE and NE should be added to the fuel/vehicle pathway integration and hydrogen production technical teams, and FE and NE should be linked closely with systems analysis efforts in the Hydrogen, Fuel Cells and Infrastructure Technology program.

Recommendation. The committee believes that significant development efforts should be directed to distributed hydrogen production, including natural gas reforming and electrolysis, as well as exploratory work for other distributed generation options.

Coal and Carbon Sequestration

Coal is a viable option for producing hydrogen in very large, centralized plants. The United States has enough coal to make all of the hydrogen that the economy could need for more than 200 years. U.S. estimated recoverable reserves of coal are about 270 billion short tons (EIA, 1999). In addition, new and very promising gasification technology is under development that can lead to high-efficiency hydrogen manufacture at costs comparable with those of gasoline.[3] It would require over 100 million metric tons a year of hydrogen to fuel the

---

[3]The *Hydrogen Economy* estimated the current cost of coal gasification to produce hydrogen at about $2.10/kg H and the future cost at about $1.70/kg H, whereas the gasoline-efficiency-adjusted cost of gasoline for a gasoline-fueled hybrid electric vehicle was estimated to be $2.12/kg H, assuming a petroleum cost of $30/barrel (NRC/NAE, 2004).

entire U.S. light-duty vehicle fleet by 2050, assuming that hydrogen fuel cell development is successful and meets current goals.

However, in a CO_2-constrained world, managing the increased CO_2 emissions from coal could become a significant barrier to its use. Since it is hard at the present time to imagine a long-term hydrogen economy in the United States without coal as a major hydrogen feedstock, the CO_2 issues must be overcome. This is particularly true when the volume of hydrogen required to significantly reduce oil imports and CO_2 emissions is considered.

To put the CO_2 issue into perspective, 6 gigatons (Gton)/year of CO_2 are emitted in the United States, 1.7 Gton of it from transportation. If coal is used to supply the hydrogen-fueled fuel cell vehicle fleet in 2050 (NRC/NAE, 2004), the CO_2 produced for transportation alone would be 6 Gton/year. Thus, CO_2 must be reduced or removed from coal plant emissions. This could potentially be accomplished through CCS—also called carbon sequestration—which involves separating and capturing the CO_2 at the plant, transporting the CO_2 to a disposal location, and storing it underground in depleted reservoirs, coal seams, or saline aquifers.

There are many questions to answer about CCS technology and its environmental impact before it can be concluded that CCS will be successful in managing the CO_2 produced when coal is the source of hydrogen for transportation. For example, the mass of CO_2 that would have to be transported by pipeline would be twice the mass of natural gas transported today. This presents huge infrastructure issues. Also, CO_2 transport and storage present safety issues. In 1986, an 80 million cubic foot eruption of CO_2 in Cameroon killed 1,800 people. Another issue is that there must be tremendous subsurface capacity to be able to handle the high volumes of CO_2 that will be generated over the next several millennia, and they must be able to trap the CO_2 for hundreds of years. While most oil and natural gas reservoirs probably have sufficient trapping capability, they probably have a CO_2 capacity of only a few decades to about 100 years.[4] Saline aquifers and/or deep ocean storage will most likely be required, and very little is known about their suitability. Finally, the costs associated with high-volume CCS are completely unknown. CCS therefore has many issues to resolve before it can be concluded that coal is a viable feedstock for hydrogen in a carbon-constrained world (NRC/NAE, 2004).

---

[4]Estimates of oil and gas reservoir capacities vary, but some estimate a range from about 25 billion tons to 40 or 50 billion tons of carbon in the United States (Beecy et al., 2002); also, G. Hill, "CO_2 capture project: Hydrogen production with geologic sequestration," Presentation to the Committee on Strategies and Alternatives for Future Hydrogen Production and Use on April 23, 2003. Emissions from light-duty vehicles are about 400 metric tons carbon/year in 2000, projected to increase to 700 metric tons carbon by 2050 assuming conventional gasoline-fueled vehicles only (NRC/NAE, 2004). Thus, light-duty vehicle CO_2 emissions might be sequestered in U.S. oil and gas reservoirs for anywhere from a few decades to about 100 years.

The CCS program at DOE is managed by FE. It is funded at \$50 million to \$60 million per year, with more long-term funding planned but not appropriated. The program contains core R&D programs and regional partnerships with industry that include field tests, environmental impact studies, public education programs, and systems studies.[5] The core R&D program includes capture and geological, terrestrial, and ocean sequestration. In addition, the program has a number of milestones, including a significant 2012 goal of predicting CO_2 storage capacities with a precision of ±30 percent. Thus, the program is broad and long term.

The CCS program appears to the committee to encompass most if not all of the areas required to make CCS successful or at least to determine if CCS could prevent high volumes of CO_2 from being added to the atmosphere by the use of coal. However, it is difficult to identify any ties between the CCS and HFCIT programs.

It is very important that a CCS systems team develop an understanding of how the CO_2 delivery infrastructure will be developed and ultimately configured. For example, what would be the best location for a coal plant relative to the associated hydrogen filling stations, the CO_2 sequestration sites, and the coal supply? Successfully dealing with the need for carbon sequestration is critically important to making coal and natural gas acceptable energy sources in a carbon-constrained world. Research in this area should be an integral part of the program.

Recommendation. DOE should create a CCS systems subteam (under the hydrogen production team) in the FreedomCAR and Fuel Partnership and make it part of the overall HFI.

Terrestrial and ocean environments are options that may provide effective carbon storage over long time periods. The HFI will have to understand the real capacity and trapping integrity of hydrocarbon reservoirs and coal seams by 2010 to 2012 in order to determine if funding for the hydrogen from coal program should continue.

Recommendation. The goal of ±30 percent precision in estimating CO_2 capacity should be focused on geological storage.

Recommendation. DOE should strengthen the ties between managers of the CCS effort at HFCIT and managers at FE by developing a specific CCS program for hydrogen within FE. In addition, DOE should increase the shared management responsibility of the CCS program between EERE and FE.

---

[5]L. Miller and S. Klara, "Carbon capture and storage," Presentation to the committee on March 21, 2005.

However, unique storage issues in the chain from production to tank are likely to be found, which could lead to high costs and energy losses. (See Chapter 3, section on hydrogen storage, for a discussion of onboard hydrogen storage for vehicles.)

As discussed in Chapter 2, the learning demonstration programs are very important to validate current component and systems concepts and to uncover previously unknown issues. They will establish many system and engineering parameters for a complete operating hydrogen supply and fuel cell transportation system, especially for addressing the interfaces between the vehicle and the hydrogen fueling appliance, and between the appliance and the on-site production and/or refueling system.

Recommendation. The technical teams working on hydrogen production, delivery, dispensing, and storage should identify the unique R&D needs for hydrogen storage for production, as well as for delivery and dispensing, that are not being adequately addressed by the current project portfolio.

REFERENCES

Beecy, D.A., V.A. Kuuskraa, and C. Schmidt. 2002. "A perspective on the potential role of geologic options in a national carbon management strategy." *Journal of Energy and Environmental Research* 2(1).

DOE (U.S. Department of Energy). 2004. *Hydrogen, Fuel Cells and Infrastructure: Multi-Year Research, Development and Demonstration Plan*, DOE/GO-102003-1741. Washington, D.C.: U.S. Department of Energy, Energy Efficiency and Renewable Energy. Available on the Web at <http://www.eere.energy.gov/hydrogenandfuelcells/mypp/>.

DOE. 2005. Department of Energy FY2006 Congressional Budget Request. DOE/ME-053. Available on the Web at <http://www.mbe.doe.gov/budget/06budget/Start.htm>.

EIA (Energy Information Administration). 1999. *U.S. Coal Reserves: 1997 Update* Available on the Web at <http://www.eia.doe.gov/cneaf/coal/reserves/front-1.html>.

NRC/NAE (National Research Council/National Academy of Engineering). 2004. *The Hydrogen Economy: Opportunities, Costs, Barriers, and R&D Needs*. Washington, D.C.: The National Academies Press.

5

Overall Assessment

This chapter presents an overall assessment of the FreedomCAR and Fuel Partnership, summarizing not only its main achievements thus far, but also the main barriers that remain before the goals of the program can be achieved. Many of these issues are discussed in the preceding chapters. The chapter ends with the committee's observations on the adequacy, balance, and funding of the program.

The committee believes that research in support of the Partnership's vision is justified by the potentially enormous beneficial impact for the nation. At this early stage, no insurmountable barriers to achievement of this vision have been identified, but several critical components of the program have been noted. Specific, quantitative technology and cost goals for 2010 and 2015 have been established by the technical teams. These goals bear on each important element of the program, and the current status of the program relative to these goals is discussed in the body of this report. In view of the large number of unknowns and the need for breakthroughs, the committee does not feel that it is appropriate or useful at this time to speculate on the probability of this program achieving its long-term vision according to its current plan. Funding levels and the consequent research results during the next few years should allow future reviews to make a more firmly based assessment.

MAJOR ACHIEVEMENTS AND TECHNICAL BARRIERS

Major Achievements

Identifying the major achievements associated with the FreedomCAR and Fuel Partnership is challenging, primarily owing to two factors that make this program different from most other government-sponsored programs: (1) many of the technical activities are continuations of activities that began under the Partnership for a New Generation of Vehicles (PNGV) program (or, in some cases, even before the PNGV) and (2) the program is envisioned to be a multidecade program involving not only many technologies but technological challenges ranging from those that will probably be solved soon to those that may never be solved. Even so, there are many noteworthy achievements, both technical and nontechnical, that should be acknowledged.

Nontechnical Achievements

Nontechnical achievements are extremely important, because they provide the mechanisms for pursuing hopefully successful outcomes to the technical challenges. Among the more significant of the nontechnical achievements are these:

- *The overall strategy and implementation plan.* The plan is well thought out and well executed.
- *Active and continuing participation by both energy companies and automobile manufacturers.* Such participation is essential for any hope of identifying and solving the most critical problems and, ultimately, reaching the long-term goals.
- *The formation of numerous expert technical groups (technical teams).* Experts from government and industry are working together to identify the needed research and help advance specific technologies.
- *The creation of a priority activity to minimize the potential negative impact of inconsistent, or nonexistent, codes and standards.* This often-neglected activity is essential for the success of any pathway to the widespread production, transportation, storage, and utilization of hydrogen as a transportation fuel.
- *The development of a well-considered, well-organized, and comprehensive plan.* The plan includes short-term, mid-term, and long-term goals as well as roadmaps—some complete, some still in draft form—for pursuing these goals.
- *The emergence of more comprehensive cost models.* Cost is a barrier to the widespread acceptance of virtually every new technology being pursued, so realistic, viable cost models are extremely important.
- *The decision to create hydrogen storage centers of excellence that are expected to be working this year (2005).*
- *The establishment of the International Partnership for the Hydrogen Economy.* This partnership is a worldwide collaboration on hydrogen technologies involving 15 countries and the European Commission.
- *The convening of workshops to address essentially all of the more challenging technologies in the program.*

- *The establishment of an independent hydrogen storage test facility at the Southwest Research Institute.*
- *The initiation of basic research programs.* Research has started on the direct production of hydrogen from biological systems and from solar energy, and work on high-temperature nuclear heat processes for hydrogen production is being expanded.

Technical Achievements

Modest evolutionary achievements are evident in every area of technology being pursued. The technology areas include those associated with the advanced internal combustion engine (ICE), hybrid electric vehicles (HEVs), and fuel cell vehicles, all of which loosely correlate with the short-term, mid-term, and long-term goals of the program. The achievements include these:

- Gaining a better understanding of low-temperature combustion in ICEs and of the processes that produce emissions through sophisticated experimentation. This better understanding could lead to higher efficiency as well as lower NO_x and particulate production for both advanced ICE and hybrid vehicles. Early tests at 20 percent power have shown NO_x and soot reductions of 90 percent and 20 percent, respectively.
- Developing effective computer codes for vehicle systems analysis. Clearly, the ability to perform virtual evaluations and comparisons is a desirable alternative to building and testing actual hardware and systems. Continuing improvement in modeling and systems analysis will benefit every aspect of the program.
- The TIAX cost model shows the projected cost of compressed hydrogen automotive fuel cell systems (high-volume production) has been lowered from $275/kW (2002) to $175/kW (2004). Note, however, that the committee has not been able to validate these TIAX projections and that questions exist about the absolute values.
- Lowering the projected cost of baseline 25-kW lithium ion battery systems from $1,750 and $70/kW (1999) to $1,200 and $48/kW (2003).
- Advances in many of the longer-term technologies associated with fuel cell components, onboard hydrogen storage, and electrochemical energy storage. Unfortunately, none of these advances is a breakthrough, and breakthroughs are clearly needed at some point for these technologies to become viable.

Technical Barriers

Technical barriers are difficult to quantify. The program is very broad: Efforts range from hydrogen production and distribution to mass-produced, fuel-cell-powered vehicles. Further, the timescale for achieving various program goals range from relatively near term to 20 years or more into the future. The truly formidable technical barriers are those associated with two long-term goals: the extensive use of hydrogen as a transportation fuel and making fuel cells a viable option for powering transportation systems. It should be noted that the long-term goals of the FreedomCAR and Fuel Partnership are "energy freedom," "environmental freedom," and "vehicle freedom" rather than hydrogen fuel and fuel cell power systems per se, but thus far, these are the only options being pursued to achieve those goals. The technical barriers arise in almost every aspect of achieving widespread distribution of affordable hydrogen and in almost every aspect of devising fuel cell technologies for eventual commercialization. However, impressive progress has already been made in these areas, and the timescale is such that much more progress can be expected.

Technical Barriers for Nearer-Term Goals

The nearer-term technical barriers are mainly those associated with improving ICE vehicles and greater market penetration for hybrid vehicles. Among the more significant of these barriers are the following:

- *Affordable lightweight, high-strength materials.* There have been several candidates to replace steel. The continuous casting of aluminum sheet has not been commercialized, and the cost of carbon fibers for the reinforcement of composites remains unacceptably high.
- *Improvements in the thermal efficiency of ICEs, with concurrent reductions in emissions and particulates.* Current experimental efforts are expanding our understanding of the low-temperature combustion processes applicable to both spark ignition and compression ignition engines, but the ultimate goals have not been met.
- *Lower cost, more compact electrochemical energy storage.* Batteries are essential components of most hybrid vehicles and are likely to be for fuel cell vehicles as well. They add weight and increase costs relative to nonhybrid vehicles. Current efforts are directed primarily at lithium batteries, which seem to have the most promise for the needed advancements.
- *Lower cost, more compact electric drive motors and power electronics.* Advances are needed to gain wider acceptance for hybrid vehicles and to hasten the deployment of fuel cell vehicles.
- *Hydrogen production technologies and infrastructure for potential transition to a widespread system featuring hydrogen as the fuel for transportation.*

Technical Barriers for Longer-Term Goals

Although it is possible that other alternative fuels and energy conversion technologies might emerge, the present vision is to realize the widespread environment-friendly production of affordable hydrogen for vehicles powered by affordable, driver-friendly fuel cell systems. The realization of the hydrogen vision appears to be necessary but not sufficient to ensure the emergence of mass-produced fuel cell vehicles. On the other hand, the hydrogen vision could conceivably be realized even if the goal of fuel-cell-powered vehicles is not, since hydrogen can be used to produce power in combustion engines. That makes the infeasibility of widespread and affordable hydrogen a primary (but certainly not the only) barrier to the success of fuel cell vehicles.

Barriers to Fuel Cells Other barriers to the successful deployment of fuel cell systems for transportation include cost and performance. The most recent cost projection, which assumes mass manufacture, is about \$125/kW (\$175/kW including compressed hydrogen storage), roughly four times the cost goal of \$30/kW (not including hydrogen storage) (TIAX, 2004). A number of areas are being pursued for potential cost savings, including the reduction or replacement of precious metal catalysts; less expensive, more durable membrane materials; and reductions in other material and production costs for membrane-electrode assemblies.

Performance barriers generally include factors that would result in operational behaviors inferior to competing ICE vehicles. Among these are start-up times of 1 minute instead of 30 s; possible damage when vehicles are parked at subfreezing (–40°C) temperatures; slow transition times from idle to full power; and unacceptably short expected operational life.

While overcoming the barriers associated with costs and performance is a formidable task, there is no reason to believe they cannot be overcome. Indeed, research activities are under way and progress has already been recorded in each area. Further, the program timescale is long enough that carefully considered approaches to resolving the various issues can be undertaken.

Barriers Associated with Hydrogen At this point, virtually everything associated with the production, distribution, and onboard storage of hydrogen for personal transportation use faces significant barriers. Since hydrogen does not exist naturally in significant concentrations on Earth, it must be produced by extraction from other substances containing hydrogen, such as (but not limited to) hydrocarbons (natural gas, petroleum, coal, etc.) and water. Typically, extraction ensures that the energy produced will be less than the energy value of the source material plus the energy required to produce it. Thus the cost of hydrogen at the fueling station is, and is likely to remain, a barrier. Many biological and electrochemical processes are being pursued to reduce costs, but the technologies are very immature and the probability of success is unknown. Also of great concern is the CO_2 that would be produced when using fossil fuels, especially coal, as a feedstock.

Owing to hydrogen's extremely low energy density, its distribution is difficult and expensive. In cryogenic liquid form (which requires more energy and is even more expensive to produce than compressed gas), hydrogen requires nearly four times as many tanker trucks (assuming the same volume) or pipeline capacity as gasoline to transport the same energy value. Depending on pressure, the volume of compressed hydrogen gas needing to be transported could be 10 times as large. Thus, the transportation of hydrogen adds to the cost barrier posed by its production. Efforts are under way to make localized production a viable and lower-cost alternative to centralized production and long-distance transportation.

Onboard storage of hydrogen remains a significant barrier. Compressed gas at up to 350 bar (~5,000 psi) is routinely used for demonstration vehicles. However, this either limits vehicle range or requires much larger fuel storage tanks than does gasoline to achieve equivalent range. Higher pressure tanks have been developed, up to 700 bar (10,000 psi), which increases range, but at a cost. This cost includes heavier, more costly tanks and additional energy for compressing the hydrogen. Metal hydrides, carbon materials, and other possibilities for storing and releasing higher fractions of hydrogen are being investigated and could reduce the storage barrier.

As with fuel cells, the timescale for resolving the many issues surrounding the use of hydrogen as a transportation fuel covers many years. However, the inability to resolve these issues could prove to be the most difficult barrier facing fuel cell vehicles.

ADEQUACY, BALANCE, AND FUNDING OF THE PROGRAM

DOE's total FY05 budget for hydrogen-related activities (hydrogen technology and fuel cells)—the Hydrogen Fuel Initiative—is about $225 million, while that for vehicle technologies—the FreedomCAR Initiative—is about $85 million. Thus, the total funding of relevance to the charter of the committee is about $310 million. This is depicted in Figure 5-1. The detailed allocation of these funds by major element in the hydrogen program is shown in Table 5-1.

This level of expenditure is consistent with the priorities and recommendations of *The Hydrogen Economy* (NRC/NAE, 2004) and is also consistent with the President's commitment of $1.7 billion over 5 years (FY04-FY08) in his 2003 State of the Union message. The emphasis is on R&D activities related to fuel cell materials and components, hydrogen production and delivery technologies, and hydrogen storage materials. The budget also includes $29.2 million for basic science, which is consistent with *The Hydrogen Economy*, which recommended increased emphasis on the fundamental science related to hydrogen and fuel cell technologies. The Office of Science program just getting started was not reviewed in this study, but future reviews will assess its adequacy.

The FY05 funding for the DOE FreedomCAR and Vehicle Technologies (FCVT) program activity on related programs is $85.3 million (Figure 5-1) and is allocated as shown in Table 5-2. Funding is highest for hybrid and electric propulsion (ca. $40 million), reflecting their critical importance to both advanced

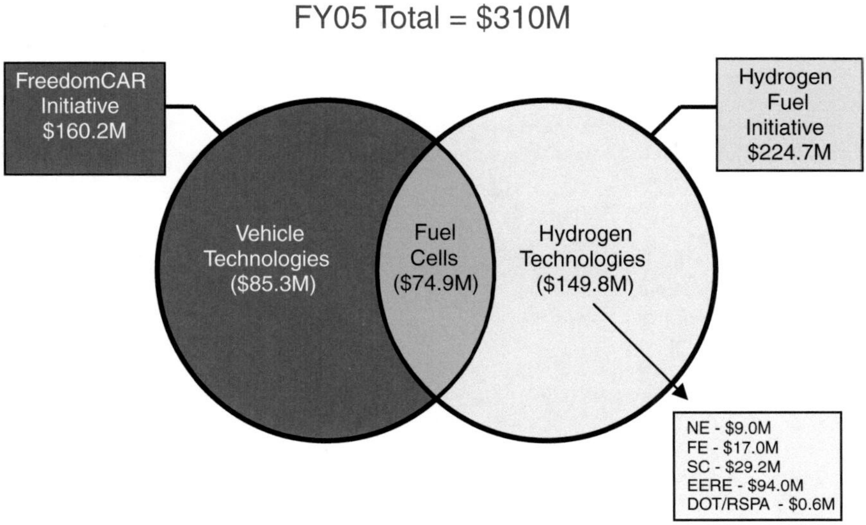

FIGURE 5-1 FreedomCAR and Fuel Partnership funding for FY05 (rounded numbers based on Tables 5-1 and 5-2). SOURCE: Adapted from DOE.

ICEs and fuel cell vehicles. The vehicle program also includes substantial funding for combustion and emissions control (ca. $19 million) and materials technologies (ca. $18.5 million). Education efforts include hydrogen education, which develops and distributes training materials to target audiences; the Graduate Automotive Technology Education (GATE) activity, creating GATE centers of excellence and multidisciplinary curricula and providing funds for research fellowships; and Challenge X, in which university teams partner with General Motors to integrate advanced vehicle technologies and appropriate fuels to minimize the use of petroleum.

The FCVT program is also responsible for the 21st Century Truck (21st CT) Partnership, a partnership similar to the FreedomCAR and Fuels Partnership but involving primarily heavy truck manufacturers. The budget for the 21st CT in FY05 is almost identical to the FCVT budget for FreedomCAR ($86 million versus $89.7 million), and the priority areas are similar, such that, overall, hybrid and electric propulsion received $45 million; advanced ICE/combustion and emissions, $54 million; and materials, $40 million.

The breakdown of the approximately $85 million for FY05 that comes from the FCVT program is national laboratories, 47 percent; industry, 40 percent; federal, 2 percent; consortia, 5 percent; universities, 4 percent; and automotive companies (OEMs), 2 percent. Funding for OEMs is exclusively for competitively selected combustion and emissions control R&D, where each OEM is part of a team that may include engineering companies, suppliers, and energy compa-

TABLE 5-2 DOE Funding Supporting FreedomCAR and Fuel Partnership Goals in the Office of FreedomCAR and Vehicle Technologies (thousands of dollars)

Program Area Request	FY04	FY05	FY06
Vehicle systems	3,659	4,486	4,800
Ancillary systems	1,155	1,268	1,300
Simulation and validation	2,504	3,218	3,500
Innovative concepts: Graduate Automotive Technology Education (GATE)	494	493	500
Hybrid and electric propulsion	38,538	39,885	43,335
Energy storage	22,338	23,073	25,700
Advanced power electronics	13,181	13,168	13,900
Subsystem integration and development	3,019	3,644	3,735
Combustion and emissions control	18,640	18,775	20,765
Materials technologies	18,980	18,437	21,000
Propulsion materials	2,766	1,972	2,000
Lightweight materials	16,214	16,465	19,000
Fuels technologies	4,104	1,367	7,000
Advanced petroleum-based fuels	3,808	0	3,000
Non-petroleum-based fuels	296	1,367	4,000
Technology introduction	889	986	1,300
Technical/program management support	854	853	1,200
Biennial peer review of FreedomCAR	494	0	500
Total	86,652	85,282	100,400

SOURCE: Provided by DOE (on April 26, 2005) in response to a request from the committee. Note that these appropriations are through the Congressional Subcommittee on Interior and Related Agencies.

While the committee endorses the overall size and relative allocation strategy in the hydrogen program budget, there are four areas of concern. First, as discussed in Chapter 2, congressionally directed activities (earmarks) account for 40 percent of the hydrogen technology budget (about $37 million out of $94 million) in FY05 and 16 percent of the total hydrogen budget. The earmarks divert funds from required R&D areas in the program, severely restricting the ability of DOE to effectively manage the program, and have delayed several critical elements of the program, including hydrogen storage and safety.

The second area of concern is carbon sequestration. While large-scale carbon sequestration is not directly within the purview of this committee's study, its feasibility will essentially determine the likelihood of sourcing hydrogen from coal and/or natural gas in a future carbon-constrained environment and consequently affects both the economics and the viability of hydrogen as a future fuel (energy carrier). DOE needs to assess the resources being devoted to this activity in the light of its criticality to the success of the program.

The third area of concern is the balance between basic research and applied development. As illustrated in Figure 5-2, 13 percent of hydrogen program spending is on basic research, with the remaining 87 percent devoted to applied R&D and demonstrations. To some extent, the latter activity will disclose areas requiring more fundamental research and drive the research agenda and funding, but the committee cautions against overemphasis on development (which is usually best performed by the private sector) at the expense of basic research.

The fourth area of concern is the process of innovation. Historically, in fields as disparate as microelectronics and medical devices, pathbreaking commercial innovations have come from start-up companies at least as often as from the industry incumbents. The FreedomCAR and Fuel Partnership should create opportunities for start-up companies to participate in the commercialization process, either independently or in partnership with one of the member companies. This would lead to a more balanced program than one relying on industry incumbents alone.

In summary, there are four areas of concern in the hydrogen program, namely congressional earmarks, carbon sequestration, spending that may be skewed too

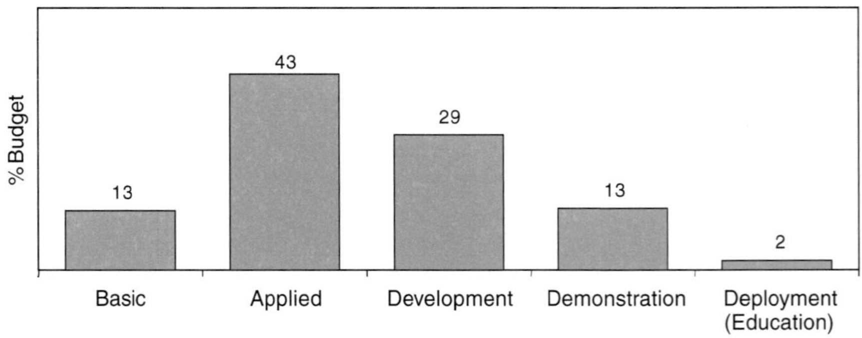

FIGURE 5-2 Distribution of funding for hydrogen technology and fuel cell activities for FY05 by RD&D category. Note that "education" supports technology transfer and adoption. SOURCE: R.F. Moorer, "FreedomCAR and Fuel Partnership peer review," Presentation to the committee on November 17, 2004.

much toward development, and the process of innovation. The committee strongly supports the HEV and advanced ICE spending but, as noted under "Structural Materials," it suggests that some of the materials spending ($19 million in FreedomCAR, $20 million in 21st CT) be reallocated to high-priority research areas.

Finally, the program involves both short-term goals that are related to hydrocarbon-fueled vehicles, used during a transition period, and much longer-term goals aimed at enabling "a clean and sustainable transportation energy future." The committee considers the current split of funding between the long-term and shorter-term goals to be appropriate. Hydrogen-related activities absorb approximately 70 percent of the funds. The remaining funds support the development of transition technologies, where in many cases cost is the most significant barrier.

REFERENCE

NRC/NAE (National Research Council/National Academy of Engineering). 2004. *The Hydrogen Economy: Opportunities, Costs, Barriers, and R&D Needs*. Washington, D.C.: The National Academies Press.

Appendixes

Appendix A

U.S. Council for Automotive Research (USCAR) Consortia

The U.S. Council for Automotive Research (USCAR) has the following consortia with the indicated missions.[1]

- Automotive Composites Consortium (ACC): To conduct joint research programs on structural polymer composites in precompetitive areas that leverage existing resources and enhance competitiveness.
- Electrical Wiring Component Applications Partnership (EWCAP): To permit and encourage cooperative research and development, which includes the joint sharing of technologies and resources to develop common electrical connection systems.
- Environmental Research Consortium (ERC): To improve understanding of the environmental impact of vehicle and manufacturing emissions.
- Low Emissions Technologies R&D Partnership (LEP): To coordinate research and development efforts on emissions control technologies through exchange of technical information and licensing of promising technical breakthroughs.
- Occupant Safety Research Partnership (OSRP): To conduct or direct precompetitive research and development on crash-test dummies and related areas such as modeling, instrumentation, data management and reduction, and subsystem safety test development.
- United States Advanced Battery Consortium (USABC): To pursue research and development of advanced energy systems capable of provid-

[1]Information on USCAR consortia available on the Web at <http://www.uscar.org/consortia&teams/consortiahomepages/cons&ttINDEX.htm>.

ing future generations of electric vehicles with significantly increased range and performance.

- United States Automotive Materials Partnership (USAMP): To conduct vehicle-oriented research and development in materials and materials processing to improve the competitiveness of the U.S. auto industry.
- Vehicle Recycling Partnership (VRP): To promote an integrated approach to the technical and economic feasibility of recycling for vehicles built in North America for the global marketplace.

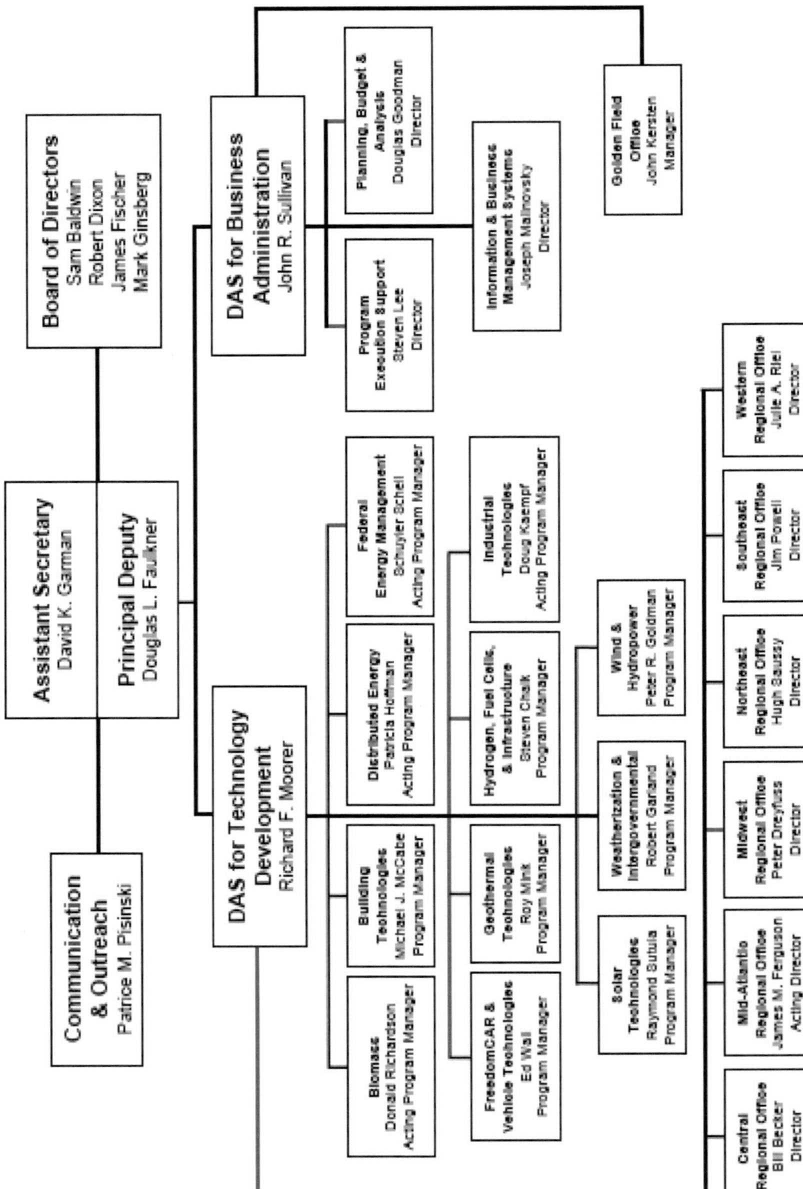
Board of Directors
Sam Baldwin
Robert Dixon
James Fischer
Mark Ginsberg
Assistant Secretary
David K. Garman
Principal Deputy
Douglas L. Faulkner
Communication & Outreach
Patrice M. Pisinski
DAS for Business Administration
John R. Sullivan
Planning, Budget & Analysis
Douglas Goodman
Director
Program Execution Support
Steven Lee
Director
Information & Business Management Systems
Joseph Malinovsky
Director
Golden Field Office
John Kersten
Manager
DAS for Technology Development
Richard F. Moorer
Biomass
Donald Richardson
Acting Program Manager
Building Technologies
Michael J. McCabe
Program Manager
Distributed Energy
Patricia Hoffman
Acting Program Manager
Federal Energy Management
Schuyler Schell
Acting Program Manager
FreedomCAR & Vehicle Technologies
Ed Wall
Program Manager
Geothermal Technologies
Roy Mink
Program Manager
Hydrogen, Fuel Cells, & Infrastructure
Steven Chalk
Program Manager
Industrial Technologies
Doug Kaempf
Acting Program Manager
Solar Technologies
Raymond Sutula
Program Manager
Weatherization & Intergovernmental
Robert Garland
Program Manager
Wind & Hydropower
Peter R. Goldman
Program Manager
Central Regional Office
Bill Becker
Director
Mid-Atlantic Regional Office
James M. Ferguson
Acting Director
Midwest Regional Office
Peter Dreyfuss
Director
Northeast Regional Office
Hugh Saussy
Director
Southeast Regional Office
Jim Powell
Director
Western Regional Office
Julie A. Riel
Director

Appendix C

Biographical Sketches for Committee Members

Craig Marks (NAE), *chair*, is chairman of the board of trustees of Altarum, a not-for-profit scientific R&D organization engaged in the application of advanced information technology to solve national defense, health care, and environmental problems. For 27 years he worked at General Motors in engineering and management positions. He subsequently became vice president of Engineering and Technology for the TRW Automotive Sector and then vice president of Technology and Productivity for the Allied Signal Automotive Sector. In the latter position he headed an automotive research and development center and was responsible for the staff functions of manufacturing; quality, health, safety, and environment; and communications. After retiring, Dr. Marks became an adjunct professor at the University of Michigan, with a joint appointment in the College of Engineering and the School of Business Administration, where he helped found the Joel D. Tauber Manufacturing Institute. He has served on a number of NRC committees, including as chairman, Committee on Review of the Research Program of the Partnership for a New Generation of Vehicles and Committee for the Review of the Intelligent Vehicle Initiative—Phase 2. He is a member of the National Academy of Engineering and a fellow of the Society of Automotive Engineers (SAE) and the Engineering Society of Detroit. Dr. Craig holds B.S., M.S., and Ph.D. degrees in mechanical engineering from the California Institute of Technology.

Peter Beardmore (NAE) was director, Ford Research Laboratory, Ford Motor Company, prior to his retirement in August 2000. His primary research interests are in the deformation and fracture of materials, including extensive research experience in metals, polymers, and composites, and he has published over 83 technical articles. He is a recognized international authority on composite materials and on the application of new materials to automotive structures. His management responsibilities at Ford covered a wide area of research activities relative to the automotive industry, including materials, environmental chemistry, sensor technologies, automotive catalyst development, and the application of modern analytical techniques. He is a member of the American Society for Materials (ASM), The Metallurgical Society (TMS) of AIME, and the Engineering Society of Detroit (ESD). He was elected a fellow of ASM in 1989 and a fellow of ESD in 1991. In 1992, he was elected a member of the National Academy of Engineering. He holds a B.Met. in metallurgy from the University of Sheffield and a Ph.D. in metallurgy from the University of Liverpool.

David L. Bodde serves as a professor and senior fellow at Clemson University. There, he directs innovation and policy at the International Center for Automotive Research. Prior to joining Clemson University, Dr. Bodde held the Charles N. Kimball Chair in Technology and Innovation at the University of Missouri in Kansas City. Dr. Bodde serves on the board of directors of several energy and technology companies, including Great Plains Energy, the Commerce Funds, and EPRI Solutions. His executive experience includes vice president, Midwest Research Institute; assistant director of the Congressional Budget Office; and deputy assistant secretary in the U.S. Department of Energy. He is a member of the NRC's Board on Energy and Environmental Systems and recently served on the Committee on Alternatives and Strategies for Future Hydrogen Production and Use. He was once a soldier and served in the Army in Vietnam. He has a doctorate in business administration from Harvard University, M.S. degrees in nuclear engineering (1972) and management (1973), and a B.S. from the United States Military Academy.

Glenn A. Eisman is director of the Center for Fuel Cell and Hydrogen Research and professor of materials science and engineering at Rensselaer Polytechnic Institute. Dr. Eisman is also principal partner of Eisman Technology Consultants, LLC. His previous positions include chief technology officer, Plug Power, Inc.; technical leader, The Advanced Materials Program, Central Research and New Businesses, Dow Chemical Company; and project leader, discovery research R&D and inorganic chemical research, Dow Chemical Company. Dr. Eisman has extensive experience in R&D and product development for fuel cells, hydrogen technologies, electrochemical engineering, physical and inorganic solid state chemistry, and new technology commercialization and business development. He received the Inventor of the Year Award from Dow Chemical Co. (1993) and is a member of the Electrochemical Society. He received a B.S. in chemistry from Temple University and a Ph.D. in physical inorganic chemistry from Northeastern University.

David Foster is the Phil and Jean Myers Professor of Mechanical Engineering at the University of Wisconsin, where he has been a faculty member since completion of his Ph.D. He teaches and conducts research in thermodynamics, fluid mechanics, and chemical kinetics and emission formation processes in internal combustion engines. He is an active member of the Engine Research Center (ERC), of which he served as the director from 1994 through 1999. The ERC has won multiple Center-of-Excellence competitions for engine research and is world renowned for its accomplishments and its extensive facilities for research on internal combustion engines. He is now codirector of the General Motors–ERC Collaborative Research Laboratory, which was established in 2003.

Professor Foster's research activities focus on the experimental investigation of engine combustion systems and the incorporation of simplified or phenomenological models of emission formation processes into engineering simulations. Most recently he initiated work on understanding the impacts of fuels and combustion processes on the detailed characterization of diesel engine particulate matter, an important basis for understanding and mitigating the health effects of engine emissions and for developing exhaust gas aftertreatment systems.

Professor Foster is an active consultant for industries both here and abroad. He has published extensively in journals of the professional societies serving the combustion and internal engine communities. He is a recipient of the Ralph R. Teetor Award, the Forest R. McFarland Award, and the Lloyd L. Withrow Distinguished Speaker Award of the Society of Automotive Engineers. Dr. Foster is a registered professional engineer in the State of Wisconsin and has won departmental, engineering society, and university awards for his classroom teaching. He served as a member of the NRC Committee on Review of the Research Program of the Partnership for a New Generation of Vehicles for 6 years. He is a fellow of SAE. He received his B.S.M.E. and M.S.M.E. from the University of Wisconsin, Madison, and his Ph.D. in mechanical engineering from the Massachusetts Institute of Technology (MIT).

John B. Heywood (NAE) is Sun Jae Professor of Mechanical Engineering at the Massachusetts Institute of Technology (MIT) and director of the Sloan Automotive Laboratory. Dr. Heywood's research has focused on understanding and explaining the processes that govern the operation and design of internal combustion engines and their fuels requirements. Major research activities include engine combustion, pollutant formation, and operating and emissions characteristics and fuel requirements of automotive and aircraft engines. He has served on a number of NRC committees, including the Committee on Review of the Research Program of the Partnership for a New Generation of Vehicles. He has consulted for many companies in the automotive and petroleum industries and for government organizations. He has received many awards, from the American Society of Mechanical Engineers, the British Institution of Mechanical Engineers, and the Society of Automotive Engineers for his research contributions. He has a Ph.D. in mechanical engineering from MIT, an Sc.D. from Cambridge University, and honorary doctorates from Chalmers University of Technology (Sweden) and City University (U.K.).

John G. Kassakian (NAE) is a professor of electrical engineering and director of the MIT Laboratory for Electromagnetic and Electronic Systems. His expertise is in the use of electronics for the control and conversion of electrical energy, industrial and utility applications of power electronics, electronic manufacturing technologies, and automotive electrical and electronic systems. Prior to joining the MIT faculty, he served in the U.S. Navy. Dr. Kassakian is on the boards of directors of a number of companies and has held numerous positions with the Institute of Electrical and Electronics Engineers (IEEE), including founding president of the IEEE Power Electronics Society. He is a member of the NAE, a fellow of the IEEE, and a recipient of the IEEE's William E. Newell Award for Outstanding Achievements in Power Electronics (1987), the IEEE Centennial Medal (1984), and the IEEE Power Electronics Society's Distinguished Service Award (1998). He has an Sc.D. in electrical engineering from MIT.

Harold H. Kung is professor of chemical and biological engineering at Northwestern University. His areas of research include surface chemistry, catalysis, and chemical reaction engineering. His professional experience includes work as a research chemist at E.I. du Pont de Nemours & Co. He is a recipient of the P.H. Emmett Award and the Robert Burwell Lectureship Award from the North American Catalysis Society, the Herman Pines Award of the Chicago Catalysis Club, the Cross-Canada Lectureship of the Catalysis Division of the Chemical Institute of Canada, the John McClanahan Henske Distinguished Lectureship of Yale University, and the Olaf A. Hougen Professorship at the University of Wisconsin, Madison. He has a Ph.D. in chemistry from Northwestern University.

James J. MacKenzie is a senior fellow in the World Resources Institute's (WRI's) Climate, Energy, and Pollution program. Prior to joining WRI, MacKenzie was a senior staff scientist, Union of Concerned Scientists; senior staff member for energy, President's Council on Environmental Quality (CEQ); and a member of the joint scientific staff of the Massachusetts and national Audubon societies. Much of his recent research and analysis has focused on transportation technologies and the impact of the transportation system on the environment. He is coauthor (transportation chapter) of *Frontiers of Sustainability: Environmentally Sound Agriculture, Forestry, Transportation, and Power Production*; author of *Climate Protection and the National Interest*; *Oil as a Finite Resource: When is Global Production Likely to Peak?*; and *The Keys to the Car: Electric and Hydrogen Vehicles for the 21st Century*. He is also coauthor of *Car Trouble*, a book on the impacts of cars on the American scene, and of several major WRI reports, including an analysis of the subsidies to motor vehicles in the United States, the impacts of global motor vehicle use on climate change, and the effects of multiple air pollutants on U.S. forests and crops. He has also completed a policy report exploring the linkages among the problems of climate change, air pollution, and national energy security. Dr. MacKenzie received his Ph.D. in physics from the University of Minnesota and completed postgraduate work at the Los Alamos and Argonne National Laboratories and MIT before joining the Audubon Society.

Christopher L. Magee (NAE) has been with MIT since January 2002 and has a joint appointment as a professor of the practice in the Engineering Systems Division and Mechanical Engineering. He also directs a multidisciplinary research center (the Center for Innovation in Product Development). Before Dr. Magee joined MIT, he had more than 35 years of experience at Ford Motor Company, beginning in the Scientific Research Laboratory and progressing through a series of management positions to executive director of Programs and Advanced Engineering. The latter position had global responsibility for all major technically deep areas involved in Ford's Product Development Organization and consisted of about 7,000 people located in the United States, the United Kingdom, and Germany.

At the beginning of his career, Dr. Magee made major contributions to understanding the transformation, structure, and strength of ferrous materials. He was internationally recognized for this work and in his early 30's won the Howe Medal (best paper of the year in Trans-ASM/AIME) and, one year later, the Alfred Nobel Award (given by the ASCE for the single best contribution by someone under 33 from candidates from all five "founding" engineering societies (IEEE, ASME, AIME, AIChE, and ASCE). Dr. Magee then made contributions to lightweight material development, including dual-phase and other HSLA steels (including new stamping technology), and led their implementation at Ford. Simultaneously, Dr. Magee pioneered experimental work on high-rate structural collapse aimed at vehicle crashworthiness. During this latter period, Dr. Magee also led the work that initiated Ford's computer-aided engineering for structural and occupant simulation for crashworthiness. Dr. Magee has led (from 1981 on) efforts at Ford to adapt systems engineering to the modern automotive design process. In addition, he was instrumental in developing new approaches to the program creation process at Ford and from 1987 through 1999 had the technical lead for all major Ford product concept efforts.

Dr. Magee is a member of the National Academy of Engineering (since 1997), a fellow of ASM, and a participant in major NRC studies, ranging from design research to materials research. Dr. Magee received his B.S. and Ph.D. from Carnegie Mellon University and an M.B.A. from Michigan State University.

Michael P. Ramage (NAE) is retired executive vice president, ExxonMobil Research and Engineering Company. Previously he was director, executive vice president, and chief technology officer of Mobil Oil Corporation. He held a number of positions at Mobil, including general manager of exploration and producing research, development, and technical services; vice president of engineering; manager of process development; and president of Mobil Technology Company. He has broad experience in many aspects of the petroleum industry, including R&D, chemical processes, and capital project management. He is a director of the American Institute of Chemical Engineers. He served on a number of university visiting committees and is a member of a number of professional societies. He recently served as chairman of the NRC Committee on Alternatives and Strategies for Future Hydrogen Production and Use. He is a member of the National Academy of Engineering and serves on the NAE Council. He has B.S., M.S., and Ph.D. degrees in chemical engineering from Purdue University.

Vernon P. Roan is retired director of the Center for Advanced Studies in Engineering and professor of mechanical engineering at the University of Florida, where he has been a faculty member for more than 30 years. Since 1994, he has also been the director of the University of Florida Fuel Cell Research and Training Laboratory. Previously, he was a senior design engineer with Pratt & Whitney Aircraft. Dr. Roan, who has more than 25 years of R&D experience, is currently developing improved modeling and simulation systems for a fuel-cell bus program and working as a consultant to Pratt & Whitney on advanced gas-turbine propulsion systems. His research at the University of Florida has involved both spark-ignition and diesel engines operating with many alternative fuels and advanced concepts. With groups of engineering students, he designed and built a 20-passenger diesel-electric bus for the Florida Department of Transportation and a hybrid electric urban car using an internal combustion engine and lead-acid batteries. He has been a consultant to the Jet Propulsion Laboratory, monitoring its electric and hybrid vehicle programs. He organized and chaired two national meetings on advanced vehicle technologies and a national seminar on the development of fuel-cell-powered automobiles and has published numerous technical papers on innovative propulsion systems. He was one of the four members of the Fuel Cell Technical Advisory Panel of the California Air Resources Board, which issued a report in May 1998 on the status and outlook for fuel cells for transportation applications. Dr. Roan received a B.S. in aeronautical engineering and an M.S. in engineering from the University of Florida and a Ph.D. in engineering from the University of Illinois.

Bernard Robertson (NAE) is president of BIR1, LLC, an engineering consultancy specializing in transportation and energy matters that he founded in January 2004, upon his retirement from DaimlerChrysler Corporation. During the latter part of his 38-year career in the automotive industry, Mr. Robertson was elected an officer of Chrysler Corporation in February 1992. He was appointed senior vice president coincident with the merger of Chrysler Corporation and Daimler-Benz AG in November 1998 and was named senior vice president of Engineering Technologies and Regulatory Affairs in January 2001. In his last position, he led the Liberty and Technical Affairs Research group; Advanced Technology Management and

FreedomCAR activities; and hybrid electric, battery electric, fuel cell, and military vehicle development. In addition, he was responsible for regulatory analysis and compliance for safety and emissions. Mr. Robertson holds an M.B.A. from Michigan State University, a master's degree in automotive engineering from the Chrysler Institute, and a master's degree in mechanical sciences from Cambridge University, England. He is a member of the National Academy of Engineering, a fellow of the Institute of Mechanical Engineers (U.K.), a chartered engineer (U.K.), and a fellow of the Society of Automotive Engineers.

R. Rhoads Stephenson is currently a technology consultant. Previously, he held a number of positions at the Jet Propulsion Laboratory (JPL), the National Highway Traffic Safety Administration (NHTSA), and Martin Marietta Corporation. At JPL, these included deputy director and acting director, Technology and Applications Programs; manager, Electronics and Control Division; deputy manager, Control and Energy Conversion Division; and manager of the Systems Analysis Section. He also served as associate administrator for R&D at NHTSA and while at Martin Marietta worked on energy conversion devices for space power. He has been a consultant to the Motor Vehicle Fire Research Institute, has been providing peer reviews of automotive safety issues, and recently published a number of papers on crash-induced fire safety issues with motor vehicles, including hydrogen-fueled vehicles. He brings extensive expertise in vehicle safety analysis, advanced technology systems, energy conversion technologies, and energy and environmental analysis. He has B.S., M.S. and Ph.D. degrees in mechanical engineering from Carnegie Mellon University.

Kathleen C. Taylor (NAE) is retired director of the Materials and Processes Laboratory at General Motors Research and Development and Planning Center. Dr. Taylor was simultaneously chief scientist for General Motors of Canada, Ltd. Earlier, she was department head for physics and physical chemistry and department head for environmental sciences. Currently Dr. Taylor serves on the board of the North American Catalysis Society, the Advanced Photon Source scientific advisory committee at Argonne National Laboratory, the advisory board for the Nanoscale Science and Engineering Center at Columbia University, the board of directors of the National Inventors Hall of Fame, and the DOE Basic Energy Sciences Advisory Committee. Dr. Taylor was awarded the Garvan Medal from the American Chemical Society. She is a member of the National Academy of Engineering. She is a fellow of SAE International and the American Association for the Advancement of Science. She has been president of the Materials Research Society and chair of the board of directors of the Gordon Research Conferences. She has expertise in R&D management, fuel cells, batteries, catalysis, exhaust emission control, and automotive materials. She received an A.B. in chemistry from Douglass College and a Ph.D. in physical chemistry from Northwestern University.

Appendix D

Presentations at Committee Meetings

1. COMMITTEE MEETING, SOUTHFIELD, MICHIGAN, NOVEMBER 17-18, 2004

Program Vision: Auto Industry Perspective
William L. Peirce, General Motors

Energy Partner Perspective
Joe Kaufman, ConocoPhillips

FreedomCAR and Fuel Partnership
Steve Chalk, Richard Moorer, Ed Wall, U.S. Department of Energy

Vehicle Systems Analysis Technical Team
Lee Slezak, U.S. Department of Energy
Larry Laws, General Motors

Advanced Combustion Engines and Emission Control (ACEC) Activities
Ken Howden, U.S. Department of Energy
Richard Peterson, General Motors

Electrochemical Energy Storage
Tien Duong, U.S. Department of Energy
Ahsan Habb, General Motors

Fuel Cell Technical Team
Valri Lightner, U.S. Department of Energy
Fred Wagner, General Motors

Electrical and Electronics Technical Team
Susan Rogers, U.S. Department of Energy
Vijay Garg, Ford

Hydrogen Storage Joint Technical Team
Sunita Satyapal, U.S. Department of Energy
Scott Jorgensen, General Motors
Farshad Bavarian, ChevronTexaco

Materials Technical Team
Joe Carpenter, U.S. Department of Energy
Bob McCune, Ford

Hydrogen Production Tech Team
Peter Devlin, U.S. Department of Energy
Steve Schlasner, ConocoPhillips

Hydrogen Delivery Tech Team
George Parks, ConocoPhillips
Mark Paster, U.S. Department of Energy

Fuel/Vehicle Pathway Integration Tech Team
Fred Joseck, U.S. Department of Energy
Don Gardner, ExxonMobil

Codes and Standards Tech Team
Patrick Davis, U.S. Department of Energy
Brad Smith, Shell Hydrogen

2. COMMITTEE MEETING, THE NATIONAL ACADEMIES, WASHINGTON, D.C., JANUARY 24-26, 2005

Systems Analysis Introduction
Steve Chalk and Ed Wall, U.S. Department of Energy

Systems Analysis: Model Utilization and Integration
Fred Joseck, U.S. Department of Energy

Macro-System Model
Dale Gardner, National Renewable Energy Laboratory

Moving Toward Consistent Hydrogen Analysis: H2A
Margaret K. Mann, National Renewable Energy Laboratory

HyTrans: A Dynamic Optimization Model of Market Transitions to Hydrogen Use by Light-Duty Vehicles Integrating Fuel Supply, Vehicle Production and Consumer Demand
David L. Greene and Paul N. Leiby, Oak Ridge National Laboratory

Well-to-Wheels Analysis of Advanced Vehicle/Fuel Systems—Application of the GREET Model
Michael Wang, Argonne National Laboratory

Systems Analysis: PBA's Market Modeling and Analysis
Phil Patterson, Jeff Dowd, Randy Steer, Brian Unruh, Scott Hassell and Tien Nguyen, U.S. Department of Energy

Vehicle Systems Analysis Technical Team
Lee Slezak, U.S. Department of Energy

Systems Analysis Progress
Fred Joseck, U.S. Department of Energy

PSAT (Powertrain Systems Analysis Toolkit)
Lee Slezak, U.S. Department of Energy

Systems Integration
Dale Gardner, National Renewable Energy Laboratory

Codes and Standards Technical Team
Patrick Davis, U.S. Department of Energy
Brad Smith, Shell

Ending the Energy Stalemate: A Bipartisan Strategy to Meet America's Energy Challenges
Drew Kodjak, National Commission on Energy Policy

Hydrogen Production for a Sustainable Energy Future
C. Lowell Miller, U.S. Department of Energy

Basic Research Needs for the Hydrogen Economy: New Research Activities in DOE's Office of Basic Energy Sciences
Harriet Kung, U.S. Department of Energy

NHTSA's Hydrogen, Fuel Cell, and Alternative Fuel Vehicle Safety Research Plan
Barbara C. Hennessey, Department of Transportation

Nuclear Hydrogen Initiative: Programmatic Overview
David Henderson, U.S. Department of Energy

Hydrogen Codes and Standards: SNL Project Overview
John Keller and Chris Moen, Sandia National Laboratories

PEMFC Cost Slides for NAS Review
Eric Carlson, TIAX LLC

FreedomCAR and Fuel Partnership Peer Review: Overview Discussion
Steve Chalk and Ed Wall, U.S. Department of Energy

3. COMMITTEE MEETING, THE NATIONAL ACADEMIES, WASHINGTON, D.C., MARCH 21-23, 2005

NAS PEER Review
Ed Wall and Steve Chalk, U.S. Department of Energy

National Hydrogen Vehicle/Infrastructure "Learning Demonstration"
Steve Chalk, U.S. Department of Energy

Carbon Capture and Storage
C. Lowell Miller and Scott M. Klara, U.S. Department of Energy

Appendix E

Participants in Hydrogen Storage Projects for FY04 and FY05

INDUSTRY

Air Products
ATK/Thiokol Propulsion
Carnegie Institution of Washington
Ceralink
Gas Technology Institute
General Electric
HRL Laboratories
Intematix Corporation
Millennium Cell
Mo-Sci Corporation
NexGen Fueling
PoroGen, LLC
Research Triangle Institute
Rohm and Haas
State Scientific Research Institute
Superior Graphite Co.
TIAX LLC
TOFTEC, Inc.
US Borax

UNIVERSITIES

Alfred University
California Institute of Technology
Drexel University
Duke University
Michigan Technological University
Northern Arizona University
Northwestern University
Penn State University
Rice University
Stanford University
State University of New York
University of Alabama
University of California-Berkeley
University of California-Davis
University of California-Los Angeles
University of California-Santa Barbara
University of Connecticut
University of Florida
University of Hawaii
University of Illinois
University of Michigan
University of Missouri

University of Nevada-Reno
University of North Carolina
University of Oklahoma
University of Pennsylvania
University of Pittsburgh/
Carnegie Mellon
University of Utah
University of Washington
Yale University

NATIONAL LABORATORIES

Brookhaven National Laboratory
Idaho National Engineering and Environmental Laboratory (INEEL)
Jet Propulsion Laboratory
Lawrence Berkeley National Laboratory
Lawrence Livermore National Laboratory
Los Alamos National Laboratory
National Institute of Standards and Technology
National Renewable Energy Laboratory
Oak Ridge National Laboratory
Pacific Northwest National Laboratory
Sandia National Laboratories
Savannah River National Laboratory

Appendix F

Acronyms and Abbreviations

21st CT	21st Century Truck
ACC	Automotive Composites Consortium
ACEC	advanced combustion engines and emission controls (technical team)
AEMD	automotive electric motor drive
AIPM	automotive integrated power module
ANL	Argonne National Laboratory
ANSI	American National Standards Institute
ASCE	American Society of Civil Engineers
ASM	American Society for Materials
ASME	American Society of Mechanical Engineers
BES	Basic Energy Sciences program (DOE)
BMW	Bayrische Motor-Werken
BNL	Brookhaven National Laboratory
CAFE	Corporate Average Fuel Economy
CAR	Cooperative Automotive Research
CBO	U.S. Congressional Budget Office
CCS	carbon capture and storage
CEQ	Council on Environmental Quality
CFRP	carbon-fiber-reinforced polymer
CIDI	compression ignition direct injection
CLEERS	crosscut lean exhaust emission reduction simulation

CNG	compressed natural gas
CO_2	carbon dioxide
CRC	Coordinating Research Council
DARPA	Defense Advanced Research Projects Agency
DLC	double layer capacitor
DOD	U.S. Department of Defense
DOE	U.S. Department of Energy
DOT	U.S. Department of Transportation
E85	85 percent ethanol
EE	electrical and electronics (technical team)
EERE	Office of Energy Efficiency and Renewable Energy (DOE)
EPA	U.S. Environmental Protection Agency
ERC	Environmental Research Consortium
ERC	Engine Research Center
ESD	Engineering Society of Detroit
EV	battery electric vehicle
EWCAP	Electrical Wiring Component Applications Partnership
FC	fuel cell
FCHEV	fuel cell hybrid electric vehicle
FCVT	FreedomCAR and Vehicle Technologies (program)
FE	Office of Fossil Energy (DOE)
FPITT	fuel pathway integration technical team
FRP	fiber-reinforced polymer
FY	fiscal year
GATE	Graduate Automotive Technology Education
GREET	Greenhouse Gas, Regulated Emissions, and Energy Use in Transportation (model)
Gton	gigaton
H	hydrogen
HAMMER	Hazardous Materials Management and Emergency Response (facility)
HCCI	homogeneous charge compression ignition
HEV	hybrid electric vehicle
HFCIT	Hydrogen, Fuel Cells and Infrastructure Technology (program)
HFCV	hydrogen fuel cell vehicle
HFI	Hydrogen Fuel Initiative
HSRP	Hydrogen Safety Review Panel
HSS	high-strength steel

OEM	original equipment manufacturer
ORNL	Oak Ridge National Laboratory
OSRP	Occupant Safety Research Partnership
PBA	Office of Planning, Budget and Analysis
PHMSA	Pipeline and Hazardous Materials Safety Administration
PEM	proton exchange membrane
PM	particulate matter
PNGV	Partnership for a New Generation of Vehicles
PNNL	Pacific Northwest National Laboratory
PV	photovoltaic
R&D	research and development
RFG	reformulated gasoline
RSPA	Research and Special Projects Administration
SAE	Society of Automotive Engineers
SBIR	Small Business Innovation Research
SC	Office of Science (DOE)
SCR	selective catalytic reduction
SiC	silicon carbide
SNL	Sandia National Laboratories
SRI	Stanford Research Institute
SUV	sports utility vehicle
TMS	The Metallurgical Society
TRB	Transportation Research Board
ULSAB-AVC	Ultralight Steel Auto Body-Advanced Vehicle Concept
USABC	United States Advanced Battery Consortium
USAMP	United States Automotive Materials Partnership
USCAR	U.S. Council for Automotive Research
VRP	Vehicle Recycling Partnership
VSAT	vehicle systems analysis team
WRI	World Resources Institute

ISBN: 9780872332461

Library of Congress Cataloging-in-Publication Data

Names: Counter, S. Allen.
Title: North Pole promise : Black, White, and Asian friends / Dr. S. Allen Counter.
Description: Peterborough, New Hampshire : Bauhan Publishing, 2017.
Identifiers: LCCN 2017000370 | ISBN 9780872332461 (softcover : alk. paper)
Subjects: LCSH: Peary, Robert E. (Robert Edwin), 1856-1920—Friends and associates. | Henson, Matthew Alexander, 1866-1955—Friends and associates. | Explorers—United States—Biography. | African American explorers—Biography. | Henson, Anaukaq, 1906- | Peary, Kali, 1906- | Inuit—Greenland—Biography. | North Pole—Discovery and exploration.
Classification: LCC G634 .C38 2017 | DDC 910.92/2—dc23
LC record available at https://lccn.loc.gov/2017000370

Book Design by Cassie Baron
Cover Design by Henry James
Printed by Versa Press

BAUHAN PUBLISHING LLC
PO BOX 117 / 44 MAIN STREET PETERBOROUGH NEW HAMPSHIRE 03458
603-567-4430
WWW.BAUHANPUBLISHING.COM

Manufactured in the United States

North Pole Promise

Black, White, and Inuit Friends

Dr. Counter placing hearing protectors in the ears of ten-year-old Inuit rifle hunter Magssanguaq Henson in Moriussaq, in northwest Greenland

To

Navarana K'avigak
Aviaq Henson
Olivia Counter

Contents

Part Four: Keeping Up with Friends

Part Five: The Centennial Approaches

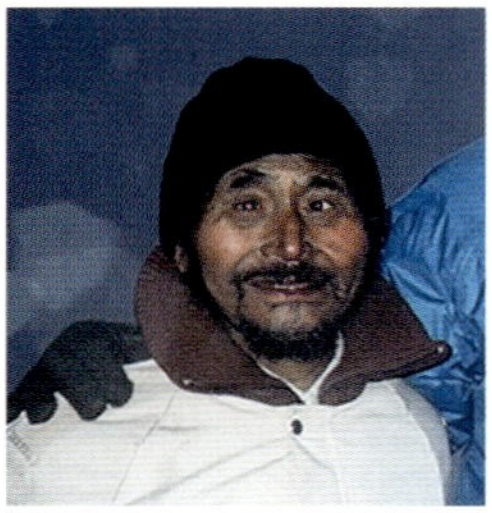

Anaukaq Henson, the son of Matthew Henson and the Polar Inuit woman Akatingwah , might be called an Afro-Inuit or Amer-Eskimo. After his father left Greenland when he was three, Anaukaq grew into a cheerful, funny, and fiercely intelligent master hunter, or *peeneeahktoewah*. The father of five sons, each of whom became an outstanding hunter in his own right, Anaukaq was known to his people as a widely-respected leader. For most of his life, he nursed a desire to see the land of his father.

Kali Peary, the younger son of Robert Peary and Ahlikasingwah, was a Euro-Inuit or Amer-Eskimo. A witty, smart, and friendly man, and, like his "cousin" Anaukaq, a renowned *peeneeahktoewah*, Kali fathered two sons and three daughters who grew into outstanding hunters and village leaders, respectively. Kali had dreamed since he was a boy of meeting his American half-brother, whom older Polar Eskimos saw in the 1930s visiting a stone monument erected in Peary's honor.

Magssanguaq Henson, the son of Avataq Henson, Anaukaq's oldest son, was his grandfather's favorite. An outstanding hunter who killed his first seal at age ten, Magssanguaq (pronounced *mah-SHING-wahk*) learned the traditional skills of the Inuit from his father and brought down his first *nanook*—the Inuit word for the ferocious polar bear—by the time he was twelve.

Aviaq Henson is Anaukaq's granddaughter, and was named for his wife. When she was a teenager, her family moved to the township of Nuuk in southern Greenland so she and her siblings could receive an education. Aviaq spearheaded the project that culminated in the Henson–Peary North Pole Centennial Commemoration.

The grandchildren of the sons Henson and Peary left behind, although not pictured here, are Avataq, Malina, Ussarqak, Ajako, Vittus, and Kitdlaq Henson, and Peter, Talilinquaq, Paulina, Massauna-Henrik, Marta, and Mikkisuk Peary.

Ole Peary is the son of Talilinquaq Peary, Kali's second son. Charming and friendly, he has been one of the best dogsled drivers in Greenland since his youth. Ole left his village for the first time at age eighteen to lay a wreath on his great-grandfather's grave in Arlington National Cemetery.

Jorgen Ootah is the great-grandson of the legendary Polar Inuit hunter and explorer for whom he is named. Like his ancestor, "Little Ootah" is known as a great hunter and Arctic navigator.

Navarana K'avigak, a descendant of the famous guide Ootah, acted as our translator in both Greenland and America.

S. Allen Counter is a neuroscientist and professor of neurology at Harvard University, as well as a professional explorer and a member of The Explorers Club. He is also the Consul General of Sweden in Boston and New England, and adjunct professor at Sweden's Karolinska-Nobel-Institute. Counter has explored northwest Greenland, where he found the Eskimo descendants of Robert Peary and Matthew Henson in 1986.

Olivia Counter was a sixteen-year-old junior at Cambridge Rindge and Latin High School in 2009. A Swedish-American, Olivia traveled with her father on his voyage to Greenland, where the Eskimo teenagers welcomed her and taught her how to use the dog-driving whip and sled. The Harvard North Pole Centennial Commemorative Committee presented Olivia with a special medal for traveling farther north above the Arctic Circle than any high school student in New England.

Ussarqak Henson (right) and son Massauna-Henrik preparing a kayak for a narwhal hunt

Preface

North Pole Promise tells the amazing story of two journeys that spanned some four thousand miles and four generations of people from different cultures and ethnic backgrounds.

At the turn of the last century, explorers from all over the world were trying to reach the North Pole, to become the first human to reach the top of the earth. Working against them were the bitter cold, enormous ice ridges, and impassable open water breaks in the ice. In spite of incredible odds, and armed with only the rudimentary technology of that era, six men—one white, one black, and four Polar Inuit Eskimos of Asian ancestry—worked together to accomplish one of the greatest feats in human history, reaching the top of the earth on foot and dogsled in 1909. One of those men, Robert E. Peary, is known to many; his right-hand man, Matthew A. Henson, is known to some; their four Inuit assistants are almost unknown.

Over the years, I have lectured on the discovery of the North Pole at middle schools, high schools, and colleges throughout the nation. Many young people have written to ask me for information they could use in their class papers. Beyond the known facts of the expedition, I have tried to provide them with insight into how newly discovered indigenous people were viewed and treated during the era of exploration at the beginning of the twentieth century. I have also tried to describe the struggle against racial discrimination, and the search for equal representation and recognition, for the African-American and Inuit explorers who worked side by side with Peary.

Dr. Counter meeting Anaukaq Henson (left) and Kali Peary for the first time in Moriussaq, in northwest Greenland

What many of the students find most fascinating is that this is a story not only of a major historical achievement accomplished by a multicultural group, but also features two mixed-race boys of American paternity who grew up in the isolation of the Arctic. Born in secrecy to indigenous Polar Inuit women in northwest Greenland, the boys were abandoned by their fathers who returned to America and never saw them again. Left behind at the top of the planet, they were the human collateral of the North Pole expeditions—never to be spoken of and, so the expectation went, never to be found.

In 1986, when I was about a thousand miles above the Arctic Circle on a medical science trip associated with my research at Harvard University and Karolinska Institute, I discovered two of the sons still living there, both eighty years old. One of them part black and one of them part white, Anaukaq Henson and Kali Peary were raised in a place where the inroads of modern technology were slow to take hold. In many ways, they grew up like young boys anywhere on earth, with the same interests, desires, and teenage anxieties, and yet they were quite different: Anaukaq and Kali were dogsled hunters who built and lived in igloos while on the icepack, surviving on a diet of raw meat, including seal, polar bear, birds, walrus, and narwhal. They grew up to be the greatest teenage hunters in their remote Arctic settlement, and they traveled around northern Greenland bringing back wild game to feed their community.

As adults, each fathered five children who, with their families, still lived in the surrounding settlements, hunting for food, building igloos, and hunting narwhals and walrus by kayak. The men told me that, although they had always felt loved and accepted by their Inuit community, throughout their youth they had felt as if they had been abandoned in the unreachable far north by their American families, and had dreamed of one day traveling to the land of their fathers to "touch the hand of a relative." I determined to help them achieve that dream, to help honor their fathers, and to celebrate the centennial of the 1909 North Pole discovery with their families.

I'm happy to report that one hundred years after Peary, Henson, Ootah, Seegloo, Egingwah, and Ooqueah reached the Pole, the United States and Greenland celebrated the centennial of their accomplishment. The ceremony again united people of different races, working together at the top of the earth, but this time the contributions of each were acknowledged. I'm especially pleased to say that I was able to bring more recognition for Matthew A. Henson, who has been a hero of mine since long ago when my grandmother gave me a copy of his 1912 book, *A Negro Explorer at the North Pole*.

I hope readers will find *North Pole Promise* an uplifting story. As a member of The Explorers Club, I also hope that this book will inspire young explorers to travel to distant places and come to know and preserve indigenous cultures, from which we can learn much about our planet, about conservation, and about the commonality of humanity.

— S. Allen Counter, Jr.
Harvard University

Magssanguaq peering out of an igloo he built with his father, Avataq Henson

Matthew Henson, age 60, exhibiting photo of his hero Robert Peary in Greenland

1
The Beginning

In 1887, two men met in a fur storage shop in Washington, DC. One was a thirty-year-old United States Navy lieutenant who had never been to sea, the other a nineteen-year-old clerk who had worked on merchant ships for five years.

Matthew Alexander Henson, a native of Charles County, Maryland, and resident of Washington DC, and Robert Edwin Peary, a Pennsylvania native who had been raised in Maine, could not have been more different: Peary was white, college-educated (at Bowdoin College in Maine), and deskbound; Henson was

US Navy Lieutenant Robert E. Peary (left) and Matthew A. Henson

black, smart but little schooled, and had traveled the world as a cabin boy during his teens. But each recognized the other's talents and worth.

Matthew Henson in New York City, circa 1930

Henson's intelligence, experience, and skills were so impressive that Peary brought him along as his personal assistant when he traveled to Nicaragua, where the navy had assigned him to survey a site where a canal could be cut to link the Atlantic and Pacific oceans. (The site was deemed unfeasible and a different one was chosen in Panama—hence, the Panama Canal.) By the time the men returned from the Nicaraguan jungle, they had forged a sincere, mutually respectful friendship. Peary found Henson a position as a messenger in the naval shipyard, but he had him firmly in mind for a bigger job: an expedition to find the North Pole.

In the late 1800s, the North Pole was the ultimate prize for explorers. In Europe and the United States, some people believed that anyone who reached it would plunge into a vast abyss; others thought it was inhabited by people whose fires lit the Aurora Borealis—the Northern Lights. Adventurers from many nations traveled north to gain glory for themselves and their homelands, sailing in the open water lanes to the northern tip of Greenland or Ellesmere Island in Canada, only to have their ships freeze in place once winter set in. From there, they would set out on foot and by dogsled in an attempt to "strike" the North Pole, some 475 nautical miles from the northernmost strip of land. So far, all the other explorers had failed. Peary and Henson would not.

2
Eighteen Years in the Arctic

Robert Peary made his first expedition to Greenland during a six-month leave from the navy in 1886, and traveled nearly one hundred miles inland before his team's food supplies ran short. When he led his second expedition in 1891, the experienced seaman Henson and Peary's wife, Josephine Diebitsch Peary, went with him. Josephine Peary gave birth to the first known European-American or white child born at such a northern latitude, a girl named Marie Ahnighito Peary. The natives called her "Snowbaby."

Of all the team members, Henson and Josephine Peary were closest to the expedition's leader. Henson admired Josephine and spoke very highly of her; she treated him with the same respect she showed the white men on the team. At one point during the trek she threw a surprise dinner party to celebrate Henson's twenty-fourth birthday, an act of kindness he never forgot. Josephine later said Henson was the only explorer to remain loyal and committed to her husband's expeditions over many years of exploration.

Matthew Henson on his first expedition to northwest Greenland in 1891

On the 1891 expedition, Peary and Henson used the state-of-the-art navigation tools of their time: celestial

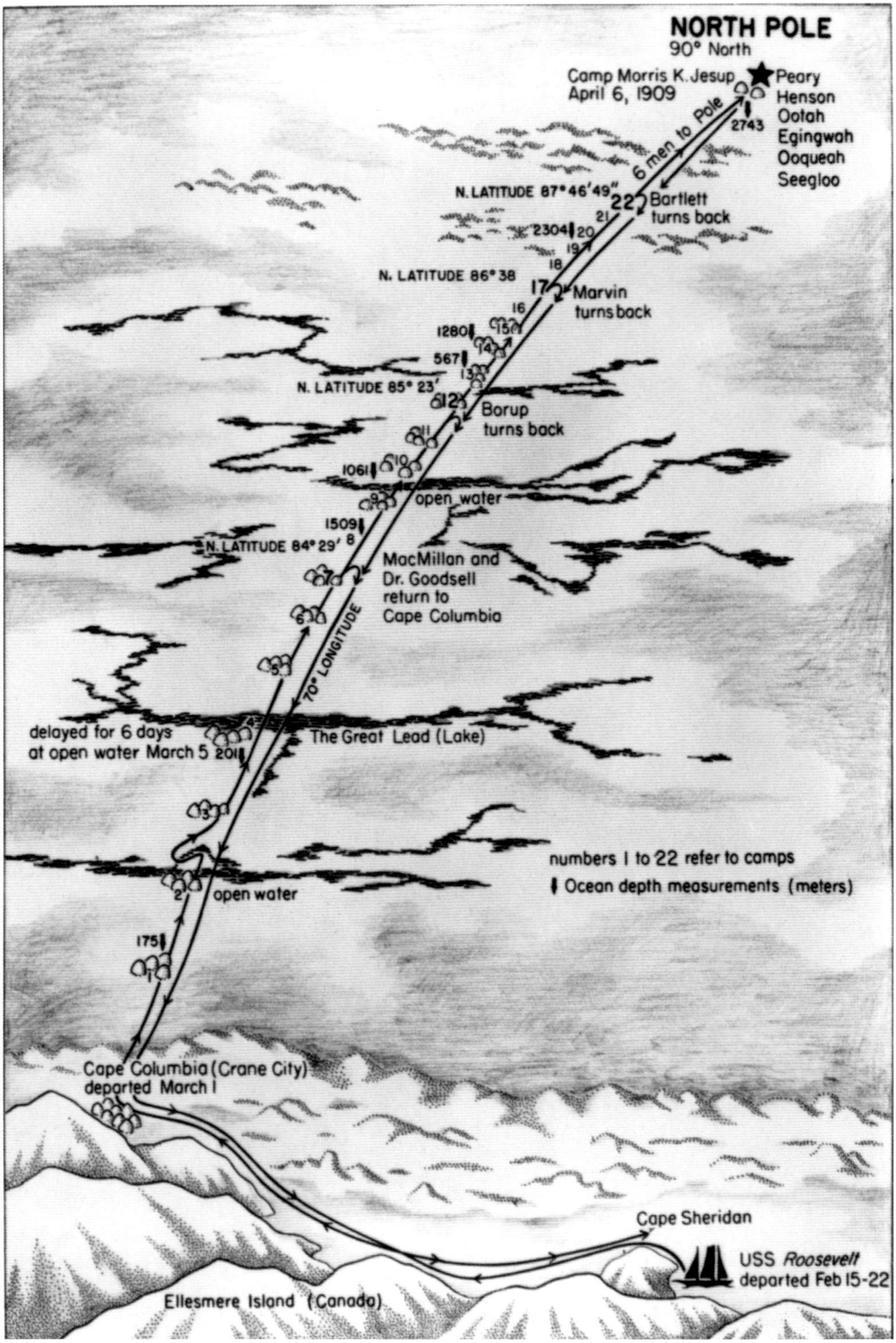

Peary's route to the North Pole on the 1909 expedition

Inuit women wearing traditional Polar Eskimo clothing onboard Commander Peary's ship during the 1909 expedition

observation, trays of mercury to estimate angles of reflected sunlight, a sextant, and dead reckoning (a method for calculating one's current position by taking a previously determined position and advancing it by a known or estimated speed over a certain amount of time and in a certain direction).

The explorers pushed farther north this time and made their first contact with the indigenous population, whom Peary called "Polar Eskimos." The natives (descendants of people who probably crossed over a land bridge from Asia to Alaska, and then through what is now Canada, around 500 BCE) called themselves *Inyooweesh* or "the humans." These nomadic people in the lonely northernmost latitudes had believed that they were the only human beings on the earth, and referred to themselves as such. *Inyooweesh* was later translated as "Inuit" by Europeans and North Americans. "Eskimo" reportedly comes from indigenous Cree, one of the first tribes to have contact with the *Inyooweesh*; "Eskimo" simply means "man who eats raw meat." (The indigenous people of northwest Greenland accept this label, and even refer to themselves as Eskimo. They do not consider the term derogatory.)

Members of the 1891 expedition to northwest Greenland, including Commander Peary's wife, Josephine

Like the first expedition, the 1891–1892 trip fell short of the Pole, stopping at Navy Cliff in Greenland's far north. But this time Peary's team built camps and food-storage igloos as supply and rest stations, learned to rely on the Inuits as hunters and dogsled drivers, and studied Inuit life, learning how to build igloos and fashion fur clothing. Henson, especially, put himself out to befriend the native people and learn their ways, and eventually even mastered their language—the only member of Peary's team to do so.

On a third expedition, from 1898–1902, Peary and his men reached far enough to sight Greenland's northern tip, which Peary named Cape Morris Jesup. The American Geographical Society and Royal Geographical Society of London honored Peary for discovering Cape Jesup, and the American banking scion George Crocker gave him a $50,000 grant. Peary used the money to buy the USS *Roosevelt*, which would become his team's home on the journey of 1905–1906 and on the final expedition of 1908–1909.

In all, Henson and Peary explored the Arctic together for over eighteen years between 1891 and 1909. The only black man on any of the expeditions,

3
Striking the Pole

On March 1, 1909, Henson and his Inuit assistants headed out to break a trail northward for the other teams to follow. Reaching the geographic North Pole out in the Arctic Sea, more than five hundred miles from their ship, the USS *Roosevelt*, and their last land base at the northern tip of Ellesmere Island, Canada, was a daunting and challenging task. The expedition had to be well planned, down to the rations of food for the team and the number of hearty dogs to pull the sleds filled with supplies. Using five relay teams to ferry supplies, each with an American and three Inuit assistants to build igloo supply houses along the way, the expedition crept northward. The last of the American-led ferrying teams turned around at latitude 87° north and headed back to the ship. From that point on, only Peary, Henson, and four Inuit assistants traveled northward.

On April 6, 1909, after eighteen years, seven unsuccessful expeditions, and a trek of roughly five hundred miles over the Arctic sea by dogsled

Commander Peary wearing traditional polar Eskimo clothing

Matthew Henson dressed in traditional polar Eskimo clothing

The four Polar Inuits assistants who reached the north pole with Peary and Henson in 1909: (left to right) Seegloo, Egingwah, Ootah, and Ooqueah

and on foot, Peary, Henson, Ootah, Egingwah, Seegloo, and Ooqueah finally reached latitude 90° north, the geographic top of the world. Matthew Henson planted the American flag in the ice.

Peary claimed victory for the United States, but was quickly challenged. Frederick Cook, who had been a surgeon on Peary's second expedition, declared that he had reached the Pole with his own team in 1908, a year earlier. Cook's claim was swiftly rejected by a commission of geographic scientists from the University of Copenhagen, but it cast a cloud on Peary's accomplishment.

In the years since other doubters have made names for themselves by asserting that Peary and Henson never actually reached the geographical North

Matthew Henson (center) planting the American flag in the ice at the North Pole on April 6, 1909, along with Seegloo, Egingwah, Ootah, and Ooqueah

Pole. Each of their claims has been challenged and proven wrong. Modern evidence—including mathematical measurements of the angles of sunlight in the photographs Peary's team took at the Pole, and testimony from other explorers who have reached the Pole using GPS—supports Peary's claim. A collective of some of the best navigators in the United States, including the US Navigation Foundation and the adventurer and mountaineer Terris Moore of The Explorers Club, now affirm that on April 6, 1909, Matthew Henson, Robert Peary, Ootah, Egingwah, Seegloo, and Oqueah were the first humans to stand at the attic of the earth.

4
Acclaim—and Anonymity

When the expedition returned to the United States, Peary was widely celebrated by government leaders and high society. The National Geographic Society hailed his accomplishments, he was promoted to rear admiral in the Navy Civil Engineer Corps, and given the thanks of Congress by special act in 1911. He gained fame, wealth, and a permanent place in the history books.

Henson, of whom Peary had said, "I cannot make it without him," and whom MacMillan had acknowledged as "of more value to Commander Peary on the Polar Sea than all white men combined," was virtually ignored. Despite his colleagues' commendations and President William Howard Taft's recognition of his achievements, it took thirty years before Henson was admitted to the prestigious all-white, all-male Explorers Club of New York—and even that did not happen until the famous, and white, adventurers Peter Freuchen and Vilhjalmur Stefansson nominated him for membership. (Freuchen, the famous Danish explorer, wrote that, to the Inuits, Henson was "a hero magnified by Eskimo legend to immortal proportions. To the Eskimos, who loved him, Matt was the greatest of all the men. . . .") Henson was at last admitted in 1939. Another fifteen years passed before he and his second wife, Lucy Ross Henson, were invited to the White House, where they met President Dwight David Eisenhower.

Unlike the financially well-off Peary, Henson went from a vital position on Polar expeditions to work in minor jobs. He spent thirty years as a messenger at the New York Customs House, a pensioned post for which he was recommended by Peary and approved by Taft. Henson took pride in the

Matthew Henson (front row, second from left) with fellow Explorers Club members. In 1939 he became the first African-American admitted to the club

work, which supported him and his wife, who worked as a bank clerk, in a respectable apartment in Harlem.

Peary died in 1922 at age sixty-three, and was buried in Arlington National Cemetery in the shadow of a magnificent, globe-shaped monument donated by the National Geographic Society.

Henson died in 1955 at age eighty-eight, acknowledged as one of the greatest explorers of his time—but only by black society. His funeral was held at the Abyssinian Baptist Church in Harlem, where his pastor, the Reverend Adam Clayton Powell, Jr., described him to more than a thousand mourners as "the black Magellan."

Commander Robert E. Peary with daughter, Marie, in northwest Greenland in 1898

Polar Inuit woman Akatingwah with her son, Anaukaq Henson, fathered by Matthew Henson in 1906

the boys' paternity was a fiercely kept secret. Not only did Peary and Henson hold their tongues, so did the rest of the men on the expeditions, all of whom knew of the Inuit offspring but held to the code of silence subscribed to in that era by white men who had affairs with "colored" women. Having fathered a half-Inuit child would have destroyed Peary's reputation in the United States, and overshadowed his achievement with Henson at the North Pole.

Peary and Henson left the Arctic for good in September of 1909. It was the last time the boys saw their fathers.

Polar Inuit woman Ahlikasingwah with her son, Kali Peary, fathered by Robert Peary in 1906

Dr. Counter conducting hearing test on eighty-year-old rifle hunter, Anaukaq Henson, in Moriussaq, Greenland

Part Two

New Discoveries

Dr. Counter with eighty-year-old Anaukaq Henson (left) and Kali Peary in Moriussaq, Greenland

6
The Secret Sons

In 1986, as part of my job as a professor at Harvard University and a neurophysiologist at Massachusetts General Hospital who studies nerve and brain function, I traveled to the villages of Morriussaq and Qaanaaq in Northwest Greenland to conduct health-related research among the indigenous Polar Inuits.

I had other plans as well: Matthew Henson was one of my boyhood heroes. I wanted to learn firsthand from the Inuits about his life in the Arctic and, perhaps, further promote his reputation.

I had been told by a number of elderly African-Americans, including my grandmother, that Henson had contributed much more to the discovery of the North Pole than was in the history books. When Peary and Henson set foot on the Pole in 1909, it was a feat as momentous as the moon landing sixty years later. But while Peary gained all the fame of Neil Armstrong, the majority-white society of the twentieth century never accorded Henson the recognition and respect given Buzz Aldrin, the second man to walk on the moon.

He was, however, known and celebrated in the African-American community. On October 19, 1909, the black establishment in New York City held a formal dinner in his honor at the prestigious Tuxedo Club in midtown Manhattan, attended by black dignitaries from all over the eastern United States, who praised his achievements and gave him a gold watch. In Henson's time, his community understood the significance of his world-class achievement, but without mention in the official record, his fame dimmed with the years.

One of my aims in Greenland was to experience the Arctic as my hero had known it. While also testing the hearing of Inuit hunters (most of whom have

Minik Wallace, Polar Eskimo boy from Northwest Greenland who was adopted by the head of the New York Museum of Natural History after his father died from American microbes.

would someday return and take them to visit what they called "the land of our fathers." Like most Inuits, they knew nothing of the United States except what the tribe had heard (and believed to be exaggerations) from Minik, one of a group of six Eskimos Peary had brought to America in 1897 for the purpose of anthropological study and proof of his expeditions near the top of the earth.

Anaukaq told me that while Peary's elder son had died, his second son, Kali, whom he called "cousin" was still alive, and some weeks later I traveled north to a small settlement on Qeqertarsaaq (Herbert Island) in the Smith Sound Bay to meet him. This Amer-Inuit was pale-skinned, and looked far more Caucasian than ethnic Eskimo. Kali, too, readily mistook me for a relative of Maripaluk as he welcomed me into his igloo for raw seal meat and coffee. He shared with me loving memories of his mother, Ahlikasingwah, but he had only a few memories of his American father, such as his habit of moving his head from side to side when deep in thought.

In time, I was able to arrange for Kali and his forty-year-old son, Talilinquaq, to be helicoptered to Morriussaq,

Minik Wallace, Polar Eskimo boy from northwest Greenland who was adopted by the head of the New York Museum of Natural History, with his first bicycle

Tragically, four of the group, including Minik's father Qisuk quickly died from exposure to bacteria and viruses found in the American population, and the American Museum of Natural History in New York displayed some of their bodies as "mummies" or skeletons. Minik was adopted by the museum's curator, William Wallace, and lived with the family for a number of years as Minik Wallace. He was devastated when he learned that his father's remains had been mounted and displayed for public viewing in the museum—he had taken part in a staged burial with what turned out to be a fake body—and he demanded the return of his father's remains for proper burial. In his late teens Minik found his way back to the Arctic on whaling ships, and described for his people the marvels he had seen in New York, such as streetcars, electric lights, and telephones. Unable to readjust to Inuit life, Minik eventually returned to the United States and worked as a common laborer. He died in the influenza pandemic of 1918 and is buried in Pittsburgh, New Hampshire.

where the cousins who had grown up together met for the first time in many years. Anaukaq had married a "full-bloodied" Inuit woman, Aviaq, with whom he had five sons (and a daughter who died at age six—a devastating loss that still pained the family). Kali too had married a full-bloodied Inuit, Eqariusaq, with whom he had fathered three daughters and two sons.

I spent weeks interviewing both men and getting to know them. Kali had adopted the Peary surname and Anaukaq the Henson surname in late adulthood, after they were "Christianized" through baptism by Danish Lutheran missionaries. They told stories of their parents, their childhoods, and their wives, each now dead. Anaukaq said, "Kali's father Peeuree (the Inuit for Peary) had two wives . . . Mrs. Peary in his homeland and Ahlikasingwah in Greenland." At this, both men laughed uproariously.

While eating their favorite delicacy, raw bird, Anaukaq talked for the first time about race. He said the elders "always said that it is true . . .

Dr. Counter and Anaukaq Henson heading out by dogsled to examine seal traps

Enjoying the summer sunshine in Moriussaq:
(left to right) Kali Peary, Dr. Counter, and Anaukaq Henson

Anaukaq Henson eating raw birds,
one of his favorite foods

you are the son of Matthew Henson. But they never isolated me from everyone because the people up here are very nice, even though they knew I was black." He said the only time his race was mentioned was "sometimes when I was a child, if I was too rough with the other kids, they shouted, 'You are black, you are black,'" he said with a laugh. "But it was my fault for being too rough."

Kali and an elderly woman in the settlement named Mikkisuk joked that Anaukaq and his sons glided instead of walked, and could dance better than the other Eskimos in the region. "Everyone up here knows that about the Hensons," Mikkisuk said humorously.

Poignantly, there came a day when Anaukaq and Kali walked over to my "igloo" (in this case, a wooden shack built for Danish technicians who visited the area periodically) and asked to speak with me. We sat down together and I served them oatmeal and hot tea. Then Anaukaq said that all their lives, he and Kali had felt abandoned by their fathers and neglected by the relatives they

Dr. Counter telling Anaukaq Henson about finding his American relatives and their desire to meet him and his children

7
Finding Family

I left Morriussaq with a promise to return. However, I had a feeling that though my new Inuit friends were happy about my visit, they did not believe I would be able to unite them with their extended families in the United States. But I was determined to search for their kin, though I knew of no relatives of either man at the time, or where to begin. Luck and academic training came to the fore, and by the fall of 1986, I had done it. Starting at the Library of Congress, looking for Hensons in Maryland, I was able to locate Olive Henson, my hero's grandniece. While I was doing this, *The Boston Globe* ran a piece about my Arctic adventure, and the story led the explorers' New England-based descendants to contact me.

Kali Peary (left) and Anaukaq Henson expressing their happiness after learning they will travel to the United States to meet their relatives and see their fathers' graves

The American Hensons were delighted to learn of Anaukaq's existence. The American Pearys, however, denied Kali almost entirely. In fact, some of the immediate family members refused to acknowledge the admiral's Inuit progeny, even though Peary's own wife, Josephine, had known of her husband's infidelity. A spokesperson for the Pearys (the son of

THE EXPLORERS CLUB

'Pole secret': *Peary (above center), son Karree (above)*

Peary's clan: *Grandson (left), great-grandson (above)*

PHOTOS COURTESY OF DR. S. A. COUNTER

Polar Conquests

On April 6, 1909, Cmdr. Robert Peary reached the North Pole accompanied by his longtime associate Matthew Henson and four Eskimos. When the expedition returned to the United States several months later, the accomplishment was met first with disbelief, then grudging acceptance and finally public acclaim. Part of the controversy surrounding the trip centered on Peary's choice of Henson as the only other member of the team to make the final leg to the pole—the rest of the expedition was left behind, 133 miles from the goal. "Henson must stay with me. I cannot get along without him," Peary explained. At the time, some thought it inappropriate that a Navy officer would afford a black man the honor of planting the American flag at the top of the world.

Now a Harvard University professor, Dr. S. Allen Counter, has shed light on the bond between Peary and Henson. Counter traveled to the northwest reaches of Greenland searching for descendants of the Eskimo members of the 1909 expedition. They in turn led Counter to discover what he calls the explorers' "North Pole secret": two men fathered by Peary and Henson with Eskimo women on an earlier trip. Both Karree Peary and Ahnahkaq Henson were born in 1906, says Counter, and have always been known by the polar Eskimos as the sons of the two explorers whose visit 80 years ago is still a subject of conversation. According to Counter, Ahnahkaq Henson and his family had never seen another black man before and greeted the professor, through an Innuit interpreter, saying, "You must be a Henson and you've come to find me." Counter, who directs a foundation at Harvard devoted to intercultural studies, is arranging to bring members of the Henson and Peary clans to the United States, where he hopes they can meet some of their North American relatives.

BETTMANN ARCHIVE

COURTESY OF DR. S. A. COUNTER

Explorer's family: *Counter (second from left) with Ahnahkaq (with cane), grandsons, great-grandaughter; Henson (inset)*

1986 Newsweek magazine article about Dr. Counter's discovery of the "secret sons" of Robert Peary and Matthew Henson in northwest Greenland.

Peary's daughter, Marie) told me that the family did not wish to meet with Kali or his children. The family seemed to fear that publicly acknowledging their patriarch's out-of-wedlock son could bring new attention to the controversy sparked by Frederick Cook's long-ago claims of reaching the Pole first, or otherwise blemish their family name; some even called Kali an illegitimate child. I believe there is no such thing as an illegitimate child, only illegitimate parents.

Nonetheless, with a promise of help from the United States Air Force, I cheerfully announced to the Amer-Inuit cousins that I had arranged for them and some of their respective families to fly to America to meet their relatives and tour the places that had shaped their fathers' lives. Anaukaq and Kali were overjoyed.

Kali Peary (standing) being welcomed by the congregation of the historic Abyssinian Baptist Church in New York City

8
An Inuit-American Family Reunion

In June 1987, I traveled back to Northwest Greenland on a US Air Force C-141 jet to escort Anaukaq and Kali on their first trip to America. Traveling with Anaukaq were his five sons, Avataq, Ussarqak, Ajako, Vittus, and Kitdlaq; his grandson Henrik, son of Ussarqak; and his granddaughters, Malina, the daughter of Ajako, and Aviaq, the daughter of Avataq. With Kali were his son, Talilinquaq, and grandson, Ole. Navarana K'avigak, my Inuit translator from Qaanaaq, agreed to join us for the American tour.

Helicopters flew us over mountains and glaciers from the settlements of Morriussaq and Qaanaaq to Thule Air Base where we boarded the massive C-141 that took us directly to McGuire Air Force Base in New Jersey. Once there, we climbed aboard a chartered bus that took us to Harvard University in Cambridge, Massachusetts, for a formal celebration of what I had dubbed the "North Pole Family Reunion."

As we settled the Inuits into dormitories, I learned that Kali's half-brother, eighty-three-year-old Robert Peary, Jr., was alive and well in Augusta, Maine. I reached the branch of the Peary family there and asked his wife, Ruth Peary, if Kali could visit them. She agreed that Kali should be able to meet his brother for the first time, and warmly invited us to "come on up" to their home.

When we arrived, the Pearys and their son, Robert Peary, III, welcomed us with affection. Robert Peary, Jr. was clearly delighted to meet his half-brother, showing him around the house and pointing out souvenirs of their father's Arctic adventures, including a dogsled, furs, and walrus tusks. The half-brothers exchanged stories and laughs as they played on the old player piano preserved

from Peary's ship, the USS *Roosevelt*. At one point Robert asked, "Do you have the classic Peary gap between your two front teeth?"

"I think I used to," said the denture-wearing Kali.

Kali was visibly happy all through the visit. He later said that for the first time in his life, he felt like a real member of the Peary family.

Back in Cambridge, we attended a special Sunday morning service at Harvard's Memorial Church, where the Reverend Professor Peter J. Gomes honored the men "who came to us from very far distant places," and apologized for the weather. "We hope you will forgive us if it is too hot," Gomes said. He praised Anaukaq and Kali for "taking the risk to travel to America to find their relatives."

The next evening, then-university president Derek C. Bok welcomed Anaukaq and Kali at a formal Harvard banquet held in their honor. He talked about the discovery of the North Pole, and presented each man with an engraved silver plate in appreciation of his father's contributions to Arctic

Reverend Samuel Proctor addressing the congregation of the Abyssinian Baptist Church

Anaukaq Henson presenting a gift from Greenland to a young relative in Boston, Massachusetts

Kali Peary (left) and Anaukaq Henson presenting a gift to Matthew Henson's grandniece Olive Henson Fulton in Boston, Massachusetts

Dr. Counter speaking with Anaukaq Henson at Matthew Henson's original gravesite in Woodlawn Cemetery in the Bronx, New York

Part Three

Honor the Fathers

Kali Peary placing a wreath at his father's gravesite in Arlington National Cemetery, with eighteen-year-old grandson Ole Peary (left) and Anaukaq Henson (with cane)

9
Two Different Resting Places

From New England, we traveled to Washington, DC, so Kali could see his father's grave in Arlington National Cemetery. On a bright afternoon, under the auspices of the president of the United States, we conducted a magnificent wreath-laying ceremony at Robert Peary's gravesite in Arlington National Cemetery. The chaplain of the US Navy was there, and from the White House Ronald Reagan sent a splendid statement of tribute to Matthew Henson, Robert Peary, and their visiting Greenlandic progeny. Chase Untermeyer, the assistant secretary of the navy, read the tribute aloud, then Kali laid a small wreath he had brought with him from Greenland on his father's grave. Anaukaq and the other Inuits joined Kali in placing a larger wreath at the foot of the globe-shaped monument that marks the site.

Later that evening, the mayor of Washington, DC, Marion Barry, hosted a ceremony in the city's convention center to proclaim Wednesday, June 3, 1987, "Matthew Henson Day." The Inuits cheered when Navarana explained the mayor's proclamation. Henson relatives from all over the Washington, DC/ Maryland area came to the ceremony and the reception that followed. Dressed in jackets and ties we had shopped for earlier, Anaukaq and Kali greeted the guests like seasoned politicians. Representative Walter Fauntroy arranged for Washington civic leader Kent T. Cushenberry to present Anaukaq and Kali with American flags that had flown over the United States Capitol—an honor the Inuit Hensons and Pearys fully understood and appreciated.

After visiting more of Anaukaq's relatives from the Georgetown area, where Matthew Henson had lived with his sister, Eliza Carter, one hundred years earlier,

Chase Untermeyer, Assistant Secretary of the US Navy, (center) presenting gifts and a letter from President Ronald Reagan to Kali Peary (left) and Anaukaq Henson

we traveled to Maryland. Maryland takes pride in honoring its sons. African-Americans from the Charles County Historical Society escorted Anaukaq and his family to the ruins of the cabin where Henson had been born in 1866. As Anaukaq stood at the site of his father's birth, he stared pensively at the old red bricks scattered around the cabin's disintegrated foundation. One of the members of the Historical Society gave him one of the bricks as a memento, and for the rest of the day Anaukaq did not let go of this small piece of his father's beginnings. The faded brick went with him when he returned to Greenland.

Our last stops in Maryland were a middle school named for Matthew Henson and the State House in Annapolis, where a bronze plaque dedicated to him rests on the entry wall. Finally we headed to New York City, home of The Explorers Club. The club is private and only rarely opens its doors to non-members, but its president, John C. D. Bruno, welcomed our group for an extensive tour guided by the noted archivist George Michanowsky. At the end, both sons signed the club's official guest registry—a proud and meaningful moment for both men, who could barely write. As I watched them sign their names with dignity and

Kali Peary (left) and Anaukaq Henson with Marine Corps Honor Guard at Matthew Henson's original gravesite in Woodlawn Cemetery, Bronx, New York

Anaukaq Henson and his sons receiving a Proclamation from Mayor Marion Barry, declaring June 3, 1987 Matthew Henson Day in Washington, DC

Anaukaq Henson (with cane), Dr. Counter, and Kali Peary with Henson's grandsons (back row, left to right) Ajako, Ussarqak, Kitdlaq, Avataq, and Vittus, and great-grandchildren (middle row, standing) Malina, Aviaq, and (front row, left) Massauna-Henrik

Anaukaq Henson and Kali Peary visiting the Abyssinian Baptist Church for a special service in their honor. The photo in the background is of the church leader, the Rev. Adam Clayton Powell.

monument in the National Cemetery. I did not know how to explain the disparity to these unsuspecting Greenlandic Inuits, nor the American racial injustice that had burdened Henson in life and in death. Instead I promised Anaukaq and his family that I would do everything in my power to right this wrong, and have Henson reburied alongside his comrade in the nation's most prestigious cemetery. Thus, I made my second promise to the Inuit sons of the North Pole expeditions.

And then I made a third: Anaukaq, Kali, and their offspring asked that I help their families have Henson and Peary remembered at the North Pole on the one hundredth anniversary of its discovery. Anaukaq was very keen on acknowledging his father's historical role and having it appreciated by his fellow Inuits, and others. I promised them that in 2009 I would travel to the North Pole with some of their sons and daughters to commemorate the centennial of their fathers' achievement. This is the vow I call my North Pole Promise.

Navarana reading a proclamation as Washington, DC, civic leader Kent T. Cushenberry (right) presents Anaukaq Henson and Kali Peary with United States flags that had flown over the US Capitol Building

10
By Order of the President

Shortly after Anaukaq's and Kali's departure from the United States, I began a campaign to have Matthew Henson disinterred from his common grave in Woodlawn Cemetery and moved to a site of prominence in Arlington National Cemetery. I wrote to President Reagan about the disparity in Peary and Henson's burial sites, and in their treatment by American society, and I appealed to him to issue a presidential order to re-bury Henson among the heroes in Arlington. Secretary of Housing and Urban Development Samuel R. Pierce, Jr., helped by transmitting my letters directly to the president. In late 1987, the president granted my request. By presidential decree, Matthew Henson would be buried in Arlington National Cemetery on April 6, 1988, the seventy-ninth anniversary of day he reached the North Pole.

Avataq accepting ceremonial flag that drapped Matthew Henson's casket

I returned to Arlington to meet with the cemetery's superintendent, Raymond Costanzo, and his staff to ask that Henson be buried next to Robert Peary. Costanzo, who had a deep appreciation for Henson and his accomplishments, was very supportive and granted my requests for both the

The Reverend Peter J. Gomes (left) presiding over Matthew Henson's 1988 re-interment at Arlington National Cemetery as the US Navy Honor Guard prepares the ceremonial flag

Commander Guion S. Bluford, Jr., America's first African-American astronaut in space, addressing attendees at Matthew Henson's re-interment ceremony

The new headstone at Matthew Henson's gravesite in Arlington National Cemetery

The headstone at Robert Peary's nearby gravesite in Arlington National Cemetery

Matthew Henson's grandniece, Audrey Mebane, (left) and niece, Virginia Carter Brannum, posing proudly at the new headstone for Matthew Henson's gravesite

(left to right), Avataq Henson, Dr. Counter, and Ebony magazine founder John H. Johnson standing beside the new headstone for Matthew Henson's gravesite in Arlington National Cemetery

With plans for the re-interment ceremony well in hand, the next task was to arrange to bring some of the Greenland Hensons back to the United States to participate. In March 1988, a US Air Force plane carried them to the United States. Anaukaq had died a year before, so his eldest son, Avataq, took the lead in representing the family. Avataq's youngest son (and Anaukaq's favorite grandson), the ten-year-old hunter Magssanguaq, traveled with him to play a key role in the ceremony. The other Inuit Hensons attending were Anaukaq's youngest son, Kitdlaq; his second-oldest son, Ajako; and Ajako's son, Jens.

On April 6, volleys of rifle fire echoed through the hills as Matthew Henson was interred in Arlington National Cemetery with full military honors. The Reverend Professor Peter J. Gomes of Harvard University's Memorial Church and Commander Charles W. Marvin, the navy chaplain, presided. Together, Magssanguaq and I uncovered the monument, which had been draped with the same veil used to cover the monument to the astronauts who died in the

Ajako, Magssanguaq, Dr. Counter, and Avataq standing at attention while a US Navy Honor Guard carries the remains of Matthew A. Henson to his new gravesite in Arlington National Cemetery

Challenger space shuttle disaster.

The ceremony included speeches by Commander Guion S. Bluford, Jr., the first African-American astronaut to fly in space, who extolled Matthew Henson as "an Arctic explorer and a black American. May all future explorers follow in his footsteps"; the Civil Rights leader Dorothy Height; John H. Johnson, the *Ebony* founder, who co-chaired the re-interment committee; and General Julius W. Becton, Jr., who represented the president of the United States. As the red and white flag of Greenland fluttered on a flagpole next to the American flag, Kitdlaq Henson delivered a speech on behalf of the Inuit Hensons. He expressed the family's pride and their appreciation that their forebear would rest in a place of great honor in the national cemetery—and next to his fellow Arctic explorer, Robert E. Peary.

"Now the old friends of the North Pole are back together again," he said.

11
Launching the USNS Henson

While waiting for the North Pole centennial to arrive, it became my aim to resurrect the name and contributions of Matthew Alexander Henson, a great American patriot, a courageous explorer, an honorable man, and a compassionate humanitarian. Henson was also a pioneer in race relations. At a time in America when relations between white and black were deplorable and perilous, he was able to reach across racial lines and establish respect and admiration for persons of his ethnic background. It is most revealing that Henson's character, qualities, and leadership inspired Donald MacMillan, the white navy admiral who accompanied him on the 1909 expedition, to write the US Congress in 1931:

> *"In consideration of the very valuable work of Matthew A. Henson, assistant to Robert Peary for 18 years, (1891–1909), I would suggest a special medal to be awarded by Congress. . . .*

Henson went to the Pole with Peary because he was better than any man in Peary's party," MacMillan wrote. "And in fact, he was of more real value to Commander Peary on the Polar Sea than all white men combined . . . race, creed, or color should not stand in the way of recognition."

I became determined to follow MacMillan's lead. Along with retired Admiral W. Norman Johnson, then dean of students at Boston University, I wrote letters to the secretary of the navy and other United States officials to recommend that a ship be named for Matthew Henson. I believed this gesture would honor not only Henson, but the navy as well, and would be appreciated by African-

Matthew Henson's grandniece, Audrey Mebane, christens the USNS Henson by smashing a ceremonial bottle of champagne against the hull

The US Naval Ship Henson, a five-thousand-ton, T-AGS-63 class, multipurpose oceanographic survey ship at sea

Americans and Americans of all races who believe that the achievements of all who serve our country admirably and with distinction should be recognized, regardless of their background, race, religion, or gender.

The navy agreed. In 1996, it commissioned the US Naval Ship *Henson*, a five-thousand-ton, T-AGS-63 class, multipurpose oceanographic survey ship designed to scan and map the ocean floor, measure sea depth, and acquire marine data that would be used to update nautical charts and develop new ones.

At the navy's invitation, I traveled to Moss Point, Mississippi, for the official launch of the USNS *Henson*. At my request, Henson's grandniece Audrey Mebane christened the ship. On the afternoon of October 21, 1996, the retired Washington, DC, schoolteacher proudly and forcefully struck the bow of the massive ship with a bottle of champagne, initiating its slow slide from wooden moorings into the sea.

12
The National Geographic Hubbard Medal

There was one other important recognition that I wanted to secure for my boyhood hero. In 1906, the National Geographic Society had voted to give its prestigious Hubbard Medal to Robert Peary for traveling farther north than any other human; three years later they awarded the medal to one of his white assistants, Robert Bartlett, who admitted that he never came closer than 125 miles to the geographic North Pole. The African-American Henson, who besides Peary was the only American of any ethnicity to *actually* reach the Pole in 1909, was not so honored. In fact, the National Geographic Society virtually ignored Henson's achievements during his lifetime.

Matthew Henson's grandniece, Audrey Mebane, accepting Matthew Henson's the Hubbard Medal in Washington, DC

For several years, I wrote letters to the Society appealing for it to right this historic wrong and posthumously present the Hubbard Medal to Matthew Henson. Finally, in 2002, it did. Invited to participate in the program, I again escorted Henson's grandniece Audrey Mebane to the ceremony at the newly named Matthew Henson Earth Conservation Center in Washington, DC, to accept the award on behalf of the Henson family.

The American Hensons were deeply appreciative of this recognition. I felt that my hero had finally received the respect of the National Geographic Society, and that he would have been pleased with this honor for his contributions to Arctic exploration, ninety years after it was bestowed on his white partner in adventure.

Avataq showing off the head of a walrus that he killed

Ten-year-old rifle hunter Magssanguaq Henson, great-grandson of Matthew Henson, displaying birds that he shot in the hunting grounds outside Moriussaq, Greenland. His father, Avataq Henson, and Dr. Counter join him in celebrating his rifle hunting skills

13
Life in the Arctic

In 1996, I made another trip to Greenland, this time with a film crew in tow. In the years since Henson had been buried at Arlington National Cemetery, I had written a book, *North Pole Legacy,* that documented my discovery of Peary and Henson's secret sons. I was now making a film to accompany it and illustrate the lives of the Polar Inuit families I'd come to know.

After landing at Thule Air Base I boarded a helicopter to travel to Morriussaq, the tiny settlement where I had first met Anaukaq Henson, to see my old friends. I was warmly greeted at the landing site by Avataq, Malina, Aviaq, and Ajako Henson, and Avataq's son, Magssanguaq—a joyous reunion, with lots of laughter and hugs. The family looked great: healthy, cheerful, and in good form. They walked me to Anaukaq's old igloo, now occupied by Avataq, and made up a bed for me in Anaukaq's old room.

Anaukaq and Kali welcoming Dr. Counter to Morriussaq

Avataq told me that he was still vigorously hunting—walrus (his favorite meat), seal, rabbits, ducks, and even polar bear. However, polar bears eat garbage as they migrate across the Arctic from Alaska to Canada and Siberia, and from that they can contract the parasitic worm *Trichinella spiralis*. When humans eat the raw meat of an animal infected

Danish Lutheran Church
in northwest Greenland

Avataq Henson conducting Sunday
religious service for Polar Inuit
congregation in northwest Greenland

with *T. spiralis* they can contract trichinosis, a painful, potentially fatal disease in which the worm parasites invade muscles and organs. Polar Eskimos consider raw polar bear meat a delicacy, and trichinosis was a common illness among them until modern times, when Danish doctors began to treat it. Avataq himself had been one of their patients.

Although he still prided himself on being a *peeneeahktoewah*, Avataq had entered a new phase of his life. He had become religious. In fact, he had adopted the Danish Lutheran faith and become the minister of a makeshift church in the Morriussaq settlement, delivering weekly sermons in a room in the village schoolhouse. In some ways, it was hard to believe that one of the greatest Eskimo hunters was now a Christian minister. I attended one of Avataq's services and watched him trade his bearskin pants and deerskin anorak for a black pullover shirt and white clerical collar. From his simple but impressive pulpit, Avataq spoke softly, but clearly and distinctly, and with admirable credibility as he read scripture and led the small congregation through recitation and songs.

I was surprised by the maturity of

his son, Magssanguaq, the little hunter whom I first met when he was nine years old. He had grown into a tall, muscular, handsome man of eighteen. One of the best soccer players in the area, he looked forward to advanced schooling and said that in the future he would train as an electrician so he could help his village. But for now, Magssanguaq proudly told me that he wanted to become a master hunter to honor his grandfather.

Ajako, the businessman of the Henson family, was still managing his variety store—the only one in the small settlement of about sixty Inuits. He complained that he and his wife, Puto, would soon have to close their doors because so many people were leaving Morriussaq to move to the larger settlement of Qaanaaq, some sixty miles away by dogsled, where there were about seven hundred inhabitants. He did not look forward to closing his little store, which he and Puto had run for more than twenty years. Ajako talked with great sadness of the loss of his second son, Nukka, an eighteen-year-old hunter who had died in a kayak accident while tracking walrus a year earlier. Kayak accidents are all too common among the Polar Eskimos. Ajako was grateful

Magssanguaq Henson, great-grandson of Matthew Henson, crawling into the new igloo he and his father built in the hunting grounds outside Moriussaq

Eighteen-year-old Magssanguaq Henson aims rifle at prey during a hunt outside Moriussaq

Eighteen-year-old Massauna-Henrik, great-grandson of Matthew Henson, in kayak with sealskin buoy and harpoon, preparing to hunt narwhals

Ajako's daughter, Aviaq, carrying
her favorite puppy in Moriussaq

that his first son, Jens, was still alive and well. His nineteen-year-old daughter, Aviaq, was his joy, and he was happy that she had gone to Qaanaaq and later the southern township of Nuuk to advance her education. (I still remembered her as a lively, carefree nine-year-old, laughing as she ran through the village with her puppies at her heels!)

The Hensons and I reminisced about Anaukaq, the patriarch they all missed dearly: his contagious laugh, his calming words, his stories, and his always-appreciated wisdom. Among the Polar Eskimos, Anaukaq was one of the most admired and respected men of his time. He was popular and welcomed in every village in northwest Greenland. He, like Kali, had even gained celebrity status after their trip to the United States in 1987. But Anaukaq was a modest man, and only wanted the history of his Greenlandic Inuit and American families to be acknowledged, connected, and remembered.

Magssanguaq was especially fond of his grandfather, and spoke lovingly of

Little Ootah readying his sled dogs for a hunting trip

him and the great times they had had together. Avataq recalled with pleasure his trips to the United States for the family reunion in 1987 and with Magssanguaq for the re-interment ceremony at Arlington a year later. Magssanguaq spoke of the three-ring circus he had attended as part of that trip, where the massive elephants that had mesmerized him. We laughed remembering how his innate hunting instincts had kicked in, and he had exclaimed that the beasts' enormous poops would make them easy to track! The Inuit Hensons wanted to visit the United States again so their children and grandchildren could see their American relatives, especially Matthew Henson's grandnieces Olive Henson Fulton and Audrey Mebane.

One pleasant surprise was to learn that Jorgen Ootah and his wife, Inaluk, were living in Morriussaq. "Little Ootah," as I called him, is the namesake and great-grandson of the Inuit guide who traveled to the North Pole with Peary and Henson in 1909, whom the American explorers credited with saving their

Ten-year-old Magssanguaq proudly displaying the many birds he shot in the hunting grounds outside Moriussaq

Strips of seal and walrus meat drying on a high rack, outside the reach of sled dogs, in Moriussaq, northwest Greenland

lives on their long, grueling trek. The Inuits are proud that today there is an island at the top of northwest Greenland named for Ootah, and they still tell stories of the legendary explorer, who became a revered leader in his community.

The Ootahs invited me to their igloo to share some food from a recent hunt. We ate fresh, raw seal meat that Ootah pulled from a wooden rack that stood about six feet high outside his igloo, well above the reach of his dogs. During our meal, Inaluk sat near us sewing sealskin snow boots that she planned to wear the next winter. Ootah talked about the disappearing animals in his area, where game had become scarce. He said that most of the other hunters, many his old friends, had moved north for better prospects, and that he and his wife would soon have to follow, possibly moving to Qaanaaq.

The sacred Inuit site, Upside-Down Tea Cup Mountain or Uummannaq, now part of Dundas Air Force Base in Thule, Greenland

Little Ootah also spoke about Uummannaq (pronounced *OO-mah-nahk*) or "Upside-Down Tea Cup Mountain," an area many of the Inuit people in northwest Greenland consider sacred, believing that their ancestors' spirits reside inside the mountain. The Inuits living in the area had been forced to abandon it when Dundas Air Base was built at Thule in the 1950s. In the 1980s, they tried to reclaim it, arguing that it was their tribe's historic home and that the mountain was their most sacred site, but they were not successful.

Ootah's eyes snapped with excitement when told me that another hunter about seven and a half miles north of Morriussaq had spotted a polar bear that he believed was still in the area, searching for seal. Ootah said he would head out on the ice in a few days to hunt it down, and his wife Inaluk would go with him and use an ulu (an indigenous knife) to begin skinning and preparing the meat from whatever he shot. If he was lucky enough to catch a *nanook,* the Ootahs would share the meat with the entire village, and it would provide food for about a week. The hide would be used to make new pants for Ootah and one

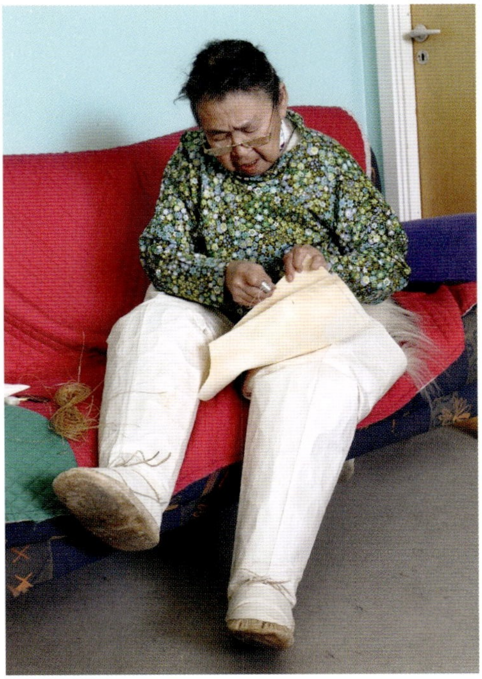

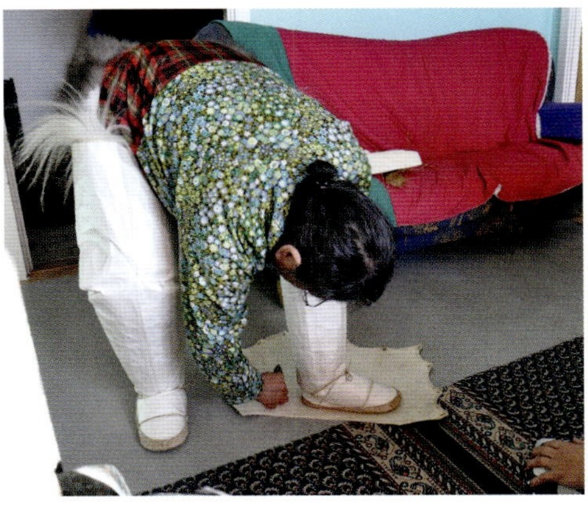

Above: Matthew Henson's granddaughter Malina using an ulu (an indigenous knife) to remove flesh from a polar bear hide

Left: Avataq Henson's wife, Simisuk, making boots from a polar bear hide

Dr. Counter practicing his sled dog whip-cracking skills as Avataq and his wife, Simisuk, look on

of his relatives. To my delight, he said that Avataq and Magssanguaq would be a part of the hunting party—and even better, he invited me along.

As it has been for centuries, the ferocious and aggressive *nanook* remains the favorite food of the Polar Eskimos. Outsiders call this huge, dangerous beast the "lion of the Arctic." It is also hard to hunt, requiring dogs to chase it down and tire it out before it can be killed, and hunters who can kill it are especially admired.

A few days later, we headed to the hunting grounds on a convoy of six dogsleds, each pulled by a team of eight to ten dogs. Magssanguaq raced his team in front of the others, laughing as he passed the older hunters. The others smiled as the young upstart showed off his budding alpha-male swagger. They understood and appreciated his excitement.

After some hours, we stopped to rest the dogs and have some tea and slices of walrus heart, and feed the dogs to refuel them for the rest of the trip. The men threw large chunks of seal or walrus innards and blubber at each wolf-like animal, which the dogs snatched mid-air and gulped down without chewing.

A ten-year-old Magssanguaq Henson with his first dog sled, preparing to hunt outside Moriussaq

Avataq, Magssanguaq, and Dr. Counter planting the Explorers Club flag at their newly completed igloo

Then we were on the move again, traveling across the vast snow and ice fields at remarkable speed, the dogs yelping and the hunters' whips snapping in the air above their heads.

We reached the edge of the hunting grounds at the end of the day. The teams untied their dogs, attaching the long leashes to stakes in the ice. As we unloaded our gear, Magssanguaq talked about how excited he was to build an icehouse, or "igloo," for our shelter. He was still learning this traditional Eskimo skill from his father, a master igloo builder. On the ice, knowing how to create shelter can save a life, especially during snowstorms or blinding whiteouts. I, too, wanted to learn igloo building, and offered to assist the Hensons in this task.

Avataq taught Magssanguaq how to cut large blocks of hardened snow from a nearby ice mound to make the base of the igloo. The lower blocks had to be larger and perfectly shaped, with angled sides and rounded edges so they would fit the next block. Magssanguaq enthusiastically joined in, carrying the heavy blocks to the spot where we would begin construction. Slowly the igloo—wide enough to sleep three persons, with a tunnel entrance—began to take shape. It was tough work and we were soon dripping with sweat.

As we worked, I marveled that I was witnessing the grandson and great-grandson of one of the greatest Arctic explorers in history, the intrepid African-American who also learned to build igloos for shelter and food storage on expeditions to the North Pole. In many respects, I was watching "black Eskimos" build an igloo. How unlikely is that?

In a little more than an hour, Avataq was putting the "sky roof" on the igloo. This delicate task required special skill, but Avataq was more than up to it. He skillfully carved a rounded piece of ice and snow while Magssanguaq studied his technique. Once the igloo was finished, we crawled inside where the Eskimos spread animal skins across the floor as sleeping mats. In no time, the walls began to melt slightly from our body heat and Avataq used his hand to smooth the inside of them to create a tight seal. Soon the igloo was so warm we could take off our outer clothing and settle in for a comfortable sleep.

Ajako, grandson of Matthew Henson, washing the skin of a polar bear that he killed on the hunting grounds

Six hours later, we were awakened by Little Ootah sticking his head into the entrance and shouting "*Poo ee hee, poo ee hee!*"—seals! We popped outside to watch him scale a one-hundred-foot mountain of ice to scan the horizon for animals. Again he shouted "*Poo ee hee!*" as he pointed to the west. He could see seals in the distance. To the Eskimos, seal meat has been a primary source of protein for centuries. Just as we eat cow, pig, or chicken, they eat seal.

The hunters tied their dogs back up to the sleds, checked and loaded

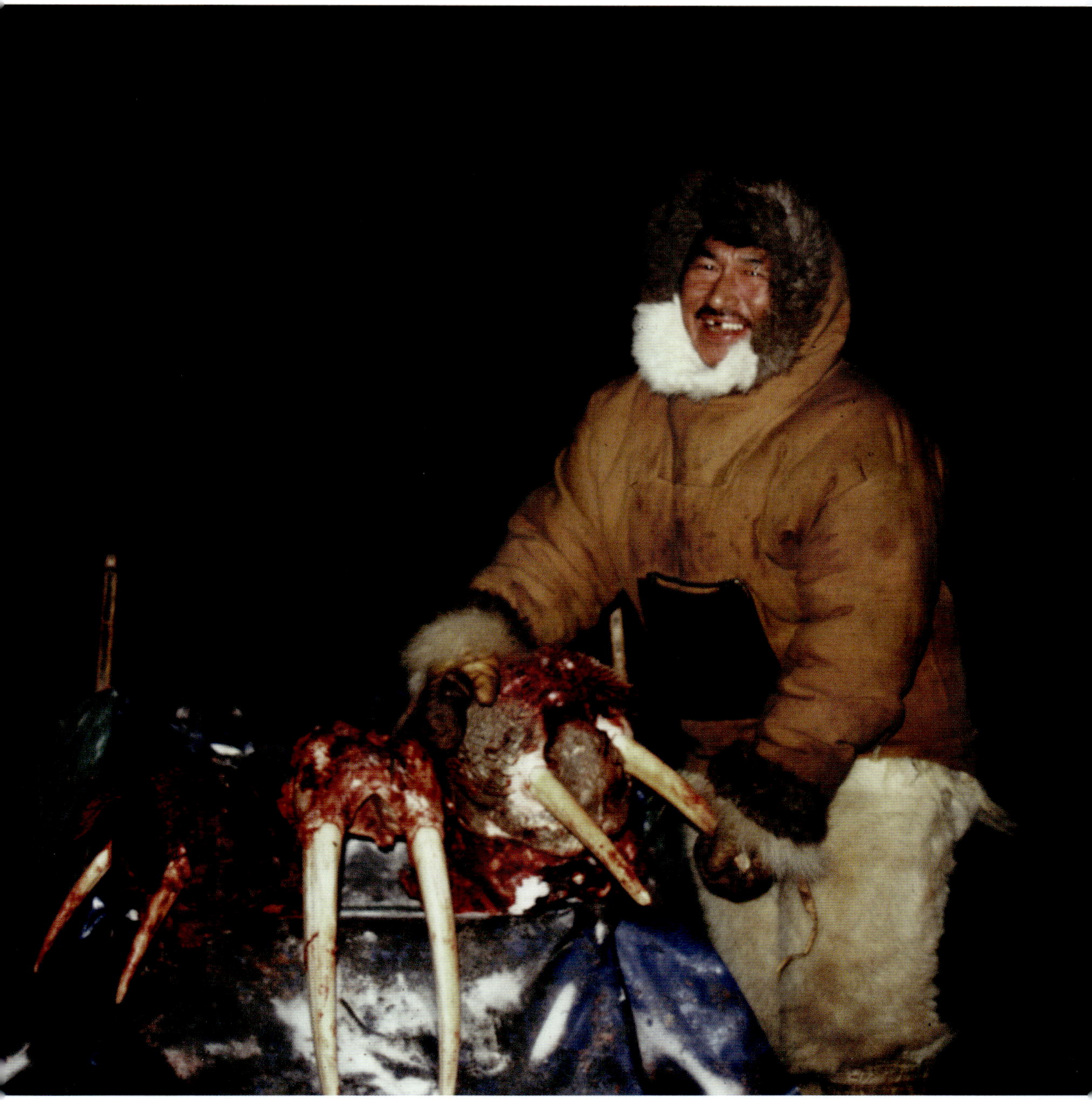

Avataq displaying the heads and tusks of three walruses that he killed on the hunting grounds

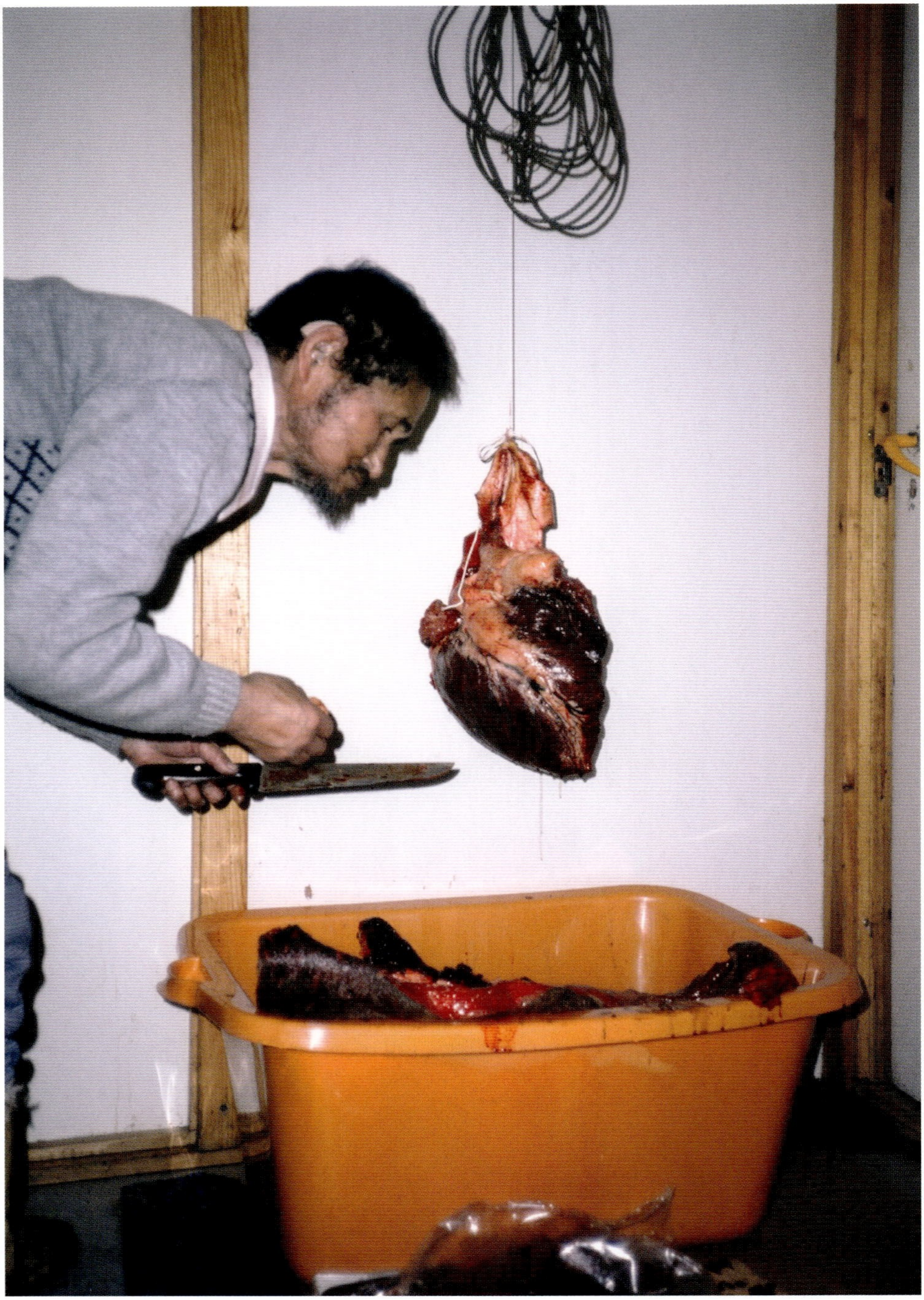

Anaukaq Henson savoring slices of raw walrus heart, one of his favorite delicacies

their rifles. Avataq and Magssanguaq's faces were intense. The hunt was on!

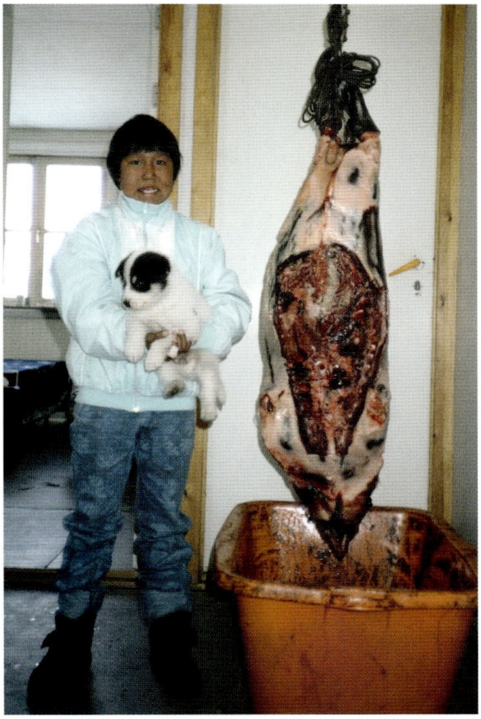

Ten-year-old Magssanguaq Henson with his puppy, displaying a seal that he shot during a hunt outside Moriussaq

The hunters halted their dogs a short distance away from the seals—big animals, basking near holes in the ice. The men grabbed their rifles and snow screens to hide themselves from view as they crept closer. Avataq and Magssanguaq quickly dropped to their knees and ordered me to do the same. Hiding behind their screens, rifles carefully mounted in the center, they inched slowly forward. I crawled after them, camera in hand. At thirty yards, Avataq and Magssanguaq took aim and fired almost simultaneously. Both shots struck their prey.

At the sound of the gunfire, the other seals, seven or so, scampered for the ice holes as the rest of the hunters started shooting. Through my binoculars, I could see several animals collapse and roll over, shot in the head.

Avataq and Magssanguaq rushed back to the sleds, untethered their dogs, and sped toward the dead seals. The dogs barked wildly, as if they understood that the kill meant there would be fresh meat for them. Magssanguaq was thrilled. He had fired one shot, and it had found its mark. He and his father could feed their entire family and dogs for at least a month with the meat from their kills.

The other hunters were jubilant as well. We had bagged a total of five big seals (but no *nanook*), each about the size of a small cow. As the men compared their trophies, a mild competition began—and it turned out Magssanguaq had shot the largest. He was ecstatic. He beamed, grateful to be acknowledged by his elders as a *peeneeahktoewah*, a master hunter. Little Ootah had shot the second-largest seal, and he too was proud. Avataq later told me quietly that he was

Anaukaq Henson, Kali Peary, and friends enjoying tea in the summer sunshine outside a wooden "igloo" in Moriussaq

immensely pleased with his son's progress as a hunter. We loaded the seals on the dogsleds, tied them in place, and headed back to the igloos, where we would spend another day before traveling back to the settlement.

The next day, we arrived back at Morriussaq. The children of the settlement ran to meet the teams and see what the hunters had killed. They knew it meant tasty, fresh meat. The hunters tied up their dogs and took their kills to special areas behind their respective igloos, where they would skin and gut the seals and remove their edible organs. I followed Avataq and Magssanguaq, running behind their dogsled. Avataq's daughter, Malina, and Ajako's daughter, Aviaq, welcomed us back and joined me in watching the men eviscerate the big game. As they carved and quartered the animals, each part was placed in a separate spot. The muscles, liver, heart, even the flippers were cut off and saved, to be used as food for both humans and dogs; the dogs alone would get the intestines, lungs, tail, and head. All of the meat was placed on a high wooden storage rack, well above the reach of the jumping dogs.

That evening, Avataq and Puto prepared a big dinner for the entire family:

Dr. Counter bidding farewell to Anaukaq Henson and Kali Peary
as he prepared to return to the United States

fresh, raw seal meat, and kelp taken from the stomach of the seal. The raw seal heart was flavorful, with the consistency of meat, but with a fishy smell. We drank hot Constant Comment tea—its citrusy flavor a new and welcome one for the Inuits—that I had brought as a gift from the United States. Avataq pulled out a bottle of red wine, and we all had a sip to toast the hunt's success. The dinner was delicious.

The Eskimos are sociable people. From time to time, village families travel many miles by dogsled to larger settlements, where they visit relatives and friends and indulge in one of their favorite pastimes, gambling with cards. Occasionally, there are weddings and other celebrations that give them reason to travel from one settlement to another. And inevitably, there are deaths and funerals.

All too soon it was time for me to travel back home. I was sad to leave these Eskimos, and they, too, were melancholy about this third separation. In our brief times together, I had grown as close to these remote families as I was to my own. The evening before I left, we sat on the floor around a table telling stories. Even in this remote settlement at the top of the earth, the people led interesting

Eighteen-year-old Magssanguaq Henson, great-grandson of Matthew Henson, proudly displaying his hunting rifle to Dr. Counter at door of newly built igloo

Ootah taking aim at prey during
a hunt outside Qaanaaq

Part Five

The Centennial Approaches

Hunting grounds outside Qaanaaq

14
North Pole Promise

In the spring of 2008 I was starting to worry. A letter from President Barack Obama commemorating Robert Peary and Matthew Henson's discovery of the North Pole was on its way, and April 6, 2009—the date of the long-awaited commemoration of their expedition—was rapidly approaching. Yet all the plans were not yet in place.

Getting to the North Pole is not easy. In the 1980s and '90s, the US Air Force, First Space Wing, took me to the top of Greenland onboard a massive Lockheed C-141 Starlifter that "rotated" supplies and personnel back and forth to Thule Air Base. C-141s are unusual: big, green, lumbering jets whose wings sweep back from the top of the fuselage. The seats face the rear of the plane, and there is none of the comforting interior insulation found on commercial airplanes; passengers are given earplugs to muffle the roar of the engine during the eight-hour flights from McGuire Air Force Base in New Jersey to Greenland. The planes are so massive that they include bunk beds in their cockpits where relief pilots can nap during long flights. The only food on board comes in a small box you buy at the base canteen before takeoff, containing a cold salami or baloney sandwich, a packet of mustard, and a juice box. But traveling in the company of friendly airmen and officers is enjoyable.

In 2008, I made plans to fly back to Thule on the rotator and from there kick off my plans to reach the North Pole. To my surprise—and distress—the base commander informed me that the C-141s had been retired, and the rotator program was no longer in service. Instead, flights to Thule had been contracted out to private companies, so even military personnel flew in on commercial aircraft.

US Air Force C-141 Starlifter that "rotated" personnel and supplies from McGuire Air Force Base in New Jersey to Thule, Greenland

Beyond losing the adventure of the military flight, getting to the northernmost American air base would now be harder and more expensive than anticipated.

I explained my project to the new base commander at Thule. I told him I wanted to travel to the North Pole and back on April 6, 2009, aboard a military aircraft with the part-Inuit descendants of the men who conquered the Pole, along with a centennial commemorative case and capsule that would be placed on the icepack at precisely latitude 90° north during a ceremony to salute their achievement. The commemorative case would then be returned to the United States to be exhibited at American schools as part of an educational initiative about Arctic exploration, the story of Peary and Henson's Amer-Eskimo progeny, and the culture of the Polar Inuit people.

The commander agreed to help. In the spring of 2008 he sent two air force officers from Thule to Cambridge to help organize the project. The officers said

US Air Force C-130 Hercules scheduled to fly the Hensons, Peary and Dr. Counter to the North Pole

that the US Air Force would support my project with military aircraft. They recommended that we fly on a military C-130 outfitted with skis from Thule directly to the North Pole, where we would land on the ice and the Henson-Peary party and I would then conduct the commemorative ceremony.

I was thrilled. The massive C-130, which had flown missions to the South Pole, could accommodate many members of the Greenlandic families, as well as Navarana, our translator from earlier years. At my request, Navarana contacted members of the Inuit Henson and Peary families from all over Greenland and invited them to travel with me to participate in the centennial ceremony.

15
The Capsule

I began working on creating the North Pole discovery centennial commemoration with Matthew Henson's remarkable great-granddaughter, Aviaq. The daughter of Henson's fourth grandson, Vittus, Aviaq lives in Nuuk, Greenland, and speaks fluent English. On earlier visits, there had been little or no telephone communication in the tiny, isolated settlements; by 2008, I could reach Aviaq and other Inuits by cellphone.

We decided the commemorative capsule should be made of thick Plexiglas, and contain items associated with Henson and Peary's achievement. Among these were a folded American flag; a bible from Harvard's Memorial Church signed by its pastor, the Reverend Professor Gomes (Henson took his bible with him to the Pole, snuggled inside his anorak as his *vade mecum*, or "guidebook kept at hand"); Peary's 1910 book, *The North Pole*; Henson's 1912 book, *A Negro Explorer at the North Pole*; a letter from President Reagan recognizing Peary and Henson's Arctic achievements and their sons' visit to America in 1987; a letter from President Obama honoring the North Pole centennial; my own book, *North Pole Legacy: Black, White, and Eskimo*; sacred Inuit symbols such as polar bear claws; sealskin and polar bear fur gloves worn by Henson's son, Anaukaq; photos and an insignia cap from the launching of the USNS *Henson*; and photographs, letters, articles, poems, and other items associated with or touched by the story of Peary, Henson, and their Greenlandic Inuit descendants.

16
The Centennial

Twenty years after Matthew Henson was interred in Arlington National Cemetery, I was finally ready to keep my North Pole Promise. I had packed my clothing and equipment for the frozen north, including compasses, tents, heaters, layers of clothing, and other items for Arctic exploration. I had chosen two people to join me in documenting the centennial commemoration: my technical assistant of twenty-five years, Anthony Jacobs, who would serve as technician and sound recordist, and the expert filmmaker, Anna Schoenfeld, who would be the cinematographer.

My then-sixteen-year-old daughter, Olivia, helped me organize the expedition and agreed to come along as my assistant. Everything was in place for us to travel with the US Air Force to northwest Greenland, pick up members of the Inuit Henson and Peary families, and fly directly to the North Pole with the commemorative case and a planned program of commemorative speeches. Our passenger list would be Kitdlaq, Vittus, Ussarqak, and Aviaq Henson; Talilinquaq, Marie, and Ole Peary; Jorgen Ootah; Navarana K'avigak; Olivia and myself.

But in early March 2009, our plans were crushed. One of the air force officers I had met with in Cambridge contacted us to say that warming conditions at the North Pole made the ice unsafe for landing a large aircraft. The air force had determined that our well-planned trip on the C-130 with skis was imprudent, and refused to go forward with it.

We had to find another way to get to northwest Greenland to meet up with the Inuit families, and from there, reach the North Pole. We contacted the Danish and Swedish air forces who both thought it was a worthy project, but

The Beech King Air 200 aircraft that flew Dr. Counter and his team from Canada to Qaanaaq

they said they could not help with flight support. Soon we found a Canada-based company that transports scientists to the Arctic, and chartered a Beech King Air 200 that could fly us to Greenland and then to the Pole. The only drawback was the plane's seating capacity: it was designed to carry only ten passengers. Because we would have to remove some of the seats to load extra drums of fuel to enable us to reach the Pole and return, that meant limiting the number of Henson-Peary family members that could make the journey.

Olivia, Tony, Anna, and I made the trip to northern Canada. With several stops along the way, we eventually reached Iqaluit Bay, and from there headed north to the Inuit village of Pond Inlet. We flew from there to Thule, landing just north of the base on the newly created airstrip at Qaanaaq where the Hensons, Pearys, and Ootahs greeted us cheerfully. It was a tremendous pleasure to see Navarana, and Anaukaq's second son, Ussarqak, "the great kayaker." It was like coming home to my own family. The teenagers welcomed Olivia joyfully and

Olivia Counter holding the one-year-old great-great-grandson of Matthew Henson

The Peary and Henson centennial cake, baked by the Inuit families of Qaanaaq

spirited her away to introduce her to their culture and learn about hers. Though initially shy, Olivia quickly grew attached to her new friends. They saw her as family, handed her their babies, and proudly taught her how to use a whip to drive the dog teams and other traditional Inuit skills.

After speaking with the families, we decided to bring all of the Inuit Hensons and Pearys together in Qaanaaq, one of the northernmost inhabited settlements on earth, to celebrate and commemorate the centennial of their ancestors' discovery there. From Qaanaaq, a smaller group of descendants of the 1909 expedition would then fly to the Pole early on April 6.

Henson's great-granddaughter, Aviaq, took the lead in organizing the commemorative ceremony. She met with members of the families and organized the food, the music, and other aspects of the celebration. She and some of the women in the settlement also baked a special cake for the occasion, writing the names of Peary and Henson in icing. Beneath the names, Aviaq wrote "*100 ÅR*"—"100 years" in Danish.

Everyone in Qaanaaq was excited about the celebration. They decorated

Talilanquaq Peary, grandson of Robert Peary, and Ussarqak Henson, grandson of Matthew Henson, cutting the centennial celebration cake

Aviaq Henson, great-granddaughter of Matthew Henson, holding a letter of congratulation from President Barack H. Obama and a picture of the First Family

The Henson and Peary families gathering in the Qaanaaq school to celebrate the one-hundredth anniversary of the North Pole discovery

the local school with bright ribbons and photographs of Henson and Peary, and hung Greenlandic and American flags from the ceiling. The elders told stories about "Matthew the Kind One" and the quiet but determined Peary, and their struggle to attain the Pole. They spoke proudly of the important contributions of the Polar Inuits, especially the senior assistant, Ootah.

Beneath the two flags, I presented President Obama's letter to Kitdlaq Henson, the grandson of Matthew Henson, and Talilinquaq Peary, the grandson of Robert Peary. I also presented them each with a framed photograph of the First Family. The men and their families were deeply honored that the president of the United States would acknowledge their grandfathers' contributions to Arctic exploration and the discovery of the Pole. With his own and Peary family members assembled for the centennial, Kitdlaq expressed his gratitude and with a big smile joked, "I want you to know that all of us Eskimo people up here voted for President Obama."

The excitement continued as we waited for April 6 to arrive. Qaanaaq is at a high enough latitude that the sun is up nearly all day and night in early

Olivia Counter displaying her Polar Inuit gloves, made of sealskin with polar bear fur

Olivia learning to use the dog whip in Qaanaaq

Aviaq Henson, Matthew Henson's great-granddaughter, chipping ice for drinking water in Qaanaaq

Dr. Counter and Aviaq Henson, now a schoolteacher, in Qaanaaq, Greenland

Eskimo tradition, the whip never touches the dogs' head, but simply makes a popping sound above their ears. Olivia quickly mastered the technique, and I was proud of her efforts to learn the Inuit culture and techniques. She was taking her first steps in Arctic exploration.

In early April, the pilots of the charted aircraft reluctantly informed me that an attempt to reach the North Pole on the King 200 aircraft was unrealistic, and possibly foolhardy. Even with the extra drums of fuel, it was not certain that we could reach the North Pole and return safely to Thule. Without a backup plane in case of a fuel shortage or mechanical failure, we could not risk it.

I was deeply disappointed. I wanted more than anything to fulfill my promise to Anaukaq, Kali, and their progeny. But I could not put the Inuits or my daughter at risk by trying to take our small plane to the Pole and back.

It looked bleak.

But then Olivia reminded me of what I had always taught her—the explorer's motto: *Aut inveniam viam aut faciam*—I shall find a way or make one.

Hunting group outside Qaanaaq:
(left to right) Olivia, Simisuk Henson, Ussarqak,
Dr. Counter, Little Ootah.

17
Fulfilling the Promise

The United States Navy had supported all of my efforts in connection with the Peary-Henson legacy, from the first visit of their Inuit descendants to the United States in 1987, to Henson's re-interment, to the christening of the USNS *Henson*. I was also mindful of the explorers' historical association with the navy. If we could not get to the North Pole by air, perhaps it was possible to get there by sea.

Knowing that the navy had ships in the Polar Arctic, I had contacted naval staff earlier about the possibility of transporting me, along with my group and the commemorative case, to the North Pole. I now reached out to Vice Admiral Melvin G. Williams, Jr., commander of the Second Fleet, which controls America's air, surface, and submarine forces in the Atlantic, to see if he could help us. He liked my project, and said he hoped that something could be worked out "to support this historic and patriotic venture." He referred my request to the commander of submarine forces, and in a short time I was informed that the navy would provide the nuclear submarine USS *Annapolis* to transport the centennial commemorative case to the Pole. The *Annapolis* was then under the command of Michael Otto Brunner, a decorated officer and admired leader, and a man who demonstrated respect for persons of all backgrounds. Commander Brunner was proud to help honor Peary and Henson's accomplishments. My task was to deliver the capsule to the sub when it reached the Arctic.

At first I hoped to travel out on the ice with Olivia by dogsled to meet the *Annapolis*, but that proved unrealistic. We would have had difficulty locating the submarine in the vast icy Arctic sea, and there was the issue of time: an we were rapidly approaching the critical date of April 6, 2009. I made the decision

Commander Michael Brunner displaying the centennial capsule at the North Pole on April 6, 2009

Commander Michael Brunner of the USS Annapolis (right) and fellow officers reading a letter from Dr. Counter as part of the centennial commemoration held at the North Pole on April 6, 2009

of the *Annapolis* to the surface, where they erected a metal sign with the sub's insignia.

That afternoon, with his crew surrounding him, Commander Brunner read this statement aloud:

> On April 6, 1909, Admiral Robert Peary, then a commander, Matthew Henson, and their four Inuit assistants reached the North Pole and claimed it for the United States of America. Today, we are here to commemorate that historic journey. But we also celebrate the progress that our society has made in its ability to judge a man by the content of his character, and the greatness of his achievements, rather than the color of his skin, because back then, in 1909, Admiral Peary alone received credit for that journey to the North Pole. . . . Matthew Henson, who was an African-American, was ignored by the press and left out of all the history books. Our crew is a shining example of how far we have come—for we live amongst each other and place our lives in each other's hands every day, no matter what our race or religion.

His remarks were followed by those of the chief weapons officer of the USS *Annapolis*, Lt. Cmdr. Howard Craig, who read, with great emotion, these lines from a poem by Horatius Bonar, which Matthew Henson had kept tucked inside his diary:

Up and away, like the dew of the morning
That soars from the earth to its home in the sun;
So let me steal away, gently and lovingly,
Only remembered by what I have done.
My name, and my place, and my tomb all forgotten,
The brief race of time well and patiently run,
So let me pass away, peacefully, silently,
Only remembered by what I have done.
Gladly away from this toil would I hasten,
Up to the crown that for me has been won.
Unthought of by man in rewards or in praises-
Only remembered by what I have done.
. . . .
Not myself, but the truth, that in life I have spoken—
Not myself, but the seed that in life I have sown
Shall pass on to the ages—all about me forgotten,
Save the truth I have spoken, the things I have done.
(Horatius Bonar, "The Everlasting Memorial")

In a communiqué dated that same day, Vice Admiral Williams confirmed to me that the case had been delivered to the North Pole by the *Annapolis* and "the Harvard North Pole Discovery Centennial Commemorative Project [was] a success." With help from him and Commander Brunner, my two-decade-old North Pole Promise to Anaukaq Henson and Kali Peary had been fulfilled. The memories of Matthew Henson and Robert Peary and their achievements had been honored, and their names had echoed through the frigid air at latitude 90° north, one hundred years to the day after they first struck the Pole. They and their Polar Eskimo friends would have been pleased by this singular recognition.

Some months later, onboard the USS *Annapolis* at the Naval Submarine Base New London, in New London, Connecticut, the US Navy held a ceremony

The crew of the nuclear submarine USS Annapolis posing with the centennial capsule at the North Pole on April 6, 2009

under the auspices of Vice Admiral Williams, Commander Brunner, Rear Admiral Bruce Grooms, and the *Annapolis's* crew, to return the commemorative capsule to me for exhibition at Harvard and in middle and high schools throughout the United States. Aboard the *Annapolis*, Williams said:

> "Honoring the exploits of Admiral Peary and Matthew Henson, and all who achieved this venture back 100 years ago . . . typifies the importance of diversity—that people from different backgrounds can achieve greatness. So, as we look forward to the future, there are no obstacles. People with different backgrounds can come together to accomplish greatness."

Onboard the USS Annapolis at Naval Submarine Base New London in Groton, Connecticut: (left to right) Vice Admiral Melvin G. Williams, Jr., Rear Admiral Bruce Grooms, and Commander Michael Brunner (far right) presenting Dr. S. Allen Counter with the commemorative capsule that traveled to the North Pole on April 6, 2009

The nuclear submarine USS Annapolis in port at Naval Submarine Base New London in Groton, Connecticut

18
One Last Honor

In June 2009, in response to my petition to the US Navy requesting an appropriate rank for a patriotic American who had served on a United States naval field expedition for nearly twenty years as assistant to a naval officer, Matthew Alexander Henson was officially named an Honorary Command Master Chief in the United States Navy.

Certificate of Appointment
The Master Chief Petty Officer of the Navy
with Great Pride and Pleasure
Hereby Appoints

Matthew A. Henson

an

Honorary Command Master Chief

The appointee is granted the full honors incumbent to this office, and will carefully and diligently perform the duties of an Honorary Command Master Chief. Therefore, by the authority vested in me, and with enthusiastic approval of the Chief Petty Officers of the Navy, I declare this appointment effective.

Given under my hand, this 2nd day of June in the year of our Lord, two thousand nine.

R. D. WEST
Master Chief Petty Officer of the Navy

Certificate from the Master Chief Petty Officer of the Navy appointing Matthew A. Henson Honorary Command Master Chief in 2009

Matthew Henson at age thirty in New York City

Kali Peary and Anaukaq Henson in Moriussaq, Greenland, preparing to travel to the United States to meet their American relatives and visit their fathers' gravesites

Thoughts for Further Discussion

❧ A close friendship between a white man and a black man was unusual for Robert Peary's and Matthew Henson's time, but the different welcomes they received when they returned from the North Pole were quite typical. Do you think similar accomplishments by people of different races are celebrated equally today?

❧ Most of Robert Peary's family did not want to meet Kali, his son by Ahlikasingwah, when he came to the United States. Some of them referred to Kali as "illegitimate." What do you think of this way of describing a child's relationship to a parent?

❧ Minik, his father Qisuk, and four other Polar Inuit Eskimos (Nuktaq, Atangana, Uisaakassak, and Aviaq) were brought to the United States in 1897 by Robert Peary, who delivered them to the American Museum of Natural History. There they were treated as living specimens to be studied by scientists and put on exhibit for visitors. Such "ethnological exhibits" or "human zoos" were not uncommon; examples can be found even in the late twentieth and early twenty-first centuries. What do you think of such a practice? Can you make assumptions about the people who organize such exhibits? What about the people who are put on exhibit?

Little Ootah on dog sled heading for the hunting grounds outside Qaanaaq, Greenland, with other hunting teams

~ Little Ootah spoke of the Inuits' beliefs about the sacredness of Uummannaq ("Upside-Down Tea Cup Mountain")—the massive rock formation more commonly referred to as Thule Mountain. In the 1950s, the Danish government relocated the Inuit people who lived near it so that the area could be leased to the US Air Force to build Thule Air Base, which hosts, among other things, a global network of sensors providing missile warning, space surveillance and space control to North American Aerospace Defense Command (NORAD) and Air Force Space Command (AFSPC). Given the importance of "Upside-Down Tea Cup Mountain" to both the Inuits and the American military, how would you decide whether to return the land to the original inhabitants or leave it under foreign control?

Suggested Study Questions

Today we use GPS technology to reach the North Pole. At the beginning of the twentieth century, however, Commander Peary and Matthew Henson used the state-of-the-art tools of their time to travel more than four hundred miles over land and sea ice to reach the Pole. Can you name some of the tools they used?

When they visited the United States in 1986, Anaukaq Henson and Kali Peary described what, for each of them, were the greatest moments of their trip to America. What were those moments?

Although he explored the Arctic with Robert Peary for over eighteen years as an assistant to a field naval commander, Matthew Henson was never given a proper military title during his lifetime. In 2009, following a petition by Dr. S. Allen Counter, the United States Navy granted Matthew Henson what title?

The names of the four Inuit assistants who traveled to the North Pole with Henson and Peary were also left out of most history books, including Ootah who saved Peary and Henson's lives on a number of occasions when they fell through the ice. Where are their names now permanently engraved?

The Counters and Hensons saying goodbye after the centennial commemoration:
left to right, Olivia, Kitdlaq Henson, Dr. Counter, and Vittus HensonOlivia, Kitdlaq, Allen, Vittus

Matthew Henson proved to be of invaluable help to Robert Peary during their eighteen years searching for the North Pole. Can you name at least three things he contributed to make their expeditions successful?

What is the name of the metal utility device invented by the indigenous Polar Inuit people and used mainly by women and girls to carve game and scrape blubber and fat from the animals killed by hunters?

Answers to Study Questions

~ Celestial observation, dead reckoning, a sextant, and a tray of mercury for measuring angles of reflected sunlight.

~ Anaukaq said that the greatest moment of his life was meeting his American family and the friends of his father. He said the second greatest moment was visiting his father's humble grave in New York. For Kali, the greatest moment was meeting his American brother for the first time, and the second was finally feeling accepted as a member of the Peary family.

~ Honorary Command Master Chief.

~ On the monument at Matthew Henson's gravesite in Arlington National Cemetery.

~ He learned the Inuit language and could speak it like a native; he earned the trust of the Inuit people; he had seafaring experience; he could build and repair the equipment the explorers needed; he was an expert dogsled driver; he convinced the Americans to dress in native garb to stay warm.

~ Ulu (pronounced OO-loo).

Navarana in Polar Inuit garb cheerfully throwing snow as she enters her wooden "igloo" in Qaanaaq

DEAR FUTURE EXAM SUCCESS STORY

First of all, **THANK YOU** for purchasing Mometrix study materials!

Second, congratulations! You are one of the few determined test-takers who are committed to doing whatever it takes to excel on your exam. **You have come to the right place.** We developed these study materials with one goal in mind: to deliver you the information you need in a format that's concise and easy to use.

In addition to optimizing your guide for the content of the test, we've outlined our recommended steps for breaking down the preparation process into small, attainable goals so you can make sure you stay on track.

We've also analyzed the entire test-taking process, identifying the most common pitfalls and showing how you can overcome them and be ready for any curveball the test throws you.

Standardized testing is one of the biggest obstacles on your road to success, which only increases the importance of doing well in the high-pressure, high-stakes environment of test day. Your results on this test could have a significant impact on your future, and this guide provides the information and practical advice to help you achieve your full potential on test day.

Your success is our success

We would love to hear from you! If you would like to share the story of your exam success or if you have any questions or comments in regard to our products, please contact us at **800-673-8175** or **support@mometrix.com**.

Thanks again for your business and we wish you continued success!

Sincerely,
The Mometrix Test Preparation Team

Need more help? Check out our flashcards at:
http://MometrixFlashcards.com/TExES

TABLE OF CONTENTS

Introduction

Thank you for purchasing this resource! You have made the choice to prepare yourself for a test that could have a huge impact on your future, and this guide is designed to help you be fully ready for test day. Obviously, it's important to have a solid understanding of the test material, but you also need to be prepared for the unique environment and stressors of the test, so that you can perform to the best of your abilities.

For this purpose, the first section that appears in this guide is the **Secret Keys**. We've devoted countless hours to meticulously researching what works and what doesn't, and we've boiled down our findings to the five most impactful steps you can take to improve your performance on the test. We start at the beginning with study planning and move through the preparation process, all the way to the testing strategies that will help you get the most out of what you know when you're finally sitting in front of the test.

We recommend that you start preparing for your test as far in advance as possible. However, if you've bought this guide as a last-minute study resource and only have a few days before your test, we recommend that you skip over the first two Secret Keys since they address a long-term study plan.

If you struggle with **test anxiety**, we strongly encourage you to check out our recommendations for how you can overcome it. Test anxiety is a formidable foe, but it can be beaten, and we want to make sure you have the tools you need to defeat it.

Secret Key #1 – Plan Big, Study Small

There's a lot riding on your performance. If you want to ace this test, you're going to need to keep your skills sharp and the material fresh in your mind. You need a plan that lets you review everything you need to know while still fitting in your schedule. We'll break this strategy down into three categories.

Information Organization

Start with the information you already have: the official test outline. From this, you can make a complete list of all the concepts you need to cover before the test. Organize these concepts into groups that can be studied together, and create a list of any related vocabulary you need to learn so you can brush up on any difficult terms. You'll want to keep this vocabulary list handy once you actually start studying since you may need to add to it along the way.

Time Management

Once you have your set of study concepts, decide how to spread them out over the time you have left before the test. Break your study plan into small, clear goals so you have a manageable task for each day and know exactly what you're doing. Then just focus on one small step at a time. When you manage your time this way, you don't need to spend hours at a time studying. Studying a small block of content for a short period each day helps you retain information better and avoid stressing over how much you have left to do. You can relax knowing that you have a plan to cover everything in time. In order for this strategy to be effective though, you have to start studying early and stick to your schedule. Avoid the exhaustion and futility that comes from last-minute cramming!

Study Environment

The environment you study in has a big impact on your learning. Studying in a coffee shop, while probably more enjoyable, is not likely to be as fruitful as studying in a quiet room. It's important to keep distractions to a minimum. You're only planning to study for a short block of time, so make the most of it. Don't pause to check your phone or get up to find a snack. It's also important to **avoid multitasking**. Research has consistently shown that multitasking will make your studying dramatically less effective. Your study area should also be comfortable and well-lit so you don't have the distraction of straining your eyes or sitting on an uncomfortable chair.

The time of day you study is also important. You want to be rested and alert. Don't wait until just before bedtime. Study when you'll be most likely to comprehend and remember. Even better, if you know what time of day your test will be, set that time aside for study. That way your brain will be used to working on that subject at that specific time and you'll have a better chance of recalling information.

Finally, it can be helpful to team up with others who are studying for the same test. Your actual studying should be done in as isolated an environment as possible, but the work of organizing the information and setting up the study plan can be divided up. In between study sessions, you can discuss with your teammates the concepts that you're all studying and quiz each other on the details. Just be sure that your teammates are as serious about the test as you are. If you find that your study time is being replaced with social time, you might need to find a new team.

Secret Key #3 – Practice the Right Way

Your success on test day depends not only on how many hours you put into preparing, but also on whether you prepared the right way. It's good to check along the way to see if your studying is paying off. One of the most effective ways to do this is by taking practice tests to evaluate your progress. Practice tests are useful because they show exactly where you need to improve. Every time you take a practice test, pay special attention to these three groups of questions:

- The questions you got wrong
- The questions you had to guess on, even if you guessed right
- The questions you found difficult or slow to work through

This will show you exactly what your weak areas are, and where you need to devote more study time. Ask yourself why each of these questions gave you trouble. Was it because you didn't understand the material? Was it because you didn't remember the vocabulary? Do you need more repetitions on this type of question to build speed and confidence? Dig into those questions and figure out how you can strengthen your weak areas as you go back to review the material.

Additionally, many practice tests have a section explaining the answer choices. It can be tempting to read the explanation and think that you now have a good understanding of the concept. However, an explanation likely only covers part of the question's broader context. Even if the explanation makes perfect sense, **go back and investigate** every concept related to the question until you're positive you have a thorough understanding.

As you go along, keep in mind that the practice test is just that: practice. Memorizing these questions and answers will not be very helpful on the actual test because it is unlikely to have any of the same exact questions. If you only know the right answers to the sample questions, you won't be prepared for the real thing. **Study the concepts** until you understand them fully, and then you'll be able to answer any question that shows up on the test.

It's important to wait on the practice tests until you're ready. If you take a test on your first day of study, you may be overwhelmed by the amount of material covered and how much you need to learn. Work up to it gradually.

On test day, you'll need to be prepared for answering questions, managing your time, and using the test-taking strategies you've learned. It's a lot to balance, like a mental marathon that will have a big impact on your future. Like training for a marathon, you'll need to start slowly and work your way up. When test day arrives, you'll be ready.

Start with the strategies you've read in the first two Secret Keys—plan your course and study in the way that works best for you. If you have time, consider using multiple study resources to get different approaches to the same concepts. It can be helpful to see difficult concepts from more than one angle. Then find a good source for practice tests. Many times, the test website will suggest potential study resources or provide sample tests.

Practice Test Strategy

If you're able to find at least three practice tests, we recommend this strategy:

Untimed and Open-Book Practice

Take the first test with no time constraints and with your notes and study guide handy. Take your time and focus on applying the strategies you've learned.

Timed and Open-Book Practice

Take the second practice test open-book as well, but set a timer and practice pacing yourself to finish in time.

Timed and Closed-Book Practice

Take any other practice tests as if it were test day. Set a timer and put away your study materials. Sit at a table or desk in a quiet room, imagine yourself at the testing center, and answer questions as quickly and accurately as possible.

Keep repeating timed and closed-book tests on a regular basis until you run out of practice tests or it's time for the actual test. Your mind will be ready for the schedule and stress of test day, and you'll be able to focus on recalling the material you've learned.

Secret Key #4 – Pace Yourself

Once you're fully prepared for the material on the test, your biggest challenge on test day will be managing your time. Just knowing that the clock is ticking can make you panic even if you have plenty of time left. Work on pacing yourself so you can build confidence against the time constraints of the exam. Pacing is a difficult skill to master, especially in a high-pressure environment, so **practice is vital**.

Set time expectations for your pace based on how much time is available. For example, if a section has 60 questions and the time limit is 30 minutes, you know you have to average 30 seconds or less per question in order to answer them all. Although 30 seconds is the hard limit, set 25 seconds per question as your goal, so you reserve extra time to spend on harder questions. When you budget extra time for the harder questions, you no longer have any reason to stress when those questions take longer to answer.

Don't let this time expectation distract you from working through the test at a calm, steady pace, but keep it in mind so you don't spend too much time on any one question. Recognize that taking extra time on one question you don't understand may keep you from answering two that you do understand later in the test. If your time limit for a question is up and you're still not sure of the answer, mark it and move on, and come back to it later if the time and the test format allow. If the testing format doesn't allow you to return to earlier questions, just make an educated guess; then put it out of your mind and move on.

On the easier questions, be careful not to rush. It may seem wise to hurry through them so you have more time for the challenging ones, but it's not worth missing one if you know the concept and just didn't take the time to read the question fully. Work efficiently but make sure you understand the question and have looked at all of the answer choices, since more than one may seem right at first.

Even if you're paying attention to the time, you may find yourself a little behind at some point. You should speed up to get back on track, but do so wisely. Don't panic; just take a few seconds less on each question until you're caught up. Don't guess without thinking, but do look through the answer choices and eliminate any you know are wrong. If you can get down to two choices, it is often worthwhile to guess from those. Once you've chosen an answer, move on and don't dwell on any that you skipped or had to hurry through. If a question was taking too long, chances are it was one of the harder ones, so you weren't as likely to get it right anyway.

On the other hand, if you find yourself getting ahead of schedule, it may be beneficial to slow down a little. The more quickly you work, the more likely you are to make a careless mistake that will affect your score. You've budgeted time for each question, so don't be afraid to spend that time. Practice an efficient but careful pace to get the most out of the time you have.

Secret Key #5 – Have a Plan for Guessing

When you're taking the test, you may find yourself stuck on a question. Some of the answer choices seem better than others, but you don't see the one answer choice that is obviously correct. What do you do?

The scenario described above is very common, yet most test takers have not effectively prepared for it. Developing and practicing a plan for guessing may be one of the single most effective uses of your time as you get ready for the exam.

In developing your plan for guessing, there are three questions to address:

- When should you start the guessing process?
- How should you narrow down the choices?
- Which answer should you choose?

When to Start the Guessing Process

Unless your plan for guessing is to select C every time (which, despite its merits, is not what we recommend), you need to leave yourself enough time to apply your answer elimination strategies. Since you have a limited amount of time for each question, that means that if you're going to give yourself the best shot at guessing correctly, you have to decide quickly whether or not you will guess.

Of course, the best-case scenario is that you don't have to guess at all, so first, see if you can answer the question based on your knowledge of the subject and basic reasoning skills. Focus on the key words in the question and try to jog your memory of related topics. Give yourself a chance to bring the knowledge to mind, but once you realize that you don't have (or you can't access) the knowledge you need to answer the question, it's time to start the guessing process.

It's almost always better to start the guessing process too early than too late. It only takes a few seconds to remember something and answer the question from knowledge. Carefully eliminating wrong answer choices takes longer. Plus, going through the process of eliminating answer choices can actually help jog your memory.

Summary: Start the guessing process as soon as you decide that you can't answer the question based on your knowledge.

How to Narrow Down the Choices

The next chapter in this book (**Test-Taking Strategies**) includes a wide range of strategies for how to approach questions and how to look for answer choices to eliminate. You will definitely want to read those carefully, practice them, and figure out which ones work best for you. Here though, we're going to address a mindset rather than a particular strategy.

Your odds of guessing an answer correctly depend on how many options you are choosing from.

Number of options left	5	4	3	2	1
Odds of guessing correctly	20%	25%	33%	50%	100%

You can see from this chart just how valuable it is to be able to eliminate incorrect answers and make an educated guess, but there are two things that many test takers do that cause them to miss out on the benefits of guessing:

- Accidentally eliminating the correct answer
- Selecting an answer based on an impression

We'll look at the first one here, and the second one in the next section.

To avoid accidentally eliminating the correct answer, we recommend a thought exercise called **the $5 challenge**. In this challenge, you only eliminate an answer choice from contention if you are willing to bet $5 on it being wrong. Why $5? Five dollars is a small but not insignificant amount of money. It's an amount you could afford to lose but wouldn't want to throw away. And while losing $5 once might not hurt too much, doing it twenty times will set you back $100. In the same way, each small decision you make—eliminating a choice here, guessing on a question there—won't by itself impact your score very much, but when you put them all together, they can make a big difference. By holding each answer choice elimination decision to a higher standard, you can reduce the risk of accidentally eliminating the correct answer.

The $5 challenge can also be applied in a positive sense: If you are willing to bet $5 that an answer choice *is* correct, go ahead and mark it as correct.

Summary: Only eliminate an answer choice if you are willing to bet $5 that it is wrong.

Which Answer to Choose

You're taking the test. You've run into a hard question and decided you'll have to guess. You've eliminated all the answer choices you're willing to bet $5 on. Now you have to pick an answer. Why do we even need to talk about this? Why can't you just pick whichever one you feel like when the time comes?

The answer to these questions is that if you don't come into the test with a plan, you'll rely on your impression to select an answer choice, and if you do that, you risk falling into a trap. The test writers know that everyone who takes their test will be guessing on some of the questions, so they intentionally write wrong answer choices to seem plausible. You still have to pick an answer though, and if the wrong answer choices are designed to look right, how can you ever be sure that you're not falling for their trap? The best solution we've found to this dilemma is to take the decision out of your hands entirely. Here is the process we recommend:

Once you've eliminated any choices that you are confident (willing to bet $5) are wrong, select the first remaining choice as your answer.

Whether you choose to select the first remaining choice, the second, or the last, the important thing is that you use some preselected standard. Using this approach guarantees that you will not be enticed into selecting an answer choice that looks right, because you are not basing your decision on how the answer choices look.

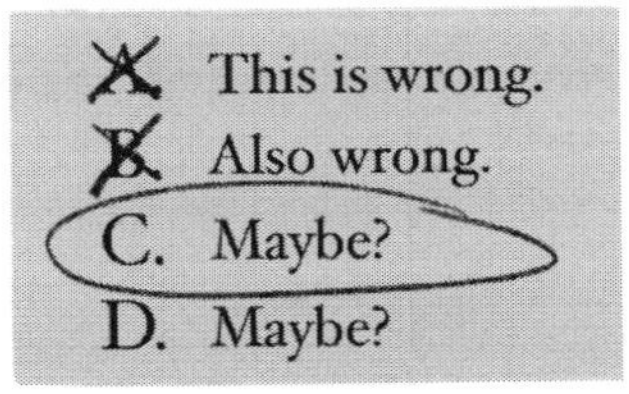

This is not meant to make you question your knowledge. Instead, it is to help you recognize the difference between your knowledge and your impressions. There's a huge difference between thinking an answer is right because of what you know, and thinking an answer is right because it looks or sounds like it should be right.

Summary: To ensure that your selection is appropriately random, make a predetermined selection from among all answer choices you have not eliminated.

Test-Taking Strategies

This section contains a list of test-taking strategies that you may find helpful as you work through the test. By taking what you know and applying logical thought, you can maximize your chances of answering any question correctly!

It is very important to realize that every question is different and every person is different: no single strategy will work on every question, and no single strategy will work for every person. That's why we've included all of them here, so you can try them out and determine which ones work best for different types of questions and which ones work best for you.

Question Strategies

✓ Read Carefully

Read the question and the answer choices carefully. Don't miss the question because you misread the terms. You have plenty of time to read each question thoroughly and make sure you understand what is being asked. Yet a happy medium must be attained, so don't waste too much time. You must read carefully and efficiently.

✓ Contextual Clues

Look for contextual clues. If the question includes a word you are not familiar with, look at the immediate context for some indication of what the word might mean. Contextual clues can often give you all the information you need to decipher the meaning of an unfamiliar word. Even if you can't determine the meaning, you may be able to narrow down the possibilities enough to make a solid guess at the answer to the question.

✓ Prefixes

If you're having trouble with a word in the question or answer choices, try dissecting it. Take advantage of every clue that the word might include. Prefixes can be a huge help. Usually, they allow you to determine a basic meaning. *Pre-* means before, *post-* means after, *pro-* is positive, *de-* is negative. From prefixes, you can get an idea of the general meaning of the word and try to put it into context.

✓ Hedge Words

Watch out for critical hedge words, such as *likely, may, can, often, almost, mostly, usually, generally, rarely,* and *sometimes.* Question writers insert these hedge phrases to cover every possibility. Often an answer choice will be wrong simply because it leaves no room for exception. Be on guard for answer choices that have definitive words such as *exactly* and *always.*

✓ Switchback Words

Stay alert for *switchbacks.* These are the words and phrases frequently used to alert you to shifts in thought. The most common switchback words are *but, although,* and *however.* Others include *nevertheless, on the other hand, even though, while, in spite of, despite,* and *regardless of.* Switchback words are important to catch because they can change the direction of the question or an answer choice.

⊘ BENCHMARK

As you read through the answer choices and you come across one that seems to answer the question well, mentally select that answer choice. This is not your final answer, but it's the one that will help you evaluate the other answer choices. The one that you selected is your benchmark or standard for judging each of the other answer choices. Every other answer choice must be compared to your benchmark. That choice is correct until proven otherwise by another answer choice beating it. If you find a better answer, then that one becomes your new benchmark. Once you've decided that no other choice answers the question as well as your benchmark, you have your final answer.

⊘ PREDICT THE ANSWER

Before you even start looking at the answer choices, it is often best to try to predict the answer. When you come up with the answer on your own, it is easier to avoid distractions and traps because you will know exactly what to look for. The right answer choice is unlikely to be word-for-word what you came up with, but it should be a close match. Even if you are confident that you have the right answer, you should still take the time to read each option before moving on.

General Strategies

⊘ TOUGH QUESTIONS

If you are stumped on a problem or it appears too hard or too difficult, don't waste time. Move on! Remember though, if you can quickly check for obviously incorrect answer choices, your chances of guessing correctly are greatly improved. Before you completely give up, at least try to knock out a couple of possible answers. Eliminate what you can and then guess at the remaining answer choices before moving on.

⊘ CHECK YOUR WORK

Since you will probably not know every term listed and the answer to every question, it is important that you get credit for the ones that you do know. Don't miss any questions through careless mistakes. If at all possible, try to take a second to look back over your answer selection and make sure you've selected the correct answer choice and haven't made a costly careless mistake (such as marking an answer choice that you didn't mean to mark). This quick double check should more than pay for itself in caught mistakes for the time it costs.

⊘ PACE YOURSELF

It's easy to be overwhelmed when you're looking at a page full of questions; your mind is confused and full of random thoughts, and the clock is ticking down faster than you would like. Calm down and maintain the pace that you have set for yourself. Especially as you get down to the last few minutes of the test, don't let the small numbers on the clock make you panic. As long as you are on track by monitoring your pace, you are guaranteed to have time for each question.

⊘ DON'T RUSH

It is very easy to make errors when you are in a hurry. Maintaining a fast pace in answering questions is pointless if it makes you miss questions that you would have gotten right otherwise. Test writers like to include distracting information and wrong answers that seem right. Taking a little extra time to avoid careless mistakes can make all the difference in your test score. Find a pace that allows you to be confident in the answers that you select.

⊘ Keep Moving

Panicking will not help you pass the test, so do your best to stay calm and keep moving. Taking deep breaths and going through the answer elimination steps you practiced can help to break through a stress barrier and keep your pace.

Final Notes

The combination of a solid foundation of content knowledge and the confidence that comes from practicing your plan for applying that knowledge is the key to maximizing your performance on test day. As your foundation of content knowledge is built up and strengthened, you'll find that the strategies included in this chapter become more and more effective in helping you quickly sift through the distractions and traps of the test to isolate the correct answer.

Now that you're preparing to move forward into the test content chapters of this book, be sure to keep your goal in mind. As you read, think about how you will be able to apply this information on the test. If you've already seen sample questions for the test and you have an idea of the question format and style, try to come up with questions of your own that you can answer based on what you're reading. This will give you valuable practice applying your knowledge in the same ways you can expect to on test day.

Good luck and good studying!

Foundations of Gifted Education

Characteristics of Gifted Students

Characteristics and Needs of Intellectually Gifted Students

Some qualities that educators have found intellectually gifted students to exhibit include:

- Curiosity and an inquisitive nature (i.e. wanting to find out and know things, to understand processes, and to answer questions)
- Complexity in their thinking, personality, and behavior; possessing an extraordinary capacity for learning; more diversity among their interests in learning
- A preference for learning subjects holistically rather than in parts or analytically
- A tendency to learn more intuitively than methodically or logically
- Tendencies toward perfectionism; and a corresponding tendency to exhibit fear regarding risk-taking
- The need to reflect on their own thinking and learning (self-reflection)
- The need to achieve an understanding and an acceptance of the nature of giftedness; a need to develop healthy and positive skills for social interactions and relationships
- A need to solve problems in real life as a way to find meaning and connections in life

Emotional Characteristics

Gifted children often display emotional intensity by experiencing and expressing extremes of emotions. They likely exhibit more emotional sensitivity than other children. They may feel more anxiety, guilt, and sense of responsibility than other children their age. Despite their superior abilities, they can feel inadequate or inferior due to their extreme sensitivity, high standards, perfectionism, and tendency to internalize. They may be shy and timid, and as a result they may be lonely. However, they are also likely to show concern for other people. Their sense of right and wrong is often heightened, prompting strong reactions against dishonesty, hypocrisy, and injustices. They tend to retain stronger memories of emotional experiences. They may have difficulty adjusting to changes. Because of their emotional sensitivity, they are susceptible to depression. They tend to have a need for security. They may suffer from physiological symptoms of emotional states, such as getting a stomach ache from feeling anxious.

Social Characteristics

Because intellectually gifted students are proficient in understanding abstract concepts, they often show an interest in social and philosophical issues. They are typically concerned about seeing justice done and tend to object to actions or decisions they perceive as unfair. Developmental psychologists such as Jean Piaget and Lawrence Kohlberg would say that such children display advanced moral reasoning. Because they have high standards for themselves and others as well as the ability to do many things better than others, even achieving perfection in certain tasks, they tend to be perfectionists. They often have more internal motivation than external motivation to achieve. Desires such as wanting good grades and teacher approval are external, whereas desires such as wanting to learn about a subject or to beat one's own personal best on tests are internal. Other prevalent characteristics include: good relational skills with parents, teachers, and age peers; well-developed senses of humor; energetic approaches; and emotional and physical sensitivity.

Review Video: Piaget's Cognitive Development Theory
Visit mometrix.com/academy and enter code: 100376

Aesthetic Characteristics in Terms of Intensity of Sensory Experiences

The Polish psychiatrist and psychologist Kasimierz Dabrowski (1902-1980) proposed intensities he termed "overexcitabilities" or "supersensitivities" in gifted children. His concept of sensual intensity involves heightened awareness and perception of the senses of vision, hearing, touch, smell, and taste. Gifted toddlers with this sensitivity may hate the feel of grass on their bare feet, gifted children may feel nauseated at the smells of certain foods, and/or they may overeat other foods that taste especially good to them. This characteristic is consistent with a craving for pleasure and a need or desire for comfort, two additional characteristics of gifted children with sensory sensitivities. These children often are more sensitive to air pollution and/or to the discomfort of labels/tags in clothing. Another aspect of this sensitivity involves the aesthetic appreciation of beautiful jewelry, visual art, music, literature, and the beauties of nature. Children with this sensitivity often can be moved to tears by beauty.

Characteristics of Imagination

Dabrowski identified the following sensitivities gifted children: psychomotor, sensual, emotional, intellectual, and imaginational. Many gifted children have very vivid imaginations. They are likely to have vivid dreams and to recall them. Because their imaginations allow them to conceive of diverse possibilities, to see things from different viewpoints, to look at things in new/different ways, to be inventive, and to generate original and unusual ideas, these children are likely to develop and utilize good senses of humor. They also are likely to daydream and have imaginary friends. They tend to love fantasy, poetry, drama, and music. They are more likely to have strong visualization skills as well as mental images with more detail than the mental images of other children. Because they can imagine all sorts of possibilities, they can envision the worst-case scenarios for any situation. This characteristic can promote fear of the unknown, of taking risks, and of new situations.

Intellectual Characteristics

Many gifted children have stronger reasoning abilities than do other students, and gifted children are better able to understand abstract concepts at earlier ages. Also, they often do not need to reinforce what they have learned with as much repetition as other children need. They should not simply be asked to recite or repeat something they have learned. Instead, they should be asked to use the higher cognitive skills they possess. For example, for comparison-contrast, you can ask the child the similarities and differences in words, familiar people, characters in a story, familiar songs, holidays, seasons of the year, birds and butterflies, etc. For categorizing, ask the child to group things, e.g. books, foods, toys, clothes, friends, or feelings. For sequencing, ask the child what happened in the beginning, middle, and end of a story. For summarizing, ask the child the most important events during a party, trip, or other activity with varied experiences.

Profiles of Types of Gifted Students

Challenging Gifted Students

Gifted students described as "challenging" are "divergently gifted." In other words, they are highly creative and unconventional. Schools often do not identify such students for gifted programs unless such programs have been implemented for a sufficient number of years and teachers have received enough in-service training. These children do not learn to use the system to their own benefit as more "successful" gifted students do. They tend to question authority and challenge teachers in class. Consequently, they are rarely rewarded for their abilities and often experience conflict. Those children who challenge their peers are not accepted, while others are liked for their originality and humor. They can be disruptive in class due to their spontaneous behaviors. Frustration and low self-esteem are risks for these students. Additionally, they may risk delinquency, drug use, and dropping out of high school if they do not receive suitable intervention by the time they reach middle school/junior high school.

Dropout Gifted Students

Researchers identify a salient characteristic of the "dropout" gifted student as anger. These students have experienced long-term failure of the system to meet their needs. Disappointment with themselves and with adults along with their feelings of rejection can lead to depression. In these students, this depression may manifest as social withdrawal or as defensive acting-out behavior. These students often have abilities and interests in unusual areas not included in school curricula, so they are not affirmed or rewarded for these abilities and interests. School may seem irrelevant or even antagonistic to those students with very different gifts. Many dropout students were not identified as gifted until later, e.g. in high school. This long-term lack of recognition and appropriate educational programming engenders resentment, bitterness, and low self-esteem. Educational experts recommend that these students need individual and group counseling, a close working relationship with a trusted adult, and diagnostic testing for possible remedial areas. Experts also state that traditional programming is not indicated for "dropout" gifted students.

Successful Gifted Students

Researchers have estimated that about 90 percent of students identified as gifted fit the profile described as "successful." These children are most often identified as gifted because they have learned which behaviors adults prefer; as a result of this learning, these children demonstrate the adults' preferred behaviors. . They typically achieve high scores on IQ, aptitude, and achievement tests. They are not likely to display problem behaviors because they crave adults' approval of and know which behaviors will obtain this desired result. Although many people expect such children to succeed independently, such success is often not the case. The researchers find these children tend to depend on adults for direction. Bored with regular curriculum, these children often "work the system" in order to achieve with minimal effort, not developing the autonomy necessary to pursue their own interests and gifts. Liked by peers and rewarded by adults, these students grow up socially well-adjusted but not prepared for life's changing demands; as adults, they may be underachievers.

Underground Gifted Students

Commonly researchers have found that middle school females often make up a group of students classified as the "underground" type of gifted student. In pre-adolescence and early adolescence, girls' need for group belonging and peer acceptance increases exponentially; as a result, many girls will hide their gifts or deny them to fit into a group. The researchers note that gifted boys in this category tend to go underground later, typically in high school, as a reaction to the pressure to join athletic teams and events and to their desire for identification with more popular "jocks" rather than less popular "nerds." Anxious and insecure, "underground" students have needs that conflict with adult expectations. Adult pressure only worsens their denial and resistance. Experts find that they benefit more from acceptance. As such, experts advise adults neither to push nor to permit abandonment of all prior interests in order to fit in; instead, adults should help these students find alternatives to meet their academic needs during what is considered a period of transition.

Double-Labeled Gifted Students

Researchers identify "double-labeled" gifted students as those with both giftedness and physical, emotional, or learning disabilities. Although dual diagnosis is possible with thorough evaluation, most educational programs for the gifted neither identify such students nor afford them differentiated instructional programming that integrates and meets their specialized needs. Because of their combinations of gifts with disabilities, schools can frequently overlook these students, as many of them display behaviors that school personnel do not associate with giftedness. Students with learning disabilities can display poor organizational skills, problems with memory, difficulties with reading, poor handwriting, and etc. Those students with emotional or behavior disorders may exhibit disruptive behavior. Due to the contradictions of their abilities and disabilities, these students are often confused and stressed and suffer from low self-esteem. Often these students are either perceived as average and therefore ignored, or targeted for remediation of their disabilities while their respective gifts are overlooked.

Autonomous Learners

Educational researchers find that parents may observe "autonomous learner" types of gifted children exhibiting signs of this style at home, but these children typically demonstrate it somewhat later in school once they have learned the system. Unlike the "successful" type of gifted student, who seems to have high self-esteem but depends on adult direction and reinforcement, autonomous learners have strong internalized self-concepts. Rather than meeting the system's demands with as little work as possible as do "successful" gifted students, autonomous gifted learners employ the system to create new opportunities to meet their own needs. Self-directed and independent, these students are characterized by self-acceptance and risk-taking ability. They feel secure about forming personal and educational goals. They exhibit a strong sense of self-efficacy (belief in their ability to do things) and personal power. They are active rather than passive, and they express their needs, emotions, and goals with freedom, but also within appropriate limits.

Personality Characteristics of Persons with Various Creative Gifts

Jane Piirto (Understanding Those Who Create, 1998) found that when compared to others with creative gifts, artists are more impulsive and spontaneous. The author found that writers tend more toward nonconformity than do the other creatively gifted types. In addition, Piirto found that in comparison to other creative types, musicians tend more toward introversion. She discovered that architects are not as flexible as other creative persons. Furthermore, she concluded that creative engineers and inventors were likely to be more well-adjusted overall than were other creative types. These findings imply that specific personality traits may need to be considered in order to make predictions, since different personality traits are associated with different domains of creativity. Kerr and Gagliardi (2009) find this domain-specific approach more useful than "...seeking one creative personality type that fits all creative occupations." Indeed, the latter seems doomed to failure in light of the many types of creativity.

Creativity in Gifted Students as a Function of Cognition or Personality

Some researchers view creativity as a strictly cognitive process; others view it as a group of personality traits. Kerr and Gagliardi (2009) assert that while some students might be identified as creative in their thinking, without personality characteristics such as independence and persistence, these students may not produce any creative work. Hence they recommend that both aspects be considered. Divergent thinking is the most prevalent cognitive attribute studied in creativity research. With respect to personality, creativity has been most associated with the traits of autonomy, introversion, and openness to experience. (The latter two traits are included in the "Big Five" personality traits identified by McCrae & Costa and others.) Researchers have noted (e.g. King and Pope 1999, Feist 1999) that creative types are likely to work independently on their creative products and to avoid much social stimulation and group influence. Openness to experience affords the stimulation of new ideas, questions, and topics.

Individual Differences Within the Population of Gifted Students

Social contexts such as geographic residence, socioeconomic status, and cultural background affect personal, family, and community definitions of giftedness. Some groups place high value on

black parents of gifted students who underachieved in school also were found to have less self-confidence in their parenting skills than parents of higher-achieving students in this population.

Approaches to Development of Gifted Students

Research on Effective Practices in Gifted Education

The National Research Center on Gifted and Talented (NRC/GT) is seeking to find out "what works in gifted education" through a five-year plan beginning with the 2008-2009 school year. Their plan involves integrating the examination of systems of identification, reading and math curricula based on theoretical models, and assessments of giftedness. Researchers are concentrating on three areas:

- Developing a solid system of identification that will build upon prior research
- With students identified via both traditional criteria and expanded criteria, analyzing the effects of certain units of reading and math curricula
- Measuring student outcomes using standardized achievement tests, extended standards-based assessments, or structured performance assessments

Researchers aim to create a system of identification for students across socioeconomic and cultural groups. Such groups will feature "talent pools" of elementary-school students, identified using standardized tests and teacher ratings, participating in randomly assigned reading and math curricula based on models.

Benefits to School Districts for Participation

Research studies regarding effective teaching methods in gifted education often develop curriculum units to test in actual practice at schools. Such units are designed to challenge gifted students, differentiated for all student levels in general education classrooms and based upon national standards. Such units educationally benefit more students at more levels than non-differentiated instruction. Participating school districts often receive tools for the identification and the assessment of gifted students free of charge, which represents a financial benefit. Major research centers also provide support for educators' professional development, benefiting school districts financially and benefiting teachers, other school personnel, and students. This support often includes online technical assistance, an additional benefit addressing the increasing use of online resources and tools. Participation in a research study conducted by a center offering professional development courses for educators also gives priority to some members of school district staffs in registering for these courses.

Characteristics of Gifted Students Relative to Achievement Motivation

Researchers have discovered that when gifted students consistently bring high skill levels to schoolwork, externally imposed limitations that represent threats to their senses of self-determination are particularly likely to affect them negatively. Gifted children frequently know they have the ability to exceed their non-gifted peers in performance; therefore, they do not rely as much on external rewards, the contingencies for earning them, or feedback from teachers. A more important need for them involves preserving their internal motivation, which cannot be forced or learned, but can too easily be quashed. While research finds gifted students as typically highly motivated, it also finds that motivation in classrooms is problematic for them. This finding is attributable to gifted students not responding well to rigidly structured, specific tasks or assignments, which non-gifted students often need. Since gifted students are self-motivated instead of teacher-motivated, gifted students perform better when given choices, flexibility, and unstructured tasks.

Intrinsic Motivation Principle of Creativity

While what Renzulli has termed "schoolhouse giftedness" is more easily identified via IQ test scores and more amenable to academic success than what he called "creative-productive giftedness, " which is not as easily identified via IQ test scores. Social psychologists conducting research into creative production over 25 years have found it associated with students' motivational orientations, and those orientations are influenced by environmental factors. They find that internal motivation encourages creativity, while external motivation discourages creativity—this finding is called the Intrinsic Motivation Principle of Creativity. Intrinsically motivated students do tasks for their own sake because they enjoy them, so their rewards come from within them. In contrast, extrinsically motivated students do tasks for a reward coming from outside them.

Influences of Environmental Factors on Expression of Creativity

The influence of environmental variables can be seen historically through creative proliferation at certain times and places. For example, Florence, Italy in the 15th century; the Harlem Renaissance in 1920s and 1930s New York in America; and San Francisco, California in 1960s America all exhibited unusual amounts of creative inspiration and production. Some of the environmental factors that allow people with creative gifts to express them in meaningful and socially valuable ways include:

- Social support for a subculture consisting of creative persons
- Availability and accessibility of materials and resources for creative production
- A political and cultural climate that allows freedom of expression
- The existence of patrons with the means and interest to lend financial, material, and moral support to creative endeavors

Influences impeding creative expression include the race, gender, and socioeconomic status of some gifted individuals due to others' stereotyping and lowered expectations because of such environmental characteristics.

Nurturing Students' Areas of Giftedness

Teachers

Teachers can observe student behaviors in order to identify their personal interests and abilities. When they notice a student excelling in a certain subject area, teachers can provide enrichment activities for that student. Even in a general education classroom with no gifted programs, teachers can provide additional activities in preferred areas to students who show evidence of giftedness by asking for additional activities. Teachers can vary their instruction by offering activities and assignments at higher grade levels to gifted students not challenged or even bored at their class' grade level. Teachers pressed for time can arrange peer tutoring by older and/or more advanced students to allow younger gifted students more 1:1 interaction and individualized learning at the students' own pace. Teachers can help gifted students plan and develop special projects in their areas of interest, allowing them to learn about these areas in more depth and with more complexity.

Parents

Parents can begin reading to children regularly as soon as the children are able to attend. Furthermore, as children develop, parents can encourage children to read to them as well. As children grow, parents can observe and encourage their interests/aptitudes. Providing books for verbally precocious children is crucial (libraries offer a good option if funds are limited), since good writers begin as good readers. Parents can give children with mathematical talents numerical puzzles and games. For children interested in producing reactions by mixing substances, junior chemistry sets or household materials can be provided. Beginner telescopes and trips to the local planetarium are good for children fascinated with stars and planets. Parents can provide interested children with introductory geology sample sets and take them on local field trips to discover rocks, minerals, and/or fossils. Basic art supplies are important for children with artistic abilities. Parents should allow and encourage free expression, exploration, discovery, learning, skills development, and new insights.

Content Skills and Process Skills

Content skills involve acquiring and retaining the information presented in a specific instructional subject area. Examples include learning the correct spelling of words; knowing the correct meaning of vocabulary words; knowing factual material such as places, names, and dates in history or social studies; remembering mathematical equations or formulas, etc. In contrast, process skills involve more interpretation of information. For example, a student not only knows what reportedly happened in a historical event, but s/he also can interpret that event in terms of its historical context, such as the frame of reference and viewpoints affecting participants. Process skills include analyzing information through comparisons-contrasts; generalizations; finding cause-and-effect relationships; classifying and sequencing; summarizing; and drawing inferences, conclusions, and predictions. Such skills include applying critical thinking skills and systematic methods of inquiry, especially in sciences. Content and process skills are related because content must be learned to process it. Integrating the two is most effective.

Enrichment Activities to Develop Skills in Forming and Testing Hypotheses

Gifted children typically exhibit advanced abilities in conceptual thinking. Their skills should be developed with activities wherein they manipulate information rather than merely memorizing or recognizing it. To stimulate and practice hypothesizing, you can ask questions of a gifted child beginning with "What would happen if we...?" For example, what if we put a block into a glass filled with water? What if we put a glass over a lit candle? If we put tennis balls into a shoebox, how many could it hold?, etc. Then you can test these hypotheses by following through and performing the hypothetical action to see what happens. These activities suit children in Piaget's Concrete Operations stage. For children who can think abstractly without concrete objects, activities can be extended to imaginary events and fantasies, e.g. "What would happen if you could fly?... if you were invisible?... if snow were ice cream?... if you were twelve feet tall?", etc.

Developing Skills in Making Inferences, Math, Reading, and Language

One way young gifted children can be encouraged to exercise their cognitive abilities is through inferences, which require them to move beyond literal and concrete perceptions to more abstract thinking involving speculation based on existing information. For example, you can show a child a picture of people wearing uniforms and ask what their jobs may be, or show a picture of people with various facial expressions and ask what they might be feeling. In both cases, ask what could have happened before and after what is in the picture. For children with mathematical gifts, it is important to incorporate fun, game-like math activities into daily living routines. For developing reading readiness skills in young gifted children, you can label common household objects with their printed names on 3x5 cards. Start open-ended stories; let them make predictions and hypotheses about what will happen next and what will happen if that occurs.

Language Experience Stories to Develop Language Skills

Younger gifted children with advanced abilities in language often will enjoy activities using language experience stories. You ask the child to make up a story on any subject the child prefers. For example, the story can be about the child's pet; the child's best friend or circle of friends; a favorite relative; a trip they took on vacation that they especially enjoyed; a day trip to a museum, zoo, planetarium; a party they had or attended; or any other subject of the child's choice. The child tells the story aloud and you write it down. You have the child draw illustrations to accompany the story. You then read the story back to the child, aloud from the printed version. Record your reading. Play back the recording and listen together, having the child follow the printed words to associate spoken and written language. Make multiple stories into a book for the child.

Affective Development of Gifted Individuals

Need for Further Research

Educational researchers specializing in gifted education find that many aspects of the social and emotional development of gifted individuals have not been researched nearly as much as the intellectual and academic aspects of their development. They recommend that researchers, school psychologists, guidance counselors, and other professionals can contribute to exploration of the affective domain. Asynchronous development in cognitive vs. affective domains has long been observed in the gifted, but it demands further study, as does perfectionism in the gifted. Scientists also want more research into underachievement, depression, eating disorders, self-injurious behaviors, and substance abuse relative to giftedness, as well as responses to life events such as divorce, loss and grief, serious illness, injury, and relocating. Additionally, research into giftedness is lacking related to obsessive-compulsive disorder, sexual abuse, physical disabilities, Asperger's syndrome, serious conflicts between parents and their gifted children, and problematic developmental transitions.

Support

Gifted education programs, both inside and outside of general education classrooms, should incorporate curriculum elements designed to support emotional, social, and career development in gifted students. At all age/grade levels, school curricula can be proactive by including psychoeducational knowledge regarding the way in which giftedness influences these developmental domains. When teaching social sciences and literature, teachers can give assignments related to the psychosocial aspects of these subjects. Classes can have group discussions about developmental challenges. Career and talent development often present issues for gifted children at much younger ages than for other students, so these topics should be addressed in the curriculum. Viewing and addressing underachievement and high achievement from a developmental perspective can help both students and educators. Teachers should affirm student resiliency and personal strengths, which can be eclipsed by performance/underperformance. Advocates for legislation, funding, and services for gifted students also should emphasize affective considerations in their efforts.

Review Video: Early Childhood Developmental Domains
Visit mometrix.com/academy and enter code: 100380

Influence of Common Characteristics of Gifted Children

The developmental tasks for gifted children are the same as for all children. For example, all children must develop such concepts as a sense of identity, a sense of self-efficacy, differentiation of self, relationships with peers, autonomy in life and in school, and paths toward careers. But gifted children can experience these developmental milestones quite differently from other children. Researchers find gifted children are commonly more emotionally sensitive, intense, and overly excitable. They also are commonly more perceptive about human behaviors and personalities; along with their talents, their moral development is precocious, so they tend to have different concerns at higher cognitive levels and younger ages than other children. Even the faster cognitive processing of gifted youngsters can intensify their emotional responses to external stimuli. These common gifted characteristics may sometimes impede developmental processes. Therefore educators, parents, counselors, psychologists, and psychiatrists working with gifted students should possess adequate and accurate knowledge regarding these children's affective development.

Attitude of Experts

As many experts in gifted education assert, gifted students deserve for adults to attend not only to their gifts, talents, and performance, but also to their overall wellbeing. Gifted children go through developmental processes and achievements universal to all children, although they may experience these processes and achievements differently. They also have other developmental experiences unique to each individual. Because educators and other adults so frequently focus on the academic and creative potential of gifted students and their performance or lack thereof, they may not focus enough on the affective dimension of their development. Adults need to remember that gifted students' developmental progress is not only cognitive, but also emotional, social, and career-related. Moreover, progress in cognitive and academic areas also has emotional and social effects. Thus, organizations such as the National Association for Gifted Children take the position that teachers, administrators, and counselors in schools should deliberately, proactively nurture the emotional and social development of gifted students.

Stages in Team Problem-Solving

If a group of gifted students wants to solve a problem as a team, the first stage they should address involves understanding the nature of the specific problem. That accomplished, in the second stage they would brainstorm ideas for potential solutions to the problem. Once they had a number of possible solutions, the third stage would involve deciding what to do. Dr. E. Paul Torrance, who developed the most widely used standardized tests of creative thinking, founded the Future Problem Solving Program (FPSP) in 1974. Today it is used with hundreds of thousands of students, nationwide and in several other countries. The FPSP model contains three sections and six steps: Under the first section, Understanding the Problem, steps 1 and 2 are identifying challenges in the problem, and selecting an underlying problem. The second section, Generating Ideas, has step 3, producing solution ideas. The third section, Planning for Action, includes steps 4, 5, and 6: generating and selecting criteria, applying criteria, and developing an action plan.

Skills Developed Through Creative Problem-Solving Process

Participating in creative problem-solving develops the higher-order cognitive skills of gifted students, which helps them to apply their knowledge for solving problems. According to experts, there is more emphasis in this process on learning "how to think" than on acquiring information. Creative problem-solving helps gifted students enhance their analytical skills as well as their creative abilities. It builds their skills in spoken and written communication and encourages gifted students to develop skills for conducting research and improving these skills. When gifted students participate in groups to solve problems creatively as a team, the process contributes to their developing responsibility as group members. The creative problem-solving process provides gifted students with guidance that enables them to develop not only responsibility, but also more self-direction. This process also furnishes gifted students with a model for problem-solving they can incorporate into their lives. Moreover, it engages gifted students' interest in the future.

Defining Giftedness

Defining the Concept of Giftedness

Educators and advocates agree that no universal consensus exists regarding the definition of giftedness. The National Association for Gifted Children (NAGC) asserts that the concept of giftedness, like the concepts of intelligence and talent, is fluid in nature. Depending upon culture and other contexts, such concepts can be manifested and perceived differently. Moreover, the term "gifted" has a variety of meanings and subtleties within each meaning, even in school systems and within individual schools. In any given school, individual educators likely hold a range of different personal beliefs regarding this term. The NAGC finds that individuals showing "...aptitude (...exceptional ability to reason and learn) or competence (documented performance or achievement in top 10% or rarer) in one or more domains" (2008) are gifted. They define domains as structured activity areas within symbol systems such as language, math, or music, and/or within sensorimotor skill sets such as art, dance, or sports.

Differentiation Between Definitions of Giftedness and Talent

Canadian psychology professor/researcher Dr. Francoys Gagné (not to be confused with American educational psychologist Dr. Robert M. Gagné), in his "A Differentiated Model of Giftedness and Talent (DMGT)" (1985, updated 2000), clearly distinguishes giftedness from talent using the criteria of competence/ability/aptitude versus performance/skill/achievement. He finds that "...superior natural abilities (called aptitudes or gifts), in at least one ability domain..." that constitute giftedness are "untrained and spontaneously expressed." He defines talent as "...superior mastery of systematically developed abilities (or skills) and knowledge in at least one field of human activity..." The differences are (1) that talent requires methodical development and learning, while giftedness needs no training; and (2) that gifts are in "ability domains" while talents are in "fields of human activity." He designates both as placing an individual in the top 10% of age peers in a given domain or field. His ability/aptitude domains include Intellectual, Creative, Socioaffective, and Sensorimotor.

Roles of America's Federal and State Governments in Defining Giftedness

While America has federal laws governing special education services for students with disabilities, these laws do not contain specific provisions that mandate each state to offer special education services for gifted and talented students. As a result, each American state can create its own definitions of giftedness and talent and its own programs for gifted and talented students. A state's definition will inform how it identifies such students; how it decides eligibility for services; and the kinds of educational programs it develops. Indiana, Nebraska, and Washington use the term "high ability student." Eighteen states use "gifted," while the other twenty-five states use "gifted and talented." Four states—Massachusetts, Minnesota, New Hampshire, and South Dakota—have no state definition of giftedness or talent. Giftedness and talent are defined by the state legislatures in 25 states, while the remaining 21 states have each mandated or authorized their state Boards of Education to define these terms.

Characteristics Necessary to Demonstrate Gifted Behavior

Dr. Joseph Renzulli and colleague Dr. Sally Reis, both from the University of Connecticut, created a Schoolwide Enrichment Model (SEM) (1976, 1977, 1985, 1997) for academically gifted and talented students and for all schools to develop students' strengths and talents. Renzulli focuses more, as do other behaviorists, on students' actions—i.e. gifted behavior—than on students' abilities/potentials. He proposes that three groups of characteristics must interact to produce gifted behavior: (1) above average abilities, either general or specific; (2) high levels of motivation; and (3) high levels of creativity. He identifies any child who succeeds either in having or in developing this combination of characteristics and applying it to any "potentially valuable" field of human activity as gifted or talented. Renzulli states in the SEM that gifted behaviors exist "in certain people (not all people), at certain times (not all the time), and under certain circumstances (not all circumstances)."

Domains of Ability or Aptitude

Domains of aptitude (identified by F. Gagné) are Intellectual, Creative, Socioaffective, Sensorimotor, and "others." A student who is gifted in the Intellectual domain might demonstrate this giftedness through superior learning, academic and/or otherwise, as evidenced by extent, breadth, depth of knowledge and/or speed and ease of learning; through superior achievement in school grades and standardized tests; and/or facility beyond grade level in reading or mathematics, etc. Gifts in the Creative domain might be demonstrated by advanced learning and understanding of principles in music, art, etc.; superior musical interpretation, performance, and/or composition; superior composition and execution in visual arts; originating mathematical/scientific formulas; or producing inventions. Socioaffective gifts might appear in a student's heightened ability in making friends, helping peers resolve conflicts, disarming bullies/hostility, and empathy for others. Sensorimotor gifts can emerge in athletic abilities or physical abilities for certain crafts and arts. "Others" includes extra-sensory perceptions such as precognition, clairvoyance, and communicating with spiritual realms.

Individual and Group Differences Among Gifted Children

Researchers have pointed out that such individual factors as genetic composition, life experiences, personality, and personal development, as well as the influences of families, interpersonal relationships, and educational experiences result in individual variations in the ways in which gifted children respond to and express their special abilities. While the research has acknowledged individual differences among gifted children for at least 50 years, less work has been done to identify various groups among the gifted.

Roeper (1982) theorized five types of gifted children based on their approaches for coping with and expressing their feelings. These types include: (1) the Perfectionist; (2) the Child/Adult; (3) the Winner of the Competition; (4) the Self-Critic; and (5) the Well-Integrated Child.

Betts and Neihart (1988) identified six "profiles" of gifted students based on their needs, feelings, and behaviors: (I) The Successful; (II) The Challenging; (III) The Underground; (IV) The Dropouts; (V) The Double-Labeled; and (VI) The Autonomous Learner.

Fluency, Flexibility, Originality, and Elaboration in Creativity

Fluency, flexibility, originality, and elaboration are four qualities considered by many experts as key components of creativity. The original version of Torrance's creativity scales measured these four qualities. Fluency refers to an individual's ability to generate a large number of concepts or of alternate solutions to a problem. Flexibility refers to an individual's ability to approach a problem from a variety of perspectives, therefore enabling the individual to come up with ideas and solutions to problems in a variety of categories. Originality refers to an individual's ability to generate unique or uncommon ideas. Fluency, flexibility, and originality are hallmarks of creative thinking. These qualities also are reflected in divergent thinking, a key characteristic of creativity involving generating many different possibilities rather than narrowing ideas down to one conclusion, as in convergent thinking. Elaboration in creative thinking involves the ability to add details to embellish one's ideas or results.

Convergent Thinking and Divergent Thinking

Convergent and divergent are opposites. In convergent thinking, different thoughts narrow/come together (converge) on one conclusion. An example is forensic detective work. If a suspect's fingerprints are found on the murder weapon, his DNA is found on the victim, and eyewitnesses place him at the scene of the crime at the time of death, then these pieces of evidence converge to show that suspect's guilt. Considering whether this suspect was framed or whether the evidence was circumstantial would involve some divergent thinking in which different thoughts

data as well as more integration of assessment techniques and economy in analyzing data are indicated to promote both learning and accountability.

Needs in Program Management of Gifted Education Programs

Educational researchers have observed that in some gifted programs, little time and few resources are left for the crucial components of curriculum design, program evaluation, and staff development, because so much time and effort are disproportionately devoted to the identification of gifted students. Nonetheless, researchers have clarified that this observation reflects not a criticism of program managers but instead an insight about the unrealistic expectations and insufficient investment in quality programs for students with high ability. Because trends increasingly emphasize curriculum planning and student assessment for such quality programs, greater program management capacities are needed. Progress includes greater recognition and provision of a variety of service options at the program level to meet diverse student and community needs, program developers' responses to the National Association for Gifted Children's standards for conveying program expectations, and the positive perceptions by constituents of the quality of staff in good gifted programs.

Javits Act

The Javits Act, passed in 1988, after Public Law 94-142, aka the Education for All Handicapped Children Act (EHA) passed in 1975, renamed the Individuals with Disabilities Act (IDEA) in 1997, most recently reauthorized in 2004, with final regulations released in 2006. IDEA provides for all children to be educated, but it does not specify services for the gifted as for people with disabilities. The Javits Act, initiated by then-Senator Jacob Javits of New York, funds educational programs for gifted children from low-income families. It states, "The term gifted and talented student means children and youths who give evidence of higher performance capability in such areas as intellectual, creative, artistic, or leadership capacity, or in specific academic fields, and who require services or activities not ordinarily provided by the schools in order to develop such capabilities fully." This legislation meets the need for special services to gifted students in order to fulfill their special abilities.

Challenges Teachers Face with Gifted and Talented Students

An ever-increasing challenge in our schools is that due to economic considerations, school budgets are being cut, and special or supplementary services to gifted students often are eliminated. As a result, general education teachers find they have more gifted and talented students in their classrooms. Their challenge involves meeting the educational needs of students with average ability, students with high ability, students with disabilities whose inclusion in regular classrooms is mandated by law, and students with both gifts and disabilities. Teaching the same curriculum to all students, i.e. teacher-centered instruction, is no longer feasible. Information technology recently has become popular for use with gifted students. As such, information technology can replace some existing delivery methods and be added to others to enhance them. Assistive technology also aids students with vision, hearing, physical, and other disabilities. Those students with both gifts and disabilities receive multiple benefits from technology. Students can work at their own paces, access multiple modalities, select preferred modalities, and learn in more depth through enriched learning experiences.

Review Video: Teaching Through Multiple Modalities
Visit mometrix.com/academy and enter code: 100371

Procedural Safeguards Regarding Education of Gifted Students

While the federal government has mandated procedural safeguards for the educational placement of students with disabilities, federal laws do not specifically address the placement of students who are gifted without disabilities. Each state Department of Education issues its own set of procedural safeguards for the placement of gifted students. They typically derive these standards from the federal requirements for students with disabilities in that all students must receive a free, appropriate public education (FAPE). Some common safeguards across states include the following standards: districts must give parents prior notice of initiation or refusal to identify, evaluate, place, or provide a FAPE to a gifted student, in the parents' primary communicative mode. Parents must be given copies of procedural safeguards. Written informed parent consent is required before providing individual formal evaluation or initial service to a gifted student, and such consent is voluntary and revocable before the action. Parents have the right to view their child's educational records and participate in educational planning meetings.

Each state Department of Education writes its own procedural safeguards. Specifically, many states require that if parents obtain independent evaluations of their gifted children at their own private expense, the state's school districts must consider those evaluation findings in any educational decisions they make for those students. In addition, states commonly require that if an administrative law judge requests an independent evaluation as part of a hearing, the expense becomes a public responsibility. Usually the district must finance the evaluation. Another standard of state procedural safeguards, adopted from Federal disability laws, is the right of gifted students and their parents to a due process hearing regarding the student's identification, evaluation, or educational placement. States also commonly stipulate that students remain in their current placements while an administrative or judicial proceeding is pending.

Legal Issues Concerning Gifted Students

Federal laws protect the legal rights of students with disabilities but not the legal rights of gifted students without disabilities. Even the Jacob K. Javits Gifted and Talented Student Act (1994), which funds research and projects related to giftedness, does not protect gifted students' legal rights. Each US state has jurisdiction over its educational procedures. However, discrimination based on giftedness is equally as unlawful as discrimination based on race, color, national origin, gender, or disability. Despite the lack of specific federal mandates for gifted education, the US State Department of Education's Office of Civil Rights protects the educational rights of gifted students participating in government-funded programs and activities. From 1985-1995, this office has ruled in 86 cases involving gifted students regarding discrimination, identification, and program admissions. The American Civil Liberties Union has also filed discrimination suits; for example, against a school district for offering fewer Advanced Placement programs in schools with African-American and Hispanic populations with lower socioeconomic status.

Chapter Quiz

Ready to see how well you retained what you just read? Scan the QR code to go directly to the chapter quiz interface for this study guide. If you're using a computer, simply visit the online resources page at **mometrix.com/resources719/texesgtsupp** and click the Chapter Quizzes link.

Assessment, Curriculum, and Instruction

Research and Statistics in Gifted Education

Research Methodology

Popular conceptions of research methodologies characterize quantitative methods as using hard data, numbers, objective information, and findings that can be generalized to the larger populations represented by study samples; and qualitative methods as using "soft" data, words, subjective impressions, and findings more specific to the study's subjects but not generalizable to larger groups. Each method has its own strengths and weaknesses. Rather than choosing one or another method by personal preference, many researchers (cf. Neill, 2006), believe the nature of the research question should dictate the most applicable methodology. Neill also points out that qualitative and quantitative methods are not opposing paradigms; rather, each has its use, and the existence of mixed-methods studies proves they can be used together. Researchers are more likely to use qualitative methods earlier in a study to learn more comprehensively about a topic, using quantitative methods later on to seek precise answers to more specific questions.

For educators who want to determine the grade level at which a gifted student is reading, for example, quantitative methods are best for measurements that can be represented numerically and somewhat discretely. If a gifted child in a regular second grade is tested as reading at a sixth-grade level, this finding helps teachers select reading material for that child that is intellectually more suitable, yet still age-appropriate emotionally, socially, and experientially. Quantitative measures are used in standardized tests of IQ, knowledge, and academic achievement. Qualitative methods are more helpful for determining which content areas are an individual gifted student's favorites. Such methods also are useful for discovering a gifted student's particular learning style as well as the student's academic strengths and learning needs. Qualitative data collection methods such as interviews, observations, and self-reports can yield thorough profiles of each gifted student as a whole person, incorporating intellectual, psychological, emotional, social, and behavioral characteristics.

Qualitative Research

One type of qualitative research is the case study. This method seeks to gain insights into a phenomenon such as giftedness by studying one individual case of the phenomenon in great depth and detail. The case could be an individual gifted student, a specific class of gifted students, a school for gifted students, or a particular event involving gifted students. Phenomenology is another type of qualitative research. Phenomenology refers to the process wherein the researcher describes the structures of his or her experiences as they are apprehended by the researcher's consciousness, without using theories, assumptions taken from other fields/disciplines, or deductive reasoning to influence that description. Another kind of qualitative research is represented by ethnography. In ethnography, the researcher makes detailed field observations of sociocultural events or characteristics, typically within a particular community or population. This type of research can lend itself to studying groups of gifted students and gifted educational programs.

Methods for Collecting and Analyzing Data

Qualitative research is descriptive, experiential, and interpretive rather than statistical or numerical, as is quantitative research. Qualitative research demands rich, detailed descriptions to reveal the research process and to communicate implicit knowledge. As such, qualitative research uses three methods of data collection and analysis: observations, interviews, and self-reports. In observation, the researcher observes the nonverbal and verbal behaviors of study subjects (avoiding influencing the subjects' behavior insofar as is possible) and records descriptions of observed behaviors. Interviews are interactive with the subjects: the researchers first ask the participants to describe specified events or phenomena orally, and then the researcher records the participants' responses. Researchers also may ask participants to write down descriptions of their experiences; these are self-reports. Researchers analyze the initial data they collect to inform and direct further data collection. All these methods can be used to obtain data from gifted students, their parents, teachers, non-gifted peers, etc.

DISADVANTAGES AND ADVANTAGES

Because quantitative forms of research using statistical analysis are the predominant methods used in educational and psychological research as well as in other social sciences, critics of qualitative types of research often regard these methods as more subjective. However, as Myers (2002) states, "Since we maintain our humanity throughout the research process, it is largely impossible to escape the subjective experience…" This humanity includes intuitive realizations and "aha moments." Qualitative research demands extensive time and effort. Furthermore, because of the subjective nature of qualitative research results, these results obtained with samples cannot be generalized to larger populations in the same manner as can the results from statistical studies. However, Myers and other researchers (e.g. Neill, 2006) find qualitative methods to transcend those drawbacks with other redeeming qualities, such as in-depth descriptions with sufficient detail for readers to realize the idiosyncrasies of the individual research situation, and providing the researcher's perspective on that situation.

QUANTITATIVE RESEARCH

Qualitative methodologies tend to be inductive, i.e. they proceed from the specific to the general and generate new theories. Quantitative methods tend to be deductive, i.e. they proceed from the general to the specific and test existing theories. Qualitative methods incorporate more subjective aspects, while quantitative methods restrict findings to objective facts and figures. For example, if you wanted to find out which of two curricula is preferred by individual gifted students in a general classroom, you likely would use qualitative methodology and ask students which curricula they liked better—thus introducing a subjective element. If you wanted to find out which curriculum resulted in higher scores on standardized achievement tests taken by the gifted students in the general classroom, you likely would use quantitative methodology to determine differences in test scores and to discover whether a correlational or causal relationship (or both) exists between the curricula and the test scores.

One quantitative research design is descriptive, which seeks to obtain more information about a specific attribute or phenomenon within a field. Descriptive designs can be used to defend existing practices, identify problems, develop theories, develop opinions, or identify others' work in the same field. Descriptive designs do not look for causal relationships or manipulate variables. Experimental designs represent the paradigm for using the scientific method. As such, main elements of experimental designs are randomization, manipulation, and control. Experimenters randomly select participants, i.e. each has an equal chance of selection. Experimenters manipulate some variable(s) in the study, and some participants receive some kind of intervention/treatment. Experimenters also control some variables. A common form of experimental design is the before/after or pre-test/post-test design. Quasi-experimental designs are used when control is unfeasible and/or to protect validity when randomization, manipulation, or control is missing. Correlational/ex post facto designs look for relationships (NOT causality) without manipulating variables.

Qualitative and Quantitative Research of Giftedness and Gifted Education

Quantitative research methods use statistical procedures to measure or quantify phenomena in giftedness and gifted education, while qualitative research methods do not use such measures. Qualitative research is interpretive in nature. Implicit in the qualitative approach is the assumption that one must investigate giftedness as a whole in order to understand it. Quantitative research typically analyzes components of giftedness/gifted education. Proponents of qualitative research argue that such piecemeal examinations overlook important parts of the holistic appreciation they advocate. Rather than an absolute reality posited in quantitative research, qualitative methods propose multiple realities, which themselves vary across different times and places. Holistic understanding of the complexity, richness, and depth of giftedness/gifted education often reveals meanings that help researchers gain new insights. While all quantitative research is conducted according to an overall structure, qualitative research methods vary in their individual theoretical orientations relative to the specific topic or phenomenon under study.

One kind of qualitative research is grounded theory. In this approach, a theory is developed inductively (rather than deductively, which uses the reverse process of formulating a hypothesis and then collecting data to test it) from a body of data. While simpler quantitative statistical tests such as analysis of variance yield only the main effects of one variable upon another, grounded theory takes a case-oriented rather than a variable-oriented approach, assuming that variables interact in more complex ways and function as a whole within a case. Grounded theory also has a comparative orientation, based on John Stuart Mill's methods of similarities and differences, to uncover the causes. Another type of qualitative research is historical. The researcher collects data about past events, and objectively evaluates these events to test his/her hypotheses about their trends, causes, and/or effects, which can help to explain current related events and predict future related ones.

Mean, Median, and Mode

Statistically, **mean**, **median**, and **mode** are measures of central tendency, meaning they quantify trends wherein the majority of values fall near the center/middle. The mean/arithmetic mean is an average of all values obtained. For example, if in a group of 11 gifted students, four score 100% on a test, three score 96%, two score 94%, one scores 92%, and one scores 90%, the mean = 96%. If one student scored 70%, this would corrupt or skew the mean. The median represents the midpoint of all scores. In this example, with this distribution: 100+100+100+100+96+96+96+94+94+92+90, the midpoint is 96. Median does not inform the other values; if one student scored 70%, the median would not change. The mode is the most frequent value; in this example, 100. Mode does not reflect non-central, extreme values. If this example had four scores of 100% and four of 96%, it would have a bimodal distribution.

Bell Curve and Normal Distribution

A **bell curve**, aka a normal curve or Gaussian curve, graphically visualizes a normal distribution. Most values obtained fall near the median (middle/center), and fewer fall higher or lower, resulting in a curve with a bell shape - higher in the center, sloping downward toward the edges, and symmetrical, with equal numbers above and below the middle. With normal distributions, the mean (average of all values), median (the middle of all values) and mode (the value occurring most often) should be equal or nearly so. If a standardized IQ test is administered to a general education class including a few gifted students and the scores are plotted on a graph, the overall distribution of all scores is likely to be normal; however, the few gifted students' scores are likely NOT to be found near the median with most other student scores, but at one extremity, among the fewest and highest values.

When data from a group can be graphed as a bell curve, they have normal distribution. The largest number of scores will fall on or around the mean or average. In normal distributions, approximately 68% of the values fall somewhere within 1 standard deviation above or below the mean; about 95% of the values fall within 2 standard deviations above or below the mean; and about 99% fall within 3 standard deviations above or below the mean. A steep curve with most values near the mean shows a small standard deviation; flatter curves show larger standard deviations with more values spread away from the mean. If the average IQ score in a regular classroom is 100, with one gifted student having a score of 150, that score represents 1% of the population, as it is more than 3 standard deviations (typically 15 points with IQ tests, x 3 = 45) above 100.

STANDARD DEVIATION AND VARIANCE

In statistics, **standard deviation** shows how much the numbers in a given set vary within their distribution and how close to or spread away they are from the mean/average. Standard deviation is the square root of variance. **Variance** is calculated by squaring each variation from the mean and averaging the squares. Standard deviation indicates how much variation around the average is expected; greater variations are statistically significant. In IQ scores, the most common standard deviation is 15 points; variations of + 2 standard deviations, i.e. 30 points above or below the mean, are statistically significant. Thus, a student scoring 134 does not necessarily have a higher IQ than one scoring 130 on the same IQ test. However, a student scoring 160 compared to another scoring 130 on the same test shows a significant difference of 30 points, or two standard deviations. This difference is sometimes used for labeling levels or degrees of giftedness.

Review Video: Standard Deviation
Visit mometrix.com/academy and enter code: 419469

RELIABILITY, VALIDITY, INTERNAL VALIDITY, AND EXTERNAL VALIDITY

In scientific research, reliability refers to whether a measurement can be repeated with consistent results, regardless of the administrator. For example, a reliable standardized achievement test given to the same gifted student should yield similar scores each time, even when different teachers/personnel administer the assessment. Validity refers to whether an instrument or procedure measures what it claims to measure. For example, a standardized achievement test is not a valid instrument for measuring IQ; conversely, a standardized IQ test is not a valid instrument for measuring academic performance. Internal validity refers to whether the findings of an experimental study are attributable to the intervention or treatment used in the study and not to any other variables. For example, students could test better due to being taught a different curriculum or to having a different teacher. External validity refers how much a study's findings can be generalized to populations beyond the study's sample.

TEST CEILING

A **test ceiling** means a test's upper limit, i.e. the highest score it is capable of yielding. Most intelligence and achievement tests used in schools cannot accurately measure gifted student levels if they exceed the limits the test can measure. For example, many IQ and achievement tests direct administrators to stop when the student gets three consecutive items wrong. The disadvantage: a student can get two wrong, one right, one wrong, etc., never meeting the stopping criterion of three wrong consecutively. The student does miss a number of items, yet never hits the test's ceiling. An additional disadvantage: the student never reaches the point when items become too difficult to answer, which represents the true ceiling. Therefore, that student's score may be accurate, a bit too low, or much too low, but this is impossible to determine. Only certain is that the student's score is his/her lowest possible score.

Safeguards for Identifying, Assessing, and Finding Services for Gifted Students

While federal law provides that students with designated disabilities are guaranteed a free appropriate public education (FAPE), it does not address gifted students. Thus each individual determines the procedural safeguards for gifted students' identification, assessment, and eligibility for services, just as each state also determines its definition of giftedness and gifted programs. Some examples of procedural safeguards include: requiring school districts to give parents prior written notice before identifying, evaluating, placing, changing, or providing a FAPE to their child; obtaining written parental informed consent before initially evaluating their child to determine eligibility and before initially providing services for gifted children; parents' right to examine their child's school records and participate in educational planning meetings for their child; parents' right to due process hearings to resolve disputes related to identification, evaluation, or placement of their child or provision of a FAPE; and many others.

Measuring Intelligence

Naglieri Nonverbal Ability Test®, 2nd Edition (NNAT®-2)

According to the test's author, the NNAT®-2 is "culturally neutral," thereby making it suitable for testing culturally diverse student populations. It is nonverbal, making it a good choice for ESL/ELL and nonverbal students. It can be used with ages 5 through 17 years and grades K through 12. It has seven levels corresponding to grades K; 1; 2; 3-4; 5-6; 7-9; and 10-12. It takes 30 minutes and is easy to administer. Students do not need good reading skills, math skills, well-developed vocabularies, or factual knowledge to respond to the test items. For identifying gifted/talented students whose socioeconomic backgrounds have limited their development of verbal skills and acquisition of information, this test is a useful instrument. Among tests of ability, the NNAT®-2 has the most recent norms, from 2008. Its administration and instructions are facilitated by pictures. Its graduated difficulty range allows identification of advanced as well as gifted and talented students.

Online Version

The NNAT®-2 offers both pencil-and-paper and online administration options. The online version saves teachers and administrators a great deal of time compared with the time required for the pencil-and-paper version. It also saves money, as schools do not have to pay for printed paper tests. Furthermore, this test saves time, space, money, and human energy by relieving schools of the need for storing, distributing, organizing, and shipping test booklets. It has minimal system requirements, enabling more users and computers to access it. The test's publisher, Pearson, uses a secure browser, protecting the safety of student information. Paperless, this test offers a more environmentally friendly option than does the traditional paper-and-pencil test. The online version yields instant results, allowing teachers to spend more time with students requiring additional, more focused instruction. Online reporting allows teachers to tailor data disaggregation and filtering to their district's needs. The online format is appealing and motivating to students, promoting test completion for fuller measurement. One item per screen helps students focus.

WISC-IV, 2003

Subscales

According to a number of educational researchers who have analyzed its content and results, the Wechsler Intelligence Scales for Children, 4th edition (WISC-IV) is a useful diagnostic tool for giftedness. They note that although it seldom gives scores above the 140s and has a score ceiling of 160, it also indicates ability beyond these limits. WISC-IV's ten required subtests include: Similarities, Vocabulary, Comprehension, Matrix Reasoning, Picture Concepts, Block Design, Letter-Number Sequencing, Symbol Search, Digit Span, and Coding. Optional supplementary subtests

and Processing Speed scores were not lower enough than Verbal Comprehension and Perceptual Reasoning scores to lower their Full-Scale IQ (FSIQ) averages. In contrast, the gifted sample averaged 27.4 points difference between highest and lowest subscale scores; the greatest variation was 69 points—over four standard deviations. Almost 60% of gifted students had 23-point differences between Verbal Comprehension and Processing Speed Composites. Gifted students scored 25 points higher than the norm group in Verbal Comprehension, but differed by below 2 points in Processing Speed. This trend implies gifted students are superior in verbal abstract reasoning but not processing; as a result, the WISC-IV's FSIQ should not be used for gifted identification.

Using the DWI-1 and Flanagan and Kaufman's GAI

At the Gifted Development Center, Silverman, Gilman, and Falk (2004) found that gifted students' WISC-IV Composite Scores in Verbal Comprehension and Perceptual Reasoning were typically high enough to qualify them for gifted educational services, but their scores on Working Memory and Processing Speed subscales were typically below qualifying levels. Additionally, since the Full-Scale IQ (FSIQ) score average incorporates these lower scores, they drag the average down, and also are individually not good indicators of giftedness. Consequently, the FSIQ identifies neither gifted strengths nor relative weaknesses. The Dumont-Willis Index-1 (DWI-1) computes a score combining the Verbal Comprehension and Perceptual Reasoning Composites using only six WISC-IV subscales, which these researchers find an excellent option for schools to identify gifted students. Flanagan and Kaufman's General Ability Index (GAI) averages the same two Composite scores; Harcourt Assessments (PsychCorp) trainers support this. Another solution involves basing identification on assessments of reasoning, delivering gifted services/accommodations, and adding accommodations for relative weaknesses.

WPPSI-III

The **Wechsler Preschool and Primary Scales of Intelligence** (WPPSI-III) is for ages 2.6 to 7.3 years. It includes 14 subscales: Block Design, Information, Matrix Reasoning, Vocabulary, Picture Concepts, Symbol Search, Word Reasoning, Coding, Comprehension, Picture Completion, Similarities, Receptive Vocabulary, Object Assembly, and Picture Naming. By various combinations of subtests, measures can be computed for verbal IQ, performance (fluid) IQ, processing speed quotient, general language composite, and full-scale IQ. Verbal IQ, performance IQ, and full-scale IQ are computed from core subtests. The WPPSI-III features short activities to accommodate attention spans of the youngest children in the age range. Activities are colorful and game-like, appealing to and interesting young children. Varied samples, second chances, and layered scoring enabling partial credit help children do their best. With children aged 6.0-7.3, the WISC-IV, for ages 6:0-16:11, is recommended for identifying giftedness, the WPPSI-III for general academic reasons. Because they are similar, these should not be administered successively to avoid practice effects.

Disadvantages

The WPPSI-III is a standardized instrument with proven validity and reliability. However, it does have some disadvantages. In testing the youngest children, the WPPSI-III's subscales reflect an expectation for IQ scores to jump significantly every 2-3 months. To qualify children for gifted services, experts advise scheduling testing optimally for ruling in/out such qualification. They also warn that a retest can take up to two years, so if a child has taken the WPPSI-III and needs re-evaluation sooner, they should use a different test. Another consideration about the WPPSI related to the youngest children is that IQ results are not as stable at earlier ages. Therefore, administering the WPPSI-III near the end of the kindergarten year is recommended to allow greater maturation and pre-academic/academic skills development. An exception is when earlier testing is indicated, e.g. a child displays signs of extreme giftedness at a very young age.

MDSD

The **Multi-Dimensional Screening Device** (MDSD) (Kranz, 1978) is based on the idea that intelligence is multidimensional in nature, not reflected by any one measurement. The author designed this instrument to facilitate better initial identification of giftedness among "the less accepted school population." A staff development program for teachers required to participate in the screening process is part of this instrument. In fact, the first step in its implementation is staff development. Other steps include rating students individually and forming a local screening committee to select individual students for screening. The MDSD's categories of giftedness are: visual arts ability; performing arts ability; demonstrated creative/productive thinking; discipline-specific academic ability; general intellectual ability of 1 in 100 or more; leadership characteristics, organization, and decision-making; psychomotor history and ability; history and use of spatial and abstract thought; wide difference between performance and general intellectual ability; and talent related to cultural heritage.

GRS

The **Gifted Rating Scales** (GRS) are norm-referenced, based on a multidimensional model of giftedness, and informed by current giftedness theories as well as federal and state definitions and guidelines. The GRS-P form, for ages 4 years to 6 years, 11 months, includes domains of intellect, academic readiness, motivation, creativity, and artistic talent. The GRS-S form, for ages 6:0 through 13:11 and grades 1-8, includes the same five domains as the GRS-P, except for an academic domain rather than academic readiness and the additional sixth domain of leadership. Administration takes 5-10 minutes. Relative strengths and specific gifted areas can be identified. Statistical validity studies positively correlate the GRS with the Wechsler Preschool and Primary Scales of Intelligence (WPPSI-III), Wechsler Intelligence Scales for Children (WISC-IV), and Wechsler Individual Achievement Test (WIAT-II). These standardized scales are designed to help qualify students for placement in gifted/talented programs. Domain-specific identification guidelines are included. Teacher form completion is easy and quick.

Measuring Creativity

Difficulty in Measuring Creativity

Although many researchers agree about common elements of creativity such as the originality, applicability, and value to society of creative work, they find it much more difficult to agree about which methods to use to make their definitions operational, and about which instruments of measurement to use. Creativity is such a complex quality that most measures cannot sum it up alone. Experts advise using multiple measures, but even this tactic can fail to achieve a comprehensive representation. Forty years ago, researchers found that the absence of any one unified and broadly accepted theory of creativity caused problems with understanding creativity's relationship to other abilities, the ramifications of different testing instruments and administrations, and operational definitions. Researchers find the same problems today. Meanwhile, so many creativity tests have been developed with varying degrees of psychometric credibility that difficulties in focusing on criteria have only increased.

Guilford Battery of Tests for Creative Thinking

Psychologist Joy Paul Guilford (1897-1987) formulated his Structure of the Intellect model in 1962, identifying 180 different kinds of thinking, many of them divergent. He found that the types of divergent thinking most pertinent to creativity were the abilities to generate new information from existing information and to transform knowledge or experience into new patterns and configurations. Based on this model, Guilford developed a battery of tests for creative thought. This battery's ten tests each measure divergence in producing the following:

- Names for Stories - semantic units
- What to Do with It - semantic categories
- Similar Meanings - semantic relationships
- Writing Sentences - semantic systems
- Kinds of People - semantic implications
- Make Something Out of It - figural units
- Different Letter Groups - figural categories
- Making Objects - figural systems
- Hidden Letters - figural transformations
- Adding Decorations - figural implications

The Guilford Battery contains ten tests of different types of divergent thinking, a cognitive process associated with creativity. Half of the tests are verbal measures, and half are nonverbal/figural measures. The verbal measures portion of the battery asks students to offer as many solutions as they can in the following categories: names for stories, verbal classes/categories, synonyms, sentence structures, verbal inferences about categories, figures, categories of figures, systems of figures, transformations of figures, and inferences about the figures. Each test is timed and scored for fluency by number of responses and for originality by statistical infrequency of responses. Not much validity research exists regarding these tests. Follow-up studies by Meeker (1978) found that students identified as creative by Guilford's tests in elementary school also had high creativity scores in high school. Michael and Bachelor (1990) found limited confirmation of Guilford's findings through conducting factor analyses.

Identifying and Predicting Aspects of Creativity

Researchers find that many instruments designed to measure creativity can identify the characteristics of fluency in ideation and divergence in thought, commonly accepted as components of creativity. However, they find that these measures do not predict future creative behaviors in a student identified as creative. Some scientists attribute this characteristic not to psychometrics but poor methodology—e.g. some studies were not continued long enough, data not having normal distributions did not receive adequate statistical procedures, or the criteria for outcomes in longitudinal studies were not well operationalized. Some researchers also note that while personal definitions and theories of creativity are not included, they should be. These researchers have proposed that individuals acting creatively are directed by their personal beliefs about the nature of creativity, how to nurture and measure creativity, and that these beliefs may differ substantially from the theories that experts in the study of creativity have developed.

Intelligence vs. Creativity Measures for Indicating Giftedness

Traditionally, standardized intelligence tests were used to determine whether a student was gifted. IQ scores above certain numbers (e.g. above 120, 130, etc.) qualified a student as gifted. However, IQ scores are an indication of intellectual giftedness rather than of creativity. Traditional standardized intelligence tests typically feature questions or problems with only one right answer. When taking this kind of intelligence test, the student must use convergent thinking to narrow down the possibilities to the correct one. In contrast, many measures of creativity test divergent thinking, which generates multiple ideas and possible responses. During the 1960s, J. P. Guilford and E. P. Torrance both developed standardized tests of divergent thinking that are still popular today. Because the traditional standardized intelligence tests require convergent thinking and little or no divergent thinking, and because divergent thinking is identified as a key characteristic of creativity, scientists have come to hypothesize that intelligence and creativity are separate constructs.

Torrance Tests of Creative Thinking

In order to operationalize creativity for research purposes, E.P. Torrance focused on its problem-solving aspect. His tests include nonverbal and verbal forms. "Thinking Creatively with Pictures" features three activity groups: namely, drawing lines elaborating on one shape, drawing lines finishing an incomplete picture, and drawing as many different pictures as possible using the same shape. "Thinking Creatively with Words" features six activities involving generating questions, alternate uses for things, and guesses. All these activities are timed and scored for fluency, flexibility, and originality. In addition, the nonverbal forms are scored for elaboration. Torrance's tests apply to grade levels from kindergarten through graduate school and are the most widely used creativity tests. They also have the most research supporting their validity worldwide, as they have been translated into many languages. While their longitudinal predictive validity is not high (.62 for males, .57 for females), Torrance indicates they match or exceed rates for intelligence measures.

Challenges Faced by Gifted Students

Strengths and Potential Problem Behaviors

Intellectually gifted students often have advanced organizational skills. This advancement benefits them academically, since it facilitates learning. However, this advancement also can manifest in these students exhibiting a need to organize things and people beyond their own academic subjects. Positively, this characteristic contributes to leadership qualities; negatively, it contributes to making rules too complex for others and to being perceived as bossy. Intellectually gifted students often possess knowledge ahead of their peers in both level and breadth; they frequently have larger vocabularies and more facile use of their respective vocabularies. Problems associated with these strengths include boredom with school and classmates and the potential for using language to manipulate others. Intellectually gifted students tend to have high expectations both for themselves and others. While this strength is a motivational quality, its drawbacks include perfectionism, a lack of tolerance for their own and/or others' shortcomings, and a potential for developing depression due to disappointment over unmet expectations.

Intellectually gifted students often learn and remember new information more quickly than other students. A potential side effect of this facility involves impatience with other less gifted, as well as with routine procedures they find superfluous and limiting. The curious/inquisitive nature of intellectually gifted students as well as their need to seek meaning can result in questions that embarrass others; others' perceptions that the gifted students are nosy and/or rude; and perceptions that the gifted students go overboard in pursuing their interests. Another strength involves internalized motivation; however, as a downside, these students can seem overly willful and resistant to direction from others. Students with superior capacities for synthesis and conceptualization tend to enjoy solving problems and dealing with abstract ideas. However, these students are more likely to question/challenge teaching practices and routines. They also tend to focus on truth and fairness, so they may have excessive humanitarian worries and/or difficulty accepting inequitable realities in life.

Often intellectually gifted students are sensitive and promote empathy for others. The problem associated with this area of giftedness involves the students' need for the acceptance of others, as well as an accompanying oversensitivity to rejection and/or criticism. Creative and inventive students enjoy finding new ways to do things. This gift proves a boon to society in terms of innovation, but it also can be perceived as disruptive and abnormal. Intellectually gifted students often have longer attention spans than other students, can concentrate on one subject more intensively, and are more persistent in the areas interesting them. Many important accomplishments are attributable to these qualities. However, these qualities also can result in the neglect of other academic subjects and/or persons and relationships, ignoring interruptions, and being thought stubborn. Many intellectually gifted students are highly alert, enthusiastic, and eager to learn; in addition, these students may possess high energy levels. These assets can be misperceived as hyperactivity/overactivity. These students can also become frustrated when inactive.

Many intellectually gifted students are academically independent. Superior abilities for learning, knowing, and understanding afford them both the skills and the confidence for autonomy and self-direction in learning. These skills and confidence, plus their thinking and learning differently from their age peers, promote self-reliance in such students. They are likely to prefer individualized tasks. Because of these factors, these students may be seen by others as not conforming to accepted norms. They also may reject input from their peers, teachers, or parents, and these other individuals likely will not receive such rejection well. Greater amounts of versatility and diversity in their abilities and interests give these students more options, but these characteristics also can cause time-management issues and cause a disorganized impression. Strong senses of humor can be misconstrued by age peers, and as a result these students may evolve into "class clowns." Their tendencies to seek causal relationships can make them uncomfortable with irrational/illogical feelings or traditions.

Uneven Development in Intellectually Gifted Children

Educational researchers have found that intellectually gifted children do not always develop evenly across all domains of ability. While some gifted students excel in all areas, particularly in the higher grades, many more of these students demonstrate comparatively greater gifts in certain domains than in others. One example of uneven development very common among younger gifted children occurs when their cognitive and conceptual abilities develop far in advance of their physical motor skills. When this uneven development happens, while a young child is capable of more sophisticated ideation, he or she is incapable of executing his or her ideas physically due to less advanced motor abilities. This frustrating inequity proves frustrating and produces tantrums in many children. Another example of uneven development occurs when an intellectually gifted child uses his or her superior cognitive skills to interact with the environment almost exclusively, leaving his or her emotional and social skills underdeveloped from lack of practice.

Gifted Students' Motivation for Subjects at Odds with Their Actual Talents

Some students are gifted in certain areas, yet their interests lay elsewhere, where they are less talented and even exhibit weaknesses. For example, a parent and retired educator had a gifted child with talent in mathematics and playing violin. However, his passions were for soccer and singing, where he was not gifted or average, but comparatively weak. His parents did not press him to focus on his gifted areas or give up on what he loved because he did not excel. Despite shortcomings in favored fields, his passion and persistence eventually won him a place on his high school soccer team and he realized his goal of making All-State choir in senior year. He teaches and directs voice and coaches soccer as an adult, happy with his choices. This illustrates that gifted students need not confine their efforts to gifted areas and may succeed in other fields with passion and perseverance.

Contributing Factors for Underachievement in Some Gifted Students

Some gifted students express their gifts during the earlier parts of their childhood by pursuing areas of special interest to them, excelling in certain school subjects, and working on advanced projects. Some students suddenly drop all these activities, deny their previous interests, or claim they have lost interest or outgrown them. Experts have found girls most likely to exhibit this trend during middle school/junior high school, while boys are more likely exhibit this trend in high school. Researchers attribute the change to social and emotional needs, especially in school. Girls approaching and entering adolescence experience a great increase in their needs for peer acceptance and for a sense of belonging. Consequently, some gifted girls will hide or deny their giftedness in order to fit in with a non-gifted group of peers. Boys in high school often experience pressure to play sports, a celebrated skill, while academic excellence without athleticism is ridiculed.

Not identified as gifted earlier in school, some gifted students reach their high school years without any special educational interventions or programs designed for their unique abilities. These students may have been overlooked in the school system if their gifts exist outside of their schools' areas of curricula. While these gifted students may have the ability to perform well in certain subjects, teachers might find these students' different methods of achieving such performances unacceptable or invalid. Furthermore, such students may attend a school in which personnel lack sufficient training or experience with giftedness. By the time students receive neither acknowledgement nor support for their gifts in high school, they have felt rejected for years and are usually angry and/or depressed. Without support or appropriate curriculum, they will not achieve and may drop out of school.

Some gifted students figure out early on which behaviors are desired and rewarded by adults, and then these students use their superior abilities to produce these behaviors. They are successful according to adult standards and appear to have high self-esteem because of the adult reinforcement they receive. They are not identified with behavior problems because they comply with adults' wishes in order to succeed. However, such students, especially when placed in regular classes with mainstream curricula, often secretly feel bored with school. These students use their expertise to get through school by expending the minimal effort possible. As a result, they do not develop the independence and skills needed for self-direction, nor do they explore subjects of personal interest to them. Instead, they rely on adults for direction, guidance, and feedback. They often develop into adults who display competence but not originality or imagination. When they continue these behaviors in adulthood, they can become underachievers in relation to their ability.

Some gifted students have superior abilities that they can express congruently within the traditionally accepted framework of academic performance and achievement. Other gifted students have superior abilities best expressed outside of this framework. These students often think in more markedly different ways than other students, who often fail to understand them. They may also have talents not included in school curricula. They do not receive rewards for their gifts in the same manner as students with more traditionally acceptable gifts. As a result, the gifted students become frustrated. Due to the unconventional nature of these students' gifts, educators often do not identify them as gifted, so they receive little instructional support. They also are often highly creative. Their spontaneity can be perceived as disruptive in school. With their self-esteem compromised and their gifts not nurtured, such students likely underachieve in school and exhibit increased risk for dropout if they do not receive specialized instructional attention.

Contributing Factors to Underachievement in Minority Gifted Students

Low self-esteem in general hampers student achievement. Researchers also find that poor self-concepts specific to academics and social interactions interfere. Some researchers have emphasized the need to consider the strength and/or positivity of a gifted minority student's sense of racial/ethnic identity. Students whose racial identities are not positive can be more susceptible to counterproductive peer pressure. They also may associate academic achievement with the dominant culture, hence with betraying their minority culture, i.e. "selling out" or trying to "act white." This view results in less effort, which results in less achievement. Psychologically, with a high internal locus of control, a student attributes his or her achievement to ability and effort and is more likely to achieve. Minority students who expect external factors such as discrimination and injustice to block their success demonstrate a high external locus of control and are less likely to achieve up to their potentials.

Researchers have found that among gifted black students, for example, underachieving students have reported experiencing less positive teacher-student interactions, less supportive classroom climates, insufficient time to understand lesson material, and feeling uninterested and unmotivated toward school. They also attributed part of their lack of interest to their schools' lack of multicultural education. Teachers often have lowered expectations of minority and low-income students, especially for teachers with inadequate preparation in both gifted education and multicultural education. Such teachers are less likely to refer minority students for gifted education services, and with lower expectations they are also less likely to identify these students as underachievers. Without access to suitable educational services that address their individual strengths and needs, minority gifted students can become frustrated, bored, lose interest in school, and become underachievers.

Review Video: Multiculturalism/Celebrating All Cultures
Visit mometrix.com/academy and enter code: 708545

Underachievement Due to Mismatches Between Learning and Teaching Styles

The predominant cultural values in America favor individualism, competition, and excellence in individual achievement, both in society and in school. In contrast, many other countries have cultures that value collectivism, cooperation, and placing the good of the group ahead of one's individual concerns. Therefore, many students belonging to minority groups in the United States have cultural values that more closely reflect the values of other countries of origin than the majority American values. These students tend to be more social and cooperative in their attitudes and behaviors, and less individualistic and competitive. Teaching styles in America tend to place emphasis on competing with others to be the best. Too much emphasis on competition can interfere with minority gifted students' achievement: rather than increasing their motivation, competition can decrease motivation by provoking anxiety and by negatively influencing students' social and academic self-images.

The majority of teaching methods in public schools tend to use the verbal modality, to deal in abstract concepts, and to teach concepts and skills in isolation, i.e. taken out of context. These methods contrast markedly with the learning styles of many minority students. For example, research finds that black students are more likely to be visual rather than verbal learners, i.e. they learn better using visual images than words. In addition, black students are more likely to be field-dependent rather than field-independent learners, meaning that they rely on the surrounding context to understand the material presented and will not grasp it as well (or at all) when it is presented out of context. Finally, the research indicates that black students tend to learn more concretely rather than abstractly, so teaching abstract concepts through using concrete objects rather than presenting such concepts only in abstract terms would enhance their learning. Mismatched teaching methods produce confusion and frustration in students with these learning styles, thereby leading to underachievement.

students do not address concomitant disabilities, while conversely most programs for disabled students do not address giftedness. Researchers are endeavoring to remedy these errors. At times teachers might overlook these students' areas of giftedness and instead focus on the behaviors produced by the students' disabilities. Teachers may not consider that a child with messy handwriting or disruptive classroom behavior, for example, might exhibit giftedness in areas such as art, science, music, dance, or athletics.

Characteristics and Challenges for Students Gifted in Math

Often children with mathematical gifts can grasp new concepts very quickly. In the general education classroom, they end up without recourse to additional intellectual exercise while they wait for the rest of the students to understand and master the new math ideas presented. Some consequences of letting gifted students go unchallenged are boredom, less than optimal thinking skills, and undesirable study habits. One way to offer them challenges is through differentiated instruction. The teacher can present math problems with graduated levels of difficulty and complexity. When students have a choice of problems at different levels, they have the opportunity to try to solve problems just within their current cognitive capacity. Such methodology challenges students with problems that are not too easy, without overly frustrating them by being out of range of their ability; it cultivates their passion for math; and it stimulates their cognitive growth and development.

Many mathematically gifted students enjoy challenges that engage their superior abilities for calculating and solving problems. However, gifted students who only have experience with effortless school work become uncomfortable confronting challenges and/or frustration inherent in learning more difficult material. In addition to teaching advanced students, teachers are charged with the job of helping them realize that confining their education to easy pursuits deprives them of many rewards, and a life without struggle leads to less achievement and fulfillment. Teachers can give students examples illustrating challenges to adults. One educator (Zaccaro, 2008) would tell his students about the time he did not calculate whether his rope was long enough to rappel off a cliff and reach the ground, and he had to jump 15 feet. He also gave them problems to work where mistakes were common so they could see that making errors was not the end of the world.

Things to Avoid in Order to Not Interfere with Development of Gifts

Parents

In general, toddlers are prone to tantrums due to frustration. Gifted toddlers are even more likely to demonstrate this characteristic when their cognitive development outpaces their physical development and they can conceive of actions but cannot perform them. Parents should not punish such tantrums as they allow emotional venting and reduce tension. Researchers have found that parents who are overly directive and controlling tend to influence children to become less exploratory and curious, less independent and self-directed, less internally motivated, and less creative. Total permissiveness is not ideal, since children need appropriate limits, boundaries, structure, and guidance. Detached or uninvolved parents are not helpful. Parents should avoid punishing or disapproving of gifted children's actions as abnormal just for being unusual or unfamiliar. Parents should not avoid activities with their gifted children from insecurity regarding their own preparation or competence. Activities confer important bonding experiences as well as children's learning and positive reinforcement.

Teachers

When a child exhibits giftedness in certain areas, the teacher should not prohibit or curtail the child's interest in preferred subjects and activities. If a gifted child prefers some subjects to the exclusion of other, required curriculum, the teacher can arrange behavioral contracts or contingencies; e.g., completing an assignment in a non-preferred area can be rewarded with more time exploring a preferred interest. Teachers should not punish or disapprove extensive interest in certain subjects, a characteristic of gifted students. Teachers should not force gifted students to follow traditional procedures for learning and performing if they learn best with different procedures. As a student, Albert Einstein reputedly was failed in math by a teacher for not following the prescribed procedures, even though he got the right answers. Fortunately, he overcame this deterrent through his intellect, curiosity, and persistence. Teachers should not ignore gifted students who grow bored or who clamor for more material.

General Mental Health Considerations

Research finds gifted students no more prone to experience mental health problems than are other students. Nonetheless, if educators and counselors have limited knowledge and experience of gifted students, their perceptions can be influenced by exclusively positive stereotypes of the gifted. Such a perception can be dangerous, in that adults may not recognize, and hence not address, counseling needs and/or developmental issues in gifted students. Some gifted characteristics can become liabilities in this sense. Also, adults unfamiliar with gifted students' specific characteristics may incorrectly interpret some divergent behaviors as pathological symptoms. Another consideration is that gifted students, wanting to preserve their competent reputations and avoid disappointing invested adults, often do not request needed assistance. Adults who are both uninformed regarding the complexity of gifted students' concerns and also dazzled by the gifted students' abilities are less objective, can overlook students' vulnerabilities, and cannot work effectively with them. Adult respect for gifted students is more appropriate than awe. As in curriculum and instruction, differentiation in counseling services is indicated for the gifted.

Prevailing View of Student Gifts and Talents as Assets

While superior intellectual abilities and/or creative talents may seem obvious assets, a number of factors have contributed to this view's overshadowing the recognition of less beneficial aspects of giftedness. Accordingly, samples selected for quantitative research studies have not covered a wide enough range of the gifted population. By not including sufficient socioeconomic, cultural, and performance factors in sampling, studies do not reflect the true range of issues for gifted students. In addition, such research findings inform the development of instruments for assessment and intervention, which in turn will not adequately represent "non-asset" aspects of giftedness. Also, due to the paucity of qualitative studies on gifted students, researchers are limited in understanding feelings and thoughts that gifted students have not expressed to others. Other factors include our society's emphasis on equality, and legislation mandating educational accountability and achievement, both of which are not conducive to addressing gifted students' emotional and social needs.

Review Video: Equality vs Equity
Visit mometrix.com/academy and enter code: 685648

Preparation Needed by Counselors

Giftedness has significant impacts on the emotional and social development of young people. They encounter the same universal developmental transitions and challenges as all children, yet they are likely to experience them quite differently. In addition, they experience other, individualized developmental issues. They will be concerned with moral, academic, and career issues at younger ages than other students. They may also have concerns that other students do not experience at any age. The issues that concern these gifted students have ramifications for their overall wellbeing. Thus school counselors and outside, private counselors must possess knowledge regarding the characteristics of gifted students. Knowing about the emotional intensity and hypersensitivity of many gifted students keeps counselors from misinterpreting these as abnormal. Knowledge of gifted students' characteristics can explain classroom and/or social problems and afford more perspective on developmental issues. This knowledge also should inform counselors' case studies and intervention plans.

Models of Instruction

General Goals of Good Gifted Education Programs

Good gifted education programs seek to facilitate the gifted student's mastery of basic reading and math skills at paces and depths suitable to the student's abilities; to foster abilities of reasoning and critical thinking; to create an environment conducive to divergent thought; to promote inquiry; to promote challenging attitudes relative to learning; to develop higher level skills in spoken and written language; to develop skills and methods for doing research; to cultivate students' comprehension of issues, problems, themes, and systems of knowledge that structure the extrinsic environment; to foster student development of self-understanding; to enable learning experiences outside of school that fit the individual student's educational needs; to provide or improve opportunities for future student development and planning; to develop creative thinking and problem-solving skills; to foster effective social skills for interpersonal interaction and coping; and to develop the student's skills of metacognition for self-direction and independent learning.

Reading Curriculum Units to Research Gifted Education

Three instructional models are: Carol Ann Tomlinson's Differentiation of Instruction model, which focuses on varying teaching methods; Sandra N. Kaplan's Depth and Complexity model, which focuses on attributes of deeper understanding and more complex knowledge; and Joseph S. Renzulli and Sally M. Reis' Schoolwide Enrichment Model (SEM), which uses educational enhancements that simultaneously benefit all students and address the special needs of gifted students. The National Research Center on Gifted and Talented (NRC/GT) has a longitudinal study basing curricula on these instructional models. By sampling school districts nationwide that implement gifted and talented programs, these studies' findings represent diverse populations, socioeconomic levels, and urban, suburban, and rural settings. These sample schools are selected from nominations by educators who develop curriculum models; state gifted education directors; and representatives from state and national organizations. Outcomes are measured by:

- state standards extended for advanced reading and math achievement
- performance-based measures of investigative and problem-solving skills

Research of Curricula Used in Gifted Education

Some research teams, including teams on a nationwide level, are interested in examining comparisons of different criteria for identifying gifted students, the effects of such criteria on assessment results and on under-represented populations, and comparisons of the effects of general education curricula and model-based curricula on gifted students. They ask research questions such as: When comparing standardized and performance-based measures of reading and math, whether the criteria for identifying giftedness interact in any ways with the type of instructional method (i.e. general education curricula vs. theoretical model-based curricula) delivered. They want to discover if students not identified as gifted are helped or harmed in achievement by model-based curricula, and if non-identified students in their non-treatment and treatment classes perform the same or differently on standardized reading and math achievement tests. These researchers are also interested in how teacher implementations of model-based curricula are affected by personal and environmental variables.

Experimental Research Design to Investigate Teaching Methods

Introduce two teaching methods, neither one normally used with the students. Divide students into two experimental groups and one control group by random sampling; i.e. they all have an equal chance of being in any group, preventing biased selection. One experimental group gets a lecture on a new topic followed by a question-and-answer/discussion session; the other gets a hands-on activity on the same topic followed by a question-and-answer/discussion session. The control group contains comparable students receiving no treatment; e.g., they receive the same teaching method normally used in your school, on the same topic as the experimental groups. The control group helps rule out the influences of extraneous/confounding variables on differing results between experimental groups, supporting the experimental treatments' differing influences. All groups receive the same pre-test and post-test on the topic. The teaching method is the independent variable, which was manipulated. The dependent variable is post-test score.

Collaborative Consultative Model of Education

The collaborative consultative model of education posits three component entities: the Consultant, the Mediator (the consultee), and the Target (usually the student). This model views the consultant and consultee as equal partners. They bring varying expertise to the collaboration and have mutual responsibility for identifying problems, planning strategies for interventions, and carrying out recommendations through combining their respective knowledge and skills. Disagreements between consultant and consultee are regarded as constructive opportunities to extract the most helpful information. This model assumes that consultant and consultee will both work directly with the target or student. Educational researchers have found in surveying teachers that while time shortages and scheduling issues can interfere with the success of collaborative consultative services, most teachers find them acceptable as alternatives to resource room services; furthermore, some teachers find them very desirable as alternatives. This model addresses special student needs, including giftedness as well as disabilities.

One of the benefits of using collaborative consultative models in gifted education involves the congruence of such models with recent trends in educational reform. Another benefit affords professional development for all those participating as they contribute and share their expertise. Collaboration also allows many more ideas to be produced than any individual could generate in isolation. In addition, consultative methods make the most of the opportunity for the adult participants to use their individual differences toward achieving constructive goals. Another advantage of the collaborative consultative model is that it is amenable to the school administrator's taking the role of facilitator. Moreover, using collaboration and consultation between teachers, parents, and others from outside of the school affords the benefit of parents' satisfaction through their direct involvement, contributions, learning from others, and seeing their children's knowledge and social skills improve.

Potential Issues to Consider

While there are many definite advantages for using the collaborative consultative model to meet students' special needs, including those of gifted students, there are also a few potential issues to consider. First, teachers, parents, and other participants from outside the school may have little or no training collaboration, unless the school institutes such training. If the school does so, another consideration is the time, expense, and human energy involved in providing training. Another very real consideration is that any or all of the potential collaborators may not have sufficient time to interact with one another. In addition, some educators prefer working with students, and would rather not work with other adults, especially if they lack relevant experiences. Moreover, implementation of collaborative consultative methods requires firm support from the school administrator. Finally, while collaborative consultation is effective, visible results take time, so patience is needed.

Curriculum Development

Selecting Curriculum Materials

Curriculum Design Features to Consider

Gifted students need greater breadth and depth of knowledge and processes than are afforded by school textbooks. When selecting curriculum materials for gifted students, teachers should consider the following curriculum design features:

1. The purpose and rationale of the curriculum design should be clear.
2. Selected materials should address objectives for each lesson.
3. Materials should include appropriate, stimulating, and challenging activities for gifted students.
4. Materials should include instructional strategies suitable for gifted students that teachers can use.
5. The materials should include valid and reliable assessment procedures teachers can use to assess the effectiveness of instruction using those materials.
6. References for additional materials and resources should be provided.
7. Ideas for extension of the curriculum, activities, and materials should be given.
8. Technology should be incorporated in the materials. Today many materials are available on websites, which are most likely to adhere to the preceding guidelines.

Special Needs of Gifted Students to Consider

When teaching gifted students, teachers will find that while these students often master textbook content easily, they also demand broader and more in-depth knowledge of a subject than textbooks afford. As they seek additional curriculum materials for gifted students, teachers should consider their special needs. Specifically, students with higher levels of ability need to be presented with more sophisticated ideas than others and more challenging activities than those usually offered in general education classrooms. Choosing activities that can be individually tailored to any specific student will give teachers more options for optimally meeting individual gifted students' educational requirements. Curriculum materials chosen for gifted students should require more use of higher-order cognitive skills than for other students. Teachers should select materials that incorporate relevant major themes, contain high levels of abstraction, allow student exploration according to their interests, and enable them to create products.

Aspects Within Areas of Content and Process to Consider

As Renzulli (2000) defines them, content equals authentic knowledge and process equals teaching techniques. Within the area of content, when selecting curriculum materials, teachers should consider: the organization of the content, the depth in which the content covers its subject, the accuracy of the content, and the degree to which the content moves from concrete topics to abstract concepts. Within the area of process, teachers should consider: opportunities to use/learn problem-solving skills, opportunities to use/develop communication skills, activities that exercise skills of reasoning, relevant connections among parts of the subject area, providing multiple representations of concepts to be taught, opportunities for students to verify their speculations when indicated, opportunities for students to work both individually and in groups, and opportunities for students to use divergent thinking in responding to a lesson, exercise, or activity.

Common Difficulties in Writing Curricula for Gifted Students

Curriculum writing can be quite difficult and time-consuming. Difficulties include trying to comply with state curriculum guidelines, feeling pressured to include activities currently popular in the educational community, trying to achieve a balance between content (information/knowledge) and process (teaching/learning), and juggling all these demands. Moreover, the results of these efforts often do not further education or impart meaning to it. Seeking to improve curriculum writing and believing that teachers need time and tools to create meaningful teaching units based on desired outcomes named in a curriculum guide, Renzulli et al (2000) of the University of Connecticut's National Research Center on the Gifted and Talented produced the Multiple Menu Model of curriculum development. Differentiating this model from more traditional approaches are its stronger emphases on encouraging original inquiry by students, on examining the organization and interrelatedness of knowledge, and on arriving at a balance between content and process.

Multiple Menu Model of Curriculum Development

While its authors originally conceived of the Multiple Menu Model of curriculum development as a means for differentiating instruction to address the needs of gifted and talented students, Renzulli remarks that teachers can also use it to promote original inquiry and creativity in students at all levels. This model uses the term "menu" to indicate that teachers are able to choose from among many options within each menu. The six component menus are: the Knowledge Menu, addressing the specific subject areas; the Instructional Objectives and Student Activities Menu, Instructional Strategies Menu, Instructional Sequences Menu, and Artistic Modification Menu, all of which address various aspects of pedagogy, i.e. instructional techniques; and the Instructional Products Menu, consisting of the interrelated Concrete Products and Abstract Products Menus, which address the kinds of products the student may create based on the knowledge acquired, and how students as primary inquirers construct this knowledge.

Components of Curriculum Writing

In curriculum writing, the stated instructional objectives and the specified student activities are concerned with the processes students use in constructing knowledge (for example, analysis, synthesis, and application). These processes are both cognitive and affective in nature. When curriculum writers include both variety and balance among these processes in the activities they design, they enable gifted students to practice using a fuller range of the encoding, decoding, and recoding processes involved in learning new ideas, principles, and information. Instructional strategies represent another area wherein curriculum writers should provide variety. Teachers can use class discussions, student dramatizations of topics, and independent study projects, for example, affording gifted students different ways to learn and utilizing broader ranges of abilities and learning styles. The Multiple Menu Model of curriculum development's Instructional Sequences Menu is a rubric teachers can use with any teaching technique/strategy to help them present lessons in the most effective order.

Traditional vs. Progressive Models of Curriculum Development

In developing curricula, teachers must select topics that illustrate a subject's fundamental principles and characteristic concepts. Both traditional models of curriculum development and more progressive ones, such as the National Research Center on the Gifted and Talented's Multiple Menu Model of curriculum development (Renzulli et al, 2000) take into account the students' ages, maturity levels, prior knowledge, and levels of experience. These methods differ in that traditional instructional models typically require teachers to cover a whole textbook with their classes by the end of the term or year; however, the Multiple Menu Model asks teachers instead to narrow down, from all potential texts, chapters, or other sources of information, those few that best represent the concepts and principles of the subject area. This model's authors recommend using a three-phase approach to choosing content: intensive group coverage, extensive group coverage, and intensive individual or small group coverage.

Curriculum Units with Differentions for Academically Diverse Students

Some research groups interested in determining which methods of identification, testing, curriculum, and instruction are most effective for gifted students develop curriculum units to use in their studies. They then compare test scores of students taught using different curricula. Some researcher-developed curriculum units feature differentiated instruction in order to tailor teaching to the individual needs of "academically diverse students." These units place emphasis on problem-solving skills, conceptual thinking, and "real-world disciplinary inquiry" (e.g. using the scientific method to investigate subjects in various fields applied to real life). Curriculum units also function as tools for assessing individual students' learning needs. In addition, they help students attain progressively higher levels of expertise. Researchers believe that their work will foster both fairness and achievement in school systems nationwide and will nurture student gifts by focusing on gifted identification, model-based math and reading curriculum units, and traditional and performance-based assessments.

Program Evaluation

Evaluation plans are crucial elements in the provision of services and programs for gifted students. Formative evaluations are made during the implementation of a program for educators to see that their programs accomplish their intended goals. Summative evaluations are made following full program implementation to assess the extent to which the program attains its goals and objectives. To enable both formative and summative assessment of a gifted education program, the element of evaluation should be built into the original program plan. A common error made in some programs involves waiting to plan the evaluation until the program has been in use for a year or two. Educators may wait this period in order to ensure their program is fully implemented prior to assessment, but planning should be done initially to allow both formative and summative assessments and to assure that the services offered can be evaluated in both these ways.

Review Video: Formative and Summative Assessments
Visit mometrix.com/academy and enter code: 804991

Guidelines

Some of the guidelines given by the National Research Center on the Gifted and Talented for program evaluation involve the following. In program development, educators should incorporate evaluation procedures into the earliest planning stages. They should develop a specific plan for how they will use their evaluation results. They should use multiple sources such as students, teachers, parents, administrators, and school board members, for information to develop clear program goals and descriptions. They should allow enough time and funds for evaluations. They must also prepare/train their school staff to conduct assessments and analyze the results. Educators should identify clearly all parties needing, and/or interested in, evaluation findings, and engage them in the process of evaluation. Because the results of gifted education programs are complex, educators must find or create appropriate assessment instruments. Data collection methods should reflect the organization and objectives of gifted programs, e.g. portfolio assessments, out-of-level testing, product ratings with inter-rater reliability, etc. Educators should report evaluation findings timely to all those involved, including follow-up recommendations.

Acceleration

Positive Effects and Forms of Acceleration

Although acceleration must be decided for each individual student, research finds many benefits of acceleration for gifted students. Those students who exhibit advancement for their age far beyond their grade can even develop poor study habits, behavioral problems, become bored and daydream in school, or avoid attending school if they are forced to stay with their age-grade level. Especially in math, science, and English, some gifted students are more gifted than in other subjects. Such students may attend class with higher grades in their most gifted subject(s), but stay with their age's grade level for other subjects. Another form of acceleration is tutoring, either individually or in small groups. Some high school students gifted in math, for example, might be in their school for all other subjects, but a group of them would attend advanced math classes with a local university's math professor a couple of days each week.

Addressing Concern That Acceleration Can Hurt Students Academically

For those who worry that accelerating a gifted student will cause academic harm, they should be reassured that the majority of research shows that accelerated students get higher grades than gifted children who are not accelerated. Also, their grades compare positively with grades of the older students in the class. Researchers find that gifted students who accelerate report greater enthusiasm for and an interest in school as well. Another academic concern is that students who skip grades in a subject or skip whole grades will have gaps in their knowledge. This does occur, but less often than people think, because regular curricula contain so much redundancy. Gaps also do not present substantial problems for gifted children as they learn more quickly and thoroughly and can catch up more easily. Accelerating gifted students should not be penalized for unfamiliarity with any missed material, but should be allowed to cover it.

Addressing Concerns Regarding Effects of Acceleration on Child's Wellbeing

The most common worry of parents and educators about accelerating a gifted student to a higher grade is the impact it might have on the student's emotional and social development and wellbeing. They fear that while the child's intellectual development is undoubtedly at the higher grade level, his or her social and emotional development might lack the same levels of advancement. They worry about the child interacting with older children. However, researchers find children who are emotionally well-adjusted and socially comfortable before acceleration will report after accelerating that they have two circles of friendships: one from their newer class of older students, and another from their previous age-level class and/or other children their age, with whom they maintain friendships. While gifted children who have trouble making friends may experience difficulty in older classes, many gifted students prefer friendships with older children, finding more in common with them due to their advanced development in many areas.

Attitudes of Educators Toward Academic Acceleration

The majority of research studies show that academic acceleration for gifted students affords them many benefits in both academic and social domains. However, despite these positive findings, most educators are reluctant to adopt or even consider acceleration as an option for fear that acceleration will result in problems in the students' emotional and social development. In contrast, educators who specialize in the education of the gifted student, and teachers and parents who have had personal experience with the acceleration of gifted students tend to have much more positive attitudes about accelerating and are more willing to select this option when a gifted student's profile indicates it. Pushing a student to accelerate can cause problems, as can accelerating an emotionally immature student or one continually getting negative responses from peers or educators. Having a small support group of similarly accelerated students eases the acceleration process.

Students' Perception of Their Status Relative to the Class or Group

Many gifted students in regular classes at their age level perceive that they are the smartest student in their class. When they are far enough beyond grade level to accelerate, the student often finds after acceleration that they are not the smartest in the group anymore. This change in status can require considerable adjustment, especially for students who have become accustomed to everything being too easy for them. In this situation, parents and teachers should not pressure students trying to adjust to unfamiliar circumstances. They should give them emotional support rather than pushing too much too soon for academic performance. They should also make clear to the student that a choice to accelerate can always be reversed if it seems unsuitable emotionally, socially, and/or academically for them. Adults should also help gifted students understand that reversing acceleration does not represent a failure on the part of the student.

Criteria and Process

If a student's scores on standardized achievement tests are many grades above age/grade level, or are so high that they are off the tests' charts, the student is a candidate for acceleration. If a formerly enthusiastic student begins exhibiting boredom and/or behavioral problems not due to other factors, this is another criterion. The student, parents, and school personnel ideally should all agree that accelerating would benefit the student. Early in the process, the IEP team and school psychologist should be involved. Classroom teachers should be consulted, as well as others who know the student best, e.g. the school's gifted and talented coordinator, principal, and/or guidance counselor. The school psychologist should evaluate the student relative to acceleration. Parents can consult state Department of Education coordinators of gifted education. Support from involved teachers and coordination and continuity from school officials is important. Acceleration decisions are usually made by the principal and/or IEP team/committee.

Considerations Regarding Acceleration in Only One Subject

On one hand, many educators are more receptive to accelerating a gifted student in one subject because the student learns the remaining majority of subjects with age-level classmates. These educators often feel this type of acceleration can avoid social and/or emotional problems for the student. On the other hand, certain factors must be considered when a gifted student is accelerating in only one subject. One is continuity. If educators do not carefully coordinate the student's programming, the student could learn accelerated materials one year, only to repeat these the following year. Teachers and curriculum specialists can help by defining what is taught in each grade for each subject. Another consideration is that elementary school students might need to attend a junior high/middle school, junior high/middle school students a high school, and high school students a college course, to accelerate in one subject. Transportation can present worse problems than academic or social concerns.

Examples of Testing Policy

In some districts, rules differ for a student's first and subsequent years of subject acceleration. In the first year, if a test exists at the accelerated-subject grade level, testing is required for students whose overall/age-level grade is third-grade or higher; whose accelerated-subject grade-level is fourth-grade or higher; and the accelerated subject is reading or math. Testing is more likely optional if the accelerated subject is science, social studies, or writing; and if the student's overall level is second-grade or lower and the accelerated subject is third-grade or higher. If no test exists at the accelerated grade level and one exists at the student's overall grade level, testing may be optional. In subsequent years of acceleration, if a test exists at the accelerated grade level, testing is usually required. If no test exists at the accelerated-subject grade level and one exists at the student's overall grade level, districts may prohibit testing.

Examples of Provisions in Acceleration Plan to Skip Grades

If the school has determined a gifted first-grader can do third-grade work, the written whole-grade acceleration plan should include such provisions as these: The third-grade teacher should report and discuss any signs of academic and/or emotional stress in the student to the school staff person assigned to monitor plan implementation. The monitor should check with the third-grade teacher weekly on student progress. A gifted intervention specialist and/or third-grade teacher should obtain/develop curricula and assessments that compact only the second-grade standard the student has never experienced to make up anything he will miss by skipping second grade. He will complete these standards/skills during free time at school and/or home with his parents helping him as needed to master them. Upon successful transition, he remains in the accelerated grade until additional acceleration options are indicated. By middle school, the student's program should be reviewed to see if dual-enrollment options might benefit him.

Curriculum Compacting

Compacting is a way to streamline regular curricula for gifted students who learn faster. Compacting also can fill instruction gaps for students accelerating to higher grades and for those using the time saved by compacting to pursue enrichment activities. Sequential steps in the curriculum compacting process include: identifying pertinent learning objectives for the subject or grade level; locating or creating pre-test instruments for these objectives; identifying student candidates for compacting; pre-testing these candidates to ascertain their learning levels of the identified objectives; get rid of teaching, drilling, and/or practicing time for students who have already mastered the objectives; for objectives students have not mastered, streamline their instruction for the gifted students' faster rates of mastery; offer options of enrichment or acceleration for students who have exhausted the regular curriculum; and maintain records of the entire process and of instruction options available for students participating in compacted curricula.

Virtual Learning

Thematic Units

Teachers use thematic units as a way of incorporating complex, abstract concepts into the curriculum. Thematic units are organized around a particular theme or topic. This organization provides a focus for learning abstract ideas and appeals to areas of student interest. One way that a teacher can employ both thematic units and information technology is to select the content for an integrated thematic unit and then develop a virtual learning environment (VLE) for teaching it. Today there are many Internet websites containing teacher resources, including those for thematic units. For example, the site A-Z Teacher Stuff offers thematic units on topics such as magnets, oceans, Martin Luther King Jr., Harry Potter, and many other themes, including lesson plans for teachers, class activities, etc. The website of the C.O.O.R. Intermediate School District in Michigan provides additional examples of integrated thematic units.

Websites

Once teachers have selected the content for a lesson or unit they want to deliver in a virtual (computer) environment, they can take advantage of such websites as The Educator's Reference Desk, where they can find lesson plans for any grade level or school subject. In addition to getting help creating and delivering lesson plans, teachers can incorporate hyperlinks from these sites in the virtual learning environments they create for students. Students can follow these links to access virtual lessons providing content enriching the teacher's basic lesson or unit. In addition to lesson plans, teachers can find online student projects and activities, instructional materials, and virtual "field" trips. In virtual trips, students can visit otherwise inaccessible nations, museums, galleries, institutions, organizations, and corporations all over the world. Education World, Eduscapes' Digital and Virtual Museums, and the Museum of Science and Industry are examples of websites offering virtual museum tours.

Fostering Creative Performance of Gifted Students

Educational researchers find that modifying curriculum, using alternative materials, encouraging brainstorming, and teaching strategies for thinking "outside the box" are all necessary, these strategies alone are not sufficient. Teachers, students, parents, and administrators need to collaborate to affect changes in the entire climate of our schools at every level, as well as in the environments of individual classrooms. Some gifted education programs developed in recent decades can be individualized to meet the needs and interests of each specific student. Instead of identifying only students who excel in traditional school subjects, programs taking this alternate approach acknowledge student talents and strengths across a broader range of dimensions. Researchers find the motivation of gifted/talented students is especially susceptible to classroom environmental influences, even more so among the culturally diverse and/or economically deprived. Experts call for educational reform that does not require money so much as commitment to change and willingness to collaborate to make school environments nurture internal motivation and creativity.

Self-Contained Classes

The primary value of self-contained classes for gifted students is the academic rigor possible when all students in a class are at similar intellectual levels. The general classroom teacher will not have to differentiate instruction specifically for gifted students by offering more in-depth study of subject content, materials at higher grade levels for acceleration, compacted curriculum, alternative activities, etc., if no gifted students are present. However, one inherent pitfall is that some general education teachers, relieved of the necessity to differentiate instruction for gifted students, may fail to differentiate instruction for their existing students. Such failure reflects that these teachers do not understand (or observe) that differentiation is intended to customize teaching for the needs of every individual student, not just every individual gifted student. Varying difficulty levels, paces, prior knowledge, interests, and learning styles also require differentiation among non-gifted students.

Ability Grouping

Ability grouping should not be equated or confused with cluster grouping, heterogeneous grouping, or tracking. Ability grouping involves dividing the students within one classroom and grouping them according to their ability levels, which can differ broadly within one general education classroom. This grouping by ability is not rigid but fluid, and it can be changed as needed at any time. For example, a student might be placed into a high ability group for math and participate in a group of middle ability for reading. If that student's reading improves, the student could move into the high ability group for reading. Conversely, if a student develops problems in an individual subject, s/he can move into a lower ability group to address his or her academic needs best. The flexibility of ability grouping allows gifted students more options to receive the most appropriate instruction and to adjust for needs that change over time.

Independent Study

Independent study gives gifted students encouragement and preparation for initiating, implementing, and successfully completing their own scholastic activities in their areas of interest. As such, it provides them with a structural context for making choices and decisions about what and how they learn. It also teaches them how to communicate what they have learned. When gifted students have independent study projects, these also benefit teachers by helping them to fulfill the roles of facilitating students' progress, listening to students, and clarifying ideas and objectives. Many gifted students underachieve; this can be related to lack of interest in assigned topics, frustration at not being able to study subjects in more depth, boredom with the mainstream class pace and/or types of assignments. Independent study of a topic a student is passionate about can alleviate these factors and increase task commitment. Planning and executing projects stimulates inquiry, builds motivation and self-discipline, improves study skills and habits, and promotes positive attitudes and self-concepts.

Teachers should ensure that any independent study program for a gifted student will meet their individual instructional needs as well as their particular personal interests. In self-selected studies, the gifted student chooses a topic s/he wants to study that will meet these needs and interests. In related studies or "spin-off" studies, the gifted student is acquainted with a subject in the usual teacher-directed class lessons, and then selects a topic within that subject that particularly interests her/him to investigate in more depth and specificity through an independent study project. In curricular studies, the teacher will have made a list of topics based on the subject content for the whole class to study. Gifted students can select one topic from this teacher-made list and conduct an independent study of it. Curricular topics may be studied independently by individual gifted students; or they may also collaborate on these in small group settings.

Purposes and Positive Outcomes

Independent study projects allow gifted students to explore a variety of subjects and areas in which they are interested. Planning, developing, and carrying out an independent study project can serve as the student's introduction to the methods used to organize, research, and present the results obtained to others. With guidance and support from teachers, independent study can encourage gifted students to develop individual qualities of initiative as well as their intellectual and creative abilities. Independent study activities stimulate gifted students' skills in critical thinking, logical reasoning, and continuity in focus. By working on independent study projects, gifted students develop a sense of personal responsibility for pursuing their goals. In addition, when they pursue and successfully attain their goals through independent study, gifted students attain a sense of satisfaction and achievement.

Process

The gifted student chooses a topic of special interest and reads extensively about it, developing a knowledge base. With a general grasp of the topic and knowing what s/he wants to discover, the student asks questions to answer. Students utilize and develop higher-order cognitive skills as they learn to identify and use resources in the forms of places and persons, establish goals and objectives, develop and carry out plans to accomplish these, make self-evaluations of their own work, and identify avenues whereby they can communicate their discoveries, conclusions, and knowledge. Students can use fact cards to record information they collect on the topic. By analyzing the information they gather, students can develop questions reflecting a problem to solve; for example, the student can ask what methods s/he could use to get a specific species of wildlife off the endangered list. This process demands student analysis of information, creation of novel and unusual plans, and expression of their opinions. It also involves them in authentic research, products, and audiences.

Categories of Research

One category of research that gifted students can use for independent study is historical research, which endeavors to reconstruct the past objectively. Some sources for gathering information include old newspapers, written records, oral history, baby books, folk songs, poems, and stories. Students could study the history of the school, the community; some aspect of the US Constitution or Bill of Rights, etc. Another category is developmental research, which examines trends, change, continuity, sequences, and patterns. Developmental psychology and human growth and development are examples of developmental research areas. Students can investigate such topics as human biological development, cognitive development in children, personality development over the lifespan, etc. Another category is descriptive research, which is quantitative. Numerical statistics are depicted using graphs, charts, and tables. Students can conduct opinion or fact-finding surveys, or observational or interview studies.

In experimental research, the researcher seeks to discover whether one variable causes changes in another variable. Statistical methods of multivariate analysis can also be applied to study the effects of and on multiple variables within the same experiment or set of experiments. For example, gifted students use independent study projects to research whether a plant responds differentially to various stimuli or which combinations of substances will produce a certain chemical reaction. Experimental studies involving human subjects include such research questions as whether the amount of sleep the night before an exam affects student test performance, whether one school curriculum is more conducive to student motivation than another, or whether internal or external rewards have greater positive influences on student performance. Experimenters establish a treatment group receiving the intervention being tested, and a control group receiving no treatment to control for extraneous variables. By manipulating the independent variable being tested, researchers can see if it affects the selected dependent variable(s).

Correlational research is a type of quantitative research method that compares quantitative data obtained to discover whether a relationship exists between two or more variables. Specifically, it seeks to find whether one variable affects another variable and how. A positive correlation means that as one variable increases or decreases in amount or number, so does the other variable. A negative or inverse correlation means that as one variable increases, the other decreases, and vice versa. Statistics also show the strength of the correlation, or the degree of relationship. Some examples include whether the seasons affect school attendance, whether an individual's head size and foot size are related, or whether school attendance affects achievement test scores. Correlations do not imply causations: two variables may increase or decrease together or oppositely, but this does not mean that either variable causes the other. To determine causation, experimental research must be used.

Enhancing Instruction Via Online Educational Games

Gifted and talented students tend to enjoy activities that they find stimulating and challenging. One way in which teachers can enhance their education is by offering students online educational games they can play wherein they apply the concepts they have learned in a lesson or unit. Online gaming has become so popular that it is a good choice for gifted students to extend and apply their knowledge and skills while having fun. For example, the About.com website has action games that address the content of various school subject areas. The Education World website's Online Game Archives has such games for students as Wacky Wordplay and Math Bingo. For teachers to evaluate student progress toward learning objectives they have developed for a unit, and evaluate the objectives themselves, they can find free assessment rubrics at websites like RubiStar. Teachers can customize RubiStar's rubric templates, including for assessment of virtual learning lessons/units.

Issue of Homework

Gifted children commonly start out in regular classes. Even after identification as gifted, the majority of gifted students are in regular classes most of the time with a small portion in supplementary gifted services like pull-out programs. They can end up with twice the homework: regular class assignments, plus gifted education assignments. Even students receiving accommodations in regular classrooms can receive double homework unless specific accommodations substitute gifted for regular homework. Having doubled work feels punishing, making students resent being identified as gifted. The excess can mean not doing all homework, damaging grades. Solutions include assigning more advanced work taking the same time but applying lesson knowledge instead of repeating it, which gifted students often do not need. For example, a student could do ten advanced math problems instead of 100 rote problems. Instead of writing simple sentences including spelling words, a student could write a poem or short story using them.

Individualized Education Programs

IEP for Gifted Student in General Education Classroom Setting

Gifted children do not have gifts only part of the time; they are the same individuals all the time. As a result, their educational needs also are constant. Such rationale also extends for addressing those needs throughout the school day/year. Each school district has the responsibility not only to place a gifted student in a gifted program, but also to make whatever accommodations and modifications in the regular education setting, curriculum, and instructional methods are necessary and specified in the gifted IEP (GIEP) to meet the student's individual educational needs. Parents should know that if district administrators tell them a gifted pull-out program is the "only" accommodation available, this constitutes a severe misrepresentation of the district's responsibilities and the student's and parents' rights. Parents should inform administrators they are aware of these responsibilities and rights and should hold administrators responsible for addressing and fulfilling these rights. Parents can advocate for their child at GIEP meetings and/or request due process hearings.

In general, the gifted IEP is a guide for the teacher to help focus instruction on the student's educational needs and goals. It is not a rigid directive forcing the teacher to adhere to specific activities determined months before instruction. At the beginning of the school year, the student's gifted IEP should establish the student's main educational goals and a schedule within the school year for their attainment. Because goals are more general than short-term learning objectives, they allow time for teacher and student to collaborate in determining a school subject for a given goal that is compatible with the student's abilities and preferences. After arriving at a subject, the teacher and student can collaborate further in developing individual short-term learning objectives according to the schedule previously established. These objectives specify the smaller increments/steps toward achieving a goal, permitting the teacher to adjust to the student while also assuring progress.

Considerations for Development of IEP

In part because federal law addresses special education needs for disabled students but not for gifted students, in part because of emphases on inclusive education, and for other reasons, most public schools do not have many (or any) classrooms with all gifted students. Many schools have pull-out gifted programs, which comprise a very small part (about 5%) of the gifted students' school hours. Thus, the general classroom should be the focal area for modifications or accommodations included in a gifted IEP. In addition, if any school problems arise in gifted students, such problems are likely to arise in regular classrooms. Gifted students who learn more quickly and/or with less repetition can become bored and/or impatient with the class pace and come to dislike school. Although such problems may be minor in elementary school, they still should be addressed early to prevent the development of larger problems by high school. Simple general classroom changes in gifted IEPs can make tremendous differences.

team to find ways to meet it. Parents also can advocate for courses initially not offered by the school or not at their child's grade level.

For a Gifted IEP (GIEP) meeting, educators typically will prepare some provisional educational goals, courses, and activities they deem appropriate for the student. Furthermore, educators should avoid using technical jargon so parents can understand it. Parents should bring their ideas to the meeting. They can inform educators of their child's interests, likes and dislikes, and behavior, and bring suggestions for educational programs and/or materials they want for their child. They also should ask questions about anything they do not understand or if they need more information. When educators present parents at the meeting's end with a document to sign, often called a Notice of Recommended Assignment (NORA), parents should not assume they must sign it immediately. State laws allow a number of days for them to decide. Parents can request additional and/or amended wording. They can also write exceptions to parts they do not accept, and then sign, or they can reject the entire GIEP.

PLEP and Assessment

Generally, objective assessments are needed to establish a student's **Present Levels of Educational Performance** (PLEP). While subjective teacher observations, or checklists showing student strengths and weaknesses in the curriculum, are helpful for developing a gifted IEP, they should not be used to define PLEP, which requires objective testing instruments for accuracy. Ideally, such objective measurements are taken at least annually, or more often as needed. However, in the reality of public schools, this schedule is seldom followed with gifted students. Gifted IEP teams often wrongly use report cards, which measure learning of material taught in the past rather than in the present, and do not indicate student ability—a student may get As in a class which is still three grade levels below his/her capacity, or a student may be so uninterested in this lower level as to get Cs or even Fs. Either way, the student would not be functioning at his/her present level of educational performance.

Goals and Objectives

Goals are the long-term targets for a student's education, while objectives represent short-term targets, but this is not the only difference. Goals need not be specific or measurable, whereas objectives should be both specific and measurable. While goals are more general and global, they still must be specific to the individual student rather than assigned to that student based on his or her class or group. First, identifying the individual student's educational needs and the final outcomes desired are the ways of determining the student's educational goals. Short-term learning objectives then specify the smaller steps the student must achieve toward eventually meeting the associated goal. Objectives are also the IEP team's means of measuring student progress toward a goal. Objectives quantify criteria for attainment and state when these criteria are to be met. Goals and objectives should be congruent with one another: when read together, they should "make sense" to parents as well as to educators.

Four Main Parts

When developing a Gifted IEP (or any IEP in general), the first part should indicate the student's Present Levels of Educational Performance (PLEP). Because asynchronous development is characteristic of gifted students, they have relative strengths and weaknesses as other do students, and they also may be extremely gifted in specific areas. Therefore, the PLEP must reflect school performance in all school subjects. This establishes a baseline wherefrom subsequent educational programming proceeds. The second IEP part should state the student's educational goals. Goals are more general and global; they do not include specific ways of achieving them. A third IEP part includes short-term learning objectives. These objectives do specify the time frames in which to achieve them and exactly what measures will demonstrate the learning. These are steps in achieving goals. A fourth part outlines specially designed instruction for the gifted student and explains teaching methods for attaining student short-term outcomes and long-term goals.

Above Grade Level

By itself, the phrase "above grade level" contributes no meaning to establishing a gifted student's Present Levels of Educational Performance (PLEP) or to developing that student's IEP. Some gifted advocates opine this phrase should not be included in gifted IEPs. If it does appear, both the student's current grade level in school and the grade level at which the student performs must be defined. In some US states, State Departments of Education require these determinations by law. Objective, standardized achievement tests have established norms for each grade level, so these definitions can be determined through test scores and norms. The description "above grade level" without such specific grade-level information is analogous to going clothes shopping for a child with only the information that s/he is "taller than four feet." The child's actual height must be measured; so must the student's actual performance grade level be measured, as well.

Relationship of General Education Teacher to GIEP

If a gifted student is in a general education classroom for any part of the school day, the general education teacher needs to be familiar with the content of the student's GIEP. None of the privacy laws protecting the privacy/confidentiality of student records applies to either general or gifted education teachers' access to the GIEP. Some school districts have their own privacy procedures, such as having general education teachers sign a log to read a GIEP, but teachers are still allowed to read the GIEP. They cannot align with the GIEP's goals, objectives, activities, and materials without knowing them. Familiarity with the GIEP facilitates general teachers' coordinating their instruction with gifted teachers/programs. If parents believe their child's teacher does not know about the GIEP, and/or that their child is identified as gifted, they should ask the teacher. Bringing their child's gifted written report and GIEP to parent-teacher conferences is advisable.

General Education Teacher's Involvement

Advocates observe that engaging general education teachers in the GIEP process represents the best way of developing and implementing a GIEP. The general education teacher should attend the GIEP meeting. Since a teacher familiar with the student always should attend IEP meetings, the primary general education teacher also is likely to be on the IEP team and attend the meeting. If a school/district does not include the general education teacher, parents can request the chairperson of the GIEP team or local educational agency (LEA) to include her/him. All teachers must comply with the GIEP. If a teacher refuses to implement the whole GIEP or any part(s) of it, e.g. pre-testing, giving differentiated classwork and/or homework, etc., parents either should call for another Team meeting or pursue due process proceedings. Teachers can make written objections to a GIEP on record. However, team-proposed, parent-approved GIEPs always must be implemented.

SDI Section

The **Specially Designed Instruction** (SDI) section of the GIEP is, in general, the section where classroom activities are designed to match the individual gifted student's needs. Any specific educational challenges or issues for that student, including any problems the student has in the general education classroom, should be addressed here. Any educational programming that the student and his/her parents think should become part of the school district's gifted program should also be included in the SDI section. Three considerations relative to general education teachers and GIEPs are:

- gifted education and general education for gifted students are not separate entities: gifted students' needs must be met in BOTH district gifted programs AND general education classrooms
- general education teachers cannot refuse to implement GIEPs; parents have recourse (i.e. another GIEP meeting/due process) if teachers do refuse to implement
- general education teachers have access to all parts of GIEPs

Instructional Accommodations That Can Be Requested

The only constraints on accommodations one might request in school for a gifted student are that such accommodations be reasonably expected to provide the student with meaningful educational value using the school's curriculum. Any accommodations meeting these criteria can be written into a student's gifted IEP. A common response from school/district administrators for students, parents, and advocates to consider is that no one ever has asked for a certain accommodation before, or no others are requesting a particular change. This happens more frequently in schools using only pull-out gifted services and/or those schools not normally providing accommodations to general education classrooms. Administrative responses even can give students/parents/advocates the impression that their request is inconvenient and/or unnecessary. However, rejecting a request simply because it sets a precedent is not valid. The only criteria for validity are the reasonable expectation of meaningful educational value and its educational appropriateness for the individual student.

Differentiated Homework

Since gifted students most often are placed in general education classrooms and given supplementary gifted services, they could be assigned the same homework as the regular class, plus another set of homework from their gifted program. Thus, gifted students can end up feeling punished rather than rewarded by having to complete a double amount of homework. Differentiated homework can avoid this problem by substituting homework more appropriate to the student's gifts instead of the regular class homework rather than in addition to it. Differentiated homework also can eliminate excessive rote work/repetition beyond the gifted student's needs. For example, assigning math puzzles instead of memorization lets gifted students apply their knowledge. Assigning a gifted student an additional vocabulary list, with homework only on the extra list, and/or having a gifted student write two longer stories per week instead of five shorter stories are other examples.

Independent Study Agreements

In public schools, gifted students are not commonly enrolled in completely separate gifted programs/"tracks" where all classes differ from general education classes in order to accommodate the gifted students' learning needs. Such gifted students most often are placed in general education classrooms for part/all of the school day. They may have pull-out programs for gifted instruction, but such programs represent a minority of the student's school day/week. Due to asynchronous development, some gifted students are farther ahead of the regular curriculum in some subjects than in others. Also, some gifted students still may find the regular curriculum relevant, but they simply may finish their class work sooner. Independent study agreements offer gifted students ways to occupy their free time instead of waiting for the rest of the class to catch. They can work on one big project incrementally, or on a series of smaller projects. Such projects should be designed to meet goals written in the student's gifted IEP.

Pre-Testing

Gifted students often have mastered the content of a lesson or subject at their grade level before teaching has begun in that class. Making them sit through instruction in material they already know only bores and/or frustrates them and damages damaging their motivation; moreover, they are prevented from learning new things in the subject at a higher level and at their own pace. The first section of the gifted IEP (GIEP) specifies the student's resent Levels of Educational Performance. When the GIEP team determines these, they can determine the need for pre-testing. Students whose pre-tests demonstrate mastery of material can be given alternate assignments that challenge them and allow them to learn new material. Students whose pre-tests demonstrate they have almost mastered a subject can be given only the parts they have not mastered to study, thereby saving time and redundancy. These students can use the rest of their class time for independent work on alternative assignments.

Acceleration

Acceleration is considered a more serious choice than other forms of accommodations for gifted students. Considerations for the student include the student's levels of emotional and social maturity, which can cause adjustment difficulties if a student is placed in chronologically older classes and is emotionally and/or socially unprepared. Historically, students skipped whole grades. While some cases still warrant such practices, students also can accelerate in individual subjects according to their levels of ability and performance. They can even work across grades if their learning levels dictate "straddling" two grade levels. Though accelerating is not always best, research finds gifted students more often benefit from ability-appropriate materials. Objective data on the student's current performance and potential are important to decisions to accelerate. Many school districts frown on acceleration, but they cannot have a policy prohibiting it. Attitudes against acceleration should be challenged when acceleration represents the only way to benefit the student.

Procedures If GIEP Is Not Working and/or If Student Has Problems in School

GIEP meetings exist in order to review the efficacy of the plan, not the student. Accordingly, "The Plan fails the child; the child does not fail the plan." (McIntyre, 2004) Some students may have had a GIEP for years that has not met their needs. If a student is in danger of failing in any subject or in school, the GIEP meeting should be reconvened. GIEP team members discuss the problem, appraise the circumstances, discuss available alternatives, and arrive at a solution for the individual student, changing the GIEP accordingly. The team can even write a provision into the GIEP for problems, specifying that if the student is potentially failing a subject, is unlikely to meet a GIEP goal(s), or his/her performance becomes substantially worse, the GIEP team will reconvene, including the teacher of the subject involved, and the GIEP will be modified to address the problem(s).

The main purpose of the GIEP is to address the needs of the individual gifted student. By definition, a gifted student learns differently enough from others that standard teaching methods and curricula are unsuitable. The different nature of the gifted student's learning and needs is the reason for a GIEP. The nature of the GIEP is not a class(es) or program(s) in which the gifted student participates. Instead, the GIEP offers a plan for the student's education. As such, the GIEP must be monitored: the student's educational needs will evolve/change over time, and the plan then must be adapted to continue meeting them. If problems persist after modifying a GIEP, the team should revisit the student's Present Levels of Educational Performance, ascertain what additional assessment data and other information are needed, and develop a new, more suitable/effective GIEP. Teams should obtain outside expert consultations if needed.

Review Video: Assessment Reliability and Validity
Visit mometrix.com/academy and enter code: 424680

Parental Access to School Records and About IEP Planning Meetings

Parents have the right to request and receive copies of all of their children's school records, including the results of any tests. These may not all be in one place. For example, the guidance counselor might maintain records of students' IEPs and educational programs, while IQ and other test results would be maintained by the school psychologist. Gifted advocates advise that parents of gifted students should obtain copies of current tests used to identify their child as gifted and of tests used to determine the student's achievement levels. Common misconceptions among all parents, including those with gifted children, is that the school IEP team already has created the IEP, that it cannot be altered, and that they are attending the meeting only to review and approve/disapprove its content. As such, parents must be informed of their roles as equal team members in contributing to, developing, and/or modifying their child's gifted IEP.

Math and Science

Math

Disparity Between Approaches of Math Programs for Gifted Students

Many school programs for mathematically gifted students are designed with the goal of moving them through the curriculum as rapidly as possible. However, this approach of fast advancement does not cultivate a student's interest or passion for math. Zaccaro (2008) finds this analogous to teaching aspiring musicians "all scales and no music." Students made to practice scales but never allowed to play music that moves them emotionally will not develop passion for it. Practicing the structural fundamentals is necessary for mastery, but these fundamentals should not be taught exclusively. The same is true of math; students need to be given opportunities to work with material they find exciting and wondrous. Students can experience how the earth's circumference was determined 2500 years ago using simple geometry, apply knowledge of the speed of light to realize that viewing stars is time-traveling to the distant past, calculate ship-to-shore distances using trigonometry, and so forth.

Importance of Connecting Math and Science to Real Life

Since math and science are so often heavy in calculations and facts, students are usually not taught how these subjects relate to the real world. All students, especially gifted students, need to find meaning in an academic subject and understand how it applies in life. We may not appreciate enough that math and science are not like people whose opinions we can disagree with; they provide hard, objective, unchangeable facts. Teachers can show students the consequences of ignoring facts with examples like these: mathematicians and engineers advised not launching the Challenger space shuttle, but management overruled them, thereby leading to the deadly explosion. Pop singer Aaliyah died in a plane crash after pilot and crew ignored the mathematics indicating airplane overload and flew regardless. A mathematician proved racial bias in jury selection by calculating that the mathematical probability of fair selection was approximately 1 in 1,000,000,000,000,000.

Recognition and Reinforcement

Student musicians perform in concerts and receive applause from crowds of schoolmates and parents. Student athletes compete in games or meets, also attended by cheering crowds. In addition to internal motivations to use their talents, these students obtain powerful external reinforcement from this positive public attention. In contrast, students with mathematical and scientific gifts do not enjoy the same opportunities to show their skills regularly in such public arenas. Students gifted in academic subjects do not experience much formal recognition for their abilities outside of good school grades and teacher approval. However, such students also need to receive positive reinforcement in order for them to appreciate the great value of their gifts. They may also need to receive validation for their accomplishments, as some students may attribute their successes to their natural gifts and mistakenly feel that they have not added any effort to developing them.

Students gifted in certain fields such as music and sports, enjoy positive reinforcement received for performing in public arenas like musical concerts and athletic events where large groups of people, many of whom they know as fellow students and parents, applaud and cheer them. Students with gifts in math and science do not have the same opportunities for official public recognition. Educator Ed Zaccaro (2008) recommends "Einstein Awards." He designated very difficult math/science problems as "Einstein problems." When students solved one, he presented them in class with an award featuring a picture of Einstein. He reported dramatic responses from students: One parent told Zaccaro that his daughter said the award was the best thing that had ever happened to her. Another said she had to restrict her son to two hours of math per night as he always wanted to solve more Einstein problems. Simple positive reinforcement is a powerful motivator.

Mathematical Word Problems with Increasing Levels of Difficulty and Complexity

First level: A gas tank was ¼ full. You add eight gallons and it is now ¾ full. How many gallons does this tank hold?

Second level: Sound travels one mile in 5 seconds. You yell near a rock wall and hear the echo 20 seconds later. How far are you from the wall?

Third level: A sprinkler throws water in a circle 15 feet away. It is in the center of a lawn 30 x 30 feet. How many square feet does the sprinkler miss?

Highest level: Teachers can present class awards to students solving these problems for positive reinforcement, teaching them to appreciate their gifts, building self-esteem, and enhancing motivation. Alice, Barbara, Charlotte, and Deborah are sisters inheriting money from an aunt. Alice got ½ the money. Barbara got ¼. Charlotte got ⅕. Deborah got the rest, which equaled $1,750.00. What total amount did all four sisters receive?

Merits of Working with Other Similarly Gifted Students

In recent years, the federal government has passed legislation guaranteeing students with disabilities a free, appropriate public education, including special education services as needed to accommodate disabilities in order to provide this education. This legislation has led to great emphasis on inclusion in education. However, the laws do not address students with gifts. Educators still apply the notion of inclusion to gifted students in that they should experience working with non-gifted students. This is true, but it is not a big concern as most gifted students in public schools are in regular classrooms most or all of the time. Equally important is the need for gifted students to be able to work with similarly gifted students. Such collaborative work not only promotes their cognitive development. It also contributes to their emotional and social development through sharing ideas, debates, and constructive disagreements and through decreasing the social isolation that many math-gifted children often experience.

Example of Single-Subject Acceleration Plan in Math

For a first-grader able to learn third-grade math, the school's single-subject acceleration plan could include the following provisions: The first- and third-grade teachers will schedule math at the same times. The third-grade teacher is the teacher of record for math. The student is not required to do any first-grade math assignments. A school staff member assigned to monitor the plan meets weekly with the student to assess her success and satisfaction and weekly with the third-grade teacher to discuss any areas of difficulty for the student. Throughout the transitional period, the student's parents are lent copies of the first-, second-, and third-grade math textbooks so they can help her at home with studying and homework. Until she can make new friends in the third-grade class, the student is assigned to sit next to a student in the third-grade class who is her friend and her next-door neighbor at home.

For a gifted first-grader placed in a third-grade class for math who has successfully completed the transition phase, the section of the school's plan to support her continuing progress could include provisions for the following strategies: In fourth grade, arrange for her to take an online math course at the sixth-grade level, and another at the seventh-grade level when she is in fifth grade. The plan can incorporate an option for her to complete an Introductory Algebra course when she is in seventh grade, or sooner if she is ready for it. Satisfactorily completing the online courses and Algebra/other educational options in the plan will be treated as the equivalents of completing traditional courses as prerequisites for enrollment in advanced high school-level math courses. The plan can provide that the student receives high school credits for completing Algebra and other high school math courses while she is in middle school.

Example of Single-Subject Acceleration Plan in Science

For a fifth-grader accelerating to sixth-grade science, the plan can provide: The fifth-grade teacher and/or gifted intervention specialist will find/create curriculum that compacts assessments on fifth-grade science standards unfamiliar to the student, which she completes in free time at school and/or home with parental help. The staff person assigned as monitor meets weekly with the sixth-grade teacher to discuss student progress and weekly with the student to discuss her adjustment throughout the transition. The student uses the elementary school's webcam/Skype/similar software for virtual attendance to sixth-grade science class at the middle school. The technology coordinator gives student and teachers a short tutorial to facilitate this. After transition, to assure ongoing progress, dual-credit courses in 7th and 8th grade are arranged so after one high-school semester, the student becomes a sophomore eligible to take state graduation tests that spring. High school credit is awarded for completing 9th-grade science in 8th grade.

Social Studies

Social Studies

Provisions for Acceleration from Intermediate to High School Classes and Credit

Some state Departments of Education (New York, for example) allow gifted eighth-grade students to accelerate in social studies for high school diploma credit if the superintendent or designee ascertains the student's readiness. Eighth-grade students can be awarded this credit by the high school if they attend the high school with high-school students and pass the course with the same requirements as the high-school students. Alternatively, an eighth-grade student can gain high-school credit by passing the state proficiency or Regents examination, and the credit is accepted as a transfer credit by any high school in the same state. If the state has no proficiency examination or similar assessment in social studies, an eighth-grade student can pass a course in his or her middle/junior high/intermediate school that is approved for high school credit by that district's superintendent or designee. Local examinations indicating school principal-determined high school-level performance may also be used.

Comprehension Skills Included in Enriched Social Studies Skills

In social studies, students should be able to attain and demonstrate the following: recognizing cause and effect relationships in social studies topics; making comparisons and contrasts among

Teachers can help students prepare for such understanding by instructing them to anticipate the central concept in the text. Before they begin reading, the teacher can ask students to skim the text first to develop an educated guess regarding the main idea. The teacher should point out to students such clues as the title of the book, chapter, article, or essay; paragraph headings; terms and/or names that are repeated often; etc. The teacher and class should review all of the students' predictions of the key message or concept. They can revisit these predictions after the students have read the text. Then the students can identify which clues they gleaned from skimming the text were useful and which clues were not.

Teaching Strategies Using Vocabulary, Associations, and Questions

Social studies teachers can help their gifted students prepare for a reading assignment by giving them a chance to preview the academic vocabulary or terminology used in the text. For example, the teacher can use a "word splash," a collection of key terms used in a section of text arrayed on a page with the most central concept or term in the center and related terms around it. The class can discuss these terms for comprehension before reading and for recognition during reading, and the teacher can post them on the wall or board. To make associations or connections, teachers should have students ask themselves what they think they know about the subject before reading. This achieves the dual purposes of uncovering student biases and/or misperceptions, establishing a context for the reading and making them feel familiar with the subject, and stimulating their engagement and interaction with the material.

Development of Critical Thinking Skills Through Reading of Subject Content

Gifted students need to develop their skills for analyzing, synthesizing, and evaluating subject content that they read. Students need to realize that anything in print was written by someone. Teachers can help their gifted students discover who wrote or published the essay, article, textbook, eyewitness report, or primary document they are reading. Furthermore, teachers can ask the students to ascertain the author's originally targeted audience, why the author wrote the material, and what purposes the author hoped to accomplish by writing it. Teachers can ask their students to draw inferences about the author's intended audience. Such an activity will help the students learn how to select the best research sources for their own class and independent study projects. Developing their skills in critical reading will help students develop their general critical thinking skills, as well.

Five W's, Comparisons and Contrasts, and Cause-and-Effect Relationships

When their social studies students are reading journalistic content such as newspaper or magazine articles, teachers should include in their assignments the identification of the "Five W's" of journalism: Who, What, When, Where, and Why. Answering these questions helps students focus on the key elements of persons, actions, times, places, and reasons in historical or current events. Teachers should call students' attention to comparison/contrast, showing how the author saw similarities and/or differences between/among events, actions, or situations, and why the author found these similarities/differences important. Teachers can show how authors describe cause-and-effect relationships with qualifying phrases like "as one result...", "in part because of...", "this helps to explain...", etc., and then have students compile lists of such qualifiers. Teachers can help students explain authors' causal arguments without agreeing with them and differentiate opinion from fact by asking students how the author explains the causes of an event, rather than asking them what the causes are.

Directing Attention to Questions and During-Reading Strategies

Students learn more from their reading if teachers instruct them to use what they read to answer questions. This focuses their attention on the reading and helps them to apply it. Teachers should encourage advanced readers to think of their own questions and try to answer them through their reading. Many gifted students will already have formed their own questions out of their curiosity about the subject. For gifted students who are younger and/or not as proficient in reading, teachers also can create their own questions and use them as an outline for reading. This will help students to focus their reading to identify key points in the text. Teachers can write these question outlines to focus on the subject content or on student reading and study skills. A teacher also can identify during-reading strategies specific to the textual material to help students self-monitor their comprehension.

Analyzing Writing Styles in Assigned Reading Material

Gifted students should understand and apply both the content and the writing style in their social studies reading. Teachers can show students various writing approaches in written materials. For example, some authors first will establish the context for a historical event and then recount its details in chronological order of occurrence. Others may open by relating an anecdote and then explain how it illustrates or is related to the topic. Reporters often use a common journalism technique of opening an article with a paragraph summarizing the key points of the report and then filling out these key points in subsequent paragraphs with more details and/or quotations of comments or responses from involved people whom they interviewed or those persons they consulted as expert sources. Teachers can stimulate discussions by asking students to project how each of these styles might influence different reading audiences. They can also provide challenges to gifted students by having them apply the styles in their own writing.

Sequential and Causal Relationships, Graphic Organizers, and Paraphrasing

Social studies teachers can point out to their gifted students that when authors of reading materials use terms like "...and then...", "next," "subsequently," "thereafter," "later," and "finally" to represent a string of events, this indicates a chronological sequence, but it might or might not also indicate a cause-and-effect relationship. Teachers should advise their students to seek other clues in the writing before they assume that one event or action actually caused the next one. After reading, students still may need support to figure out the author's primary message or argument. Graphic organizers such as concept maps, flow charts, or outlines help students to visualize verbalized ideas. Teachers can have students paraphrase what they read in 3-5 written or spoken sentences to show their comprehension. Paraphrases should include the subject/topic, the main idea, key terms, and the most important details.

Post-Reading Teaching Strategies to Foster Critical Thinking

After they have read assigned social studies texts, students will benefit if their teachers have them list the most significant points in the material and rank them in order of importance. This will help students clarify priorities among logical points/reasons in an argument, enhancing critical thinking skills. When the time sequence of events is significant, teachers can assign students to list 5-10 chronological events the author has cited. Teachers can help students identify an author's viewpoint by giving them statements to identify as true or false according to the author and having them cite specific pieces of text as the bases for their choices. The teacher also can use true-false statements to help students distinguish authors' opinions from facts. Regarding important issues, teachers can have students evaluate the author's argument, giving sufficient time and teacher guidance. This can motivate further reading and research: in the evaluation process, students will want to consult additional sources.

Examples of Questions to Facilitate Exploration of Multiple Perspectives

Educators advise social studies teachers against adhering to a single source or a narrow definition of sequences of historical or current events. Instead, they recommend encouraging student appreciation of a breadth of perspectives within the subject's topics. Teachers can do this by asking students to consider topics such as: from whose viewpoint an account was written; whether other views or interpretations might exist, and if so, why; whose voices are identified in the account and whose voices are not; what evidence is given in the account for its assertions; how the students can evaluate the quality of that evidence; how specific individuals and/or groups of people are depicted in the account; why they might be depicted in that way; why different accounts of historical events exist; and what influence the existence of varying accounts has on our concepts of historical accuracy and of "truth."

Scavenger Hunt

Researchers (Doty, Cameron, and Barton, 2003) indicate that "...teaching reading in social studies is not so much about teaching students basic reading skills as it is about teaching students how to use reading as a tool for thinking and learning." One way teachers can help gifted students construct meaning from text is a "scavenger hunt" through the book. For example: ask them how many chapters and/or sections the book has and how it is organized. Ask what kind of material is at the beginning of the text and the significance of this. Ask them what kinds of skills or techniques the reader may require to read and understand the text. Ask them to identify special features in the student textbook that do not appear in trade books, and how these features can help them to understand and organize the text content. Finally, ask them how their answers to the preceding questions will help them to read the text better, and why.

Enriched Social Studies Courses

Research and Writing Skills

Gifted students should acquire and demonstrate a number of skills for doing research, writing up research results, and writing in general. Students should learn how to acquire the information they need. In addition, they should learn different note-taking practices, how to organize information, and be able to identify primary and secondary sources of research and know how to use them. They should be able to read textbooks with comprehension, and they should have/cultivate the ability to look for patterns in the material they read. They should also be able to interpret the information they encounter in social studies, and then analyze, synthesize, and apply the information to different circumstances and real-world situations. They should be able to support their viewpoints using pertinent documents and facts. They should learn to create bibliographies and webographies. Finally, they should recognize what is important in research and writing.

Enrichment Skills in Interpersonal Interactions and Group Relationships

Social studies skills for gifted students that fall under the area of interpersonal and group relations skills include the following: students should learn to define social studies terms and identify fundamental assumptions made in social studies. They should learn how to identify the existence of conflicts between or among the values of the people and groups involved in social studies topics and issues. They should learn to recognize when and where stereotypes exist and to avoid being influenced by them. In addition, they also should learn how to avoid engaging in stereotypes, themselves. They should become able to acknowledge various points of view. Gifted students in social studies courses should develop empathy for and understanding of others. They should participate in group discussions and planning in their social studies classes. Finally, they should be able to collaborate in order to attain goals cooperatively and take responsibility for completing tasks.

Maps, Globes, Graphs, and Images

Gifted students participating in enriched social studies courses need to learn how to read maps and read, understand, and apply map legends, scales of miles, and mapping symbols. They need to be able to read and use compass roses, grids, and other mapping tools. They should be able to understand and apply time zones. They need to understand the concept of distance. In addition, they should be able to draw inferences from maps; to compare various maps, to analyze and interpret different kinds of maps, and learn how to create their own maps. Students in enriched social studies courses should be able to read and interpret charts, tables, and other graphics. They also should demonstrate the ability to interpret the meanings of cartoons, photographs, paintings, drawings, and other images relative to social studies topics. The ability to interpret visual images such as pictures and other graphic representations represents is also an important skill of analysis.

Language Arts

Differentiated Instruction

Approaches: Included among differentiation approaches in adapting language arts curricula for gifted students are acceleration; depth, complexity, challenge, and creativity. In differentiating the curriculum for gifted students, teachers must adjust their expectations regarding the demands of content, process, and concept to higher levels. To meet these higher expectation levels, one method is to give students access to more advanced content or curricula when they are younger while still assuring that they can meet prescribed standards at all levels below that. Students also may be allowed to accelerate through the standard curriculum. Either way, teachers must adjust their expectations for advanced student levels. To meet gifted students' needs for advancement, depth, and complexity, one example is: following a class discussion on major themes in novels, a teacher could assign a gifted student to select a novel and write an essay examining how its major themes are treated in one chapter.

Rationale for Verbally Gifted Students in K-12 Education

Traditionally, instruction in the language arts has placed emphasis on basic reading skills and on assessing those skills by testing lower-level cognitive skills such as factual knowledge rather than fostering active inquiry and learning. Gifted students who have already mastered fundamental reading skills will not be challenged by these traditional methods. Verbally gifted children frequently attain linguistic proficiency at younger ages than their non-gifted age peers. Individual gifted students may be more advanced in a variety of areas in addition to reading, such as literary analysis, writing poetry, and/or writing prose. Gifted students who have mastered basic reading skills are ready before others to apply their abilities to higher-level cognitive tasks such as reading critically, writing exposition, communicating orally with others, developing their vocabularies and language usage, and learning foreign languages. These differences dictate a need for differentiated instruction for verbally gifted students in all developmental stages.

Selection of Instructional Strategies

According to expert educators, instructional strategies are not differentiated exclusively for students with gifts in the language arts. Rather, the level and character of the particular curriculum being taught is used to determine which instructional strategies are most indicated and how they should be applied. The curriculum and the choice of teaching strategies cannot be separated. Educational researchers and groups such as the National Association for Gifted Children find the diagnostic-prescriptive approach to teaching as of value for the purpose of differentiating language arts instruction for gifted students. This approach affords a process of assessment to determine each individual student's abilities and talents in the language arts, allowing educators to adapt their instruction to meet individual student needs. This lets linguistically gifted students progress at faster paces by not requiring their instruction in skills that they have already mastered.

Techniques Matched with Advanced Curricula

Educational researchers have found that certain instructional strategies can be paired with advanced curricula for effectively teaching gifted students in the language arts. One such strategy is questioning: when students have read or viewed challenging materials, teachers can stimulate class/group discussions at higher cognitive levels by asking them questions about the material. Another technique teachers can use is to assign open-ended activities. As long as these activities are sufficiently difficult, they can promote learning and growth in gifted students. Another strategy teachers can apply in conjunction with an advanced curriculum found especially effective for gifted students involves giving them poorly structured problems, which facilitates problem-based learning and challenges advanced learners to exercise and develop their problem-solving skills while applying the information and concepts that they have learned at the same time.

Characteristics of Good Instructional Programs in Writing

For students who display gifts in the language arts, good writing programs should focus on developing the students' skills in expository writing, i.e. informational, descriptive, or explanatory writing; and in persuasive writing, i.e. argumentation, logic, and rhetoric. Teachers need to help gifted students learn the process of writing by concentrating on the components of developing ideas, opinions, and arguments regarding current issues; writing drafts, making revisions, and editing. Educators also should give their gifted students exposure and practice in writing narrative, poetry, and other forms of literature by presenting them with literary models of each form and then assigning them to write their own original work in each form. Students at the high school level can be assigned exercises wherein they copy the styles of their favorite authors. Such an activity helps them to develop control over the medium, develop flexibility and versatility, and eventually develop and identify their own individual writing styles.

Oral Communication

To attain a good balance, language arts instruction in the area of oral communication for gifted students should address equally both speaking and listening to oral language. Gifted students should be assisted in developing their skills for evaluative or critical listening—not just hearing, comprehending, and remembering spoken language, but also analyzing, assessing, and judging it for veracity, accuracy, credibility, and relative bias, and distinguishing between facts and opinions. Gifted students should learn and practice the techniques of formal argumentation and debating. In addition, gifted students should practice discussion as another element of oral communication, including asking questions, probing for additional information, and expressing and hearing ideas and then building upon these collaboratively. Students talented in the creative arts will benefit from the opportunity to develop advanced skills in these areas through instruction in oral interpretation and participation in dramatic productions.

Literature

While standard school literature courses afford a number of works of high quality for students to read, gifted students will be expected and will want to go beyond these. High school students can find reading lists of books for pre-college preparation at most public and school libraries. Some gifted students read at such a high level that some of these books may be accessible and suitable for them in middle/junior high or even elementary school. Recommendations from educators include that teachers should emphasize student development of critical reading and thinking skills and help gifted students to develop their skills for analyzing and interpreting the literature. Teachers are also advised that their gifted students should read a broad range of subject matter. They should become familiar with various authors and their contents, themes, and styles. They should also come to recognize which authors are their favorites and learn more about those authors' lives.

Language Study

Language study includes the main components of vocabulary grammar, and syntax (i.e. sentence structure and word order). Therefore, instruction in the language arts for gifted students should emphasize the development, extension, and correct use of vocabulary. Instruction should include promoting an understanding of etymology or the origins of words, including word roots, prefixes and suffixes, and their original language sources. Language arts education for gifted students should include making analogies to teach an understanding of relationships between/among concepts and words. Gifted students also should learn linguistics, or the formal study of language; the history of their language; and semantics, or the study of the meanings of words. In teaching language arts to gifted students, the preferred approach integrates all of these elements of language study into one unified program for a holistic appreciation rather than teaching the elements separately on a disconnected piecemeal basis.

Foreign Languages

Verbally gifted students can not only excel in English, but also should learn foreign languages. In fact, foreign language study is one of the areas of curriculum differentiation in language arts instruction for verbally gifted students. Learning other languages enhances understandings of the history and structure of English; exposes students to the histories and cultures of other countries; gives them access to literary, scientific, and other text in foreign languages; and enables proficient students to communicate with persons from other nations and backgrounds. Such instruction even can prepare some students for future diplomatic careers. Learning foreign languages early is advantageous for gifted students. They can and should accelerate through four years of a second language and at least two years of a third. Choosing languages spoken in the community will afford follow-up opportunities. Spanish, French, German, Chinese, and Japanese are good choices. Latin and Greek (also a modern spoken language) are invaluable sources of English word roots.

Visual and Performing Arts

Cognitive and Academic Skills Improved by Art Education

Educational research has found that in addition to developing gifted students' creative abilities, instruction in the arts also enhances their cognitive and academic accomplishments. Art education improves the student's observational skills, abstract thinking skills, analytical skills, and problem-solving skills. In creating works of art, artists must use their reasoning powers to identify and define problems, to visualize and establish their goals, select methods for gathering information, propose solutions to problems, evaluate their solutions, and use their imaginations to revise those solutions. Therefore, the creative process in the arts requires, develops, and exercises higher-level cognitive skills. Teachers can integrate the arts into their curriculum and design activities tailored to gifted students' unique abilities, needs, and interests, and challenge them with increasing sophistication and complexity in the activities. Educational researchers advise teachers designing parallel curricula or differentiated instruction to clarify their students' educational goals before developing alternative instruction. Arts learning goals and activities can be integrated into math, science, language arts, and social studies curricula.

Visual and Performing Arts

Beliefs and Goals Regarding Role of Visual and Performing Arts in Education and Life

Persons and groups dedicated to the education of gifted students believe that the visual (drawing, painting, sculpture, photography, cinematography, etc.) and performing (singing, dancing, acting, etc.) arts are essential parts of healthy, productive lives. The Arts Network of the National Association for Gifted Children, for example, finds that "...the health and productivity of a society is reflective of the degree of artistic expression among its citizens." This network thus commits to the initiation, development, and implementation of teaching resources and strategies to support student aptitudes and interests in these areas, including furthering professional and public acknowledgement of the visual and performing arts as a necessary domain of giftedness, promoting more research into the topic of artistic gifts and talents, furnishing practical teaching methods and resources to cultivate artistic expression, and raising educational and public awareness of aesthetic perception, appreciation, values, and expression.

Learning Goals and Activities to Incorporate Arts Instruction into Reading

Arts instruction reinforces spoken and written communication and gives gifted students more opportunities to use analytical thought and creative problem-solving skills. Having students visually illustrate (draw, paint, etc.) a salient aspect of a story they read and discuss these illustrated aspects in small groups enhances the students' critical thinking skills. Teachers can help students adapt a written story into a dramatic piece, choose the most important scenes, explain their selections, and perform it as actors and narrators. This type of activity stimulates not only analytical thinking, but also imaginative interpretation and reading. Teachers also can have gifted students identify a problem, conflict, or issue in their assigned reading; assign the parts of characters included in the text to different students; and then debate the topic they have identified. This activity can make gifted students more aware of various points of view and of people's different motivations for their behaviors.

Examples of Integrating Goals and Activities in the Arts into Writing Curriculum

To develop gifted students' skills for synthesizing information, teachers can have them use a variety of sources such as paintings, photographs, musical compositions, and literature or other written work, to write an essay, script, poem, or sketch about some current event they find reported in a newspaper. The teacher can assign students to write an account of this event from various points of view, such as their best friend; their mother; their father; their sister; their brother; their family's pet dog, cat, or other animal; their teacher; etc. This helps students explore both the variety of sources for news stories and the variety of perspectives afforded by different individuals, as well as helping them to synthesize disparate elements. Another writing exercise integrating the arts involves having students listen to and analyze a musical composition and then write a script for a conversation analogous to the music in tone, tempo (speed/pace), divisions, and changes in speed, pitch, etc.

Teachers can use paintings, photos, and other visual art as catalysts to stimulate their gifted students to generate novel story ideas to write. They can suggest that the students use the visuals to correspond with their story's climax, its rising action just before the climax, or its dénouement (falling action). Teachers can give gifted students practice doing investigative research by first giving them a story to read about a painting that vanished, and then having them imagine and write about how they, as "art detectives," recovered it. Teachers can assign students to write fictional pieces about uncovering art forgeries, including the telltale details. To give them experience with understanding various perspectives, teachers can have gifted students write historical fiction about the origins of famous artworks and points along their journeys to their eventual/current destinations. They can write from the viewpoint of the artist, people in possession of the art, people searching for it, or the artwork itself.

Examples of Integrating Arts Goals and Activities into Social Studies Curriculum

Teachers can help their gifted students to associate the arts with social studies and vice versa by helping students relate such things as art movements with their social and historical contexts and events. For example, a teacher can ask gifted students to imagine they are time-traveling news

of light (e.g. Rembrandt, Monet, Turner, and others) and ask students to discuss the manner in which each artist depicts light's interaction with color and with water, the directions of light in the paintings, the quality of indoor and outdoor light, the quality of outdoor light at different times of day, how this affects colors, etc. The teacher also can ask gifted students to choose some aspect of light of scientific interest to them and then to participate as artists themselves by coming up with ways in which they can express this aspect of light visually via sketches, drawings, paintings, collages, sculptures, etc.

Example of Integrating Art and Math in Curriculum

When creating visual, especially representational art, artists frequently self-assess their visual perceptions against reality by estimating objective distances, sizes, heights, widths, and areas in the physical world. Teachers can provide their gifted students with comparable experiences through real-world experiments. For example, they can go outside on a sunny day in an area with some trees, each push a stick into the ground, measure the length that is above ground, and then measure the length of the stick's shadow. Then they can measure the length of a nearby tree's shadow. The teacher then asks them how they can calculate the height of the tree, in inches and in feet, based on the information they have. Teachers should allow gifted students to come up with their own methods for solving the problem, including drawing diagrams, taking photographs, or whatever they want to try. This activity sharpens accuracy in both visual perception and numerical estimation.

Learning Environments

Relationship Between Gifted Education and Special Education

Many students with gifts in specific areas and/or global giftedness also have disabilities. This complicates the circumstances of the best educational programming for them. One consideration is that disabilities can prevent some gifted students from being identified as gifted. For example, if a student has visual impairment and does not receive appropriate classroom accommodations, it might not be discovered that this student also has superior language and reading abilities. Another consideration is that instruction must be differentiated in certain ways to accommodate the needs of students with disabilities, and in other ways to accommodate the needs of gifted and talented students. The challenge for educators is to combine and coordinate both types of differentiation for these doubly/multiply exceptional students. An additional aspect is that federal laws exist to guarantee education to disabled students but not to gifted students. Therefore, many US state regulations for gifted education are modeled upon federal laws protecting the rights of disabled students.

Relationship Between Gifted Education and General Education

One factor important to consider in both gifted education and general education, since it affects both, is that compared to the numbers of special schools for gifted and talented students, there are many more public schools attended by gifted and talented students. Additionally, while some public schools include self-contained classes composed entirely of identified gifted and talented students, many more have such students mainstreamed into general education classrooms along with non-gifted students. Many public schools have pull-out programs for gifted students wherein they attend resource rooms to participate in special, individualized activities with instruction from a gifted education teacher. However, pull-out programs only involve a portion of the students' school time, typically a half-day or one day at most. Therefore, it is crucial for general and gifted education teachers to coordinate their curricula and activities. Each must know what the other is doing to facilitate the student's educational progress.

Cognitive Process Models

Educational scientists develop theoretical models and then apply the models to instructional approaches and methods. Some models focus on a particular domain, such as affective or cognitive. Cognitive process models focus on the cognitive domain, i.e. mental processes by which students learn and think. Bloom's Taxonomy, created by Benjamin Bloom in 1956, remains one of the most famous and popularly used cognitive process models. This taxonomy contains six ascending levels of cognitive processes:

1. **Knowledge**: Students can recall (retrieve from memory) or recognize (identify upon presentation) specific information they have learned.
2. **Comprehension**: Students not only remember, but understand, information they learn.
3. **Application**: Students can learn ideas, principles, and theories and then apply them to other contexts.
4. **Analysis**: Students can break information down into component parts.
5. **Synthesis**: Students can take constituent parts and think originally and creatively to combine them into a whole.
6. **Evaluation**: Students develop standards and criteria and apply these to concepts, methods, and materials presented to judge their value.

Review Video: Bloom's Taxonomy
Visit mometrix.com/academy and enter code: 755020

Affective Process Models

The affective domain relates to the emotions, emotional responses to the environment, and interactions with the social environment, including an individual's motivation, attitudes, perceptions, and values. Some process models used in gifted education focus on this domain. Krathwohl's Taxonomy of the Affective Domain is one such model. It contains five levels:

1. **Receiving**: Students show awareness of and attention to stimuli presented by teachers.
2. **Responding**: Students commit to discovery, seek out learning activities, and feel satisfaction from participating in the process.
3. **Valuing**: Students demonstrate decision-making about a value, their engagement with it, and their commitment to it. At this level, students may endeavor to convince others to commit to a value they have chosen.
4. **Organizing**: Students construct a belief system/value system, including attitudes, which they organize by the interrelationships of its components.
5. **Characterizing by a value or set**: Students have organized and internalized their value system and can apply it to many circumstances as a philosophy of life.

Virtual Learning Environment

A virtual learning environment is a computer-based environment where learning materials can be delivered to students via the Internet. By computer information technology, teachers can differentiate their instruction for gifted and talented students. Enhancing school projects by using online databases, scanning images, and incorporating hyperlinks and video clips into them is generally more motivating to students with all levels of ability. VLEs are useful for developing cultural experiences in the visual, performing, and/or creative arts. Students can visit museums, art galleries, government agencies, various industries, and institutions using VLEs. Teachers can acquaint students with a variety of ideas through online exposure to famous and/or controversial individuals. In subjects incorporating research activities, VLEs can give opportunities to gifted students for more advanced learning. Because of the virtually limitless amounts of information accessible on the World-Wide Web and the way it is instantaneously transmitted, using VLEs is a method that students find motivational and exciting.

Environmental Constraints That Stifle Creativity

Researchers (e.g. Amabile, 1983a, 1996; Hennessey, 1996) have identified five environmental constraints that destroy internal motivation and creativity:

1. **Expected Reward** – students work to receive some externally bestowed reinforcement, rather than being internally motivated to be creative for its enjoyment.
2. **Expected Evaluation** – students limit creative exploration, working instead to score well on a test.
3. **Competition** – rather than exploring a subject creatively to see where it leads, students focus on besting other students by obtaining the highest scores. Such competition limits creativity, since students' pursuits are narrowed to traditional academic tasks and to achieving the best among typical student responses rather than anything original or different.
4. **Surveillance** – when students know they are being observed, they feel less free to engage in creative exploration or find divergent solutions. They are more likely to conform and less likely to be creative.
5. **Time Limits** – when students know they are being timed, they feel pressured. Finishing on time interferes with the additional learning and production afforded by creativity.

Environmental Influences on Motivation

Although one might expect gifted students' exceptional abilities to support them in their classroom performance, research finds that gifted students are also very sensitive to influences they encounter in the classroom environment. While enthusiasm for learning and for subject content is a common characteristic of gifted children, teachers can dampen this enthusiasm when they insist that students conform to traditional classroom procedures and behaviors. Because gifted children often think and do things differently, trying to make them conform can ruin their motivation. They lose interest and get bored. In addition, many gifted children, especially younger ones, frequently do not know how to handle adults' high expectations, how to conduct successful interpersonal interactions, or how to set suitable goals for themselves. All of these factors in combination can contribute to academic underachievement. Underachievement actually is one of the most frequent problems among gifted students.

Social psychologists and other educational researchers have amassed a body of literature over 30 years regarding intrinsic motivation, creative performance, and how to promote these attributes in the classroom. Research increasingly has produced evidence that educational reforms to make our schools more conducive to internal motivation and creativity can benefit all students, not merely gifted students. Some steps that have been suggested (Hennessey, 2005) to accomplish such reforms include the following: Teachers need to make every effort to create atmospheres in their classrooms wherein students feel in control of their learning processes. Teachers and school administrators need to attain some distance and perspective in order to make objective, critical reviews of their schools' current reward systems and the types of incentives used. In addition, when external rewards are being used, educators need to help students achieve distance from the limitations these impose. Educators must help students develop proficiency in knowing their own weaknesses and strengths.

Environmental and Personal Influences on Implementation of Curricula

A number of factors in the environment and within the individual teacher can influence the way teachers implement curriculum units in reading and math. While curriculum units based on theoretical models adhere to principles of the applicable model, their implementation can vary as it is shaped by personal and environmental variables. For example, the grade level of students being taught will make a difference in the way that teachers implement curriculum. The school subject and department also will influence what elements are emphasized or minimized. Even the school building can make a difference in how teachers carry out their curricula. Variables at the school district level also affect teachers' understanding of a curriculum and how they implement it. Teachers' personal variables include their duration of classroom experience, their beliefs about learning and teaching, their particular approach to classroom management, and the grade level(s) and subject(s) they teach.

Choosing a Program

The Quality of Gifted Programs or Schools

Parents should be able to identify the philosophy and goals of a gifted program and/or school. Because giftedness is a lifelong trait, gifted students' academic needs will have some continuity. If program goals differ by age, these differences should reflect instructional variation, which should be both age-appropriate and appropriate for gifted students. Gifted programs should include both acceleration, i.e. offering instruction at a faster pace more suitable for gifted students; and enrichment, i.e. extending the curriculum by increasing the depth of study of a subject appropriately to gifted students' abilities and needs. Another essential factor involves offering multiple options in gifted education. Students gifted in the language arts are different from those gifted in math; mildly gifted and profoundly gifted students differ from one another. Owing to individual differences such as personalities and learning styles, even students with similar levels of giftedness in the same subjects can be very different. As a result, it is critical to include multiple ways to address these multiple educational needs.

Many times, schools depend on a teacher's recommendation or results of a single standardized group test to identify a student for gifted placement. Such methods typically overlook gifted students who are underachievers, have learning disabilities, and/or belong to under-represented populations such as the economically disadvantaged and racial/ethnic/linguistic/cultural minority groups. For these reasons, for years educators have advocated using multiple assessment instruments and practices. Another factor is staff development: Teachers trained in gifted education provide more effective instruction to gifted students. A school/program's teachers of gifted students should have certifications/endorsements in gifted education, and schools should provide regular in-service trainings regarding gifted students. Because gifted students are different from other students, schools also should provide guidance services to address problems such as fitting in socially with non-gifted peers and gifted hypersensitivities to the ordinary stresses of growing up and attending school. Additionally, schools should confer equivalent honors to academic talents/achievements as to athletic talents/achievements.

Any good gifted education program or school should state its learning outcome expectations clearly for the student. Providing activities that students enjoy is important, but gifted students also must learn while having fun. Specifying learning objectives and designing instruction to facilitate reaching those objectives is an essential part of any gifted program. Gifted education programs/schools should provide curricula that stimulate and challenge their students. Gifted students often grow bored if instructional tasks are too easy and do not require them to "stretch" their minds. Flexibility is important to help gifted students receive suitable challenges and meet individual needs. For example, students who perform several years above grade level in one subject should not be made to do grade-level work. Programs should be flexible to let such students study higher-level material. Students with musical gifts could exchange some school time for participation in special music programs, or study with an "artist-teacher" musician.

Review Video: Learning Objectives
Visit mometrix.com/academy and enter code: 528458

Online Information About Community Resources

Many websites offer information on community resources for gifted and talented students. One good resource is the website of the National Association for Gifted Children (NAGC). State associations can join NAGC as an affiliate. State affiliates then can offer NAGC Parent Affiliate Memberships to parents of gifted children. On the NAGC website, there is a Gifted by State page. As of its 2008 copyright, it linked to organizations for every US state except Alaska and Wyoming. Minnesota has two gifted organizations linked. Websites of state organizations for the gifted offer such information as local chapters, publications, webinars, audio recordings, upcoming events, recent news; the organization's mission statement; the state's definition of giftedness; programming standards; legislative updates; summer teacher professional development courses, workshops, institutes, conferences, etc.; e-stores; awards and scholarships; communicators; advocacy information; job postings; and more. State gifted organizations' sites also list and link to local community resources within their state, including other gifted associations; college/university programs; academic/enrichment opportunities including competitions, activities, programs, and journals.

Program Placement Options

Regular Education with Resource Services

The term "resource services" refers to the student in the general education classroom going to a separate resource room part-time, also called pull-out program or send-out classes, available in some middle schools and more prevalent in elementary schools. Typically, groups of 10-30 students attend a resource room for classes with a gifted teacher for one day or ½ day weekly. Gifted teachers may reserve one weekday for third-graders, and designate two other days for cross-grade groups of fourth- and fifth- and/or fifth- and sixth-graders in the resource room. They might visit regular classrooms to teach model lessons on a fourth weekday, and use the fifth day for completing their administrative duties.

One less common, but valuable, way to implement a pull-out program is with a multiple-year curriculum, which affords continuity. For example, using the theme of change and adaptation, fourth-graders could study environmental changes; fifth-graders could examine political changes under oligarchies, monarchies, and democracies; and sixth-graders could learn of changing family roles in different regions of our country. Another method involves extending the regular classroom curriculum by studying the same subjects or topics in more depth. A third approach is basing the pull-out curriculum on national or regional enrichment programs, e.g. Odyssey of the Mind, Science Olympiad, or the Future Problem-Solving Program, and adding activities focused on the underlying reasoning and creative skills whereby students succeed in such programs. In a fourth approach, the resource-room gifted teacher decides the pull-out curriculum. In this case, the quality of program content depends largely on the knowledge, experience, and skills of the individual teacher.

Advantages

Pull-out programs give gifted students the benefits of weekly classroom interactions with their intellectual peers and of sharing experiences involving their individual strengths. Many gifted students look forward to this so much that they will never miss school on their pull-out day. Pull-out classes often have open-ended enough curricula to allow freedom of creative exploration, enlisting the teachers' and students' skills and imaginations to design projects of interest and value, which is another advantage. Having gifted resource-room teachers is also financially advantageous to school districts, as these teachers usually work with all gifted students at various grade levels, teach lessons in regular classrooms, conduct in-service trainings for other teachers and staff, field parents' phone calls as delegated by their school principals, and perform other functions entailing intensive communication with varied audiences.

Disadvantages

One disadvantage of pull-out programs is that school personnel may not adequately coordinate their activities with the regular classroom curriculum. Consequently some teachers—and students—wonder how some pull-out activities relate to general education subjects. Another problem occurs if resource teachers base projects on their own interests rather than students' strengths. Many teacher resources are available for activities supporting higher levels of learning, but teachers often do not avail themselves of these curriculum guides and materials. The activities they design for gifted students should not only be fun and interesting for the students, but also develop their advanced cognitive skills. Rarely, some general classroom teachers resent gifted students' weekly absence and retaliate by scheduling class parties or field trips, important tests, and/or new topic introductions on pull-out days. This unprofessional behavior is uncommon but problematic. Also, some busy classroom teachers do not give gifted students other differentiated learning choices for the rest of the week, mistakenly thinking the pull-out program is enough.

Gifted Pull-Out Programs Vs. Inclusive, Regular Classroom-Based Gifted Instruction

Suppose that a school district has two schools served by a gifted resource teacher. A total of 60 gifted students are served, with 30 gifted students in each school. Each school has two classrooms per grade, and the gifted students are in second through fifth grades. Four grades x two classrooms = eight, x two schools = 16. With pull-out programs, the resource teacher can teach one group of students per grade level, or four groups at each school (eight total). The teacher can do this in two full school days, one at each school. In contrast, if this teacher had to go to sixteen separate regular classrooms to teach the gifted students, it would take more time, and that time would be more fragmented. The teacher likely could not see students as regularly. Researchers found when a school district switched from pull-out to in-class programs, it cost three times as much to hire three times as many specialists as they had needed with pull-out teaching.

Remediating Drawbacks of Pull-Out/Resource Room Programs

To remedy a lack of coordination between pull-out activities and regular classroom curriculum, the resource teacher can circulate a newsletter to classroom teachers explaining their resource room activities and how they connect with the regular curriculum. To provide both gifted resource teachers and general classroom teachers with reciprocal insights into each other's work and experiences, they can each trade places for a day or half-day. The gifted resource teacher can offer to teach a lesson for the regular classroom teacher, who can observe the teaching strategies the resource teacher uses for gifted as well as regular students. The gifted resource teacher also can propose that the two teachers work together to team-teach a regular class that includes gifted students. Educational researchers find that pull-out programs can succeed or fail depending on whether or not the regular and gifted teachers share mutual understanding, solid communication, and positive interactions.

Self-Contained Special Education Programs

Self-contained gifted classes consist entirely of gifted students. In districts having these, they are at the elementary school level. If middle and high school students are placed in each subject via ability grouping, these classes may function as self-contained. However, true self-contained gifted classes for all subjects do not exist in regular public secondary schools due to subject departmentalization. Advantages include the ability to design more gifted-appropriate instruction overall; higher gifted student achievement levels than those in pull-out gifted programs; usually no extra expense, as the gifted class will have the same number of students as a regular class at that grade level; the greater ability of gifted students to relate to their intellectual peers; and a more supportive classroom atmosphere, more conducive to freedom of expression and more encouraging relative to individual student strengths and weaknesses, which allows gifted students to perform better and more consistently.

Cluster Grouping

One practice that schools employ to meet the educational needs of their gifted students is to use cluster grouping. This practice entails grouping all of the gifted children at one grade level inclusively in one regular classroom. For example, if a school has three separate fourth-grade classrooms and a total of six fourth-graders identified as gifted, these students would all be placed in one of those classrooms rather than being distributed among several fourth-grade classrooms. This works for students gifted in individual subjects; a student with verbal gifts can be in a different classroom than a student with mathematical gifts. Some benefits are that cluster grouping is inexpensive and that students can move fluidly in and out of cluster groups. A drawback of cluster grouping is that globally gifted students cannot attend multiple classrooms simultaneously to meet their advanced needs in multiple subjects. Another consideration is that teachers must differentiate their instruction for greater variations in ability levels within one class.

Disadvantages

One disadvantage of self-contained gifted classes that continue for several years is that gifted students lack classroom interactions with non-gifted peers. One solution is to have activities such as lunch, physical education, art, and music include all students to provide interaction. Some school districts create "teams" of several classrooms, including the self-contained gifted class, so gifted and other students interact daily. Another problem occurs when either too few or too many gifted students are identified for the school's normal class size. A smaller gifted class would provoke resentment in teachers with normal, larger class sizes; a group too large would force schools to leave out some qualified gifted students. Self-contained classes also can lead to student, teacher, and parent perceptions of school elitism and isolationism. Also, though there is no proof that other students suffer, some teachers still feel that removing the gifted students leaves the other students with no role models.

Chapter Quiz

Ready to see how well you retained what you just read? Scan the QR code to go directly to the chapter quiz interface for this study guide. If you're using a computer, simply visit the online resources page at **mometrix.com/resources719/texesgtsupp** and click the Chapter Quizzes link.

4. Stephen has a very high IQ, but he must use a wheelchair and is almost completely immobilized except for limited use of one hand, with which he can only press buttons. His paralysis affects his vocal cords so he cannot speak. Stephen excels in all academic subjects, most notably the sciences and expressive and receptive language skills. His communication is limited by his disabilities. Which would be the most appropriate adaptive equipment for him in the classroom and elsewhere?

a. An adaptive pencil holder and paper holder both attached to his wheelchair
b. A computer with a voice synthesizer that he can program to speak for him
c. An electrolarynx he can use for speech in place of his paralyzed vocal cords
d. A communication board with buttons he can press to select various words

5. Which of the following is *most* accurate about exceptional students who are gifted?

a. They typically have advanced social skills.
b. Their academic performance is superior.
c. Their IQ scores indicate their giftedness.
d. They typically have difficulty relating to peers.

6. Who, in addition to gifted and talented teachers, are required to take 6 additional hours of training annually?

a. Administrators and counselors who make programming or hiring decisions
b. Only the teachers are required to have annual training
c. Administrators who make programming or hiring decisions
d. All teachers, although only gifted and talented teachers are required to have the initial 30 hours of training

7. A class includes students with disabilities, gifted students, average students, and a variety of reading levels. To support reading, which type of instructional resources would afford the most benefits to the most students?

a. Printed materials
b. Pictorial materials
c. Multimedia materials
d. Manipulative materials

8. Differentiation is:

a. The practice of grouping students in terms of commonality; for example, grouping gifted students and average students separately
b. Modifying lessons for the unique needs of a particular student
c. Illegal in all 50 states, as it violates the 14th Amendment
d. The practice of grouping students in terms of their differences; for example, grouping gifted and regular students, or grouping native English speakers with non-native speakers

9. The steps to curriculum compacting includes understanding learning objectives, identifying students who can master objectives most quickly, pretesting, streamlining instructional periods for students who understand objectives, and:

a. Recommending acceleration opportunities
b. Recommending students be advanced a grade
c. Reviewing assessments
d. Passing these students on to a Gifted and Talented teacher

10. Curriculum focusing on social awareness and adjustment, personal awareness and adjustment, a study of attitudes, values, and the individual's perceptions is:

a. Effective Curriculum
b. Sociological Curriculum
c. Affective Curriculum
d. Developmental Curriculum

11. Mario is taking AP classes at his high school. This means:

a. Advanced Placement, in which all his high school coursework will also apply as college credits
b. Alternative Potential. These are small, experimental classes taught to students who have been identified as 'alternative learners'
c. Accountability Projects. These are independent studies in which an individual or small group works closely with a mentor, who may not be a teacher but someone with professional experience in the student(s) area of interest. Students are held accountable for their projects' designs, implementation, and successful conclusions
d. Advanced Placement, in which high schools offer coursework meeting criteria recognized by institutions of higher education. Often, college credit is earned when an AP exam in specific content areas is passed

12. Fifth-grade Esmeralda, a gifted writer, told the school counselor that it really isn't her fault that much of her school work never makes it to the teacher's hand. She perceives herself as fairly vulnerable in a difficult world. Her little brother goes through her book bag and tears up her papers. Her handwriting isn't very good no matter how hard she tries and she can't read her own notes. Her mother forces her to stay inside on nice days to do homework that is boring. Sometimes other kids on the bus take her homework and throw it away. Esmeralda is also convinced that when she does manage to turn in the work, the teacher herself misplaces it. She appears to believe her excuses, and doesn't see any way to overcome them. What's going on?

a. She has a high external locus of control
b. She has a low external locus of control
c. She has a low internal locus of control
d. She has a high internal locus of control

13. The Total Talent Portfolio is a systematic method of gathering, analyzing, and classifying students' abilities in order to:

a. Protect gifted children with learning disabilities from falling through the cracks
b. Help the teacher determine how to enrich and accelerate students' learning based upon their individually unique profiles
c. Help older students with transition and placement into jobs at which they are likely to excel
d. Give teachers visual and logistical models whereby they can decide which cross-curriculum strategies are most likely to succeed for individual students

14. How many hours of annual training are required of gifted teachers to retain eligibility?

a. 15
b. 6
c. None
d. 3

15. Ten-year-old Ming has exceptional musical abilities. She can hear a piece of music once and immediately play it on a variety of instruments, and her singing range is phenomenal. She is new to the US, and speaks little English. She has learning disabilities and is dysgraphic. She is easily excited. Her classmates find her:

a. Delightful; her excitability coupled with her musical talents are entertaining
b. Odd; her lack of English, terrible handwriting, nervousness, and bizarre musical talents make her seem abnormal
c. Acceptable. Fifth graders typically don't exclude one another
d. Fascinating; her musical talent coupled with her excitability and lack of English make her an interesting distraction to their schoolwork

Refer to the following for question 16:

EXHIBIT 1

CLASS DESCRIPTION

Mr. Hamlin is a new fourth-grade general education teacher to a class of 31 ethnically diverse students. The community in which he teaches is largely underserved, and many of his students are from socioeconomically disadvantaged backgrounds. One of his students, Lindsay, has begun displaying disruptive behavior in class, such as calling out, arguing with other students, and getting out of her seat during instruction. After speaking with the guidance office regarding her behavior, Mr. Hamlin learned that Lindsay has recently been placed into foster care.

To understand his students' needs, Mr. Hamlin has reviewed their state assessments from the previous year. The results indicate that eleven of his students have below grade-level literacy and math skills. Of these eleven students, three are intermediate-level English language learners (ELLs) and one has beginner-level English language proficiency skills.

In order to build positive relationships and learn more about the individual needs in his classroom, Mr. Hamlin wants to establish communication with his students' families, but often has a hard time getting a response when he reaches out. To encourage greater family involvement, Mr. Hamlin has created a class website in which he posts information about what the class is learning each week, pictures of student work, study tips, and important assignments.

Mr. Hamlin strives to emphasize respect and the celebration of differences in his class. However, while most of Mr. Hamlin's students get along with one another, they tend to associate primarily with classmates of their own ethnic group. To diversify his students' interactions, Mr. Hamlin leads a class meeting each morning in which students are free to share thoughts, discuss concerns, and ask questions.

Mr. Hamlin is preparing to teach a lesson on fraction equivalence. He has asked the school administration to visit his classroom to give him suggestions regarding strategies for more effectively connecting with and meeting the learning needs of diverse students. The school vice principal has decided to conduct an informal walkthrough evaluation of Mr. Hamlin's lesson and takes notes to give him feedback.

Exhibit 2

Excerpt from Mr. Hamlin's Lesson Plan

Topic: Fraction equivalence	
Standard: (NY-4.NF) • Extend understanding of fraction equivalence and ordering.	
Essential Question: • How can two fractions with different numerators and different denominators be the same?	
Lesson Objectives: • Students will be able to compare two fractions with different numerators and different denominators. • Students will be able to explain why fraction a/b is equivalent to a(n)/b(n).	
Grouping: Students will be arranged in teacher-selected pairs for portions of the lesson.	
Lesson Component	**Activity**
Introduction	As a warm-up, students will be asked to identify familiar fractions based on pie-chart visual representations. The teacher will review the warm-up questions and show a slide presentation to review familiar fractions and relevant vocabulary. The teacher will then show several examples of equivalent fractions with different numerators and denominators, accompanied by visual models. Students will then see a short video clip to further explain fraction equivalence and ordering.
Whole-Group Activity	Students will be given a handout with several blank pie charts. The teacher will lead a whole-group activity in which students color in each chart to represent a given set of equivalent fractions.
Partner Activity 1	Students will work together in teacher-selected pairs to complete a set of task cards with varying practice questions and word problems related to comparing fractions.
Partner Activity 2	Students will work with their partner to demonstrate their understanding of fraction equivalence and ordering. Students will have the choice to create five relevant word problems, draw a series of charts to represent equivalent fractions, or complete an online quiz using the classroom computers.
Homework	Students will be assigned to log into an educational math website at home to complete and submit ten practice problems.
Accommodations	• Scaffolded texts for word problems • Extra time for processing as needed • Students paired with others who can provide scaffolding • Visual representations for ELL students

Exhibit 3

Notes from Informal Observation by the Vice Principal

Classroom Environment

Mr. Hamlin's classroom is brightly decorated with inspirational posters depicting individuals of varying backgrounds, displays of student work, and bulletin boards to reinforce topics learned in class. Various learning materials are available for student use, including notebook paper, pencils, rulers, and art supplies. Many of the items in his classroom are labeled, and he has created a word wall to emphasize high-frequency vocabulary related to the curriculum. Student desks were arranged in groups of three to five. Mr. Hamlin began with a class meeting in which he asked students how they were feeling and if they had any plans for the upcoming weekend.

Lesson Activities

Students were learning about fraction equivalence and ordering. Mr. Hamlin began with a review to elicit students' prior knowledge and continued with a presentation with visual representations and a video clip on the new material. After the presentation, Mr. Hamlin asked students to indicate their level of understanding by showing a thumbs-up or thumbs-down hand gesture. Seven students indicated that they did not entirely understand the new concept, and Mr. Hamlin provided additional examples.

Students then transitioned to the lesson activities. Mr. Hamlin selected student pairs to complete a series of task cards and create a product that demonstrated their understanding of equivalent fractions. It appeared that Mr. Hamlin paired students that struggled to understand the presentation with others that could help. Most students worked well together, but some seemed uninterested in working with their partner. During the activities, Mr. Hamlin moved throughout the room to check for understanding and monitor students' progress.

Student Interactions and Engagement

Overall, students were respectful of one another during the class meeting, and seemed interested in listening to one another. Students did not seem to want to ask one another questions, but rather, to share about themselves. A few students were quiet during the discussion and chose not to share.

Most students worked well together during the partnered activities. However, a few seemed uninterested in working with their partners. One student tried to leave his partner and join another pair but was redirected by Mr. Hamlin. Students seemed most interested during the second activity, in which they got to choose how they demonstrated their understanding of equivalent fractions.

16. Which of the following could Mr. Hamlin have incorporated into his lesson to support students of varying needs and abilities more effectively?

a. Extension and enrichment opportunities for gifted and talented students
b. Individualized instruction for struggling students
c. Opportunities for frequent breaks throughout the lesson
d. The opportunity for students to select their own partners for the collaborative activities

Refer to the following for question 17:

Exhibit 1

Class Description

Ms. Jewel has been teaching third grade for several years and has a great deal of experience working with students of varying learning styles and needs. Until recently, the demographics of her school primarily included one culture, but this year has seen an increased population of ethnically diverse students. To create an inclusive learning atmosphere, the school administration has been encouraging teachers to emphasize the importance of celebrating diversity and appreciating one another's differences.

Ms. Jewel has been implementing several strategies to teach her students the value of diverse backgrounds, cultures, and experiences. To decorate her classroom, she has asked each student to bring in an item that represents their heritage to hang on the class bulletin board. In addition, Ms. Jewel has been working with the other third-grade teachers to plan a cultural celebration day in which students' families will be invited to participate.

At the beginning of the school year, Ms. Jewel reviewed her students' education records to gain insight into their backgrounds, abilities, and needs in order to plan instruction accordingly. She learned that, of her 28 students, four are from socioeconomically disadvantaged households. Two of these students are native Spanish speakers and possess beginner-level English proficiency skills according to their most recent NYSESLAT scores. Another student, Brian, has been identified as homeless. Brian is frequently absent from school and often arrives to class without necessary learning materials. Ms. Jewel has noticed that he is becoming increasingly withdrawn and struggles to make friends.

Carla has been medically diagnosed with ADHD that impedes her ability to focus in class and complete assignments. She has an individualized education program (IEP) to outline her accommodations, as well as her academic and behavioral goals. In addition, her literacy skills are below grade-level according to her score on last year's New York ELA assessment, for which she is receiving Tier 2 Response to Intervention (RTI) services. Ms. Jewel also works closely with the school reading resource teacher to develop strategies that support Carla's literacy development. Carla's parents are divorced, and her father lives in a different state, which sometimes makes it difficult for him to actively participate in her education. To remedy this, Ms. Jewel always ensures that he is included in communications related to Carla's academic progress, upcoming assignments, and important school events.

Ms. Jewel is preparing a unit on identifying character traits using details from a text. She has intentionally chosen a short story that includes several characters from diverse backgrounds, cultures, and experiences. After each lesson, Ms. Jewel uses a journal to reflect upon her instructional effectiveness as well as her students' level of engagement.

EXHIBIT 2

EXCERPT FROM MS. JEWEL'S LESSON PLAN

Topic: Character traits

Standard:

- **3R3**: In literary texts, describe character traits, motivations, or feelings, drawing on specific details from the text.
- **3RF4a**: Read grade-level text across genres orally with accuracy, appropriate rate, and expression on successive readings.

Essential Question:

- How do character traits influence one's feelings and motivations?
- What can we learn about a character from details in the text?
- How can we identify character traits?

Lesson Objectives:

- Students will be able to identify character traits using details from a text.
- Students will be able to explain how character traits influence one's actions and motivations.

Grouping: Students will work with teacher-assigned partners for a portion of the lesson.

Lesson Component	**Activity**
Introduction	Students will begin with a fill-in-the-blank activity in which they use adjectives to describe themselves. Students will be called upon to read their responses. The teacher will explain that character traits are used to describe people. The teacher will show a slideshow presentation to introduce the idea of character traits and explain how they affect one's motivations and feelings. The slideshow will include three brief passages. Students will be called upon to read the passages aloud and identify the characters' traits. Students will be given a handout that lists several character traits to reference later in the lesson.
Partner Activity	Students will work with a partner to read brief three- to five-sentence passages about various characters. Students will be asked to determine if each passage depicts positive or negative character traits. Students will then choose three characters from the passages and use in-text details to draw a picture that represents each description.
Whole-Class Activity	Students will take turns reading sections of a short story aloud to the class. Prior to reading aloud, students will have a few minutes to read the story silently to themselves. Students will independently answer a set of comprehension questions that correspond with the short story. The teacher will review answers with the whole class.
Individual Summative Activity	Students will be assigned to write a response in which they compare and contrast the character traits, feelings, and motivations of two characters from the short story.
Extension Activity	Selected students will be assigned to create original stories that include at least two characters with three different character traits each.
Accommodations	• Short stories will be scaffolded according to reading comprehension skills, and will include an English-Spanish glossary for complex terms. • Visual representations will be included in the presentation. • English-Spanish character trait vocabulary sheets will be provided. • Preferential seating/reduced distractions will be used. • Graphic organizers will be provided for the independent writing activity. • Extra time to write the response will be provided as necessary.

18. Continuum of Learning Experiences refers to:

a. Challenging experiences and opportunities that resonate with the specific interests and abilities of high-level students
b. Unplanned learning experiences which are successful and therefore incorporated into the continuum
c. The correct term is Continuous Learning Events
d. Planned learning experiences that build upon a progressively stronger foundation from year to year

19. Parental permission to assess a student for a gifted program is:

a. Required
b. Not required, in order to avoid a hardship for students who meets the qualifications but do not have supportive parents
c. Not required; one of the premises of gifted education is developing a sense of independence and personal responsibility in students
d. Required only if a student requests parental involvement

20. For most school districts, the written policy regarding student identification must: include provisions for continuing screening; include assessment procedures from a number of sources; ensure that all populations have assessment available; ensure final selections are made by at minimum three local educators with specialized training; and:

a. Include provisions about reassessment, exiting strategies, and readmission policies
b. Include provisions about furloughs, reassessment, exiting strategies, transferring students, and the district decision appeals process
c. None of the above; the district is not required to have a written policy
d. Include provisions about furloughs, reassessment, exiting strategies, transferring students, and parental involvement

21. Uneven social and intellectual development, being socially at variance with expectations of the classroom and of others at the same age, experiencing finely tuned emotions, and the resulting vulnerability is called:

a. Social inequality
b. Psychological ineptitude
c. Asynchrony
d. Serendipity

22. Both qualitative and quantitative assessments must be used when identifying elementary, middle, and high school students, with what exception (s)?

a. Teachers can choose to assess kindergarten students using only qualitative assessments
b. For students being assessed for exceptional leadership abilities, creativity, or artistic areas, teachers can chose to assess qualitatively only
c. Both a and b
d. Neither a nor b

23. SAGES stands for:

a. Standard Assessment Games for English Students, and describes informal assessments based on cooperative games
b. Sagmore Adams Gates Educational Standards, and describes gifted standards for grades 1-3
c. Screening Assessment for Gifted Elementary Students, and describes a standardized assessment
d. Screening Assessment for Gifted Elementary Students, and describes an informal assessment

24. Piaget's first stage of cognitive development involves:

a. Operations with concrete things
b. Sensory input and motor output
c. Operations with abstract things
d. Intuition and animism, not logic

25. The National Research Center on the Gifted and Talented is:

a. A federally-funded center with the charter of developing effective methods of identification and testing of Gifted and Talented students, and programming for such students
b. A federally-funded center with the charter of developing effective methods of showcasing gifted students by funding venues for student performances and presentations
c. A privately-funded center with the charter of developing effective methods of showcasing gifted students by funding venues for student performances and presentations
d. A state-funded center with the charter of developing effective methods of identification and testing of Gifted and Talented students, and programming for such students

26. A student who participates in a Gifted and Talented program at her school has been invited to visit family members in China with her parents. She will attend school there, and also be tutored so that she doesn't fall behind. Her mother has asked how a two-month leave of absence will affect her participation in the Gifted and Talented program. Which of the following is most accurate?

a. If the student takes a leave of absence, she is out of the program
b. At any point, students will automatically receive furlough status of any length of time as long as the request is in writing six weeks prior to the furlough date
c. If the student takes a leave of absence, she is out of the program. She can enroll again if the district is willing, but must go through nomination and evaluation procedures
d. A student may be granted a leave of absence for specific reasons and for a pre-established length of time; the leave must be approved

27. According to Sternberg's triarchic model of intelligence, three central qualities comprise giftedness: componential intelligence (the ability to analyze), experiential intelligence (the ability to think creatively), and

a. Abstract intelligence
b. Analytic intelligence
c. Practical intelligence
d. Essential intelligence

28. What is the difference between a raw score and a scale score?

a. Raw scores are used on specific tests to show how well a student did; this score is not necessarily consistent from test to test. A scale score translates raw scores onto a measurement common to all forms of testing for a particular assessment
b. There is no difference; both terms refer to the score a student receives
c. A raw score is shown as an error number while a scale score is indicated by a percentage. For example, if a student takes a test with 100 questions and misses 7, the raw score is -7 and the scale score is 93%.
d. Scale scores are used on specific tests to show how well a student did; this score is not necessarily consistent from test to test. A raw score translates scale scores onto a measurement common to all forms of testing for a particular assessment

29. When creating a Total Talent Portfolio, who is primarily responsible for deciding what to include, maintaining the portfolio, and establishing goals?

a. The student and the teacher working as partners
b. The teacher in the primary position, supported by the student
c. The student in the primary position, with the teacher to review and approve
d. The student is primarily responsible

Refer to the following for question 30:

EXHIBIT 1

CLASS DESCRIPTION

Mr. Morris has recently moved to a new city and begun teaching 8th-grade social studies. The school population is ethnically diverse, and Mr. Morris is unfamiliar with the backgrounds, customs, beliefs, and experiences of his students and their families. However, he is determined to build positive relationships and learn how to best meet the needs of his students and the community.

At the beginning of the school year, Mr. Morris reaches out to the administration regarding professional development opportunities that focus on methods for effectively teaching students of varying cultural backgrounds, and he self-educates regarding the priorities of his new community. He also attends the school's open-house night in order to meet his students' families and gain insight into their home lives and individual learning needs. In addition, Mr. Morris plans to involve himself in the school community by participating in such activities as PTA meetings and fundraisers and attending school sporting events. In the classroom, Mr. Morris focuses on creating a positive classroom community by conducting team-building exercises, icebreaker activities, and learning-style inventories to get to know his students. He creates a classroom newsletter and website to communicate important updates and intends to maintain frequent communication with families via conferences, phone calls, and emails.

Prior to beginning his first unit on the contributions and experiences of different immigrant groups throughout American history, Mr. Morris administers a diagnostic exam on the topic to assess his students' level of understanding. The results indicate a wide range of knowledge and abilities. Mr. Morris considers these results when planning lessons in an effort to proactively meet his students' needs by implementing the most effective instructional strategies and supports. He plans to deliver instruction through a variety of means, including presentations, class discussions, and hands-on activities, as well as opportunities for individual and small-group work in which struggling students are paired with others that can provide assistance. In addition, Mr. Morris carefully researches and seeks examples from the community that demonstrate positive contributions to today's society from varying immigrant groups. Throughout instruction, Mr. Morris conducts formative assessments to evaluate understanding and overall engagement in order to reflect upon and determine areas in which his instructional approaches may need adjustment to better support his students.

EXHIBIT 2

EXCERPT FROM MR. MORRIS'S LESSON PLAN

<table>
<tr><td colspan="2">Topic: Immigration</td></tr>
<tr><td colspan="2">Standard: Compare and contrast the experiences of different groups in the United States. Explain the contributions of specific groups to American society and culture (NYCCLS – 4.7a).</td></tr>
<tr><td colspan="2">Essential Question:
• How have different immigrant groups contributed to shaping American society and culture throughout history?
• How do the experiences of early immigrants compare to those of recent immigrants?</td></tr>
<tr><td colspan="2">Lesson Objectives:
• Students will be able to compare and contrast the experiences of early and recent immigrant groups in the United States.
• Students will be able to discuss ways in which early and recent immigrant groups have contributed to American society and culture.</td></tr>
<tr><td colspan="2">Grouping: Students will be arranged in groups of three to five. The teacher will determine student groupings.</td></tr>
<tr><td>Lesson Component</td><td>Activity</td></tr>
<tr><td>Introduction</td><td>Students will be shown a political cartoon that refers to the United States as a melting pot. Students will be asked to discuss what they think that term means and how it applies to today's society.
Students will then see a presentation on early immigrants to the United States, such as those from Ireland, Germany, and England, as well as recent immigrant groups from China, Vietnam, and Latin America. Throughout the presentation, the teacher will give examples of ways in which each group has influenced American society and culture, such as food, clothing, and music.</td></tr>
<tr><td>Individual Activity</td><td>Students will conduct independent online research regarding the contributions and experiences of a specific immigrant group that the teacher will assign. Some students will be assigned to research the same immigrant group for a later collaborative assignment.</td></tr>
<tr><td>Small-Group Activity</td><td>Students will work in groups of three to five with others that researched the same immigrant group. Working collaboratively, students will create a visual representation of their choosing (chart, comic strip, slideshow, skit, etc.) to depict their research findings. Members of each group will be assigned a specific role and present their visual representation to the class in a two- to three-minute presentation.</td></tr>
<tr><td>Whole-Group Activity</td><td>The teacher will lead a class discussion focused on comparing and contrasting the experiences and contributions of early versus recent immigrant groups in the United States. Throughout the discussion, the teacher will record key points on the chalkboard using a Venn diagram.</td></tr>
<tr><td>Extension/Enrichment</td><td>Gifted students will have the opportunity to extend their research on their assigned immigrant group by applying what they learned and imagining they are immigrating to the United States. These students will be asked to create a series of diary entries that detail their initial experiences based upon their research findings.</td></tr>
</table>

EXHIBIT 3

NOTES FROM MR. MORRIS'S FORMATIVE ASSESSMENTS

DIAGNOSTIC EXAM

The diagnostic exam results indicate that most students possess basic knowledge of recent immigrant groups, but very limited understanding of the experiences and contributions of early immigrants. Extra focus on early immigrant populations may need to be included in the introductory presentation. While most students demonstrated appropriate grade-level reading comprehension skills, many were unfamiliar with key vocabulary related to the unit, and a few students indicated below grade-level reading comprehension skills. These students will need extra supports in the form of visual aids, graphic organizers, and selective grouping during instructional activities. Three students performed extremely well on the assessment. Extra learning opportunities for enrichment will be beneficial for these students.

Introduction

The presentation included slides, images, video clips, and opportunities for students to answer open-ended questions. During the presentation, I moved around the room to assess students' level of attentiveness. Students appeared to be the most interested in viewing the images and video clips, and most were highly engaged in the open-ended questioning segment of the lesson. A few students in the back of the classroom, who indicated low reading comprehension skills on the diagnostic exam, appeared bored throughout the presentation. I used proximity and maintained eye contact to redirect their focus. Students were asked periodically throughout the presentation to indicate their level of understanding by giving a "thumbs-up" or "thumbs-down" sign. If the majority of students showed a "thumbs-down" sign, I reviewed previous presentation slides and provided extra examples. Overall, students seemed to respond well and were most enthusiastic about the real-world examples of contributions from immigrant groups in everyday life.

Individual Activity

Each student was given a color-coded card to indicate their level of understanding as they conducted independent research. One side of the cards was green, and the other was red. Students were instructed to flip their cards to the red side when they needed assistance or clarification. During the activity, I moved throughout the room and stopped to help students that flipped their card to the red side. This method seemed to work well for students that do not typically ask for help.

Small-Group Activity

Throughout the small-group activity, I moved between groups to listen to student discussions, ask questions, address misconceptions, and ensure all students were on task. Assigning specific roles to group members appears to have helped maintain group focus. Struggling students seemed to be more engaged and productive when paired with others that could provide assistance.

Exit Ticket

Students were instructed to write down three new things they learned, two questions they had for further exploration, and one way in which their culture has positively contributed to shaping American society and culture as preparation for a long-term project. Most students appeared to have a firm understanding of the key points of the lesson. A few students, however, seemed to have struggled with differentiating the experiences between specific immigrant populations. This will be reviewed in the next lesson for clarification. Questions for further exploration were primarily focused on reasons why different groups of people immigrated to the United States, which will be addressed in a future lesson.

30. By incorporating opportunities for enrichment and extension, Mr. Morris does which of the following?

a. Prevents students that finish early from engaging in off-task or disruptive behavior
b. Strengthens students' understanding of diversity and its significance in shaping American society and culture
c. Provides gifted and talented students with extra work while they wait for their classmates to catch up
d. Acknowledges the learning needs of gifted and talented students by challenging them academically and intellectually

31. According to the Javits Act, gifted and talented students are described as those who:

a. Have been assessed as having intelligence at least 30% above the national average, and are not receiving services or activities to develop those capabilities
b. Are in the top 5% of their age group in terms of intellectual, creative, artistic, or leadership areas, or in specific academic fields
c. Have achieved high accomplishment in intellectual, creative, artistic, or leadership areas or in specific academic fields, and who have not been recognized and honored in an appropriate manner
d. Have the capability for high accomplishment in intellectual, creative, artistic, or leadership areas or in specific academic fields, and who need services or activities not ordinarily provided by the school to develop those capabilities

32. A number of gifted high school students are in a Language Arts class. Is their AP teacher obliged to differentiate the program of study for the gifted students?

a. No; if the teacher is already running a student-centered classroom, allowing students choices in assignments, and using flexibility in student grouping, the teacher is already differentiating. No further modification is required
b. No. All students in Advanced Placement courses are, by definition, gifted. Therefore, differentiation is already taking place
c. No. The teacher can choose to differentiate the program, but by the time gifted and talented students have entered high school, they have internalized sufficient strategies to challenge themselves and independently promote their own education
d. Yes. If the teacher is already running a student-centered classroom, allowing choices in assignments, and using flexibility in student grouping, the teacher is already differentiating. The teacher has received additional training to allow him or her to continue to modify classroom assignments to more profoundly challenge the gifted students

33. A parent has approached his daughter's teacher and is frustrated because the child seems to be completely without motivation. The father has tried rewarding his daughter for exemplary work, punishing her for inadequate work, and reviewing her work before she turns it in to ascertain if it is complete, to no avail. The father knows the daughter is extremely capable intellectually and artistically. He feels both angry at what he sees as her laziness, and guilty for not being able to motivate her. The teacher tells him:

a. To modify his strategy. He should BOTH reward exemplary work and punish inadequate work at the same time
b. That until the child takes full responsibility for her actions, no one can motivate her. Withdrawing his involvement will likely result in the girl becoming more motivated
c. That gifted student's value learning over performance. Her lack of motivation shouldn't be an occasion for blame. Instead of focusing on grades and performance, the father might consider exploring some learning opportunities in areas of the girl's interest. For example, a visit to an art museum followed by a trip to the library for books about a particular art movement might motivate the student
d. That he is doing all he can do; his methods are fine, and if he changes his course at this point it will do more harm than good

34. "Gifted" students are federally defined as:

a. Those that score above 130 on a WISC test
b. There is no decisive federal definition; each state determines its own definition
c. Those who are academically 2 classes or more above peers
d. Those who exhibit extra-ordinary abilities in the areas of the arts, mathematics, science, business, or leadership

35. Self-efficacy, the belief in one's ability to succeed at a particular task, is influenced by past performance, observing the success of peers, verbal support and persuasion by others, and

a. A willingness to take risks
b. A creative approach to the task at hand
c. Strong self-esteem
d. Physiological cues that indicate nervousness, such as sweating hands; or cues that indicate confidence, such as a feeling of ease and control

36. Bloom's taxonomy describes the three types of learning as:

a. Cognitive, affective, and psychomotor
b. Intellectual, emotional, and social
c. Intellectual, creative, and psychological
d. Cognitive, creative, and psychomotor

37. A 4th grade teacher has three gifted students in his general classroom. One strategy for differentiating their learning is extending lessons throughout the curriculum by comparing common themes and issues, studying the associations within and between individual disciplines over the course of time and/or from more than one viewpoint. The teacher is adapting the __________ of his lessons in order to challenge his gifted students.

a. Complications
b. Complexity
c. Enhancement
d. Richness

44. Binh is in fifth grade. He was identified as gifted in leadership in third grade and entered into a Gifted and Talented program. He has always been mercurial, energetic, and moody, but until now it was not apparent that his studies were affected. Because his classroom behavior has worsened and his grades have been affected, his fifth grade teacher wanted him evaluated for Attention Deficit Hyperactivity Disorder. Should he be diagnosed with ADHD, what will happen to his gifted and talented status?

a. Because this has been a documented, ongoing problem, the district will temporarily remove him from the Gifted and Talented program, concentrate on improving his issues with attention and hyperactivity, and then re-enroll him once those things are under control
b. He will remain in the program and it will be modified to become more appropriate for his learning needs. It is illegal to discriminate against students with learning or behavioral disorders
c. It will be discussed with the gifted teacher, his general classroom teacher, his parents, and Binh. If Binh promises to monitor his behavior and work to the best of his abilities, he will be permitted to stay
d. He will be removed from the Gifted and Talented program; his learning issues preclude advanced studies. If, at some future point, he wants to return to the program, he will have to be re-evaluated for it

45. A strategy whereby the curriculum is modified in depth, complexity, and originality, and suggests the four concurrent directions of Core Curriculum, Curriculum of Connections, Curriculum of Practice, and Curriculum of Identity, is called:

a. Multiple Curriculums Modification
b. Parallel Curriculum Model
c. Creative Multiple Curriculums
d. Parallel Creative Curriculums

46. When dealing with a class with widely varied levels of skill and knowledge, what common pitfall is most harmful?

a. Giving too much attention to the weaker students
b. Permitting the stronger students to correct the weaker students in class
c. Offering more encouragement to the stronger students than the weaker students
d. Overcorrecting the mistakes of the weaker students

47. The Total Talent model considers which of the following?

a. Student interest, learning styles, and extraordinary abilities
b. Student's abilities only
c. Student's interests and learning styles
d. A student's interests and learning styles and the student's cognitive ability are equally considered

48. Authentic assessment as it pertains to gifted students:

a. Assesses emotional development, using expressive models
b. Assesses learning using portfolios, performance, observations, and other assessments that model real-world representations
c. Assesses a student's sincerity or authenticity by using a lie-detector
d. There is no longer such an assessment; popular in the early 1970s, it was abandoned in 1992 because results lacked clarity or concise information

49. According to Piaget's theory of cognitive development, ______ and ______ are processes included in the overall process of ______.

a. assimilation; accommodation; adaptation
b. adaptation; assimilation; accommodation
c. accommodation; adaptation; assimilation
d. adaptation; accommodation; equilibrium

50. Pfeiffer's Gifted Rating Scales is used to:

a. Weigh the pros and cons of entering a student into a gifted program
b. Determine what percentage of the budget should be allocated to gifted programs by considering what ratio such program(s) are scaled to
c. Measure gifted intelligence overall
d. Measure the degree of disability a gifted, learning-disabled student has

51. Dr. Acevedo has a gifted child in her fifth grade classroom. In order to keep the student engaged and motivated, which of these inclusion strategies should she permit the student?

a. Establish own learning goals; self-assess; move to an independent challenge activity when student understands the lesson; encourage student to work with peers close to his or her own intellectual level
b. Have student independently design his or her own curriculum based upon classroom assignments; self-assess and self-grade; help establish learning goals for the class
c. Help teacher design his or her own curriculum based upon classroom assignments; review student's assessments and grades with the teacher; suggest learning goals for the class
d. Help establish learning goals for the class; act as the teacher's assistant teacher; grade work of classmates for the teacher

52. Gifted students who receive a vertical score equal to or greater than the Met Standard at a grade level higher than their own have already met the standard at their grade level. Is this statement true or false?

a. True. This is one of the ways gifted students are identified
b. True, although a student might meet the standard at a higher grade level in one subject but not in all subjects
c. True. Additionally, gifted students whose vertical score in a particular subject is equal to or greater than the Met Standard at a higher grade level should be working at that higher level, either by taking those classes at that grade level or via differentiation
d. False

53. Gifted and talented teachers are required to have 30 hours of specialized training within the first semester of their teaching, followed by 6 additional hours of training per year. Who determines the material to be covered in this training?

a. The school board
b. It is legislated by the state
c. It is a district decision
d. It is federally mandated

54. Ms. Wing often groups her students by ability in order to provide them with efficient, challenging instruction that proceeds at a pace appropriate to their development. She has recently been under attack by some parents, who consider her methods elitist and potentially racist. The parents further point out that so-called 'gifted' students don't need the depth of support that failing students require. The position taken by the National Association of Gifted Children include(s) the following argument(s):

a. Athletically gifted individuals are regularly grouped by ability and have the best success in such groupings
b. All areas of professional or graduate study/ preparation involve grouping
c. Ability grouping is a means by which gifted ELLS, students with economic disadvantages, and those with learning disabilities can be challenged along with their true peers
d. All of the above

Refer to the following for question 55:

EXHIBIT 1

CLASS DESCRIPTION

Ms. Hoffman has just begun a new school year teaching four sections of tenth-grade English to a culturally and socioeconomically diverse group of students. One of her sections is comprised of 31 students with a wide range of backgrounds, learning styles, and abilities. Ms. Hoffman has reviewed her students' previous education records, including assessment scores, to gain a deeper understanding of their individual needs. She has learned that nine of her students scored below grade level on last year's state assessments in English language arts. Of these nine students, three are English language learners (ELLs) with varying levels of proficiency, and another, Samantha, has recently moved to the area and enrolled in school. Samantha comes from a socioeconomically disadvantaged background, and has moved a great deal in the past, causing continuous interruptions in her formal education up to this point.

Another student, Luke, has been medically diagnosed with autism spectrum disorder and has an individualized education program (IEP). While Luke is typically a diligent student, he struggles with social cues and often has difficulty expressing himself clearly in writing.

Ms. Hoffman has begun working extensively with the school ELL teacher and special education department to find effective resources for supporting her students' learning needs.

While most of Ms. Hoffman's students appear to be getting along so far this year, she has noticed some growing racial tension between a few students in her class. To prevent escalation, Ms. Hoffman has been conscientiously making efforts to establish a positive classroom climate that emphasizes respect and appreciation for differences. Each week, she begins class with a brief meeting in which students can share experiences, discuss concerns, and ask questions. In addition, Ms. Hoffman included her students in the process of developing class norms and expectations. She plans to regularly implement team-building exercises to maintain a positive and inclusive learning environment.

Ms. Hoffman communicates with her students' families often. In addition to attending the school open-house night, she utilizes a variety of communication tools to ensure accessibility for all families, including sending text reminders, emails, and weekly progress reports, and frequently updating her class website. She has also invited families into the classroom to visit, observe, or volunteer as aides.

Ms. Hoffman is preparing a unit on writing arguments that will consist of a final project and a class debate. Students will be asked to consider whether personal technology devices should be allowed in school. The unit will last approximately one week, and will include both individual and collaborative portions. Throughout the unit, Ms. Hoffman will conduct formative assessments to reflect on her instructional effectiveness and student engagement. She has selected a group of three students to extend their project by drafting a proposal to present to the school administration.

EXHIBIT 2

Excerpt from Ms. Hoffman's Lesson Plan

Topic: Writing and presenting an argument on a specific topic	
Standard: • **WHST1:** Write arguments focused on discipline-specific content. • **WHST1a:** Introduce precise claim(s), distinguish the claim(s) from alternate or opposing claims, and create an organization that establishes clear relationships among the claim(s), counterclaims, reasons, and evidence. • **WHST1b:** Develop claim(s) and counterclaims objectively, supplying data and evidence for each while pointing out the strengths and limitations of both claim(s) and counterclaims in a discipline-appropriate form and in a manner that anticipates the audience's knowledge level and concerns.	
Essential Question: • What makes an argument credible and effective? • What information and perspectives do I need to gather when crafting an argument? • What is the significance of including counterclaims when presenting an argument?	
Lesson Objectives: • Students will be able to research and gather evidence to prepare an objective argument on a specific topic. • Students will be able to analyze multiple perspectives to prepare an argument. • Students will be able to support their claims with credible evidence from multiple perspectives.	
Grouping: Students will work in teacher-assigned groups of three for a portion of the unit. Students will choose a side to participate in the debate portion of the unit.	
Lesson Component	**Activity**
Introduction	Students will be shown a series of statements, and will be asked to move from one side of the classroom to the other to indicate whether they are for or against the argument presented in each statement. The last statement will be about using personal technology devices in school. The teacher will then lead a class discussion in which students are asked to elaborate on their stance regarding this topic, including potential pros and cons of utilizing personal technology devices in school. Students will see a short presentation that reviews the importance of supporting a claim with evidence, as well as the details of the upcoming unit.
Individual Activity	Students will be assigned to read two articles that present opposing viewpoints regarding the use of personal technology devices in school. Students will be asked to highlight key points of each article, including supporting evidence.
Small-Group Activity	Students will work together in groups of three to research and prepare an argument to support their stance regarding whether personal technology devices should be allowed in school. Student research will culminate in a final project of their choosing to present their argument.
Whole-Class Activity	Students will participate in a whole-class debate regarding whether personal technology devices should be allowed in school. Each side will choose a speaker, and will have five minutes at a time to present and defend their argument. Students will have a class period prior to the debate to practice presenting their arguments.
Formative Assessments	• After the presentation, students will be asked to indicate their level of understanding based on a scale of 1-5, with 1 indicating no understanding, and 5 indicating complete understanding. • During the individual activity, students will be given color-coded cards to indicate a need for assistance or clarification. • Students will complete an exit ticket at the end of each lesson to reflect on their understanding and progress. • The teacher will check frequently for understanding throughout all stages of the unit.
Accommodations	• Texts will be scaffolded during the individual activity. • Extra time will be allotted as needed for the individual and small-group activities. • Graphic organizers will be provided for the small-group activity. • The introductory presentation will be accompanied by visual representations.

Exhibit 3

Excerpts from Students' Exit Tickets

Question: Does your group believe that personal technology devices should be allowed in school? What part of the unit have you enjoyed the most so far? Which portion has been the most difficult?

<u>Samantha</u>

Our group is in favor of using personal technology devices in school because they can be useful for learning. So far, my favorite part of the unit has been working in a group because my classmates have been very helpful to me. I also think that the topic is interesting because we use technology in our everyday lives. The hardest parts of this unit for me so far have been understanding the articles and doing the research for our argument, but my group members have been doing most of it.

<u>Luke</u>

Our group is against allowing personal technology devices in school because they are distracting to students, and not everyone can afford their own device. So far, I liked reading the articles about the pros and cons of using personal devices in school because I got to work alone. The most difficult part for me has been working in a group because I don't know my classmates very well and don't know what to say to them. They are mainly working with each other while I watch, and I wish I could have done this project by myself. I also did not like the warm-up activity where we had to move around the room.

<u>Isabel (ELL Student)</u>

My group is in favor of using personal devices in school because technology is a big part of everyday life. So far, my favorite part of the unit has been working in a group because I like my group members and we are doing a good job at coming up with an argument. I also like that we get to pick our final project because I can be creative. I'm nervous about the debate part of the project though because I get nervous speaking in front of big groups of people.

55. Which of the following is the most likely reason Ms. Hoffman selected three students to extend their project by assigning them to draft a proposal for the school administration?

a. To demonstrate the real-world application of their learning
b. To support the learning needs of gifted students by challenging them academically
c. To ensure these students had an additional assignment in case they finished too early
d. To encourage collaboration among diverse groups of students

56. Every Friday, a teacher works with her gifted classroom on creative problem solving. She has begun bringing in ordinary household objects and asking students to work in groups to come up with different ways they could use the items if they were shipwrecked on an island. What is the likely purpose of this activity?

a. To help students learn to think outside the box
b. To teach students how to organize their ideas
c. To help students learn how to work well in groups in the classroom
d. To activate the students' imagination

57. Mr. Santiago has grouped students with similar interest in architecture and very different learning styles. He has given them a complex project and suggested the initial step to designing the project involves a brief, intense period in which all group members suggest and consider any possible ideas without ranking or judging their value. He reminds them that a flexible mind is capable of taking in more information and making brilliant creative leaps. This creative approach to problem-solving is called:

a. Brainstorming
b. Brainraining
c. Multiple thought process
d. Creative braining

58. Bruner, one of the central figures of the 'cognitive revolution,' later became critical of his earlier position and turned his attention to the matter of how _______ shapes thinking.

a. Creativity
b. Art
c. Science
d. Culture

59. A differentiation strategy for students with high ability involves streamlining work to match ability, thereby creating a more challenging environment. This is called:

a. Curriculum packaging
b. Curriculum packing
c. Curriculum compacting
d. Curriculum packeting

60. The term 'depth' refers to a number of interconnected strategies employed when exploring an area of study in terms of content. Studying details, patterns, and cultural considerations; examining content from concrete to theoretical; moving from known and familiar to unknown and unfamiliar; and ___________ are among these strategies.

a. Weighing knowledge of the physical world and the spiritual world
b. Moving beyond theoretical concepts to real-world experience
c. Considering all previous knowledge and projecting future knowledge
d. Moving beyond fact to generalizations and principles

61. How should a teacher initially plan to manage a highly diverse classroom?

a. On the first day of class, the teacher should ask each student to complete a survey to determine the students' cultural, ethnic, and racial background.
b. A good teacher should be intimately familiar with the major cultures within the local school district.
c. A teacher should plan lessons so that new material is introduced using a variety of different approaches.
d. Guest teachers from different ethnic backgrounds should occasionally be invited to teach, so that all students will feel more comfortable.

68. The first step in having a student considered for a gifted program is nomination via a Student Observation Form submitted by whom?

a. Teachers or parents
b. Other community members
c. School administrators and other personnel
d. Any of the above

69. A student who has tremendous leadership potential has been charged with a series of tasks. She will gather a variety of types of information that describe her areas of strength and continually update this information; she will categorize this material as pertaining to ability, learning styles, social skills, and special interest; she will create a time line to consider options and guide decisions in her differentiation; she will communicate with the teacher regarding these options and decisions; and she will use the information to communicate with mentors, parents, and others about her development. What is she doing?

a. Creating a Total Talent Portfolio
b. Developing a number of assessment tools
c. Developing a differentiation plan
d. Creating a Personal Growth file

70. What type of assessment is often used to identify gifted students in leadership, creativity, or artistic areas?

a. Qualitative
b. Quantitative
c. Both a and b
d. Formal and informal

71. Gifted and Talented underachievers can best be helped by parents and teacher who _______________ in an environment that____________.

a. Share consistent expectations / is flexible, respects the student's opinions, and does not demand absolute obedience
b. Demand excellence / is firm, unbending, clear, and consistent
c. Ask questions / is invigorating, inviting, inspiring, and flexible
d. Offer support / is loving, understanding, accepting, and respectful

72. According to Jean Piaget's four-stage theory of cognitive development, the distinction between people at the concrete operational stage (approximately 7–12 years of age) and the formal operational stage (approximately 12–16 years of age) of development is that:

a. Unlike those at the concrete operational stage, those at the formal operational stage can think logically in concrete terms.
b. Unlike those at the concrete operational stage, those at the formal operational stage are highly reliant on sensory and motor skills to learn.
c. Those in the concrete operational stage are capable of abstract thought, while those at the formal operational stage can think only in concrete terms.
d. Those in the concrete operational stage can think logically only with respect to concrete experiences, while those at the formal operational stage can reason in abstract and hypothetical terms.

73. Gifted children typically get similar results with different types of I.Q. tests, such as the Wechsler Preschool and Primary Scale of Intelligence (WPPSI), the Stanford-Binet Intelligence Scale (SB), and the Wechsler Intelligence Scale for Children (WISC).

a. Neither true nor false
b. False
c. True
d. These I.Q. tests are not given to gifted students

74. Once a student has been nominated for a gifted and talented program via a Student Observation Form, he or she _______ be assessed.

a. Will
b. Might
c. Will not; nomination guarantees assessment
d. Must

75. According to Erikson's theory of psychosocial development, which ability must be achieved during the period of adolescence?

a. Identity
b. Intimacy
c. Industry
d. Autonomy

76. Students who are gifted are required by federal law to have an Individualized Education Program (IEP).

a. True. Students in all 50 states receiving any form of special education must have such a plan
b. False. No such plan for children identified as Gifted and Talented has ever been federally mandated
c. Neither true nor false. IEPs are used exclusively for students with learning disabilities or mental insufficiencies
d. At one point this was true, but it has been overturned

77. Why are open-ended responses of particular value to the teacher?

a. The encourage dialogue between teacher and student, or between student and peers
b. They give the teacher insight into how a particular student structures ideas, uses language, and demonstrates understanding
c. They are not of particular value because they cannot be measured and do not remain consistent through time
d. They support the gifted student's emotional and psychological development by demonstrating respect for the student's abilities

78. In what areas do computers have the LEAST benefit for student learning?

a. Simple drills, such as teaching multiplication tables, because this frees up the teacher to work with the students on developing higher-level skills
b. Role-playing games that develop greater social awareness and that allow the students to participate more directly in the events they are learning about
c. Simulations of mechanical activities that help the students develop motor skills and reflexes
d. Exercises that teach higher-order thinking and professional skills

79. Kingore Observation Inventory (KOI) is used to:

a. The correct name is the Kingmore Ongoing Index; it is used as a continuous informal assessment
b. Evaluate students in the gifted program to determine if they should remain in the program
c. Assess the capabilities of English Language Learners with learning disabilities that might be considered for a Gifted and Talented program
d. Identify gifted and talented students, including those from culturally diverse backgrounds or those who live at or below the poverty level

80. Who is responsible for identifying gifted and talented students?

a. Each county establishes its own guidelines based upon State Board of Education criteria, which apply to all schools in the county
b. The State Board of Education is responsible for designing a systematic plan which includes assessment, evaluation, curriculum, and planning. Each district must abide by this overreaching plan, all owing slight modifications when necessary and approved
c. Each building designs its own system for identifying and serving Gifted and Talented students. It must follow loose guidelines suggested by the district, which in turn, follows loose guidelines suggested by the State
d. Each district develops and implements a program that is based upon criteria determined by the State Board of Education, which includes identifying and serving gifted and talented students

Answer Key and Explanations

1. B: Array of learning experiences. An array of learning experiences is defined as a menu of challenging learning experiences or opportunities that fit the unique interests and abilities of advanced level students.

2. D: The student finds the work too simple, so they dismiss or ignore it. Gifted and talented students learn best when they are interested in the material, allowed to explore it independently and are challenged by it. They are not motivated by grades or products so much as by the learning process.

3. B: Curiosity; adept at using multiple resources; willing to take risks; applies key words when asking questions. According to Kaplan, belief sets are a strong indicator of learning success. Students who are curious, can move comfortably between resources, are willing to take risks, and listens for key words in questions achieve their goals at the highest level.

4. B: Adaptive pencil and paper holders would not help Stephen, as he can only press buttons with one hand. He could not manipulate a writing implement with his paralysis. Therefore, option A is incorrect.

A computer with a built-in voice synthesizer would let Stephen program all the words in his vocabulary into it. He could express himself despite his paralyzed vocal cords. He has excellent expressive and receptive language skills and a high IQ, so this would be the best solution for him. Therefore, option B is correct.

An electrolarynx is used by some laryngectomy patients whose larynxes have been surgically removed because of laryngeal cancer. It is an alternative to learning esophageal speech. This device is hand-held. Stephen could not hold it up to his throat as required. It also requires the speech system's muscles to work. It emits a tone, and the user manipulates the tongue, lips, cheeks, and throat, as in normal speech, to modulate the tone into recognizable speech sounds. Stephen could not use this device with his paralysis. Therefore, option C is incorrect.

A communication board would not be the best choice for Stephen, as it would afford simpler communication of basic needs and feelings. With Stephen's high IQ and advanced academic skills, he can program a computer with more complex ideas and complete sentences. He would undoubtedly find the communication board limited and would be frustrated. Therefore, option D is incorrect.

5. D: Some gifted students excel in all areas, including athletics and social skills (A), but not necessarily. Gifted children can have different abilities and interests than peers, making social interaction awkward. They may have difficulty relating to peers with more typical abilities and interests, resulting in peers who may not want to interact with them for being different. Some gifted children prefer academic/creative activities over social activities. Some gifted children show superior academic performance (B), while some do not. Some are bored by traditional curricula (including special education for the gifted) and lack motivation, while some use unconventional methods. As a child, Albert Einstein reportedly was failed in math by a teacher for achieving correct answers without following prescribed steps. Not all gifted children get high IQ scores (C) because giftedness also includes creativity or other forms not represented by IQ scores, not just intelligence. Some gifted children do not respond to standardized measures because they may think differently and/or be uninterested in test items.

6. A: Administrators and counselors who make programming or hiring decisions. All individuals who work with or for children identified as gifted or talented are required to complete 6 hours of training annually.

7. C: The more modalities and media the learning materials include, the more they can address a larger number of varied student needs. There are two reasons for this: (1) if one modality or medium is not effective for certain students, then another one will be; (2) presenting the same information redundantly and through different modalities makes learning more effective for all students than presenting it through only one (A), (B), (D).

8. B: Differentiate for those students who understand the concept but arrive at incorrect answers. By reviewing their steps, the teacher can determine why answers are incorrect and can demonstrate the correct approach. Concept is more important than content. If the students truly understand the 'how', they have the tools they need to arrive at the 'what'. However, if a student demonstrates the same type of error repeatedly, she may not fully understand the concept. Concept is more important than content.

9. A: Recommend acceleration opportunities. The steps to curriculum compacting include understanding learning objectives; identify students who can master objectives most quickly, pretest, streamline instructional period for students who understand objectives and recommending acceleration opportunities.

10. C: Affective curriculum. Gifted students often feel their differences; they can be more highly attuned to surroundings and nonverbal cues, and they may be reluctant to take risks that would propel them beyond their comfort zones. Affective curriculum allows them the opportunity to explore social and deeply personal aspects of their giftedness.

11. D: Advanced Placement, in which high schools offer coursework meeting criteria recognized by institutions of higher education. Often, college credit is earned when an AP exam in specific content areas is passed.

12. A: She has a high external locus of control. Locus of control is the degree to which an individual believes she can control events or actions that impact them. The more in control an individual believes herself to be, the higher her internal locus of control. The less in control an individual believes herself to be, the higher her external locus of control. In this example, the student blames other students, her brother and her teacher for missing schoolwork, believing herself to have no control in these matters. Her locus of control is external, and the degree to which she feels out of control indicates it is also high.

13. B: Help the teacher determine how to enrich and accelerate students' learning based upon their individually unique profiles. The Total Talent Portfolio is a systematic method of gathering, analyzing and classifying students' abilities, given equal weight to interests and learning styles as to cognitive skills to aid the teacher in deciding how to enrich and accelerate students' individualized learning based upon their unique profiles.

14. B: 6. Teachers working with gifted students are required to complete 30 hours of professional development by the time they finish their first semester, followed by 6 additional hours annually.

15. B: Odd; her lack of English, terrible handwriting, nervousness, and bizarre musical talents make her seem abnormal. Many gifted children are highly sensitive and can be far more easily excited that their average peers. They may not know how to relate to other children, and as a result lack social skills. Ming's lack of English, terrible handwriting, nervousness and musical talents might make her seem abnormal to her classmates.

16. A: Gifted and talented students have a unique set of abilities that are often overlooked in the classroom. These students must be challenged academically and intellectually in order to properly accommodate their learning needs to sustain engagement in learning. Incorporating opportunities for extension and enrichment would have helped create a more inclusive environment by allowing these students to build upon and strengthen their learning.

17. A: Accommodating the needs of gifted students requires that teachers include meaningful opportunities for extension and enrichment into instruction. This is necessary to provide these students with engaging and intellectually challenging learning experiences aligned with their unique abilities. By incorporating an extension activity in which selected students were assigned to create original stories, Ms. Jewel effectively acknowledges and implements learning activities that meet the needs of her gifted and talented students.

18. D: Planned learning experiences that build upon a progressively stronger foundation from year to year. A Continuum of Learning Experiences is a multi-year construct that relies upon each previous year to provide the framework for further academic and artistic development.

19. A: Required. Unless the student has been emancipated, a parent or guardian must give written permission before that student can participate in a Gifted and Talented program.

20. B: Include provisions about furloughs, reassessment, exiting strategies, transferring students, and the district decision appeals process. A district's written policy regarding student identification must include provisions for continuing screening; assessment procedures from a number of sources; insure that all populations have assessment available; insure final selections are made by at minimum three local educators with specialized training; and contain provisions regarding furloughs or leaves of absence, reassessment, exiting strategies, transferring students and the process by which appeals can be made.

21. C: Asynchrony. According to the Columbus Group (1991) "Giftedness is asynchronous development in which advanced cognitive abilities and heightened intensity combine to create inner experiences and awareness that are qualitatively different from the norm. This asynchrony increases with higher intellectual capacity. The uniqueness of the gifted renders them particularly vulnerable and requires modifications in parenting, teaching, and counseling in order for them to develop optimally."

22. C: Both A and B. Many teachers prefer to assess kindergarten students using only qualitative assessments. In areas of creativity, artistic and leadership abilities, teachers can chose to use only qualitative assessments.

23. C: Screening Assessment for Gifted Elementary Students, and describes a standardized assessment. Unlike many tools used to identify Gifted and Talented students, SAGES was designed expressly for that function. SAGES is also designed to reduce prejudice in gifted and talented testing.

24. B: Piaget's first stage of cognitive development is called the sensorimotor stage. This is when infants respond to input they receive from the environment through their sensory organs by engaging in motor actions. When they realize how the environment then reacts to some of these actions, they respond again; this is called circular reactions. Operations with concrete things (A) occur during Piaget's third stage called concrete operations. Operations with abstract things (C) occur during Piaget's fourth and final stage called formal operations. Intuition and animism but not logic (D) occur during Piaget's second stage, which he called preoperational.

25. A: A federally-funded center with the charter of developing effective methods of identification and testing of gifted and talented students, and programming for such students. The National Research Center on the Gifted and Talented is one of three components of the Javits Act, a reauthorization of The Elementary and Secondary Education Act.

26. D: A student may be granted a leave of absence for specific reasons and for a pre-established length of time; the leave must be approved. The leave of absence must all be requested at minimum six weeks prior to the furlough date.

27. C: Practical intelligence. According to Sternberg's Triarchic Theory of Intelligence, componential intelligence (the ability to analyze, think in abstract terms and efficiently process information), experiential intelligence (the ability to creatively combine unrelated information and to perceive patterns in material that, on the surface, appears to be unrelated) and practical intelligence (the ability to apply thinking skills to contextual situations by reassessing one's position in terms of strengths and weaknesses) are the three core qualities of giftedness.

28. A: Raw scores are used on specific tests to show how well a student did; this score is not necessarily consistent from test to test. A scale score translates raw scores onto a measurement common to all forms of testing for a particular assessment.

29. D: The student is primarily responsible. One of the purposes of the Total Talent Portfolio is to encourage student autonomy. In taking on responsibility in the selection of materials that will compose the TTP, updating the portfolio at regular intervals and establishing personal goals, the student assumes ownership and is therefore more committed. The teacher's role is that of a guide, primarily in the process of review.

30. D: Providing gifted and talented students with meaningful opportunities to extend their learning allows them to progress at a pace that matches their abilities and learning needs. This helps to ensure that these students remain engaged in the classroom, deepens their understanding, and encourages them to continuously seek challenging learning experiences.

31. D: Have the capability for high accomplishment in intellectual, creative, artistic, or leadership areas or in specific academic fields, and who need services or activities not ordinarily provided by the school to develop those capabilities. The Jacob Javits Gifted and Talented Students Education Act (Javits) is a federal program that addresses the needs of gifted and talented children. The purpose of the Act is to coordinate programs of scientific research, projects, pioneering approaches, and the like to enhance schools' abilities to foster the educational needs of gifted and talented students.

32. D: Yes. If the teacher is already running a student-centered classroom, allowing choices in assignments and using flexibility in student grouping, she is already differentiating. She has received additional training to allow her to continue to modify classroom assignments to more deeply challenge the gifted students.

33. C: That gifted student's value learning over performance. Her lack of motivation shouldn't be an occasion for blame. Instead of focusing on grades and performance, the father might consider exploring some learning opportunities in areas of the girl's interest. For example, a visit to an art museum followed by a trip to the library for books about a particular art movement might motivate the student.

43. B: Effectively differentiating science instruction includes incorporating activities that provide the appropriate level of challenge to align with the intellectual abilities of gifted and talented learners. Doing so sustains these students' engagement in scientific learning while promoting continuous development at a pace that corresponds with their individual needs. Differentiation for gifted and talented learners can be achieved through a variety of strategies, such as extending the scope of instruction, adding a layer of complexity to a summative assessment, or allowing for flexibility to explore personal interests. In this example, the teacher can accommodate gifted and talented students in the classroom by having them research local weather patterns throughout history. Such an activity would provide enrichment by allowing these students to explore the topic on a broader scale, identify long-term weather patterns, and ultimately gain a deeper understanding of occurring and recurring patterns in nature.

44. B: He will remain in the program and it will be modified to become more appropriate for his learning needs. It is illegal to discriminate against students with learning or behavioral disorders.

45. B: Parallel Curriculum Model. This strategy modifies the curriculum in depth, complexity and originality and indicates four concurrent directions, Core Curriculum, Curriculum of Connections, Curriculum of Practice and Curriculum of Identity.

46. C: In a highly diversified class, teachers often fall into the trap of offering too much encouragement to the strongest students while neglecting the weaker students. As a result, the weaker students become resentful and feel little incentive to work harder. Answer A is not a mistake. The weaker students need more of the teacher's attention. Answer B is not really a problem as long as the stronger students are respectful and provide accurate information. Answer D is, again, not really a mistake. The weaker students need to be corrected in order to learn. Of course, the teacher should take care not to make the students feel bad about themselves.

47. D: A student's interests and learning styles and the student's cognitive ability are equally considered. Total Talent Portfolios include attention to student interests and learning styles in addition to cognitive abilities that have traditionally contributed to educational decisions.

48. B: Assesses learning using portfolios, performance, observations, and other assessments that model real-world representations. Authentic assessment tasks are not simply practice for the sake of practice; they are rehearsals for real-world experiences. They assess a wide array of literacy skills in the context of how that skill would be applied in the real world.

49. A: Assimilation is fitting a new experience into an existing schema. Accommodation is altering an existing schema or forming a new schema to accommodate a new experience. These two processes are part of the overall process of adaptation, i.e., adjusting one's thinking to the environment via interacting with it. This adaptation process helps the individual to maintain equilibrium, or balance.

50. A: Weigh the pros and cons of entering a student into a gifted program. Pfeiffer's Scales is a time-efficient way to assess giftedness using six scales for students beyond kindergarten. It is administered by teachers and covers intellectual ability, academic ability, creativity, artistic talent, leadership ability, and motivation.

51. A: Establish own learning goals; self-assess; move to an independent challenge activity when she understands the lesson; encourage her to work with peers close to her intellectual level. Research has proven these techniques more deeply engaged gifted students that traditional methods, and contribute to such students' degree of motivation.

52. D: False. Vertical scores at a particular grade level indicate the degree of mastery of that subject at that grade level. This number is unrelated to the Standards established at a higher grade level. Vertical scores are not the same as Standards.

53. C: It is a district decision. Prior to assignment in the program, teachers who provide instruction and services for gifted students are required to have a minimum of 30 hours of staff development that includes nature and needs of gifted/talented students, assessing student needs, and curriculum and instruction for gifted students. However, how many hours to devote to each category is left up to the district to determine.

54. D: All the above. The National Association of Gifted Children (NAGC) has taken a strong position on grouping. NAGC argues that athletically gifted individuals, whether children or adults, are regularly grouped by ability and have the best success in such groupings. Furthermore, all areas of professional or graduate study/ preparation involve grouping. Yet another argument is that ability grouping is a means by which gifted ELLS, students with economic disadvantages and those with learning disabilities can be challenged along with their true peers.

55. B: Gifted and talented students possess a unique set of abilities and learning needs that often go overlooked in a general education setting. Selecting, modifying, and implementing curriculum to challenge these students academically and intellectually is imperative to creating a supportive, inclusive, and engaging learning environment. By creating an opportunity for extension, Ms. Hoffman modifies instruction to provide enrichment for gifted students, thus accommodating their unique learning needs in a meaningful way.

56. A: When teachers work on incorporating creative problem solving into a classroom by posing interesting questions or scenarios, they are helping students work on finding new solutions to problems when they arise. This helps students develop the way they think, which can help them in classes and in the outside world. This helps students learn practical skills that they can apply in the real world to help them overcome obstacles they may encounter in everyday situations. While this activity will likely activate the students' imagination, it is more likely that the teacher's main intention for this activity was to work on helping students develop their skills with creative problem solving.

57. A: Brainstorming. Brainstorming is a highly-charged creative technique to approach problem-solving, in which group members spontaneously offer a wealth of associations, ideas and possible solutions. The group considers all possibilities without initial judgment, looking for creative possibilities that might otherwise be overlooked.

58. D: Culture. Bruner's original position on education underwent a major shift by the mid 1990s. In The Culture of Education he expresses the position that it is impossible to think outside of culture; regardless of how interior or 'mental' the process, thinking requires the framework or grid of culture.

59. C: Curriculum compacting. Curriculum compacting is a differentiation strategy for students with high ability involves streamlining work to match ability, thereby creating a more challenging environment. The three steps to compacting are determining goals, identifying students, and offering acceleration and enrichment opportunities. Curriculum compacting eliminates repetition of lessons that have already been learned, and speeding up lessons to match the learner's pace.

60. D: Moving beyond fact to generalizations and principles. Depth involves exploring content within a discipline; analyzing from the concrete to the abstract, familiar to the unfamiliar, known to the unknown; exploring the discipline by going past facts and concepts into generalizations, principles, theories, laws; investigating the layers of experience within a discipline through details, patterns, trends, unanswered questions, ethical considerations.

61. C: Introducing new material using several different strategies will help make it more accessible to a larger group of students, regardless of their specific backgrounds. All three of the other answers represent a shallow approach to diversity that presumes that different cultures adhere to stereotypes. The truth is that different students think differently, and using a wider variety of teaching methods will cast a wider net.

62. B: 5%. School districts are given an annual allotment equal to a particular district's base allotment multiplied by 12 per student per year. No more than 5% of the total number of students can be considered for Gifted and Talented funding.

63. D: Are anecdotal records, observations, interviews, student products, checklists, and similar materials. In contrast, quantitative measures involve standardized testing.

64. D: Middle school is a critical time during which students develop a respect for diversity, make the decision to continue with school, and develop tolerance for those who are different. Middle school students also develop higher-order thinking skills, but this represents cognitive development rather than value development.

65. C: Behaviorist learning theory states that receiving rewards for certain behaviors reinforces the behavior, or increases the chance the behavior will recur for more rewards. A student receiving positive reinforcement for studying hard will continue to do so. Social learning theory states people respond not only to direct rewards, but also to observing others receiving rewards, and will imitate rewarded behaviors. A student studying hard who observes struggling classmates getting more teacher attention may stop studying for comparable attention (A). A student observing classmates getting rewards for studying hard imitates that behavior (B). A student observing struggling classmates getting less attention will continue to study hard (D) to avoid losing the rewards. Choice C is an example of behaviorist learning theory, the others of social learning theory.

66. B: An individual's belief or feeling about his or her ability regarding a specific undertaking. An unmotivated gifted student may have low self-efficacy, believing that a task, project or other undertaking is beyond his abilities.

67. C: No Child Left Behind (NCLB). This legislation reauthorized the Elementary and Secondary Education Act, and allocated financial support for low-income student education, teacher recruitment, professional development, technology and the like. NCLB is aimed at increasing educational success and parental contribution, and charges districts with guaranteeing state proficiency standards in mathematics and reading are reached by 2014.

68. D: Any of the above. To be considered for a gifted program, a student must be nominated. Anyone can make the nomination, including family members, teachers, school administrators and other school personnel, and community members who are familiar with the child's abilities.

69. A: Creating a Total Talent Portfolio. A TTP gathers a variety of types of information describing a student's areas of strength and interest and tracks it through time, updating as appropriate. This information is categorized (typically with a chart) and studied in order to set goals and determine differentiation. The gifted student is encouraged to develop autonomy by being primarily responsible for her TTP.

70. A: Qualitative. While it is appropriate to use both quantitative and qualitative approaches to identify 1st-12th grade students in terms of general intellectual ability or specific academic subjects, a district may choose to apply qualitative assessments only in order to identify gifted students in areas of creativity, the arts and leadership.

71. A: Share consistent expectations / is flexible, respects the student's opinions, and does not demand absolute obedience. Gifted underachievers do best when teacher and parent expectations are consistent, and in learning and home environments are not overly rigid or demanding.

72. D: According to Piaget's theory, the distinction between people at the concrete operational stage (approximately 7–12 years of age) and the formal operational stage (approximately 12–16 years of age) of development is that those in the concrete operational stage can think logically only with respect to concrete experiences, while those at the formal operational stage can reason in abstract and hypothetical terms. Children at the sensorimotor stage (birth to 2 years old) rely on their sensory perception and motor skills to learn and understand the world around them, and children at the preoperational stage (approximately 3–7 years old) think in a literal, symbolic manner.

73. B: False. While average and developmentally delayed children typically get very similar scores in a variety of I.Q. tests, gifted children more often get radically differing scores from test to test.

74. B: Might. The decision of whether to assess a child is made by a committee and is partially based on observable academic strength according. If the G/T Committee decides a particular child should be assessed, parents must first give permission.

75. A: Erikson identified a nuclear conflict to be resolved in each stage of psychosocial development. Babies confront basic trust versus mistrust, toddlers autonomy (D) versus shame and self-doubt, preschoolers initiative versus guilt, school-aged children industry (C) versus inferiority, adolescents identity (A) versus role confusion, young adults intimacy (B) versus isolation, middle adults generativity versus stagnation, and older adults integrity versus despair.

76. B: False. No such plan for children identified as gifted and talented has ever been federally mandated. While gifted children do receive special education in terms of differentiation, federal law does not require they have IEPs as learning disabled children are.

77. B: They give the teacher insight into how a particular student structures ideas, uses language, and demonstrates understanding. Because open-ended questions are the beginning of dialogue, they also give the student the opportunity to express their ideas with creativity, and exhibit the degree of their abilities to abstract, find connections, analyze and assess ideas.

78. A: While useful in some contexts, computer-monitored drills do not represent the greatest use of computers. These drills can be accomplished without computers. Even if using computers for drilling exercises, the teacher should remain active with the students and would not be more "freed up" than if the drills were administered on paper. Role-playing games, simulations, and exercises that teach higher-order thinking represent uses of computers that provide benefits that are difficult to obtain without computers.

79. D: Identify gifted and talented students, including those from culturally diverse backgrounds or those who live at or below the poverty level. The Kingore Observation Inventory is an efficient assessment and differentiation process, identifies gifted and talented students via observing behaviors in seven categories.

80. D: Each district develops and implements a program that is based upon criteria determined by the State Board of Education, which includes identifying and serving gifted and talented students.

How to Overcome Test Anxiety

Just the thought of taking a test is enough to make most people a little nervous. A test is an important event that can have a long-term impact on your future, so it's important to take it seriously and it's natural to feel anxious about performing well. But just because anxiety is normal, that doesn't mean that it's helpful in test taking, or that you should simply accept it as part of your life. Anxiety can have a variety of effects. These effects can be mild, like making you feel slightly nervous, or severe, like blocking your ability to focus or remember even a simple detail.

If you experience test anxiety—whether severe or mild—it's important to know how to beat it. To discover this, first you need to understand what causes test anxiety.

Causes of Test Anxiety

While we often think of anxiety as an uncontrollable emotional state, it can actually be caused by simple, practical things. One of the most common causes of test anxiety is that a person does not feel adequately prepared for their test. This feeling can be the result of many different issues such as poor study habits or lack of organization, but the most common culprit is time management. Starting to study too late, failing to organize your study time to cover all of the material, or being distracted while you study will mean that you're not well prepared for the test. This may lead to cramming the night before, which will cause you to be physically and mentally exhausted for the test. Poor time management also contributes to feelings of stress, fear, and hopelessness as you realize you are not well prepared but don't know what to do about it.

Other times, test anxiety is not related to your preparation for the test but comes from unresolved fear. This may be a past failure on a test, or poor performance on tests in general. It may come from comparing yourself to others who seem to be performing better or from the stress of living up to expectations. Anxiety may be driven by fears of the future—how failure on this test would affect your educational and career goals. These fears are often completely irrational, but they can still negatively impact your test performance.

Elements of Test Anxiety

As mentioned earlier, test anxiety is considered to be an emotional state, but it has physical and mental components as well. Sometimes you may not even realize that you are suffering from test anxiety until you notice the physical symptoms. These can include trembling hands, rapid heartbeat, sweating, nausea, and tense muscles. Extreme anxiety may lead to fainting or vomiting. Obviously, any of these symptoms can have a negative impact on testing. It is important to recognize them as soon as they begin to occur so that you can address the problem before it damages your performance.

The mental components of test anxiety include trouble focusing and inability to remember learned information. During a test, your mind is on high alert, which can help you recall information and stay focused for an extended period of time. However, anxiety interferes with your mind's natural processes, causing you to blank out, even on the questions you know well. The strain of testing during anxiety makes it difficult to stay focused, especially on a test that may take several hours. Extreme anxiety can take a huge mental toll, making it difficult not only to recall test information but even to understand the test questions or pull your thoughts together.

Effects of Test Anxiety

Test anxiety is like a disease—if left untreated, it will get progressively worse. Anxiety leads to poor performance, and this reinforces the feelings of fear and failure, which in turn lead to poor performances on subsequent tests. It can grow from a mild nervousness to a crippling condition. If allowed to progress, test anxiety can have a big impact on your schooling, and consequently on your future.

Test anxiety can spread to other parts of your life. Anxiety on tests can become anxiety in any stressful situation, and blanking on a test can turn into panicking in a job situation. But fortunately, you don't have to let anxiety rule your testing and determine your grades. There are a number of relatively simple steps you can take to move past anxiety and function normally on a test and in the rest of life.

Physical Steps for Beating Test Anxiety

While test anxiety is a serious problem, the good news is that it can be overcome. It doesn't have to control your ability to think and remember information. While it may take time, you can begin taking steps today to beat anxiety.

Just as your first hint that you may be struggling with anxiety comes from the physical symptoms, the first step to treating it is also physical. Rest is crucial for having a clear, strong mind. If you are tired, it is much easier to give in to anxiety. But if you establish good sleep habits, your body and mind will be ready to perform optimally, without the strain of exhaustion. Additionally, sleeping well helps you to retain information better, so you're more likely to recall the answers when you see the test questions.

Getting good sleep means more than going to bed on time. It's important to allow your brain time to relax. Take study breaks from time to time so it doesn't get overworked, and don't study right before bed. Take time to rest your mind before trying to rest your body, or you may find it difficult to fall asleep.

Along with sleep, other aspects of physical health are important in preparing for a test. Good nutrition is vital for good brain function. Sugary foods and drinks may give a burst of energy but this burst is followed by a crash, both physically and emotionally. Instead, fuel your body with protein and vitamin-rich foods.

Also, drink plenty of water. Dehydration can lead to headaches and exhaustion, especially if your brain is already under stress from the rigors of the test. Particularly if your test is a long one, drink water during the breaks. And if possible, take an energy-boosting snack to eat between sections.

Along with sleep and diet, a third important part of physical health is exercise. Maintaining a steady workout schedule is helpful, but even taking 5-minute study breaks to walk can help get your blood pumping faster and clear your head. Exercise also releases endorphins, which contribute to a positive feeling and can help combat test anxiety.

When you nurture your physical health, you are also contributing to your mental health. If your body is healthy, your mind is much more likely to be healthy as well. So take time to rest, nourish your body with healthy food and water, and get moving as much as possible. Taking these physical steps will make you stronger and more able to take the mental steps necessary to overcome test anxiety.

Write out your plan so you don't get lost as you go. Having a written plan also helps you feel more in control of the study, so anxiety is less likely to arise from feeling overwhelmed at the amount to cover.

Step 3: Gather Your Tools

Decide what study method works best for you. Do you prefer to highlight in the book as you study and then go back over the highlighted portions? Or do you type out notes of the important information? Or is it helpful to make flashcards that you can carry with you? Assemble the pens, index cards, highlighters, post-it notes, and any other materials you may need so you won't be distracted by getting up to find things while you study.

If you're having a hard time retaining the information or organizing your notes, experiment with different methods. For example, try color-coding by subject with colored pens, highlighters, or post-it notes. If you learn better by hearing, try recording yourself reading your notes so you can listen while in the car, working out, or simply sitting at your desk. Ask a friend to quiz you from your flashcards, or try teaching someone the material to solidify it in your mind.

Step 4: Create Your Environment

It's important to avoid distractions while you study. This includes both the obvious distractions like visitors and the subtle distractions like an uncomfortable chair (or a too-comfortable couch that makes you want to fall asleep). Set up the best study environment possible: good lighting and a comfortable work area. If background music helps you focus, you may want to turn it on, but otherwise keep the room quiet. If you are using a computer to take notes, be sure you don't have any other windows open, especially applications like social media, games, or anything else that could distract you. Silence your phone and turn off notifications. Be sure to keep water close by so you stay hydrated while you study (but avoid unhealthy drinks and snacks).

Also, take into account the best time of day to study. Are you freshest first thing in the morning? Try to set aside some time then to work through the material. Is your mind clearer in the afternoon or evening? Schedule your study session then. Another method is to study at the same time of day that you will take the test, so that your brain gets used to working on the material at that time and will be ready to focus at test time.

Step 5: Study!

Once you have done all the study preparation, it's time to settle into the actual studying. Sit down, take a few moments to settle your mind so you can focus, and begin to follow your study plan. Don't give in to distractions or let yourself procrastinate. This is your time to prepare so you'll be ready to fearlessly approach the test. Make the most of the time and stay focused.

Of course, you don't want to burn out. If you study too long you may find that you're not retaining the information very well. Take regular study breaks. For example, taking five minutes out of every hour to walk briskly, breathing deeply and swinging your arms, can help your mind stay fresh.

As you get to the end of each chapter or section, it's a good idea to do a quick review. Remind yourself of what you learned and work on any difficult parts. When you feel that you've mastered the material, move on to the next part. At the end of your study session, briefly skim through your notes again.

But while review is helpful, cramming last minute is NOT. If at all possible, work ahead so that you won't need to fit all your study into the last day. Cramming overloads your brain with more information than it can process and retain, and your tired mind may struggle to recall even

previously learned information when it is overwhelmed with last-minute study. Also, the urgent nature of cramming and the stress placed on your brain contribute to anxiety. You'll be more likely to go to the test feeling unprepared and having trouble thinking clearly.

So don't cram, and don't stay up late before the test, even just to review your notes at a leisurely pace. Your brain needs rest more than it needs to go over the information again. In fact, plan to finish your studies by noon or early afternoon the day before the test. Give your brain the rest of the day to relax or focus on other things, and get a good night's sleep. Then you will be fresh for the test and better able to recall what you've studied.

Step 6: Take a Practice Test

Many courses offer sample tests, either online or in the study materials. This is an excellent resource to check whether you have mastered the material, as well as to prepare for the test format and environment.

Check the test format ahead of time: the number of questions, the type (multiple choice, free response, etc.), and the time limit. Then create a plan for working through them. For example, if you have 30 minutes to take a 60-question test, your limit is 30 seconds per question. Spend less time on the questions you know well so that you can take more time on the difficult ones.

If you have time to take several practice tests, take the first one open book, with no time limit. Work through the questions at your own pace and make sure you fully understand them. Gradually work up to taking a test under test conditions: sit at a desk with all study materials put away and set a timer. Pace yourself to make sure you finish the test with time to spare and go back to check your answers if you have time.

After each test, check your answers. On the questions you missed, be sure you understand why you missed them. Did you misread the question (tests can use tricky wording)? Did you forget the information? Or was it something you hadn't learned? Go back and study any shaky areas that the practice tests reveal.

Taking these tests not only helps with your grade, but also aids in combating test anxiety. If you're already used to the test conditions, you're less likely to worry about it, and working through tests until you're scoring well gives you a confidence boost. Go through the practice tests until you feel comfortable, and then you can go into the test knowing that you're ready for it.

Test Tips

On test day, you should be confident, knowing that you've prepared well and are ready to answer the questions. But aside from preparation, there are several test day strategies you can employ to maximize your performance.

First, as stated before, get a good night's sleep the night before the test (and for several nights before that, if possible). Go into the test with a fresh, alert mind rather than staying up late to study.

Try not to change too much about your normal routine on the day of the test. It's important to eat a nutritious breakfast, but if you normally don't eat breakfast at all, consider eating just a protein bar. If you're a coffee drinker, go ahead and have your normal coffee. Just make sure you time it so that the caffeine doesn't wear off right in the middle of your test. Avoid sugary beverages, and drink enough water to stay hydrated but not so much that you need a restroom break 10 minutes into the

test. If your test isn't first thing in the morning, consider going for a walk or doing a light workout before the test to get your blood flowing.

Allow yourself enough time to get ready, and leave for the test with plenty of time to spare so you won't have the anxiety of scrambling to arrive in time. Another reason to be early is to select a good seat. It's helpful to sit away from doors and windows, which can be distracting. Find a good seat, get out your supplies, and settle your mind before the test begins.

When the test begins, start by going over the instructions carefully, even if you already know what to expect. Make sure you avoid any careless mistakes by following the directions.

Then begin working through the questions, pacing yourself as you've practiced. If you're not sure on an answer, don't spend too much time on it, and don't let it shake your confidence. Either skip it and come back later, or eliminate as many wrong answers as possible and guess among the remaining ones. Don't dwell on these questions as you continue—put them out of your mind and focus on what lies ahead.

Be sure to read all of the answer choices, even if you're sure the first one is the right answer. Sometimes you'll find a better one if you keep reading. But don't second-guess yourself if you do immediately know the answer. Your gut instinct is usually right. Don't let test anxiety rob you of the information you know.

If you have time at the end of the test (and if the test format allows), go back and review your answers. Be cautious about changing any, since your first instinct tends to be correct, but make sure you didn't misread any of the questions or accidentally mark the wrong answer choice. Look over any you skipped and make an educated guess.

At the end, leave the test feeling confident. You've done your best, so don't waste time worrying about your performance or wishing you could change anything. Instead, celebrate the successful completion of this test. And finally, use this test to learn how to deal with anxiety even better next time.

Review Video: Test Anxiety
Visit mometrix.com/academy and enter code: 100340

Important Qualification

Not all anxiety is created equal. If your test anxiety is causing major issues in your life beyond the classroom or testing center, or if you are experiencing troubling physical symptoms related to your anxiety, it may be a sign of a serious physiological or psychological condition. If this sounds like your situation, we strongly encourage you to seek professional help.

Online Resources

Due to our efforts to try to keep this book to a manageable length, we've created a link that will give you access to all of your online resources:

mometrix.com/resources719/texesgtsupp

It's Your Moment, Let's Celebrate It!

Share your story @mometrixtestpreparation

Dear Future Exam Success Story

First of all, **THANK YOU** for purchasing Mometrix study materials!

Second, congratulations! You are one of the few determined test-takers who are committed to doing whatever it takes to excel on your exam. **You have come to the right place.** We developed these study materials with one goal in mind: to deliver you the information you need in a format that's concise and easy to use.

In addition to optimizing your guide for the content of the test, we've outlined our recommended steps for breaking down the preparation process into small, attainable goals so you can make sure you stay on track.

We've also analyzed the entire test-taking process, identifying the most common pitfalls and showing how you can overcome them and be ready for any curveball the test throws you.

Standardized testing is one of the biggest obstacles on your road to success, which only increases the importance of doing well in the high-pressure, high-stakes environment of test day. Your results on this test could have a significant impact on your future, and this guide provides the information and practical advice to help you achieve your full potential on test day.

Your success is our success

We would love to hear from you! If you would like to share the story of your exam success or if you have any questions or comments in regard to our products, please contact us at **800-673-8175** or **support@mometrix.com**.

Thanks again for your business and we wish you continued success!

Sincerely,
The Mometrix Test Preparation Team

Need more help? Check out our flashcards at:
http://MometrixFlashcards.com/AAFCS

TABLE OF CONTENTS

Introduction

Thank you for purchasing this resource! You have made the choice to prepare yourself for a test that could have a huge impact on your future, and this guide is designed to help you be fully ready for test day. Obviously, it's important to have a solid understanding of the test material, but you also need to be prepared for the unique environment and stressors of the test, so that you can perform to the best of your abilities.

For this purpose, the first section that appears in this guide is the **Secret Keys**. We've devoted countless hours to meticulously researching what works and what doesn't, and we've boiled down our findings to the five most impactful steps you can take to improve your performance on the test. We start at the beginning with study planning and move through the preparation process, all the way to the testing strategies that will help you get the most out of what you know when you're finally sitting in front of the test.

We recommend that you start preparing for your test as far in advance as possible. However, if you've bought this guide as a last-minute study resource and only have a few days before your test, we recommend that you skip over the first two Secret Keys since they address a long-term study plan.

If you struggle with **test anxiety**, we strongly encourage you to check out our recommendations for how you can overcome it. Test anxiety is a formidable foe, but it can be beaten, and we want to make sure you have the tools you need to defeat it.

Secret Key #1 – Plan Big, Study Small

There's a lot riding on your performance. If you want to ace this test, you're going to need to keep your skills sharp and the material fresh in your mind. You need a plan that lets you review everything you need to know while still fitting in your schedule. We'll break this strategy down into three categories.

Information Organization

Start with the information you already have: the official test outline. From this, you can make a complete list of all the concepts you need to cover before the test. Organize these concepts into groups that can be studied together, and create a list of any related vocabulary you need to learn so you can brush up on any difficult terms. You'll want to keep this vocabulary list handy once you actually start studying since you may need to add to it along the way.

Time Management

Once you have your set of study concepts, decide how to spread them out over the time you have left before the test. Break your study plan into small, clear goals so you have a manageable task for each day and know exactly what you're doing. Then just focus on one small step at a time. When you manage your time this way, you don't need to spend hours at a time studying. Studying a small block of content for a short period each day helps you retain information better and avoid stressing over how much you have left to do. You can relax knowing that you have a plan to cover everything in time. In order for this strategy to be effective though, you have to start studying early and stick to your schedule. Avoid the exhaustion and futility that comes from last-minute cramming!

Study Environment

The environment you study in has a big impact on your learning. Studying in a coffee shop, while probably more enjoyable, is not likely to be as fruitful as studying in a quiet room. It's important to keep distractions to a minimum. You're only planning to study for a short block of time, so make the most of it. Don't pause to check your phone or get up to find a snack. It's also important to **avoid multitasking**. Research has consistently shown that multitasking will make your studying dramatically less effective. Your study area should also be comfortable and well-lit so you don't have the distraction of straining your eyes or sitting on an uncomfortable chair.

The time of day you study is also important. You want to be rested and alert. Don't wait until just before bedtime. Study when you'll be most likely to comprehend and remember. Even better, if you know what time of day your test will be, set that time aside for study. That way your brain will be used to working on that subject at that specific time and you'll have a better chance of recalling information.

Finally, it can be helpful to team up with others who are studying for the same test. Your actual studying should be done in as isolated an environment as possible, but the work of organizing the information and setting up the study plan can be divided up. In between study sessions, you can discuss with your teammates the concepts that you're all studying and quiz each other on the details. Just be sure that your teammates are as serious about the test as you are. If you find that your study time is being replaced with social time, you might need to find a new team.

Secret Key #3 – Practice the Right Way

Your success on test day depends not only on how many hours you put into preparing, but also on whether you prepared the right way. It's good to check along the way to see if your studying is paying off. One of the most effective ways to do this is by taking practice tests to evaluate your progress. Practice tests are useful because they show exactly where you need to improve. Every time you take a practice test, pay special attention to these three groups of questions:

- The questions you got wrong
- The questions you had to guess on, even if you guessed right
- The questions you found difficult or slow to work through

This will show you exactly what your weak areas are, and where you need to devote more study time. Ask yourself why each of these questions gave you trouble. Was it because you didn't understand the material? Was it because you didn't remember the vocabulary? Do you need more repetitions on this type of question to build speed and confidence? Dig into those questions and figure out how you can strengthen your weak areas as you go back to review the material.

Additionally, many practice tests have a section explaining the answer choices. It can be tempting to read the explanation and think that you now have a good understanding of the concept. However, an explanation likely only covers part of the question's broader context. Even if the explanation makes perfect sense, **go back and investigate** every concept related to the question until you're positive you have a thorough understanding.

As you go along, keep in mind that the practice test is just that: practice. Memorizing these questions and answers will not be very helpful on the actual test because it is unlikely to have any of the same exact questions. If you only know the right answers to the sample questions, you won't be prepared for the real thing. **Study the concepts** until you understand them fully, and then you'll be able to answer any question that shows up on the test.

It's important to wait on the practice tests until you're ready. If you take a test on your first day of study, you may be overwhelmed by the amount of material covered and how much you need to learn. Work up to it gradually.

On test day, you'll need to be prepared for answering questions, managing your time, and using the test-taking strategies you've learned. It's a lot to balance, like a mental marathon that will have a big impact on your future. Like training for a marathon, you'll need to start slowly and work your way up. When test day arrives, you'll be ready.

Start with the strategies you've read in the first two Secret Keys—plan your course and study in the way that works best for you. If you have time, consider using multiple study resources to get different approaches to the same concepts. It can be helpful to see difficult concepts from more than one angle. Then find a good source for practice tests. Many times, the test website will suggest potential study resources or provide sample tests.

Practice Test Strategy

If you're able to find at least three practice tests, we recommend this strategy:

Untimed and Open-Book Practice

Take the first test with no time constraints and with your notes and study guide handy. Take your time and focus on applying the strategies you've learned.

Timed and Open-Book Practice

Take the second practice test open-book as well, but set a timer and practice pacing yourself to finish in time.

Timed and Closed-Book Practice

Take any other practice tests as if it were test day. Set a timer and put away your study materials. Sit at a table or desk in a quiet room, imagine yourself at the testing center, and answer questions as quickly and accurately as possible.

Keep repeating timed and closed-book tests on a regular basis until you run out of practice tests or it's time for the actual test. Your mind will be ready for the schedule and stress of test day, and you'll be able to focus on recalling the material you've learned.

Secret Key #4 – Pace Yourself

Once you're fully prepared for the material on the test, your biggest challenge on test day will be managing your time. Just knowing that the clock is ticking can make you panic even if you have plenty of time left. Work on pacing yourself so you can build confidence against the time constraints of the exam. Pacing is a difficult skill to master, especially in a high-pressure environment, so **practice is vital**.

Set time expectations for your pace based on how much time is available. For example, if a section has 60 questions and the time limit is 30 minutes, you know you have to average 30 seconds or less per question in order to answer them all. Although 30 seconds is the hard limit, set 25 seconds per question as your goal, so you reserve extra time to spend on harder questions. When you budget extra time for the harder questions, you no longer have any reason to stress when those questions take longer to answer.

Don't let this time expectation distract you from working through the test at a calm, steady pace, but keep it in mind so you don't spend too much time on any one question. Recognize that taking extra time on one question you don't understand may keep you from answering two that you do understand later in the test. If your time limit for a question is up and you're still not sure of the answer, mark it and move on, and come back to it later if the time and the test format allow. If the testing format doesn't allow you to return to earlier questions, just make an educated guess; then put it out of your mind and move on.

On the easier questions, be careful not to rush. It may seem wise to hurry through them so you have more time for the challenging ones, but it's not worth missing one if you know the concept and just didn't take the time to read the question fully. Work efficiently but make sure you understand the question and have looked at all of the answer choices, since more than one may seem right at first.

Even if you're paying attention to the time, you may find yourself a little behind at some point. You should speed up to get back on track, but do so wisely. Don't panic; just take a few seconds less on each question until you're caught up. Don't guess without thinking, but do look through the answer choices and eliminate any you know are wrong. If you can get down to two choices, it is often worthwhile to guess from those. Once you've chosen an answer, move on and don't dwell on any that you skipped or had to hurry through. If a question was taking too long, chances are it was one of the harder ones, so you weren't as likely to get it right anyway.

On the other hand, if you find yourself getting ahead of schedule, it may be beneficial to slow down a little. The more quickly you work, the more likely you are to make a careless mistake that will affect your score. You've budgeted time for each question, so don't be afraid to spend that time. Practice an efficient but careful pace to get the most out of the time you have.

Secret Key #5 – Have a Plan for Guessing

When you're taking the test, you may find yourself stuck on a question. Some of the answer choices seem better than others, but you don't see the one answer choice that is obviously correct. What do you do?

The scenario described above is very common, yet most test takers have not effectively prepared for it. Developing and practicing a plan for guessing may be one of the single most effective uses of your time as you get ready for the exam.

In developing your plan for guessing, there are three questions to address:

- When should you start the guessing process?
- How should you narrow down the choices?
- Which answer should you choose?

When to Start the Guessing Process

Unless your plan for guessing is to select C every time (which, despite its merits, is not what we recommend), you need to leave yourself enough time to apply your answer elimination strategies. Since you have a limited amount of time for each question, that means that if you're going to give yourself the best shot at guessing correctly, you have to decide quickly whether or not you will guess.

Of course, the best-case scenario is that you don't have to guess at all, so first, see if you can answer the question based on your knowledge of the subject and basic reasoning skills. Focus on the key words in the question and try to jog your memory of related topics. Give yourself a chance to bring the knowledge to mind, but once you realize that you don't have (or you can't access) the knowledge you need to answer the question, it's time to start the guessing process.

It's almost always better to start the guessing process too early than too late. It only takes a few seconds to remember something and answer the question from knowledge. Carefully eliminating wrong answer choices takes longer. Plus, going through the process of eliminating answer choices can actually help jog your memory.

Summary: Start the guessing process as soon as you decide that you can't answer the question based on your knowledge.

How to Narrow Down the Choices

The next chapter in this book (**Test-Taking Strategies**) includes a wide range of strategies for how to approach questions and how to look for answer choices to eliminate. You will definitely want to read those carefully, practice them, and figure out which ones work best for you. Here though, we're going to address a mindset rather than a particular strategy.

Your odds of guessing an answer correctly depend on how many options you are choosing from.

Number of options left	5	4	3	2	1
Odds of guessing correctly	20%	25%	33%	50%	100%

You can see from this chart just how valuable it is to be able to eliminate incorrect answers and make an educated guess, but there are two things that many test takers do that cause them to miss out on the benefits of guessing:

- Accidentally eliminating the correct answer
- Selecting an answer based on an impression

We'll look at the first one here, and the second one in the next section.

To avoid accidentally eliminating the correct answer, we recommend a thought exercise called **the $5 challenge**. In this challenge, you only eliminate an answer choice from contention if you are willing to bet $5 on it being wrong. Why $5? Five dollars is a small but not insignificant amount of money. It's an amount you could afford to lose but wouldn't want to throw away. And while losing $5 once might not hurt too much, doing it twenty times will set you back $100. In the same way, each small decision you make—eliminating a choice here, guessing on a question there—won't by itself impact your score very much, but when you put them all together, they can make a big difference. By holding each answer choice elimination decision to a higher standard, you can reduce the risk of accidentally eliminating the correct answer.

The $5 challenge can also be applied in a positive sense: If you are willing to bet $5 that an answer choice *is* correct, go ahead and mark it as correct.

Summary: Only eliminate an answer choice if you are willing to bet $5 that it is wrong.

Which Answer to Choose

You're taking the test. You've run into a hard question and decided you'll have to guess. You've eliminated all the answer choices you're willing to bet $5 on. Now you have to pick an answer. Why do we even need to talk about this? Why can't you just pick whichever one you feel like when the time comes?

The answer to these questions is that if you don't come into the test with a plan, you'll rely on your impression to select an answer choice, and if you do that, you risk falling into a trap. The test writers know that everyone who takes their test will be guessing on some of the questions, so they intentionally write wrong answer choices to seem plausible. You still have to pick an answer though, and if the wrong answer choices are designed to look right, how can you ever be sure that you're not falling for their trap? The best solution we've found to this dilemma is to take the decision out of your hands entirely. Here is the process we recommend:

Once you've eliminated any choices that you are confident (willing to bet $5) are wrong, select the first remaining choice as your answer.

Whether you choose to select the first remaining choice, the second, or the last, the important thing is that you use some preselected standard. Using this approach guarantees that you will not be enticed into selecting an answer choice that looks right, because you are not basing your decision on how the answer choices look.

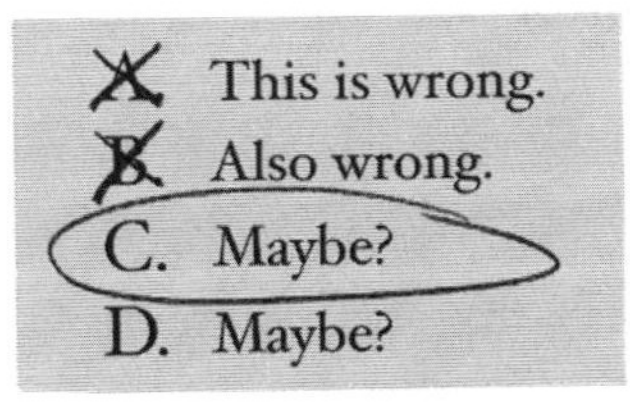

This is not meant to make you question your knowledge. Instead, it is to help you recognize the difference between your knowledge and your impressions. There's a huge difference between thinking an answer is right because of what you know, and thinking an answer is right because it looks or sounds like it should be right.

Summary: To ensure that your selection is appropriately random, make a predetermined selection from among all answer choices you have not eliminated.

Test-Taking Strategies

This section contains a list of test-taking strategies that you may find helpful as you work through the test. By taking what you know and applying logical thought, you can maximize your chances of answering any question correctly!

It is very important to realize that every question is different and every person is different: no single strategy will work on every question, and no single strategy will work for every person. That's why we've included all of them here, so you can try them out and determine which ones work best for different types of questions and which ones work best for you.

Question Strategies

✓ READ CAREFULLY

Read the question and the answer choices carefully. Don't miss the question because you misread the terms. You have plenty of time to read each question thoroughly and make sure you understand what is being asked. Yet a happy medium must be attained, so don't waste too much time. You must read carefully and efficiently.

✓ CONTEXTUAL CLUES

Look for contextual clues. If the question includes a word you are not familiar with, look at the immediate context for some indication of what the word might mean. Contextual clues can often give you all the information you need to decipher the meaning of an unfamiliar word. Even if you can't determine the meaning, you may be able to narrow down the possibilities enough to make a solid guess at the answer to the question.

✓ PREFIXES

If you're having trouble with a word in the question or answer choices, try dissecting it. Take advantage of every clue that the word might include. Prefixes can be a huge help. Usually, they allow you to determine a basic meaning. *Pre-* means before, *post-* means after, *pro-* is positive, *de-* is negative. From prefixes, you can get an idea of the general meaning of the word and try to put it into context.

✓ HEDGE WORDS

Watch out for critical hedge words, such as *likely, may, can, often, almost, mostly, usually, generally, rarely,* and *sometimes.* Question writers insert these hedge phrases to cover every possibility. Often an answer choice will be wrong simply because it leaves no room for exception. Be on guard for answer choices that have definitive words such as *exactly* and *always.*

✓ SWITCHBACK WORDS

Stay alert for *switchbacks.* These are the words and phrases frequently used to alert you to shifts in thought. The most common switchback words are *but, although,* and *however.* Others include *nevertheless, on the other hand, even though, while, in spite of, despite,* and *regardless of.* Switchback words are important to catch because they can change the direction of the question or an answer choice.

✓ Benchmark

As you read through the answer choices and you come across one that seems to answer the question well, mentally select that answer choice. This is not your final answer, but it's the one that will help you evaluate the other answer choices. The one that you selected is your benchmark or standard for judging each of the other answer choices. Every other answer choice must be compared to your benchmark. That choice is correct until proven otherwise by another answer choice beating it. If you find a better answer, then that one becomes your new benchmark. Once you've decided that no other choice answers the question as well as your benchmark, you have your final answer.

✓ Predict the Answer

Before you even start looking at the answer choices, it is often best to try to predict the answer. When you come up with the answer on your own, it is easier to avoid distractions and traps because you will know exactly what to look for. The right answer choice is unlikely to be word-for-word what you came up with, but it should be a close match. Even if you are confident that you have the right answer, you should still take the time to read each option before moving on.

General Strategies

✓ Tough Questions

If you are stumped on a problem or it appears too hard or too difficult, don't waste time. Move on! Remember though, if you can quickly check for obviously incorrect answer choices, your chances of guessing correctly are greatly improved. Before you completely give up, at least try to knock out a couple of possible answers. Eliminate what you can and then guess at the remaining answer choices before moving on.

✓ Check Your Work

Since you will probably not know every term listed and the answer to every question, it is important that you get credit for the ones that you do know. Don't miss any questions through careless mistakes. If at all possible, try to take a second to look back over your answer selection and make sure you've selected the correct answer choice and haven't made a costly careless mistake (such as marking an answer choice that you didn't mean to mark). This quick double check should more than pay for itself in caught mistakes for the time it costs.

✓ Pace Yourself

It's easy to be overwhelmed when you're looking at a page full of questions; your mind is confused and full of random thoughts, and the clock is ticking down faster than you would like. Calm down and maintain the pace that you have set for yourself. Especially as you get down to the last few minutes of the test, don't let the small numbers on the clock make you panic. As long as you are on track by monitoring your pace, you are guaranteed to have time for each question.

✓ Don't Rush

It is very easy to make errors when you are in a hurry. Maintaining a fast pace in answering questions is pointless if it makes you miss questions that you would have gotten right otherwise. Test writers like to include distracting information and wrong answers that seem right. Taking a little extra time to avoid careless mistakes can make all the difference in your test score. Find a pace that allows you to be confident in the answers that you select.

✓ Keep Moving

Panicking will not help you pass the test, so do your best to stay calm and keep moving. Taking deep breaths and going through the answer elimination steps you practiced can help to break through a stress barrier and keep your pace.

Final Notes

The combination of a solid foundation of content knowledge and the confidence that comes from practicing your plan for applying that knowledge is the key to maximizing your performance on test day. As your foundation of content knowledge is built up and strengthened, you'll find that the strategies included in this chapter become more and more effective in helping you quickly sift through the distractions and traps of the test to isolate the correct answer.

Now that you're preparing to move forward into the test content chapters of this book, be sure to keep your goal in mind. As you read, think about how you will be able to apply this information on the test. If you've already seen sample questions for the test and you have an idea of the question format and style, try to come up with questions of your own that you can answer based on what you're reading. This will give you valuable practice applying your knowledge in the same ways you can expect to on test day.

Good luck and good studying!

Integration of Foundations

Careers

Family and consumer science skills are useful in food management, financial management, human resources, public relations, tailoring, dress-making, etc. Indeed, regardless of career, an individual always finds a use for these skills in life. An individual in food management needs to know about nutrition and the proper handling and preparation of food. Financial advisors need to know how to assess resources, cut costs, and determine how much an individual needs to save before retirement. Human resource and public relations managers need social skills and training in time and resource management, human development, and psychology. Finally, tailors and dressmakers use their knowledge of textiles and textile design to create better garments.

In order to help students determine their interests and develop their skills, teachers should give them some example descriptions of various careers. Some of these examples may be family and consumer science careers, though it is not necessary for them all to be so. The class should examine a diverse sampling of different careers, especially since family and consumer science skills can be applied to virtually any setting. For example, a construction worker might not need to know about food, textiles, or housing design, but they still need to know various problem-solving techniques. It can also be extremely useful for students to get some hands-on experience applying family and consumer science concepts to the tasks associated with different careers.

Family and Consumer Education

Family and consumer education aims to improve a variety of skills that are essential for the day-to-day functioning of an individual and their family. Family and consumer education includes specific topics such as family interaction, human development, nutrition, consumer economics, types of housing and housing design, textiles, parenting, and the appropriate cooking and handling of foods. Family and consumer education covers both the physical and the psychological needs of the individual, and emphasizes appropriate social interaction between the individual and the rest of society.

Work Simplification

Work simplification is the process of discovering and implementing a series of procedures allowing an individual or a group of individuals to complete a task more easily and efficiently. Based on the particular type of task being performed, work can be simplified in a number of ways by determining the best possible way to complete a task without significantly impacting the overall quality of the work. Some of the basic methods used to make any task easier and more efficient include ensuring that individuals have access to necessary equipment, that work areas are organized, and that any steps in the work process that do not directly affect the outcome of the work are eliminated.

Goals of Family and Consumer Sciences Education

The Association for Career and Technical Education has identified nine goals commonly associated with family and consumer sciences education. These nine goals include:

1. Improving the overall quality of life for individuals and families
2. Helping individuals and families become responsible members of society
3. Encouraging healthy eating habits, nutrition, and lifestyles
4. Improving how individuals and families manage their resources

5. Helping individuals and families balance their personal, family, and work lives
6. Teaching individuals better problem-solving techniques
7. Encouraging personal and career development
8. Teaching individuals to successfully function as both consumers and providers
9. Recognizing human worth and taking responsibility for one's own actions

The first of the nine goals established for family and consumer sciences education is to improve the overall quality of life for individuals and families, which is also the primary mission of all the goals. Family and consumer sciences education teaches people about how individuals, families, and the rest of society interact with each other, along with methods for improving those interactions. These methods include problem-solving techniques, common scams and problems to avoid, methods to stay healthy both physically and psychologically, and the distribution of a variety of other information regarding how the individual, family, and the rest of society function. Ultimately, family and consumer sciences education strives to improve the quality of life by educating individuals and families in the best manner to function on a day-to-day basis. However, this goal is truly accomplished only when the other eight goals of family and consumer sciences education are met as well.

Basic Concepts of Family and Consumer Education

There are several concepts at the core of family and consumer education, but one of the most important concepts is that families form the basic unit of society. Another important concept of family and consumer education is that individuals need to be life-long learners in order to develop and function successfully. Finally, family and consumer education promotes the idea that individuals and families need to have an understanding of the advantages of experimenting with different decision-making methods and diverse ways of thinking in order to solve any given problem.

Occupational Family and Consumer Sciences Education

Occupational family and consumer sciences education is a teaching discipline that is similar to the standard discipline of family and consumer education but focuses less on the skills for day-to-day living and more on how those skills can be used in the workplace. Occupational family and consumer sciences education covers information regarding skills that can commonly be applied in fields such as health services, food service, child care, hospitality, fashion design, interior design, and many other similar fields. Occupational family and consumer sciences education places more emphasis on family and consumer skills that directly relate to a career, such as management techniques and ethical businesses practices, than the standard family and consumer education discipline. The occupational family and consumer sciences education discipline ultimately takes the skills that an individual has learned from the standard discipline and shows how those skills can be applied to a career.

Community Advisory Committees

Community advisory committees can be extremely useful to an education professional who is attempting to determine what areas of the family and consumer science discipline a teaching plan should emphasize because the committees offer insight into the concerns and demographics of the students. Each community has its own problems, concerns, and level of diversity, and it is important that a family and consumer science teacher can recognize and focus on areas of concern in the school's community. For example, a community that is having problems with widespread teenage drug abuse and teenage suicide may want the community's family and consumer science teachers to focus more on the topics of avoiding substance abuse and handling depression. The goal of a family and consumer sciences educator is to improve the overall quality of life for the students and their families, and the educator cannot do that if they do not know what problems need to be addressed.

Some of the functions that community advisory committees perform, other than offering advice to education professionals, include assessing the performance of family and consumer science programs; assessing the performance of students with special needs; and providing equipment, technology, and resources for family and consumer science programs. These resources may include raw materials, textile samples, charts and diagrams, library books, and access to computers and design software. Community advisory committees also help students improve their chances of finding better jobs and careers and act as a public relations liaisons for local family and consumer science programs. Ultimately, the primary purpose of a family and consumer sciences community advisory committee is to ensure that a family and consumer sciences program has all of the resources and training necessary to achieve the program's goals.

Laboratory Settings

A **laboratory setting** is important for a family and consumer sciences educator because it offers students an opportunity to gain hands-on experience using a variety of skills and techniques. Many important areas of the family and consumer sciences discipline center around using a combination of various skills to achieve a certain end result, and sometimes the best way to teach the appropriate way to integrate these skills is through experience. A laboratory setting offers students a place to demonstrate and improve their skills related to the family and consumer science field with the advantage of having a teacher present to answer questions and correct mistakes. Some examples of useful laboratory settings for the family and consumer sciences field include kitchens or food science laboratories, day care centers, and testing laboratories for textiles and consumer products.

Demonstrating Family and Consumer Science Concepts

There are a large number of methods that an educator can use to demonstrate concepts related to family and consumer sciences, but the best methods always involve promoting students' active participation. Some examples of active participation include allowing students to use a sewing machine; having students test the qualities of various textiles to see how soft, lustrous, resilient, absorbent, etc. each material is; and requiring students to prepare a meal. Students can also demonstrate active participation with the following activities: comparing advertisements to find the best offer for a particular product, examining common marketing tactics, visiting or working in a local daycare center, and being involved in local community service activities. Many of these activities serve not only as effective ways of teaching students about the important concepts of family and consumer sciences, but also as a means of testing the students' ability to apply the techniques, skills, and information that they have learned.

Usually, the best method an educator can use to verify whether a student understands a particular concept is to see if the student can actively apply the information they have learned to such everyday tasks as cooking, sewing, and time management; however, a student may not be able to demonstrate their understanding of certain concepts if a laboratory setting is not available. Alternative methods by which an educator can evaluate a student's level of comprehension include administering written tests, assigning projects and research papers, teaching students to design charts and diagrams, involving students in the evaluation of case studies and scenarios, and requiring students to keep a journal of their activities and eating habits. Which evaluation method an educator should use depends primarily on the curriculum being covered and the abilities of the students who are taking the class.

Professional Organizations

Professional organizations such as the American Association of Family and Consumer Sciences, also known as the AAFCS, play an important role in influencing the education of individuals in the methodology and knowledge associated with family and consumer sciences. Many of these local and national professional organizations offer seminars, courses, and publications on a wide range of topics directly to individuals and families to teach them about essential career and management skills, how to be smart consumers, the importance of following nutritional guidelines, and information about a wide range of other topics. These professional organizations also provide publications, advice, and curriculum guides to educational professionals that help these professionals teach and stay informed regarding important changes to the curriculum that result from changes in legislation, society, and the education system itself. These organizations also have a profound effect on family and consumer science education by influencing public policy and gathering support for programs that help educate and protect individuals and families from unsafe habits, business practices, products, and lifestyles.

FCCLA

The **FCCLA**, which stands for the **Family, Career, and Community Leaders of America**, is a youth organization for students in family and consumer science education. The FCCLA offers a variety of publications and programs designed to educate people about parenting, relationships, substance abuse, teen pregnancy, and teen violence, among other concerns. By focusing public attention to the problems that young people face, the FCCLA gains support for programs and laws that help protect young people and their families. The FCCLA also shows students how they can improve their family and consumer science skills and apply those skills later in life.

The FCCLA and other similar youth organizations play an important role in influencing national policy related to protecting families and consumers. Additionally, these organizations are important because they support family and consumer science educational programs, which strive to improve the overall quality of life for individuals and families by teaching people skills that will enable them to live better lives.

New Acts of Legislation

When a new act of legislation is passed, it can often have a profound impact on the types of materials used in family and consumer sciences classrooms, as well as the issues that should be addressed by family and consumer sciences education. It is important that individuals understand the legal protections and rights granted to them by the various acts put into place by state and federal governments. Since laws are constantly changing, family and consumer science educators must be able to adapt quickly and add information regarding new legislation to their curriculum.

Balancing Home and Work Roles

It is important that an individual can balance their work and home roles because it is becoming more and more common for individuals to have to act as both caregiver and provider for the family. The increasingly common presence of dual roles in society can be extremely difficult for an individual to balance, as there may be instances where work-related responsibilities and family-related responsibilities conflict with one another. Family and consumer sciences education attempts to teach individuals how to avoid and handle these conflicts through the use of successful life management tactics such as time and resources management, problem-solving and decision-making techniques, and effective communication techniques. Family and consumer sciences education also attempts to give individuals a basic understanding of what responsibilities and qualities are necessary for the successful completion of each role so that individuals can set better priorities and find better ways to plan their lives.

Family Studies and Human Services

Families, Marriages, and Parenting

Eliminating Sexual Stereotypes

Eliminating sexual stereotypes is a major concern of family and consumer sciences education. It is important for students to disregard sexual stereotypes and recognize that an individual's gender does not necessarily affect the role they play. In the early- and mid-1900s, women were commonly seen as caretakers of the home and men as providers for the family. However, these roles have changed drastically over the past fifty to sixty years, and are not entirely realistic at this point. As the cost of living increases, it becomes more difficult for a single individual to provide for an entire family. As a result, it is more common for men and women to share the caretaker and provider roles to satisfy the physiological, financial, and psychological needs of the family.

Direct and Indirect Communication

Direct communication occurs when a person who is attempting to convey a given piece of information simply states that information to the person they want to receive the information. **Indirect communication**, on the other hand, is when the person communicating the information states the information, but not to anyone in particular. For example, if a parent says, "Christine, we need to set the table," that is an example of direct communication because the parent is addressing the person they want to talk to directly. However, if the parent instead simply mutters out loud, "We need to set the table," rather than saying it to someone in particular, that would be an example of indirect communication. Direct communication is far more effective in carrying out the day-to-day functions necessary to maintain a family than indirect communication because various tasks can be assigned directly to a particular individual.

Making Communication More Effective

Families with individuals who use direct, clear communication are the most effective. These family members listen to one another, spend more time communicating, respect one another's points of view, and pay attention to the more subtle forms of affective communication. By communicating directly and concisely with other family members, each family member creates a much more effective form of communication than that which would be found in any other setting. If the individuals receiving the information listen to and respect their fellow family members and—more importantly—make the time to listen to them in the first place, the communication between family members will become much stronger. Of course, this communication can be strengthened even further if members of the family are careful to take note of emotional indicators that allow them to identify the feelings of another family member without that person having to verbally express their feelings.

Conflict Resolution Techniques

A family can successfully resolve a conflict by following steps very similar to those of the basic problem-solving model. First, the family needs to attempt to identify the problem, making sure to maintain open communication while remaining objective and minimizing hostility. After the problem is identified, the family must strive to recognize the various positions that each member has regarding the conflict while again attempting to minimize hostility. After each person involved in the conflict has made their position clear, the family must move toward a compromise that will work for everyone. Each step of the conflict resolution process requires that the people involved in the conflict remain as patient and as understanding as possible, which can often be extremely difficult when a solution or compromise cannot be determined immediately.

Social Interaction Outside the Family

Outside social interaction is extremely important for all family members, regardless of age, because it offers an opportunity for each individual to improve their social skills, learn about the world around them, and learn more about values that one might not learn from the family alone. This is especially true in the case of children. Research shows that children who have regular outside social interaction through things such as extracurricular activities are less likely to rebel or cause problems and more likely to excel in school and relationships. Outside social interaction is also necessary for the children of a family to eventually leave the household and create families of their own, as they need to seek out their own relationships. Therefore, social interaction with individuals outside of the family is necessary not only for the fulfillment of the members of the family, but also to continue the life cycle of the family.

Divorce

Divorce is the termination of the marriage union before the death of either member of the union. It has a significant impact on the stability of the family unit as a whole, and it affects the relationships and well-being of the individual members of the family. Frequently, when the marital couple decides to divorce, there has already been significant stress placed on the entire family from the difficulties the marital couple has been experiencing. However, divorce can often lead to a great deal more stress being placed on the family, especially when children are involved. As a result, individuals within and outside the marital couple may become more withdrawn or hostile as the structure of the family changes. Divorce also allows both members of the marital couple to later remarry, as their legal obligation to each other no longer exists. This can further alter the family structure by adding stepparents to the mix.

There are many factors that may influence the risk of a marriage ending in divorce, including income, education, religion, pregnancy before marriage, and whether the parents of the married couple are divorced. Couples who make over $50,000 a year are at a much lower risk of divorce than couples who make less than that amount. Couples comprised of well-educated individuals who have graduated from high school and have at least some college background also have a much lower risk of divorce than less educated individuals. Couples with no religious background or drastically different religious backgrounds have a much higher risk of divorce than couples who have religious backgrounds that do not conflict. Couples who have a baby prior to being married also have a higher risk of divorce than couples who have children after they are married. Individuals with parents who are divorced also have a higher risk of divorce than individuals from intact families.

Studies indicate that the age at which a couple marries may have a significant impact on whether they remain married for an extended period. Individuals who marry before either member of the couple is 18 will often separate within a few years of their marriage. Individuals who are in the 18–25 range will separate less frequently than those who marry before 18, but they are still at a very high risk for their marriage ending in divorce rather than death. Individuals who marry after both members of the couple are over 25 have a significantly lower risk of divorce than those who marry at younger ages. Ultimately, statistics show that the risk of divorce decreases as the age of each member of the couple at the time of the marriage increases.

Social and Economic Factors

Social and economic factors affect the overall functioning of a family. In fact, researchers use an index called the **socioeconomic status**, or SES, to measure the ability of the family to function in a healthy fashion. The SES uses the educational background of the members of the family, the family's total income, and the skill (both actual and perceived) required by the occupations of the individuals who act as providers for the family to measure the family's ability to function. Individuals who are well-educated tend to marry later in life, receive jobs with higher incomes, and have careers with a higher social status. These factors all add stability to the marriage and stability to the overall functioning of the family. Families that earn a higher income are also less concerned with obtaining basic necessities because the family consistently has the means to obtain them. As a result, there is often less stress experienced by the family.

Affective and Instrumental Communication

The two primary types of communication used by family members are **affective communication** and **instrumental communication**. Affective communication is communication in which an individual demonstrates their feelings through facial expressions, motions, gestures, or by stating their feelings outright. Instrumental communication is when an individual informs another member of the family of a piece of factual information that is necessary to carry out the normal day-to-day functions of the family. An example of instrumental communication is a mother informing her child where they can find their socks. Families that use both types of communication usually function more effectively than families that use instrumental communication more often than affective communication.

Clear and Masked Communication

Clear communication occurs when an individual explicitly states the information they are trying to convey, and there is no ambiguity as to the meaning of the statement. For example, "I am upset because Daniel is not home from the movies yet" is an example of clear communication because there is no question that the individual making the statement is upset at Daniel for not being home. On the other hand, **masked communication** occurs when an individual states the information they are trying to convey in a vague and somewhat confusing manner. For example, "I am upset" is an example of masked communication because there is no direct indication as to why the person is upset. As these examples illustrate, clear communication is always more effective in conveying a particular piece of information than masked communication.

Family Aid in the Development and Education of Family Members

One of the most important functions a family provides is developing and educating family members. Parents and grandparents pass down their heritage along with teachings of social norms and acceptable behavior to the children of the family through their customs, traditions, and ultimately their actions. Children learn about their heritage through the traditions of the family and also often learn lessons about the manner in which they are expected to behave by using the behavior of their parents and the rest of the household as a model for how they, too, should behave. Children also learn about the manner in which the world around them functions through the interactions of the members of the family with the world outside the household. This allows the child to understand more complex types of social interaction such as what goods the family needs, where the family must go to fulfill those needs, and what is needed to acquire those necessities (e.g., how much money is required to purchase an item).

Behavioral Modeling, Consumer Education, and Heritage

Behavioral modeling, when related to child development within a family structure, is the manner in which children model their own behavior after the behavior of their parents and other people with whom they interact. Children learn what behavior is socially acceptable by mimicking the behavior of the people around them.

Consumer education is the process of teaching a person about the marketplace and its goods and services, the suppliers, and the various considerations associated with searching for goods and services. These concepts are critical for family members to learn so that they can survive in a consumer society.

Heritage is anything inherited from one's ancestors, including traditions, customs, or physical characteristics. The family conveys the traditions, customs, and social norms of the previous generation to the generations that follow.

Roles That Are Essential to the Functioning of a Healthy Family

There are five major roles that are essential to the functioning of a healthy family. These roles are:

1. Provision of necessities
2. Development and education
3. Emotional support
4. Management of the family
5. Satisfaction of the married couple's needs

Individuals within the family need to provide necessities by creating income so that the family has access to food, clothing, and shelter. Family members need to teach not only customs, but also skills that will help the members of the family achieve academically and professionally. Families must provide emotional support for the family members during times of high stress. In addition, the family needs someone to take a leadership role and handle issues such as managing finances and maintaining the roles essential to the family's survival. The married couple has its own requirements, including basic necessities, sexual needs, and emotional needs that must be met for the family to continue functioning normally.

Role, Role Confusion, and Role Strain

A **role** is a collection of social rights, behaviors, and obligations that is assigned to a particular individual. For example, a mother's role might be that of a provider because she is out in the workforce earning an income for the family.

Role confusion occurs when an individual is uncertain of what role or roles they should play in a particular situation. For example, a nurse might run into a patient whom she took care of previously while out grocery shopping and be unsure of whether to act in a formal, nurse-to-patient manner or in an informal, friendly manner.

Role strain occurs when an individual is placed in a situation in which carrying out the duties of a certain role will prevent the individual from fulfilling their obligations of another role. For example, a working mother might be both caregiver and provider. If her child becomes ill, she cannot carry out both roles; she is forced to choose between working and caring for the sick child.

Married Couple

The married couple or, in some cases, the couple living together is the core of the family and therefore has a profound effect on the relationships and well-being of the family. If a marital couple is having difficulty in their relationship and the stress of that difficulty becomes apparent, the rest of the family will most likely exhibit signs of stress. For example, if the marital couple is consistently seen fighting, or even if they just become withdrawn after a fight, other family members may react to the stress and become withdrawn, upset, or even hostile. On the other hand, marital couples who are not experiencing marital difficulties and who appear warm and affectionate will foster the same feelings of warmth and affection in the rest of the family.

Marriage

Marriage is a union between two individuals that is often held as a legally binding contract in which the members of the union state their intention to live together and aid each other in maintaining a family. Even though couples who simply live together in the same household can constitute a family under the commonly used definition, the institution of marriage offers a level of stability to the family structure that is not present when an unmarried couple makes up the center of the family. This added stability is primarily a result of the societal, religious, and governmental recognition of the institution of marriage, which creates an expectation that the marriage and, more broadly, the family will remain intact. Although many married couples eventually separate and divorce, it is more difficult for a member of the marital couple to leave the family than it would be for a member of a couple who has no legal or societal obligation to remain together.

Family and Single Individual

A **family** is commonly considered a group of individuals related by birth, adoption, or marriage who reside together, usually for the purpose of raising children. However, a family can refer to any group of people who live together in the same household even if they are not related by blood or legal ties. This means that an unmarried couple who is living together or even a pair of roommates may still be considered a family. A **single** individual, though, is not considered a family because it is a person who lives alone and therefore does not regularly interact with relatives or other individuals within the household.

Family Structures

The four major types of family structures are nuclear, extended, single-parent, and blended. Each of these structures is based on the idea that a family is a group of people who participate in raising the next generation. A **nuclear** family is the traditional concept of a family in which a mother, father, and their children live in the same household. An **extended** family is an expansion of the nuclear family that includes the mother, father, and their children as well as aunts, uncles, cousins, and grandparents. A **single-parent** structure is a family in which one parent is the only one in the home caring for the children. A **blended** family, also known as a stepfamily, is one in which a parent marries or remarries when they already have their own children, and there is a parent, stepparent, and one or more children living in the household.

The typical family structure in the United States has changed dramatically in recent years as the norm moves away from the nuclear family and toward the blended family. As more people divorce and remarry, blended families are becoming much more common. In this family structure, children are cared for by both biological and stepparents. This increase in the number of blended families, which were unheard of 50 years ago, has resulted in two substructures: simple and complex. In a simple stepfamily, only one of the individuals marrying has children before the marriage. In a complex stepfamily, both parents marrying have their own children before the marriage.

FAMILY LIFE CYCLE

There are commonly nine stages in the family life cycle:

1. **Bachelor stage**: The stage in which the individual is yet to be married, and the family has not yet been established.
2. **Newly married couple stage:** The newly married couple stage in which two individuals have just married but do not have children.
3. **Full nest stage I**: The beginning of the three full nest stages, when the parents are beginning to raise children. The youngest child is under six.
4. **Full nest stage II**: The youngest child is six or over.
5. **Full nest stage III**: The stage in which an older married couple has independent children.
6. **Empty nest I stage**: The head of the household is married and still in the labor force, but the couple has no children at home.
7. **Empty nest II stage**: The same as empty nest I stage except that the head of the household has retired.
8. **Solitary survivor in labor force stage**: One member of the couple has passed away, and the survivor must continue to work to support themself.
9. **Retired solitary survivor stage**: The same as the solitary survivor in labor force stage except that the survivor has retired, and there are no longer any individuals living in the household who are still in the labor force.

PURPOSE OF A FAMILY

The primary purpose of a family is to ensure the survival of the family and to nurture the children. Families facilitate survival by sharing the work and tasks such as earning a living and taking care of the home. Family also provides emotional support to one another during stressful times. The family nurtures the children by offering social and emotional interaction, protecting them from potential danger, and educating them in social norms and customs. The family also provides the basic necessities required for the basic physical development of the children in the household, including food, clothing, shelter, and play.

MURRAY BOWEN'S FAMILY SYSTEMS THEORY

Bowen's concept of the nuclear family emotional system consists of four basic relationship patterns that determine where family problems develop. Clinical symptoms or problems typically emerge during times of intensified and protracted tension in a family. Stress levels, family adaptations to stress, and family connections with extended family and social support networks determine tension levels. In the **marital conflict pattern**, spouses project their increasing anxiety into the marital relationship. Each partner becomes preoccupied with the other's shortcomings, tries to control them, and resists being controlled. For example, a couple with a young child conceives a second child. The wife becomes anxious about meeting the needs of two children. The husband questions his wife's ability to cope in order to avoid facing his own anxieties. After the second child's birth, the husband, observing his wife's stress, helps out more at home and is more controlling of her. He starts to feel neglected and disappointed in his wife's inadequate coping. The wife, who used to drink but quit while pregnant, resumes drinking.

In Bowen theory, the relationship **pattern of dysfunction in one spouse** involves one partner pressuring the other to behave in certain ways, and the other acceding to that pressure. While both partners accommodate for maintaining harmony, eventually one does more than the other. Both are comfortable with this interaction for some time; however, if family tensions increase, the subordinate partner gives up enough self-control to become significantly more anxious, yielding to the dominant partner. Combined with other factors, this anxiety contributes to a psychiatric, social, or medical problem. For example, a couple with one young child has a second child. In the relationship pattern of marital conflict, the husband projects his own anxiety into criticizing his wife's coping abilities, taking on more household duties, and controlling her while the wife addresses her anxiety by drinking. The husband accuses her of selfishness and lack of effort. She agrees with but resents his criticism, feeling more dependent on him. Feeling increasingly unable to cope and make decisions, she escalates her drinking. He calls her an alcoholic. The wife becomes increasingly under-functional, the husband increasingly over-functional, functioning for her—all in an effort to avoid direct conflict and maintain harmony.

In the **pattern of impairment of a child or children**, parents project their own anxieties onto their child or children. They view the child unrealistically—either negatively or idealistically. The child reciprocates excessive parental focus by focusing excessively on the parents, overreacting to parental expectations, needs, and attitudes. This undermines the child's differentiation of self from family, increasing their susceptibility to either internalizing or acting out family tensions. Anxiety can disrupt the child's social relationships, school progress, and health. For example, a couple with one young child has another baby. Anxieties over the added stress of raising another child cause marital conflict and a dysfunctional relationship, developing into greater dysfunction in one spouse or parent. This causes emotional distance between spouses, who focus anxiously on the older child. She reacts by regressing, making immature demands of the parents, especially her mother. The mother externalizes her anxiety onto the child, worrying the new baby will displace her, acceding increasingly to her demands. The father avoids conflict with his wife by supporting her focus on the child, relieving her by giving the child attention when he gets home from work. Parents and child unwittingly conspire in seeing and creating dysfunction in the child.

In Bowen's family systems theory, the four basic relationship patterns are marital conflict, spousal dysfunction, child impairment, and emotional distance. Whichever pattern predominates will dictate which family members will manifest familial tensions by developing psychological, social, or medical symptoms. The **pattern of emotional distance** consistently occurs in relation to the other three patterns. When interactions between family members become too intense, they develop emotional distance to decrease intensity. However, the drawbacks of emotional distance are that distanced members can become overly isolated, and they can lose intimacy in their relationship. For example, when a couple with one child has another baby, they first project their anxieties onto each other and experience marital conflict. They then withdraw from one another emotionally to reduce the intensity of the conflict. They react to the emotional distance between them by externalizing their anxieties onto the first child, worrying she will feel left out with the new baby. The child reacts to the obsessive parental emotional over-involvement with her, reciprocating their emotional focus and overreacting to real or imagined parental withdrawal, thus creating impairment of a child. In this manner, emotional distance interacts with the other patterns.

Review Video: Bowen Family Systems Theory
Visit mometrix.com/academy and enter code: 591496

Socioeconomic and Health Variables on Parent-Adult Child Relationships

Some sociological researchers investigating relationships of parents in their mid-50s to mid-70s with their adult children found intergenerational exchanges were characterized by strong reciprocity in both the United States and Great Britain. Contrary to stereotypical views of elderly adults becoming "burdens" on adult children, researchers have seen instead that married parents who gave help and support to at least one adult child were twice as likely to receive support from another adult child as parents who did not provide such support. Investigations showed when researchers controlled for various other parent and child variables, parents who owned homes, had

continuously during sex. Though the point is not to encourage sex among immature students, educators can inform those harboring misunderstandings regarding kissing, body rubbing, masturbating, and that oral and anal sex cannot cause pregnancy without vaginal or vulvar contact with sperm. Abstaining from sex, or using both a condom and birth control continuously during sex, are the ways of preventing pregnancy. Teen couples contemplating sex should discuss birth control with each other and a parent or trusted adult, visit a Planned Parenthood center, and see a physician, nurse, or healthcare provider. Planned Parenthood's website offers a quiz to help choose a method.

Contraception

Some young (or uninformed older) people assume condoms worn by males are sufficient for contraception. However, condoms can break, leak, or slip off during or following intercourse. Ideally, foam, gel, or other spermicide should accompany condoms. Female contraception includes IUDs, diaphragms, and birth control pills. IUDs are typically inserted by physicians and worn continuously. They can periodically require removal and replacement. While effective, they can have undesirable side-effects for some women including irritation, inflammation, cramping, spotting, tissue damage, or other issues. Diaphragms are typically self-inserted by women before intercourse, often with spermicidal gel applied to the surface, and removed afterward. They are also effective, but some women have difficulty inserting them properly and/or cannot tolerate their presence. They can also sometimes shift position, impeding contraception. Birth control pills are very effective, though a very small percentage of women using them might still get pregnant. Oral hormones cause some women undesirable side effects like weight gain and symptoms resembling pregnancy. Lower-dose pills have fewer side effects; different dosages affect individual women differently. More extreme measures include tubal ligation (often reversible, but not always) and hysterectomy (irreversible) for women, and vasectomy (often reversible, but not always) for men.

STIs

Over half of Americans contract an STI (sexually transmitted infection) during their lives. According to Planned Parenthood®, there are many ways to practice safer sex, such as using condoms. Monogamous sex is generally safer, although many individuals are unaware of being infected, and others are aware but are untruthful about it, hence safer sex includes partners' getting tested regularly together. There are also sexual activities that do not transmit STIs (e.g., masturbating or mutual masturbation, fantasy sharing, cybersex, or phone sex). Other sexual activities considered low-risk include kissing; fondling (manual stimulation); "outercourse" (body rubbing); oral sex with a condom, dental dam, or other barrier; and using sex toys with partners. Vaginal and anal intercourse are high-risk activities. Without condoms, they are likely to transmit chancroid, chlamydia, cytomegalovirus (CMV), genital warts, gonorrhea, hepatitis B, herpes, HIV, human papilloma virus (HPV), molluscum contagiosum virus, pelvic inflammatory disease (PID), pubic lice ("crabs"), scabies, syphilis, and trichomoniasis. Unprotected oral sex is high-risk for transmitting CMV, gonorrhea, hepatitis B, herpes, syphilis, and HPV. Skin-to-skin contact without intercourse is risky for transmitting CMV, herpes, HPV, molluscum contagiosum, pubic lice, and scabies. Many STIs are often asymptomatic. Planned Parenthood's website has a search engine for finding health centers by ZIP code to schedule testing.

Parental Avoidance in Talking About Procreation with Their Children

Many parents feel squeamish about "The Talk" or discussing "the birds and bees" with their maturing children. This is not just discomfort over an intimate topic; parents frequently fear that discussing sex with preadolescent and adolescent children is akin to giving them permission to engage in it. However, research studies find the opposite is true: teens are more prone to sexual behaviors when their parents have *not* talked about sex with them. When uninformed of possible consequences, they are more likely to act, not knowing of any disadvantages; they may experiment to get knowledge their parents have not imparted; and/or sexual behavior may be a reaction against parental avoidance and lack of openness. Communications researchers say sex is a continuing, two-way conversation that starts when very young children see pregnant women and ask questions. They advise parents to use Socratic questions, such as "What do you think the right time is for having sex?" and sharing their own thoughts after children do. Open, receptive attitudes are critical: if children bring up sex and perceive avoidant or shocked parental reactions, they will stop approaching parents, shutting down this vital conversation.

CONFLICT IN RELATIONSHIPS

Many people try to avoid conflict at all costs because they find it unpleasant and feel threatened by confrontation. However, conflict is normal and integral to healthy relationships. Its source is rooted in differences between and among people, whether major or minor. No two (or more) people can agree about everything 100 percent of the time. Any time that people disagree, conflict results. Though some disagreements seem unimportant, any conflict that evokes strong emotions indicates some deep personal need at its core (e.g., to be valued or respected, to be closer or more intimate, or to feel safety or security). As one example, young children need to explore and take risks to learn and develop normally, while parents need to protect children's safety, and this can present a child-parent conflict. Conflicts in personal relationships can cause discord and even end the relationships when members do not understand each other's different needs. Conflicts in workplaces can ruin deals, lower profits, and end jobs. Acknowledging needs that conflict with a willingness to examine them in understanding, compassionate environments enables team-building and creative problem-solving. Both avoiding and mismanaging conflict can damage relationships, but positive and respectful conflict management can improve them.

FEMALE REPRODUCTIVE SYSTEM

Because females do not have a Y-chromosome, during embryonic and fetal development they are not affected by testosterone to develop male reproductive organs. Without testosterone stimulation, reproductive organs develop into ovaries, a uterus, and other female organs. Most internal female organs are formed by the end of the first trimester. Immature eggs (ova) form in the ovaries in utero; all of a female's eggs are produced before birth. Female infants are born with all reproductive organs formed, but these organs are immature and not functional. These do not grow much in childhood, but rapidly mature and grow during puberty. Girls typically start puberty one or two years before boys and take around four years to complete, whereas boys take around six years. The primary female sex hormone is estrogen. In the brain, the hypothalamus stimulates the pituitary gland to secrete luteinizing hormone (LH) and follicle-stimulating hormone (FSH), which stimulate the ovary to produce estrogen. (The same hormones stimulate the testes' testosterone production in males.) Estrogen stimulates uterus and breast growth; pubic hair growth; bone development; the adolescent growth spurt, which begins and ends earlier than in males; and menarche (menstrual cycle onset).

MALE REPRODUCTIVE SYSTEM

While a male fetus develops in utero, the testes begin to develop. Around two months before birth, the testes begin descending into the scrotal sacs outside of the main body cavity, allowing slightly lower temperatures to aid sperm production. The testes additionally produce hormones enabling development of secondary male sex characteristics. Puberty activates an increase in brain hormones, triggering the pituitary gland's increased production and release into the bloodstream of luteinizing hormone (LH) and follicle-stimulating hormone (FSH). In the bloodstream, LH stimulates testes cells to produce and release testosterone, which enlarges and develops the penis and other sex organs, promotes skeletal and muscular growth, and deepens the voice. Testosterone and FSH stimulate sperm production in seminiferous tubules within the testes. Each sperm cell takes 65-75 days to form; about 300 million are produced daily, stored in the epididymis, wherefrom the vas deferens carries sperm through the prostate gland below the bladder to the urethra. The male urethra releases both sperm and urine. The prostate gland and seminal vesicles—accessory sex glands—produce specialized fluids, mixing with sperm during transport, creating semen which exits from the urethra through the penis during ejaculation.

Adolescent Sexual Attitudes and Behaviors

Various studies find teens' sexual attitudes are influenced by variables including their parents' attitudes regarding teen sex, religiosity, the media, bonding in school relationships, and adolescents' perceptions of social norms among their peers. According to some experts, such research demonstrates the necessity of considering the wide range of sexual attitudes teenagers consider. Warning of negative consequences is insufficient; adults must provide information enabling teens to weigh positive and negative aspects of both engaging in and abstaining from sex to make their own best decisions as they mature and develop physically, intellectually, emotionally, and socially. While many models of teen risk behaviors emphasize perceptions of possible consequences in decision-making, studies also find positive motivations for having sex. Some investigators found teens valued sexual goals and expectations of intimacy, then social status, then pleasure in that order; but then expected sex to result in pleasure, then intimacy, then social status in that order. Male adolescents valued pleasure more; females valued intimacy more. The National Adolescent and Young Adult Health Information Center (NAHIC, 2007) found almost half of high school students reported having sex. CDC's Youth Risk Behavior Surveillance (YRBS, 2008) found sexual intercourse most prevalent in black, then Hispanic, then white, then Asian teens.

Researchers find that, although any age group can be influenced by sexual media content, teens can be especially vulnerable to media messages. Adolescence is a developmental time when individuals are forming their sexual attitudes, behaviors, and gender roles. Teenagers have recently developed the ability to think abstractly and critically; however, their cognitive skills are still not completely developed for critical analysis of media messages and decision-making that takes into account future potential consequences. This places them at higher risk for media influence. Researchers have found (Gruber, 2000) that teens viewed an average of 143 instances of sexual behavior on TV weekly during prime time. Activities between unmarried partners were depicted three to four times more often than between spouses. Network and cable TV channels show movies, an estimated 80 percent of which include sexual content. Researchers' analysis of music videos estimated that 60 percent included sexual impulses and feelings. Sexual TV messages are found to be nearly always presented in positive terms, with scarce treatment of negative consequences or risks of unprotected sex. High school students have reported substantial access to and viewing of TV and video. Over 80 percent of teens report peer discovery about sex from entertainment media.

Stages of Pregnancy

In the first trimester, a zygote transforms into an implanted embryo; organs, hair follicles, nail beds, muscles, white blood cells, and vocal cords form; and the baby starts moving around week eight. While pregnancy does not show externally, mothers awash in pregnancy hormones feel many symptoms. However, every woman and pregnancy is different; no two necessarily have the same symptoms, but most initial symptoms diminish further into pregnancy while others develop. During the second trimester, babies grow hair, begin sucking and swallowing, and their eyes and ears reposition. They have fingerprints and can hiccup and yawn by week 18. Their limbs are coordinated and their senses develop by week 21. Weight gain, capillary formation, and opening eyes occur by six months. By seven months, fetal weight doubles to two pounds. Babies perceive light and dark, taste what mothers eat, and hear their voices by week 31. Transparent skin becomes opaque by week 32; length may increase an inch during week 33. Weight reaches around six pounds by week 36; waxy vernix and hairy lanugo shed in week 38. By week 40, fetal weight is 6-9 pounds and length is 19-22 inches; babies dream, blink, and regulate their body temperatures.

LABOR AND CHILDBIRTH

In late pregnancy, symptoms can mask labor signs, or some contractions can be false labor. If contractions persist, become stronger, last longer, and occur closer together, this usually indicates labor. The "411" method is one way to judge: contractions 4 minutes apart, lasting 1 minute each, continuing for at least 1 hour. Labor's first stage is typically the longest, marked by contractions and gradual cervical dilation. The first stage has three phases: the early phase, usually comfortable, with contractions 20 minutes apart progressing to 5 minutes apart. The second, active phase generally involves 1-minute contractions every 4-5 minutes. The third, transition phase is among the shortest (1-2 hours) but hardest. Contractions are 2-3 minutes apart; some women shake and may vomit. This phase ends with complete dilation. Some women temporarily cease contractions but feel no need to push. Labor's second stage involves a need to push. It can last 3+ hours, but often less. Contractions spread out again to around every 4 minutes. This stage culminates in childbirth. Then the mother must push out the placenta, nursing the newborn aids uterine contractions to expel it. The fourth labor stage is postpartum.

Chapter Quiz

Ready to see how well you retained what you just read? Scan the QR code to go directly to the chapter quiz interface for this study guide. If you're using a computer, simply visit the online resources page at **mometrix.com/resources719/aafcsfamconsc** and click the Chapter Quizzes link.

Human Development, Education, and Services

Stress, Substance Abuse, Crises, and Decision Making

Teenage Suicide

There are a number of factors that increase the risk of teenage suicide, but studies indicate that a teenager's history, emotional and physical health, social pressures, and access to the methods necessary to carry out a suicide are the most influential factors. If a teenager has attempted suicide, has a history of drug or alcohol abuse, a history of depression or other mental illness, or another family member has committed suicide or been abused, the teenager's risk of suicide increases. Physical illness, religious or cultural pressures, and other suicides in the community can also lead to an increased risk of suicide among teenagers. Finally, if the teenager has access to guns, knives, drugs, or any other means of taking their own life, the teenager may be at heightened risk for suicide.

Although teenage suicide can be difficult to prevent, especially when teenagers have easy access to instruments conducive to committing suicide, identifying the risk factors and attempting to minimize their effects before they are allowed to escalate is the most effective way to prevent teenage suicide. Teenagers who have access to both mental and physical health facilities and have strong family, societal, and religious support are much less likely to commit suicide. In addition, teenagers who have been taught methods of solving problems and conflicts in non-violent ways have a lower risk of suicide.

Substance Abuse

Substance abuse is a disorder in which an individual begins to become dependent on or overuse a particular drug or group of drugs that ultimately has a negative impact on their health and human development. Substance abuse, especially when the individual becomes addicted to or dependent on the drug, can affect the individual's ability to interact both socially and physically. Their ability to communicate intelligibly or even to complete relatively simple tasks can be severely hindered. After an individual has become chemically dependent on a particular drug, their body develops a physical need for the drug, and the individual will experience the effects of withdrawal if they are unable to meet that need. However, substance abuse not only affects a person by causing health problems, it also severely hinders an individual's ability for social development, as the individual often has difficulty improving social skills because of their inability to control behavior, actions, or even basic speech.

Smoking Addiction

Nicotine is consistently shown to be far more addictive than alcohol; whereas only one in ten users of alcohol will eventually become alcoholics, approximately eight out of ten heavy smokers will attempt and fail to quit. The method that nicotine uses is similar to that of other addictive substances: it creates an immediate positive feeling when taken, it will cause painful withdrawal symptoms if it is not taken, and it stimulates powerful cravings in the user even after it is removed from the system. Nicotine addiction can become so strong that a heavy smoker will experience withdrawal symptoms a mere two hours after smoking. Persistent tobacco use will also lead to an increased tolerance for nicotine, and so the user will have to consume more and more to achieve the pleasure or avoid the pain.

Alcohol Abuse

There are a few guidelines that alcohol users should know so that they can avoid chronic alcohol abuse. First, never use alcohol as a medicine or as a way to escape personal problems. Always drink slowly, and if possible, alternate alcoholic and non-alcoholic beverages. It is a good idea to eat both before and during drinking so that less alcohol rushes into the bloodstream. Drinking should never be the primary reason for a social function, though individuals should also try to avoid drinking alone. At a party, it is a good idea to avoid mixed drinks, as it is often difficult to tell just how much alcohol they contain. Finally, and most importantly, every person should have the self-control to say "no" to a drink without feeling guilty or rude.

Psychological and Physical Dependence on Drugs

A psychological dependence on drugs may begin as a craving for the pleasurable feelings or relief from anxiety that the drug provides. However, this craving can soon turn into a dependency on the drug in order to perform normal mental operations. A physical dependency, on the other hand, is said to occur when the individual requires increasing amounts of the drug to get the desired effect. Many drugs, like marijuana or hallucinogens, do not cause withdrawal symptoms; others, like heroin or cocaine, may be extremely painful to stop using. Individuals with a severe chemical dependency will eventually use a drug like this simply to avoid experiencing the effects of withdrawal. Typically, an individual with a severe dependency will try to stop many times without success.

Common Causes of Conflict Within a Relationship

The many sources of conflict within a relationship are too numerous to mention, but some of the common problems include the following: setting expectations that are too high, not appreciating or respecting the other person in the relationship, not considering the feelings of the other person, being afraid of showing affection or emotion, being overdependent, being inflexible, expecting the other member of the relationship to change, and lacking effective communication. Preventing conflict can be extremely difficult. Preventing it altogether is virtually impossible, but avoiding some or all of these common sources of conflict can greatly reduce the number of conflicts that take place within any given relationship.

Problem Solving

A well-functioning family would first identify the problem itself and determine the cause of the problem. The family would then develop a list of solutions that could potentially solve the problem, and they would attempt to determine the benefits of each solution. After determining the benefits of each solution, the family would choose the solution that seems to best solve the problem and then, after putting the solution into effect, monitor the solution to make sure that it actually solved the problem. Finally, the family would decide whether the solution worked or not to determine whether it was necessary to try something else. This entire process is important to the functioning of a family because it prevents problems from being misdiagnosed early on and prevents them from getting too far out of control.

Introducing a Model for Responsible Decision-Making to Students

A health educator teaching a responsible decision-making model to middle and high school students can begin with a presentation and student worksheets with term definitions. They discuss definitions with students. Empowerment is feeling control over one's decisions and behavior, resulting in inspiration; teachers tell students they must take responsibility for their decisions to achieve empowerment. Decision-making styles determine responsibility; they explain that teens with inactive decision-making styles cannot or do not make choices, and they lack control, accountability, and the ensuing self-confidence and empowerment. They explain that teens with reactive decision-making styles let others make decisions for them; needing others to like them and being easily influenced by others also impede self-confidence and empowerment. They then identify proactive decision-making styles as those involving analyzing a necessary decision, identifying and evaluating potential actions, choosing one action, and taking responsibility for the consequences of taking that action. Teachers can then introduce students to a model for responsible decision-making as a guide for making proactive decisions.

A model for responsible decision-making is meant to make sure that student decisions result in actions that show good character; that follow guidelines which parents, guardians or other responsible adults have established for them; that demonstrate self-respect and respect for others; and that protect safety, obey the laws, and promote health. A health educator can teach students in grades 6–12 the following seven steps included in a responsible decision-making model:

1. Describe the situation requiring you to make a decision.
2. List all of the decisions you could potentially make.
3. Share this list of potential decisions with an adult you trust.
4. Evaluate what the consequences of each of the decisions could be.
5. Determine which of the potential decisions you identified is the most appropriate and responsible one.
6. Take action on the decision that you have chosen.
7. Evaluate the outcomes of the decision that you have made.

The steps in the responsible decision-making model are describing the situation wherein they need to make a decision, listing the decisions they could potentially make, sharing their list with a parent or other responsible adult, evaluating each decision's potential consequences, deciding which decision is the most suitable and responsible, acting on the chosen decision, and evaluating the outcomes of that decision. When evaluating potential consequences of each decision, students can ask themselves the following five questions:

- Will making this decision lead to taking actions that are lawful or legal?
- Will making this decision lead to taking actions protecting my and others' safety?
- Will making this decision lead to taking actions that agree with the guidelines and advice that my parents and other responsible adults have given me?
- Will making this decision lead to taking actions that demonstrate my respect for myself and for other people?
- Will making this decision lead to taking actions that are demonstrations of good character?

As they learn to make decisions responsibly, students are bound to make mistakes, as with all new learning. Teens may experience anxiety over responsibility for poor decisions with unwelcome consequences. Paralyzed by doubt and indecision, they may avoid taking responsibility and action. In the same way that many teens fear being judged, rejected, disliked, or even viewed as different, they also fear doing the wrong thing. In addition to peer pressure and desiring acceptance, fear of misusing new responsibilities can motivate inaction to avoid unintentionally doing harm and experiencing guilt. Health educators can offer four steps to take after a bad or otherwise wrong decision:

1. Admit it; take responsibility, not trying to hide the mistake, blame others, or make excuses.
2. Immediately consider things done based on the decision; avoid perpetuating actions misguided by a wrong choice.

Coping with Common Stressors in Life

Four major life skills that people can apply to cope with common life stressors are values clarification, decision-making, communication skills, and coping skills. The following are typical life stressors in early childhood, adolescence, middle adulthood, and later adulthood, along with coping mechanisms that can be applied from the life skill perspective of values clarification.

- **Early childhood**: a pet's death. According to values clarification, reviewing the pet's positive qualities (similarly to adults' celebrating the life of the deceased) and considering getting another pet can address stressors.
- **Adolescence**: unwanted pregnancy. Discussing feasible alternatives and their ramifications for the teen, unborn baby, family, and society is not only required to make decisions, but also provides positive coping.
- **Middle adulthood**: divorce. Evaluating its impact on the couple and their relatives and friends; and the roles played by marital status, social expectations outside home, and religion inform values clarification coping mechanisms.
- **Later adulthood**: retirement. When people retiring from careers or employment view retirement in terms of their values, this can facilitate their ability to choose feasible options for post-retirement living.

Among life skills that enable coping with the stress of common life events are values clarification, decision-making, communication skills, and coping skills. Common stressors in four life stages follow, accompanied by coping mechanisms utilizing the decision-making life skill.

- **Early childhood**: a pet's death. Helping the child discuss the pros and cons of each alternative for disposing of the pet's remains is a decision-making-oriented way to cope with the loss.
- **Adolescence**: unwanted pregnancy. Careful consideration and evaluation of such alternatives as abortion; carrying the baby to term and surrendering it for adoption (and open or closed, public or private adoption, etc.); carrying to term and keeping the baby, etc. are necessary decisions to make and cope proactively with stressors. Decisions about future birth-control methods are also indicated.
- **Middle adulthood**: divorce. From the decision-making perspective, considering alternatives, risks, and consequences and making choices among career options, life roles, and future social relationships have major impacts on post-divorce living.
- **Later adulthood**: retirement. The decision-making life skill enables the retiree to consider alternatives and their advantages and disadvantages (e.g., not retiring, pursuing leisure activities, embarking on a second career, volunteering, realizing a long-deferred dream).

Four major life skills are values clarification, decision-making, communication skills, and coping skills. Examples of common stressors in each life stage, plus ways to cope using communication skills, follow.

- **Early childhood**: a pet's death. When a child feels sadness and anxiety over the loss, encouraging the child to communicate their feelings and thoughts can mediate psychic distress.
- **Adolescence**: unwanted pregnancy. Teenage mothers need support from various sources including family, friends, counselors, educators, and health professionals. Effectively utilizing communication skills enables them to know how, where, and from whom to solicit help and advice to cope with their situation.

- **Middle adulthood**: divorce. Adults undergoing divorces often have to assume various new life roles, including some that their former spouses may always have addressed. As a part of the process of divorce, adults need to apply communication skills to seek out supportive friends, relatives, and professionals as they establish and adjust to these new roles and experiences.
- **Later adulthood**: retirement. When older adults retire, they may lose some of their autonomy. Using communication skills assertively can help them maintain their independence.

Decreasing Tobacco Use

WHO recognizes tobacco as "the most widely available harmful product on the market." Therefore it negotiated the first international, legally binding treaty, the WHO Framework Convention on Tobacco Control (FCTC), providing protocols and guidelines for evidence-based interventions to decrease tobacco supply and consumption. Raising tobacco prices and taxes is a documented cost-effective method that substantially increases quitting and decreases starting smoking, particularly among poor and young people. With proper implementation, enforcing smoke-free public place and workplace laws obtains high compliance levels: fewer youths start smoking; smokers are supported in quitting or reducing smoking; and smoke-free policies prevent perpetuating addiction at earlier stages, especially in youth. Informing and educating the public is another cost-effective measure. Studies in multiple countries find graphic health warnings on cigarette and tobacco packaging and creative media campaigns succeed in powerfully decreasing consumer demand, despite opposition from wealthy tobacco companies and health officials' comparatively limited resources. Another cost-effective measure is providing smoking cessation assistance, combining pharmaceutical and behavioral therapies, through primary medical care and public health providers. Though a minority of the global population has received these measures, research finds them affordable in all world nations.

Characteristics of Conflict

According to experts, a conflict is not simply a disagreement, but a situation wherein both or either party perceives a real or imagined threat. Because such perceived threats are to people's survival and well-being, conflicts continue; ignoring them does not make them go away. Confronting and resolving conflicts stop them from going on indefinitely (or until the relationship ends). People do not necessarily (or usually) respond to conflicts based on objectively considering the facts, they react to them based on their personal values, beliefs, cultural backgrounds, and life experiences. Hence individual reactions to conflict are according to individual perceptions of the situation. Conflicts naturally provoke strong feelings. Therefore, people who cannot manage their emotions under stress or who are uncomfortable with them will be unable to succeed at resolving conflicts. Another characteristic of conflicts is that they present opportunities for growth. When members of a relationship succeed at resolving interpersonal conflict, they build trust between themselves. They gain direct experience that their relationship can withstand disagreements and challenges. This proof enables them to feel more secure about their relationship's existence and future.

Unhealthy vs. Healthy Ways of Resolving Conflict

When conflict inevitably arises, one unhealthy reaction is being unable to recognize and respond to things that are most important to the other person. A healthier response is being able to identify and address things that matter most to another. Emotional reactions that are resentful, angry, explosive, or designed to hurt the other person's feelings are unhealthy. Healthier responses involve staying calm, not becoming defensive, and showing respect for the other person. When one person reacts to conflict by rejecting the other, withdrawing their affection, isolating himself or herself, saying or doing things to shame the other, or showing or expressing fears of being abandoned, these are unhealthy reactions. Healthier responses are being willing to forgive the other person; forget undesirable reactions, words, and deeds; and progress beyond the conflict without retaining anger or resentment. Being unable to see the other person's viewpoint or make any compromises is unhealthy; being able to compromise instead of punishing the other person is healthier. Fearing and avoiding conflict due to expected negative outcomes is unhealthy; believing in the mutual benefit of confronting conflict head-on is healthier.

Peer Pressure

Children and teens often have more life experience being cared for, controlled, and told what to do, and relatively less experience being on their own, making independent choices, and taking initiative. Adults should tell them that being pressured is not good for them and is not right. Many children and teens (and even adults) have difficulty resisting pressure. Motivations include because they want to be liked, don't want to alienate friends, are afraid others will reject them, do not want others to make fun of them, do not want to hurt other people's feelings, are afraid others will perceive refusal as rejection, are not sure what they actually want, or do not know how to extricate themselves from the situation. Children and teens must know they have the right to say no without needing to give any reason, and they have the right to walk away from any situation involving pressure. Some brief tips to support refusing and resisting pressure include standing up straight, making eye contact with the other person, stating one's feelings clearly, not making excuses, and standing up for oneself.

Resisting Spoken Pressure

Children and teens (and adults as well) can find it hard to resist pressure that other people exert on them through spoken words. It is normal for most people not to want to hurt other people's feelings or feel responsible for bad feelings in others. However, children and teens especially must be reminded how important it is for them to stand up for themselves in order to prevent others from verbally pressuring them into doing unsafe or unwanted things. Some strategies recommended by experts to help young people refuse to use alcohol (or do other things that they know are not in their best interests and that they do not wish to do) include the "Dos and Don'ts." Dos: do say no assertively. Do abstain from drinking alcohol. Do propose some alternate activity. Do stand up for others being pressured who do not want to drink. Do walk away from the situation. Do look for something else to do with other friends. Don'ts: don't go to a party without being prepared to resist alcohol use. Don't be afraid to say "no." Don't mumble. Don't say "no" in an overly aggressive way. Don't behave like a "know-it-all" when refusing.

Domestic Violence and Abusive Relationships

Domestic violence is domestic abuse or spousal abuse, wherein one relationship partner dominates and controls another in a way that incorporates physical violence. Some violent behaviors include having an unpredictable, bad temper; harming, threatening harm, or threatening to kill the partner; threatening to hurt or take children; threatening suicide if the partner leaves; forcing sex; and destroying the partner's belongings. Manipulative power tactics abusers employ include: dominance, humiliation, isolation, threats, intimidation, denial, and blame. The cycle of domestic violence follows a common pattern: one partner abuses the other with violent behavior to exhibit dominance. The partner appears guilty, but really fears being caught and punished rather than feeling remorse. The abuser avoids responsibility by making excuses for the violent behavior, rationalizing it, and/or blaming the other partner for it. The abuser, trying to keep the victim in the relationship and regain control, behaves contritely, "normally," or with great affection and/or charm, often fooling the victim into hoping they have changed or will change. The abuser fantasizes and plans further abuse to make the victim pay for perceived wrongs. The abuser then places the victim in a situation to justify further abuse, and the cycle repeats all over again.

Domestic violence is domestic abuse including physical violence. Physical force that injures or endangers someone is physical abuse. Physical battery or assault is a crime; police have the authority and power to protect individuals from physical attacks, whether outside or inside a family or home. Sexual abuse is an aggressive, violent act and a type of physical abuse. This includes forced sex, even by a partner with whom one also has consensual sex. Victims of physical and sexual abuse are at greater risk of serious injury and death. Even if incidents seem minor (e.g., being pushed or shoved), they are still abuse and can still cause severe injury or death. Even if incidents have only happened once or twice in a relationship, they are still abuse and are likely to continue and escalate. If physical assaults stop when the victim becomes passive, this is not a solution: the victim has given up their rights to independence, self-expression, and decision-making as a partner and as a person. Even when no physical violence exists, victims may suffer from verbal and emotional assault and abuse.

People who want to determine if they are in an abusive relationship should consider whether they think or feel the following: they feel afraid of their partner often, they avoid mentioning certain subjects for fear of making their partner angry, they feel they cannot do anything right with their partner, they believe being mistreated is what they deserve, they wonder whether they are the member of the relationship who is crazy, and they feel helpless and/or emotionally numb. To consider whether their partner engages in belittling behaviors toward them, they should consider the following: whether their partner yells at them often; whether the partner says or does things to humiliate them; whether the partner insults them or criticizes them regularly; whether the partner treats them so poorly they find it embarrassing for family, friends, or others to witness it; whether the partner dismisses, disparages, or ignores their successes and/or opinions; whether the partner blames them for the partner's abusive behaviors; and whether the partner views and/or treats them as a sexual object or property instead of a human being.

Victims in abusive relationships should consider whether their partner behaves in an overly possessive and jealous manner toward them; whether the partner controls what they do or where they go; whether the partner prevents them from seeing their family or friends; whether the partner limits their access to the car, the phone, and/or money; and whether their partner is continually checking up on what they are doing and where they are going. These are all behaviors intended to control the other person, and they are not normal or healthy. Threats of violence or violent behaviors to watch for in a partner include: the partner has a bad temper, and is unpredictable about losing their temper; the partner threatens to harm or kill them, or actually does harm them; the partner threatens to hurt their children, actually hurts them, or threatens to take them away; the partner threatens that if they leave, the partner will commit suicide; the partner forces them to engage in sex when they do not want to; or the partner takes away or destroys their personal belongings.

People often associate the idea of domestic abuse with physical battery. However, many partners are victims of emotional abuse. Without physical bruises, the victim, abuser, and other people unfortunately overlook or minimize emotional abuse. The intention and result of emotional abuse are to erode the victim's independence, control, and feelings of self-worth. Victims come to feel they have nothing without the abusive partner, or have no way to escape the relationship. Emotional abuse includes verbal abuse like blaming, shaming, name-calling, insulting, and yelling. It also includes controlling behaviors, intimidation, and isolating the victim. Threats of punishment, including physical violence, frequently enter into psychological or emotional abuse. Emotional abuse scars are less visible than physical ones, but are equally or more damaging. Financial or economic abuse is another way to control the victim. It includes withholding money, checkbooks, credit or debit cards; withholding shelter, food, clothing, medications, or other necessities; making victims account for every cent they spend; rigidly controlling the victim's finances; restricting the victim to an allowance; sabotaging the victim's job by constantly calling there and/or causing the victim to miss work frequently; preventing the victim's working or making career choices; and stealing from or taking the victim's money.

Some people observe that abusive individuals lose their tempers and/or apparently have some psychological disorder. Some also abuse substances (though others do not), and, equating their problems with the illness or disease model of substance abuse, mistakenly assume that abusers cannot control their behavior. However, experts point out that abusive behaviors and violence are deliberate choices that the abusers make to control their victims. Evidence that they can control their behavior includes that they do not abuse everybody in their lives, only those they claim to love who are closest to them; that they choose carefully where and when to abuse, controlling themselves in public but attacking the victim once they are alone; that they can stop the abusive behavior when it is to their benefit (e.g., when their employer calls or the police arrive); and that they frequently aim physical attacks to parts of the victim's body where they are hidden by clothing, so others cannot see them.

Tactics Employed by Domestic Abusers

- **Dominance**: abusers, needing to feel in control of victims and relationships, dominate by making decisions for victims and family, giving them orders, and expecting unquestioning compliance. They often treat victims as children, servants, slaves, or possessions.
- **Humiliation**: to keep victims from leaving, abusers make them feel worthless and that nobody else will want them. To make victims feel inadequate, they insult and shame them publicly and privately, making them feel powerless and destroying their self-esteem.
- **Isolation**: abusive partners make victims dependent on them by cutting off their contact with others. They may stop victims from visiting with friends and relatives, or even from going to school or work. Victims may have to ask permission to see anybody, go anywhere, or do anything.
- **Threats**: to frighten victims into dropping charges and/or prevent their leaving, abusers typically threaten to: harm or kill victims, children, other family, or pets; commit suicide; report victims to child services; and file false charges against them.
- **Intimidation**: threatening gestures and looks, property destruction or smashing objects in front of victims, hurting pets, or displaying weapons are tactics signaling violent consequences for noncompliance to frighten victims into submission.
- **Denial and blame**: abusers minimize or deny abuse or blame it on circumstances or, commonly, the victim. "You made/make me do it" is a frequent accusation used by abusers.

Warning Signs of Domestic Abuse

Warning signs of domestic abuse: the person agrees with everything the partner does and says; frequently checks in with the partner, reporting what they are doing and where they are; often receives harassing phone calls or texts from the partner; appears anxious or afraid to please the partner; and/or mentions the partner's jealousy, possessiveness, or temper. Warning signs of physical violence: the person often misses school, work, or social events without explaining; often has injuries, excusing them as "accidents" or "clumsiness"; and/or wears sunglasses indoors, long sleeves in summer, or other means of hiding injuries. Warning signs of isolation: the person never or seldom goes out in public without the partner; is unable to see friends and family; and/or has limited access to the car, money, or credit or debit cards. Psychological warning signs of being abused: someone who used to be confident displays significantly lowered self-esteem; an outgoing person becomes withdrawn; or an individual shows other major personality changes. The person appears anxious, depressed, despondent, or suicidal, verbalizing suicidal ideations or displaying suicidal behaviors.

Advice for People Who Suspect Somebody They Know Is a Victim of Abuse

Abusers are experts at manipulating and controlling victims. Victims are drained, frightened, ashamed, depressed, and confused. They need to escape the situation, but frequently have been isolated from others. Those suspecting abuse should be alert to warning signs, offer support to victims for extricating themselves, getting help, and starting the healing process. Some people may hesitate, thinking they could be mistaken; learning the victim does not want to discuss it or have them interfere; or simply be told that it is none of their business. In these cases, experts advise people to speak up regardless: expressing concern not only informs a victim somebody cares, it moreover could save that person's life. They should speak with the person privately, identifying signs they have observed and explaining why they are concerned, reassure the individual they will keep all conversation confidential, that they are there whenever they are ready to talk, and will help in any way possible. Regarding dos and don'ts, do the following: express concern, ask whether something is wrong, listen, validate the person's communications, offer help, and support the individual's decisions. Don't: wait for the person to approach you, blame or judge the individual, give advice, pressure the person, or attach conditions to your support.

Social, Emotional, Physical, and Intellectual Development

Affective, Cognitive, and Psychomotor Skills

Affective skill refers to how effectively an individual can recognize, understand, and handle emotions and relationships. Affective skills allow an individual to feel appropriate emotion in response to certain situations or stimuli, and then to respond appropriately.

Cognitive skill refers to an individual's ability to gather and understand information. Cognitive skills allow an individual to comprehend new situations and apply the knowledge that they have gathered elsewhere.

Psychomotor skill refers to an individual's ability to coordinate their physical movements. In other words, psychomotor skills are a person's control over simple and complex motor functions.

It is extremely important for an individual to be able to use a combination of their affective, cognitive, and psychomotor skills together on a day-to-day basis, as each type of skill is essential to the overall functioning of a healthy individual. An individual who has mastered their psychomotor skills may be in excellent physical health, but the individual's emotional and intellectual health will suffer if they are unable to make effective relationships and understand basic and complex concepts. The situation is the same for individuals who can only maintain effective relationships or who can only understand complex concepts, as it will be significantly more difficult for them to perform everyday functions if they have poor control of their psychomotor skills. For an individual to maintain their physical and mental health, along with that of their family, the individual must be able to use a combination of different skills.

Some of the factors that can be used to measure how well-developed an individual's affective skills are include determining how well the individual receives emotional stimuli and how well the individual responds to those stimuli. It is also important to determine how easy it is for the individual to acknowledge the worth of a particular situation, relationship, or individual and whether the individual has an organized and well-conceived value system. An individual's ability to receive and respond to emotional stimuli can be measured by how aware the individual is of a particular stimulus, how willing the individual is to acknowledge that particular stimulus, and how focused the individual is on that stimulus. An individual's ability to assign value to a situation and uphold a value system can be measured by how motivated the individual is, how the individual behaves, and how consistent that individual's behavior is. For example, a student that always comes to class and clearly always pays attention may have well-developed affective skills.

Some of the factors that can be used to measure how well-developed an individual's cognitive skills are include determining the individual's ability to retain knowledge, comprehend knowledge, apply knowledge, and evaluate knowledge. An individual's ability to retain knowledge can be measured by testing the individual's ability to remember certain facts and information through exams or simply asking questions. An individual's ability to comprehend knowledge can be measured by an individual demonstrating a concept in a different form, explaining a concept in more detail or simplifying a concept, or predicting a result based on a particular concept. An individual breaking a concept down into individual parts and demonstrating how those parts make up the whole can also show comprehension of a particular concept. An individual's ability to apply knowledge can be measured by an individual demonstrating that they can use a particular concept for a real-life purpose. Finally, an individual's ability to evaluate a particular piece of knowledge can be indicated by the individual showing the value of that knowledge.

Some of the factors that can be used to measure how well-developed an individual's psychomotor skills are include how well an individual performs physical skills and acts, how precisely the individual can perform those skills or activities, and how natural those activities seem to be for the individual. An individual's ability to use physical skills can be measured simply by how much difficulty the individual has in accomplishing a particular complex physical activity such as climbing a rope or assembling a model. How precisely the individual can perform those skills or activities can be measured by determining the quality of the result of the individual's physical activity and how long it took the individual to reach that result. For example, if the individual has constructed a model plane, does the model look like a plane, are its wings and other parts attached correctly, how long did it take to assemble, etc. Finally, an activity is natural for an individual if the individual can perform it without thinking.

Meeting the Special Needs of a Student

The first step an educator should take when determining the best way to meet the special needs of a student is to identify exactly what that particular child's needs consist of, as each student is unique in their ability to learn and comprehend. If a student is performing poorly, a teacher must determine the cause of the student's poor performance. Once the cause has been identified, the teacher can then determine how much assistance the student needs. If the student's needs can be met through such techniques as one-on-one attention or special project assignments, this is usually the best course of action. However, if the student has needs that require solutions beyond simple changes in curriculum, including potential psychological or physiological disorders, the educator has an obligation to consult with other educational professionals and to discuss other options with the child's parents.

Early Childhood Intervention and Intellectual Giftedness

Early childhood intervention is the process by which children who are experiencing or showing signs of developmental difficulties are diagnosed and treated early to allow them to continue developing in the best manner possible. Early childhood intervention services usually take place before the child reaches school age because studies indicate that the earlier a child who is experiencing difficulties receives special education, the more effective that education will ultimately be.

mass, volume, and number. For example, they can understand that two containers of different shapes that each have the same amount of liquid poured into them still contain the same amount of liquid despite their differences in appearance. The child also begins to identify and organize objects according to shape, size, and color. The child will not be able to understand more abstract concepts such as those found in calculus or algebra, however, until they reach the formal operational stage of development.

THE FORMAL OPERATIONAL STAGE

The fourth and final stage of Piaget's theory of cognitive development, the **formal operational stage**, starts at age 11 and continues until the end of an individual's life. During this stage, an individual understands more abstract concepts and develops a logical way of thinking about those concepts. In other words, an individual begins to understand ideas that are less concrete or absolute and that cannot necessarily be backed up by physical evidence or observation such as morality, advanced mathematics, and a person's state of being. It is also within this stage of development that individuals can understand all the variables in a problem and are able to determine most, if not all, the possible solutions to a problem rather than just the most obvious solutions. This stage is never truly completed; it continues throughout a person's life as the individual develops and improves their ability to think abstractly.

CHALLENGES TO PIAGET'S THEORY

Later researchers have challenged Piaget's theory of cognitive development because studies indicate that Piaget may have underestimated the abilities of younger children to learn and understand various concepts. Piaget's theory indicates that younger children are unable to understand certain concrete and abstract thoughts early within their development even if another individual teaches the child. However, this notion has been disproved. Research shows that young children can be taught how to handle and understand problems that Piaget believed only older children would be able to comprehend. Researchers have also challenged Piaget's theory because studies indicate that if a younger child is given a task like one an older child might receive, but the difficulty of the task is adjusted to compensate for age, the younger child would actually understand the concept more effectively. Also, notably, Piaget's theory ignores many of the benefits of adult learning. Piaget's theory is still important, though, because it presents the importance of active learning in a child's development.

ABRAHAM MASLOW'S HIERARCHY OF HUMAN NEEDS

Abraham Maslow theorized that there are five types of human needs that, if arranged in order of importance, form a pyramid. Maslow maintained that individuals would not be able to focus on the upper layers of the hierarchy until they were first able to meet the needs at the lower layers. The first layer of the pyramid represents the **physiological needs**, which are the basic needs required for an individual's survival such as food, water, breathable air, and sleep. The second layer of the pyramid represents the **safety needs**, which are the elements that an individual needs to feel a sense of security such as having a job, good health, and a safe place to live. The third layer of the pyramid corresponds to the **love and belonging needs**, which are needed to form social relationships such as those with friends, family, and intimate loved ones. The fourth layer of Maslow's hierarchy of human needs is the **esteem layer**, which represents the individual's need to respect themself and be respected and accepted by others. The fifth and top layer of the pyramid is the **self-actualization layer**. It represents the individual's need for morality, creativity, and trust. Maslow theorized that individuals could survive without reaching the higher levels of the pyramid but that would feel a sense of anxiousness if these needs were not met. Maslow also believed that individuals who reached the higher levels of the pyramid did not receive any tangible benefit from meeting these needs other than a feeling of fulfillment and the motivation to fulfill needs higher on the pyramid.

Maslow later added two additional layers above the self-actualization layer of the pyramid. These are the **cognitive layer** and the **aesthetic layer**. The cognitive layer is the layer that represents an individual's need to acquire and ultimately understand both abstract and concrete knowledge. The aesthetic layer, which became the final layer in later versions of the pyramid, is the layer that represents the individual's need to discover, create, and experience beauty and art. Maslow later theorized that if an individual was unable to meet the needs of any given layer of the pyramid, those needs could become neurotic needs. Such needs are compulsions that, if satisfied, would not facilitate the individual's health or growth.

Review Video: Maslow's Hierarchy of Needs
Visit mometrix.com/academy and enter code: 461825

Erik Erikson's Theory of Psychosocial Development

Erik Erikson's theory of psychosocial development breaks the process of human development into eight stages necessary for healthy functioning. The eight stages Erikson identified are infancy, younger years, early childhood, middle childhood, adolescence, early adulthood, middle adulthood, and later adulthood. During each of these stages, individuals must overcome a developmental obstacle, which Erikson called a crisis, to be able to progress and face the crises of later stages. If an individual is not able to overcome one of the crises along the way, later crises will be more difficult for them to overcome. Erikson's theory also maintains that individuals who are unable to successfully pass through a particular crisis will likely encounter that same crisis again.

Infancy

The first stage of Erikson's theory of psychosocial development is **infancy**, which spans from birth to 12 months. In this stage, a child is presented with the crisis of trust versus mistrust. Although everyone struggles with this crisis throughout their lives, a child needs to be able to realize the concept of trust and the elements of certainty. For example, a child learns that if their parents leave the room, they aren't going to abandon the child forever. If a child is unable to realize the concept of trust because of traumatic life events, such as abandonment, the child may become withdrawn and avoid interaction with the rest of society.

Younger Years Stage

The second stage of Erikson's theory of psychosocial development is the **younger years stage**, which covers ages one to three. In this stage, a child is faced with the crisis of autonomy versus shame and doubt. The child is presented with the need to become independent and learn skills such as using the toilet without assistance. If the child is able to overcome this crisis, they will gain the sense of self-pride necessary to continue fostering the child's growing need for independence. If, however, the child is unable to overcome this crisis and cannot establish their own independence, the child will develop feelings of shame and doubt about their ability to function without assistance.

Early Childhood Stage

The third stage of Erikson's theory of psychosocial development is the **early childhood stage**, spanning ages three to five. In this stage, a child is faced with the crisis of initiative versus guilt. The child is presented with the need to discover the ambition necessary to continue functioning independently. This stage is strongly linked with the moral development of the child as they begin to use make-believe play to explore the kind of person they want to become in the future. If children are unable to explore their ambitions or if they are expected to function with too much self-control, they will develop feelings of guilt as they begin to see their ambitions, dreams, and goals as unattainable or inappropriate.

MIDDLE CHILDHOOD STAGE

The fourth stage of Erikson's theory of psychosocial development is the **middle childhood stage**, which covers ages six to ten. In this stage, a child is faced with the crisis of industry versus inferiority and is presented with the need to develop the ability to complete productive tasks such as schoolwork and working in groups. If children are unable to learn how to work effectively, either alone or in a group, they will develop a sense of inferiority as a result of their inability to complete the tasks set before them that their peers are capable of completing. For example, if a child is regularly unable to complete their homework because the child does not understand the material while the rest of the child's peers are not having difficulty, this can lead the child to develop a sense of inferiority.

ADOLESCENCE STAGE

The fifth stage of Erikson's theory of psychosocial development is the **adolescence stage**, which covers ages 11 to 18. In this stage, the child is faced with the crisis of identity versus role confusion. During this stage, the child attempts to find their place in society and identify future goals and the skills and values necessary to achieve those goals. At this stage, the child also becomes more aware of how people perceive them and becomes concerned with those perceptions. If the child is unable to determine what future goals they are interested in pursuing, it can lead to confusion about what roles the child will play when they reach adulthood.

EARLY ADULTHOOD STAGE

The sixth stage of Erikson's theory of psychosocial development is the **early adulthood stage**, which covers ages 18 to 34. In this stage, the young adult is concerned with the crisis of intimacy versus isolation in which an individual needs to begin establishing intimate relationships with others. If an adult is unable to form intimate relationships with others, perhaps because of disappointing relationships in the past, this person will become more withdrawn and will isolate themself from others. Isolation can prove to be a perilous problem in the development of a healthy adult, as it prevents the individual from forming lasting relationships. The lack of social interaction can also lead to severe personality flaws, which may hinder the development of future relationships.

MIDDLE ADULTHOOD STAGE

The seventh stage of Erikson's theory of psychosocial development is the **middle adulthood stage**, occurring between the ages of 35 and 60. In this stage, an adult becomes aware of the crisis of generativity versus stagnation in which the individual is concerned with continuing their genetic line before it is too late. Generativity refers to the ability to produce offspring and then nurture, guide, and prepare that offspring for future life. At the same time, however, generativity in this context also refers to any act that gives something of value to the next generation such as teaching children how to read. If an individual is unable to contribute to the next generation in some form, the individual will feel a sense of failure resulting from stagnation, which is simply a lack of accomplishment.

LATER ADULTHOOD STAGE

The last stage of Erikson's eight stages of psychosocial development is the **later adulthood stage**, which is the period that starts at age 60 and extends to the end of one's life. In this stage, an individual is confronted with the crisis of ego integrity versus despair. During this time, an adult begins to examine the course of their life by reflecting on the kind of person that they have been. If the adult feels that they have had a meaningful life and accomplished something during it, this will lead to a strong sense of integrity. However, if the individual is unhappy with the way they have acted, this person will experience despair and will fear death as the absolute end of further achievement.

Changing Familial Roles of Men and Women

Fifty years ago, women were the primary caretakers of the family's children, and they were in charge of maintaining the household while men worked to provide for the family. This has changed, however, because of the drastic increase in the number of women entering the workforce since that time. This is partially because it has become more difficult for families to subsist on one income alone. Both members of the marital couple are often forced to work to provide for the family, which can make it difficult when trying to balance the responsibilities of caretaker and provider. Men, who were once the primary providers for the family, are still out in the workforce, but their spouses have joined them, and both individuals have to find ways to make the time to care for the family's children.

Nature Versus Nurture

The concept of nature versus nurture is the idea that of all a person's traits, some result from their genetic heritage, and some result from their environment. In this context, nature refers to any trait that an individual is born with or has acquired through genes. Nurture may be seen as the opposite of nature; it refers to any trait that an individual learns from the environment. Nurture often refers specifically to the environment created by the parents of the child, but it can refer to any environmental condition that affects the development of the child. The concept of nature versus nurture is important because it shows that individuals inherit some of their traits from their parents, but they also develop many of their traits from their environment.

Genetic and Environmental Traits

Research has shown that some traits that are almost completely genetic include eye color, blood type, and most diseases. In most cases, genetics also determines one's risk of future diseases, vision, and vision impairments. Religion and language, on the other hand, are examples of traits that researchers have proven to be almost completely environmental. These traits are all linked to specific genes or to specific environmental factors, but most traits are actually a result of both environmental and genetic influences. Traits such as height, weight, and skin color are all examples of traits that are influenced by both an individual's genes and their environment.

Robert Havinghurst's Developmental Task Concept

The developmental task concept is a theory of human development established by Robert Havinghurst that states that there are certain tasks each individual needs to go through at points during their life to continue developing into a happy and successful adult. These tasks, separated into three groups by their causes, are tasks resulting from physical maturation, personal causes, and societal pressures. A child learning to crawl is an example of a task that becomes necessary as the child matures physically. An individual learning basic first aid because they are interested in becoming an EMT is an example of a personal cause. An example of a task resulting from societal pressure is a child learning to behave appropriately in a store.

1. **Infancy and early childhood**: The period from ages zero to five, and it consists of tasks such as learning to walk, talk, and eat solid foods as well as learning right from wrong.
2. **Middle childhood**: The period of development from ages six to 12 that includes tasks such as learning to get along with others, moral values, and skills and knowledge required for day-to-day living.

3. **Adolescence**: The period from ages 13 to 18, and it requires tasks that include learning how to relate with members of the opposite sex, learning the social role of one's gender in society, and preparing for life after childhood.
4. **Early adulthood**: The period of life from ages 19 to 29, and it is the age range where tasks such as starting a long-term relationship, finding a career, and starting a family are required.
5. **Middle adulthood**: The period from ages 30 to 60 that includes tasks such as finding adult recreational activities, achieving in one's chosen career, and helping one's teenage children become healthy and happy adults.
6. **Later maturity**: The period from ages 61 to the end of a person's life. This period consists of tasks such as adjusting to the death of a spouse, adjusting to the effects of old age, and finding people in one's peer group to interact with.

Multiple Intelligences Identified by Howard Gardner

Students with high **logical-mathematical intelligence** excel at calculations and reasoning, thinking abstractly and conceptually. They detect and explore patterns and relationships, ask cosmic questions, and enjoy solving puzzles and conducting experiments. Before addressing details of a problem, they must learn and formulate concepts. Compatible teaching strategies include offering mysteries, investigations, and logic games. Students with strong **linguistic intelligence** are effective with words. Their auditory skills are highly developed. They often think in words, not pictures, feelings, or sounds. They enjoy reading, writing, and playing word games. Teaching methods and tools include reading books with them, encouraging them to see and pronounce words, lectures, books, computers, multimedia materials, voice recorders, and spoken-word recordings. Students with high **intrapersonal intelligence** tune into their own feelings and are "loners," avoiding social interaction. They are independent, confident, strong-willed, motivated, opinionated, intuitive, and wise. Introspective reflection and independent study, journals, diaries, books, creative materials, time, and privacy are useful instructional methods and materials. Students with strong **interpersonal intelligence** are social, interactive, empathetic, and "street-smart." Group activities, dialogues, seminars, telephones, audio- and video-conferencing or Skyping, and email are good teaching methods and tools.

Chapter Quiz

Ready to see how well you retained what you just read? Scan the QR code to go directly to the chapter quiz interface for this study guide. If you're using a computer, simply visit the online resources page at **mometrix.com/resources719/aafcsfamconsc** and click the Chapter Quizzes link.

Nutrition, Wellness, and Food Sciences

Nutrition and Wellness

Preventing Anorexia, Bulimia, and Obesity

Unfortunately, there is no known way of completely eliminating the risk of an individual developing an eating-related condition such as **anorexia**, **bulimia**, and **obesity**, but it has been shown that some methods may significantly lower the risk of developing these conditions. The best way to reduce the risk of developing eating disorders or becoming obese is for parents to interact more with their children and make sure that they are teaching them good eating habits early on. It is also important that parents attempt to build-up the self-esteem of their children through interactions that show high esteem in both the parents and child. Children who have been taught good eating habits by their parents and who have been taught to have high self-esteem have been shown to eat more carefully, and therefore are at a significantly lower risk of becoming obese or developing eating disorders such as anorexia and bulimia.

Anorexia is an eating disorder in which an individual views their own body as being overweight, even though they are not, which causes the person to have an extreme, unfounded fear of gaining additional weight. This fear can lead individuals to use desperate and unhealthy methods to reduce their weight below what would normally be healthy. These methods include deliberate vomiting, limiting their food intake, exercising excessively without eating enough food, and using medications to flush their system. Anorexia is commonly found to affect young women, specifically during adolescence. It can pose a serious risk to an individual's health as the obsessive attempts to reduce body weight can affect the health of the heart, brain, immune system, muscles, and other organs. Anorexia can be extremely difficult to treat as it is a complicated psychological condition, but it has been shown that psychotherapy may be able to help the individual overcome the inaccurate perceptions she or he has regarding their own body.

Bulimia is an eating disorder, similar to anorexia, in which an individual views their own body as being unattractive or overweight, but lacks the ability to control their own eating. Bulimic individuals regularly eat an unhealthily large amount of food and then attempt to flush their systems to prevent themselves from gaining additional weight. Some of the methods individuals suffering from bulimia might use to flush food from their systems include deliberate vomiting, excessive exercise, and using diet pills, laxatives, ipecac, and other medications. Bulimia is most commonly found in women, especially younger women from ages 12 to 19, and can have a significant impact on the health of the individual. It can cause problems such as anemia, weakness, muscle and heart problems, dehydration, malnutrition, damage to the stomach, and a wide range of other problems. Bulimia is difficult to treat, but a combination of group psychotherapy and low doses of anti-psychotic medications have been shown to help.

Anorexia and bulimia are very similar disorders, but the one major difference between the two conditions is control. An individual suffering from anorexia is usually already below a healthy weight, still perceives that their own weight is unacceptable, and therefore attempts to lower that weight further by limiting food intake. Bulimic individuals, on the other hand, have no sense of control over their eating habits. They instead eat excessively and then attempt to overcompensate for their excessive food intake. In short, individuals suffering from anorexia are usually below what would normally be considered a healthy weight for their age and size and attempt to control their eating to reduce their weight further. Bulimic individuals, however, are usually above a healthy weight as they have no control over their own eating.

Obesity is a condition of the body where the individual has increased their own body weight significantly beyond what is normally considered healthy, usually by excessive eating. Obesity occurs because the individual takes in more food than their body can actually use, and the excess food is stored as fat. Overeating is the primary cause of obesity, but obesity can also be tied to family history, genetic factors, stress and lack of sleep, various illnesses and conditions, and many other causes. An individual who is obese is at a significantly higher risk for certain health problems, including problems with the heart, stomach, muscles, lungs, skin, nervous system, and many other areas of the body. The best way to treat obesity is through a well-balanced diet that eliminates excessive food intake and a rigorous exercise program. In extreme cases, individuals may also use medication or even surgery to help lower their weight.

Diabetes

Diabetes is a disease that prevents the body from producing or using a hormone called insulin, which the body needs to process sugar and use it as energy for the cells of the body. Since an individual with diabetes cannot produce insulin or cannot use what is produced, the body is unable to appropriately use the sugar that the cells need to survive. Instead the sugar builds up in the body, leading to a high sugar concentration in the blood, which is a condition known as hyperglycemia. Diabetes can lead to a series of dangerous health problems including the potential failure of the heart, kidneys, nerves, and eyes as well as conditions such as high blood pressure, blindness, poor healing of wounds, and many other dangerous conditions. There is currently no cure for diabetes, but the negative effects of diabetes can be controlled through careful monitoring of blood sugar levels, the use of a specific diet to maintain the individual's blood sugar, and through the use of certain drugs and/or insulin shots.

The three types of diabetes are **type 1**, **type 2**, and **gestational diabetes**. Type 1 diabetes is a form of diabetes in which the individual's own immune system mistakes the beta cells of the pancreas, which are the cells responsible for producing insulin, as being harmful and therefore attacks and destroys them. Type 1 diabetes prevents the individual from actually producing insulin. It is treated primarily through the use of a carefully constructed diet and insulin injections or the use of an insulin pump. Both type 2 diabetes and gestational diabetes are conditions in which the individual is either unable to produce enough insulin for the body or the cells of the body are unable to use the insulin correctly. Type 2 and gestational diabetes can both be treated through the use of a carefully monitored diet along with medication to help the body use the insulin appropriately. The only major difference between type 2 diabetes and gestational diabetes is that gestational diabetes occurs specifically due to hormones present during pregnancy.

There is currently no way to completely prevent an individual from developing diabetes, but the risk of developing certain forms of diabetes can be reduced. The risk of type 2 diabetes and gestational diabetes can be reduced by eating a more healthy diet high in fiber and whole grains, exercising regularly, and lowering the amount of high-fat foods that the individual eats. It has also been suggested that the risk of a mother developing gestational diabetes may be reduced through breast-feeding as it allows the mother to release some of the excess hormones present in her body as a result of the pregnancy. The risk of type 2 and gestational diabetes can be reduced, but there is, unfortunately, no known way of preventing type 1 diabetes, as researchers know very little about the exact causes of the disease.

Review Video: Diabetes Mellitus
Visit mometrix.com/academy and enter code: 501396

Vitamin E, and its form tocopherol, is important primarily because it aids in maintaining proper brain function and eye health. It has also been suggested that vitamin E may help reduce the risk of cancer, cataracts, heart disease, and other health conditions as well as helping treat patients that have Parkinson's or Alzheimer's Disease. A significant vitamin E deficiency can result in muscle weakness, blindness, and neurological problems as a result of the body transmitting nerve impulses incorrectly. Common sources of vitamin E include peanuts, hazelnuts, coconuts, corn, asparagus, carrots, tomatoes, fish, peanut butter, and vegetable oils. There are also many multivitamins and supplements that supply vitamin E, but some studies have shown that some synthetic vitamins and supplements may actually be significantly less beneficial or even have a negative impact on the individual when taken in large doses. However, both men and women do require vitamin E, in similar amounts, to continue functioning normally.

B12, and its synthetic form cyanocobalamin, is important because it is necessary for the production of blood cells and aids in maintaining the health of the nervous system. A severe lack of B12 within the body can lead to a variety of conditions including megaloblastic anemia, a condition in which the red blood cells have less hemoglobin and therefore have more difficulty functioning properly. A significant lack of B12 can also lead to severe problems with the nervous system because the lack of B12 causes the disintegration and death of nerve cells. Individuals who lack significant quantities of B12 may develop symptoms such as numbness, tingling, and difficulty with muscle control. Some common sources of B12 include chicken, beef, pork, liver, fish, shellfish, certain breakfast cereals, milk, cheese, eggs, and yogurt. Vegetarians may have difficulty getting enough B12 since the primary sources of B12 are meat products, but B12 supplements and multivitamins can be a good option as well.

Vitamin A, and its derivative retinol, is important because it aids in bone growth, skin health, and the ability to reproduce. Vitamin A also promotes eye health and aids in the production of tears, which prevents the eyes from drying out and washes away bacteria that might cause infections. A vitamin A deficiency can cause the eyes to deteriorate and may lead to conditions such as night blindness, also known as nyctalopia, in which an individual has difficulty seeing in low light environments. Common sources of vitamin A include carrots, sweet potatoes, pumpkins, spinach, beef, pork, chicken, eggs, and broccoli. Both men and women require similar amounts of vitamin A for their bodies to function normally, but men usually require slightly more than women do.

Importance of Fiber

Fiber is an important part of an individual's diet because it helps with bowel movements, digestion, and immune responses. Fiber has also been shown to lower blood cholesterol, help prevent obesity, lower the risk of certain types of cancer including colon cancer, and lower the risk of type 2 diabetes. A significant lack of fiber in an individual's diet can lead to symptoms such as constipation and slower digestion and can cause an individual to have a higher risk of developing certain diseases. Some common sources of fiber include certain breakfast cereals, oatmeal, whole-wheat bread, beans, apples, pears, strawberries, bananas, potatoes, onions, and green beans. Both men and women require similar amounts of fiber to continue functioning properly.

Importance of Protein

Protein is important to the functioning of a healthy individual because it is necessary for the body to produce the amino acids it needs to continue functioning. Most of the amino acids that the body needs are already present in the body, but certain amino acids, known as essential amino acids, can only be produced when the body digests protein. A severe lack of protein is usually caused by malnutrition and can lead to reduced brain function, intellectual disabilities, and an overall weakening of the immune system due to a decrease in the number of white blood cells. It has also been suggested that a significant lack of protein may lead to conditions such as kwashiorkor, which causes significant weight loss, thinning and discolored hair, swelling of the organs, and weakens the responses of the immune system. Some common sources of protein include chicken, beef, wheat, rice, milk, cheese, eggs, peas, beans, peanuts, and peanut butter.

Importance of Iron

Iron is important because it aids in the proper functioning of virtually all the muscles and organs. Iron also allows the body to form hemoglobin, which is a protein in red blood cells that carries oxygen throughout the body. If an individual does not have enough iron in their diet, anemia can result, a condition in which the body is unable to produce hemoglobin. The body may also be unable to produce additional red blood cells if there is a lack of iron, and the individual will be more likely to become fatigued and develop other symptoms and conditions. Common sources of iron include oatmeal, spaghetti, whole wheat bread, sunflower seeds, broccoli, green beans, beets, peas, potatoes, green leafy vegetables, beef, pork, and chicken. Women usually require more iron than men do partially because of the regular blood loss associated with menstruation.

Importance of Calcium

Calcium is important to the development and maintenance of bones and teeth, since a lack of calcium can result in osteoporosis and other bone problems. Calcium is also important to the production of lymph, which is a fluid similar to interstitial fluid that aids in the overall operation of the body, specifically the immune system. Common sources of calcium are milk, cheese, eggs, orange juice, oranges, broccoli, rutabagas, almonds, and other nuts. Calcium can also be obtained through supplements and multivitamins, which can be a good source for people who are allergic to dairy products or have a significant calcium deficiency. Both men and women require similar amounts of calcium, but the amount of calcium required increases as an individual ages.

RDA and RDI

RDA stands for recommended dietary allowance, which is the amount of each vitamin, mineral, or other nutrient that health professionals believe an individual needs to receive daily in order to stay healthy. The RDA has been replaced by the now more commonly used **RDI** or reference daily intake. The RDI is similar to the RDA, but does not recommend different nutrient intakes based on gender and age as the RDA does. The RDI is used to determine the RDV, or recommended daily value, which is printed on virtually every food product in the United States and Canada to inform consumers of the nutritional value, or the lack thereof, that each product offers. These recommended daily values are usually based on a 2000-calorie diet, and since the RDI recommends the same nutrient intake for everyone, there is no difference shown between the intakes necessary for individuals of different age groups or genders. Both the RDA and RDI are important because they allow consumers to determine which foods are best for their diet.

Low Activity and High Activity Individuals

A **low activity** individual is usually considered to be anyone who does not exercise regularly and gets less than 30 minutes of exercise on any given day. A **high activity** individual is usually considered to be anyone who exercises regularly and gets at least 30 to 60 minutes of exercise 4 to 5 times a week. An individual can also be considered high activity if they are in a profession that is physically intense such as warehouse workers, moving company workers, gym teachers, etc. Individuals who exercise more than 30 minutes at a time, but do so only once or twice a week are usually considered mid-level activity persons rather than high or low activity individuals.

Nutritional needs change in a variety of ways as an individual ages, but the major changes involve which vitamins and the number of calories the body requires to keep functioning. Young children usually require around 1000 to 1500 calories a day, but the amount of calories a child requires increases as the child grows. Young teens usually require somewhere between 1600 and 2200 calories a day, depending on their gender and other factors. Individuals entering young adulthood usually require between 2000 and 2400 calories a day. However, after a person reaches age 30, the calorie requirement begins to decrease and continues to do so as they get older. People also require progressively more Vitamins A, C, and E as well as iron and fiber as they age. Individuals over age 50 also require significantly more calcium.

An individual that is more active requires more calories than a person who is less active. A young active child usually requires about the same number of calories as a less active child, but the difference between the calorie needs increases substantially as the children grow. Active young teens require 1600 to 2600 calories, depending on gender and other factors, and active individuals entering young adulthood usually require between 2000 and 3200 calories. As these age ranges show, the calorie needs of an active individual is usually between 400–800 calories higher than a less active person. Active people also require more protein than less active individuals, but usually require similar amounts of Vitamins A, C, and E as well as iron and fiber.

Factors Affecting Nutritional Needs

Some of the major factors that can affect an individual's nutritional needs include:

- **Age of the individual**: The age of an individual plays a major role in the nutritional needs that person has in several ways. As a child grows, for instance, the number of calories the child needs also increases, but after the individual passes a certain age, the number of calories they need decreases. A person's need for specific vitamins and minerals also increases as they get older.
- **Individual's gender**: Gender can affect an individual's nutritional needs; men typically require more protein and more calories than women do, while women usually require significantly more iron than men.
- **Individual's level of activity**: Level of activity also plays a large role in their nutritional needs because an individual who is more active will require a larger number of calories than an individual who less active.

Cholesterol

Cholesterol is a lipid, a type of fat that the body uses to produce both new cells and bile, a substance secreted by the liver that helps the body digest fat. Although the body requires a certain amount of cholesterol to continue functioning normally, excessive amounts of cholesterol can lead to heart and circulatory problems, including a condition known as atherosclerosis, a dangerous condition in which arteries are clogged by deposits of cholesterol. Clogged arteries can cause a variety of other serious problems, such as heart attacks and strokes, because the heart, brain, and many other vital organs cannot receive enough blood to function properly. One of the best ways to avoid too much cholesterol is to stay away from foods that contain high concentrations of saturated fats. Some examples of foods that are high in cholesterol and saturated fats are beef, pork, eggs, milk, butter, cheese, and most snack foods. Some examples of foods that are low in cholesterol and saturated fats include oatmeal, fish, rice, and most fruits and vegetables.

Carbohydrates

Carbohydrates, also known as saccharides, are a group of simple and complex sugars and starches that form an important class of foods in an individual's nutritional needs by supplying energy for the body. Such carbohydrates as the sugar glucose are transported throughout the body via the bloodstream and broken down into energy that keeps the body functioning. Carbohydrates are also important in the function and regulation of the body's immune system and reproductive system. In addition, because carbohydrates aid a person's blood in clotting effectively, they play an essential role in the body's ability to heal. Although many of the body's functions that carbohydrates play a role in can be carried out by proteins and fats as well, carbohydrates are easier for the body to digest and contain less cholesterol than proteins and saturated fats. Examples of foods that are high in carbohydrates include bread, beans, cereals, pasta, potatoes, and rice.

Review Video: Carbohydrates
Visit mometrix.com/academy and enter code: 601714

FATS

Fats are important to the healthy functioning of the body because they allow certain vitamins, including vitamins A, D, E, and K, to be digested and absorbed by the body. These fat-soluble vitamins can be broken down and used by the body only when enough fat is present. In addition, a lack of certain fats in an individual can lead to other vitamin deficiencies. Fats also help maintain the body's temperature; help promote hair, skin, and overall cell health; protect the body's organs; help protect against some diseases; and act as a means of storing energy for the body to use later. However, an excessive amount of fat in the body can adversely affect an individual's health, as an abundance of saturated fats and trans fats can greatly increase the cholesterol level in the blood, which in turn increases an individual's risk of certain heart conditions and other health problems. Some foods that individuals may want to avoid eating in excess because they are high in fat include butter, cream, cheese, lard, milk, and snack foods.

IODINE

Relatively small amounts of **iodine** are important to an individual's health because the element iodine is necessary for the production of thyroid hormones, which the body needs to function normally. A lack of iodine in an individual's diet or an excessive amount of iodine in the body can lead to a variety of symptoms and conditions including depression, fatigue, mental slowness or intellectual disability, excessive weight gain, and goiter, which is an enlargement of the thyroid gland characterized by an extremely swollen neck. Iodine deficiencies are fairly uncommon in the United States and other developed countries because most of the salt used in cooking has been treated with iodine to make sure that individuals ingest enough iodine on a daily basis. Other foods that have high concentrations of iodine include fish, kelp, and most dairy products.

FOOD

Food is essential for the health of an individual, because it satisfies an individual's nutritional needs. Additionally, food provides many people some degree of psychological satisfaction as well. Food is a basic necessity for the successful functioning of an individual; therefore, the body makes a person aware of its need for food by creating a feeling of hunger. When the individual satisfies this hunger by consuming various foods, the body releases hormones that result in the person's feeling a sensation of being full, an indication that the body's need for sustenance has been assuaged. Certain types of foods satisfy an individual's hunger more effectively than other foods. Foods that are high in fiber, protein, and water usually will make an individual feel more satiated, or full. If an individual's body lacks a specific type of nutrient, the body will cause that person to experience a craving for foods that will satisfy that particular nutritional need.

THE NEW PYRAMID

The New Pyramid is the current federal healthy eating guide, formerly MyPlate and the Food Pyramid. It was introduced for the 2025-2030 Dietary Guidelines for Americans. While MyPlate mainly focused on showing how a meal could be divided into different food groups, The New

Pyramid looks more closely at the kinds of foods people choose. It encourages everyone to eat real, whole, nutrient-dense foods as often as possible. This includes such foods as protein, dairy, vegetables, fruits, healthy fats, and whole grains, especially when they are in their natural or minimally processed forms. The guidelines also remind people to limit foods and drinks that are heavily processed or high in added sugars, refined carbohydrates, excess sodium, unhealthy fats, and artificial ingredients. Overall, The New Pyramid is not just about how much of each food group is on the plate. It mainly focuses on the quality of the food chosen and encourages people to build meals around choices that provide important nutrients and support better health throughout their lives.

History of American Dietary Guidelines

Early in 1977, the US Senate Select Committee on Nutrition and Human Needs, then chaired by Senator George McGovern, following extensive reviews of science and discussions, recommended dietary goals consisting of recommendations concerning food and nutrients. The first goal recommended Americans only eat as much energy from food (i.e., calories) as they expended. This goal addressed obesity through balanced consumption. Regarding nutrients, the committee recommended eating more complex carbohydrates and fewer simple (i.e., processed or refined) carbohydrates and decreasing total fat, saturated fat, cholesterol, and sodium. Food goals also included eating more vegetables, fruits, and whole grains and replacing more saturated fats with unsaturated fats. Due to controversy and skepticism from industrial and scientific groups, the US Departments of Agriculture and Health and Human Services (formerly Health, Education, and Welfare) jointly issued a nutrition and health brochure in 1980, partly based on the Surgeon General's 1979 report on health promotion and disease prevention and reflecting research findings about relationships between diet and health. However, this too met with debate.

Despite their straightforward, scientific basis, US Dietary Guidelines issued in 1980 encountered arguments from various industrial and scientific quarters. In response, Congress directed the Departments of Agriculture (USDA) and Health and Human Services (HHS, formerly HEW) to form an advisory committee to solicit formal and informal outside and private-sector advice for subsequent editions. Though the 1985 edition of the Guidelines did not incorporate extensive changes, adding non-federal scientists' input resulted in far less controversy and wider adoption. HHS and USDA formed a second scientific advisory committee in 1989 to review and revise the Guidelines. The 1990 edition reinforced earlier editions' principles, but additionally suggested numerical targets for saturated and total fat amounts, emphasizing these did not apply to a single food or meal, but to eating choices over multiple days. While these first three editions were voluntarily issued, the 1990 National Nutrition Monitoring and related Research Act first legally mandated the Dietary Guidelines report, required a new report every five years, and established a Dietary Guidelines Advisory Committee for that purpose. The 2005 committee adopted systematic methods to review scientific evidence and literature. The 2010 committee extended these, also establishing the Nutrition Evidence Library to support its work with a comprehensive, evidence-based review process.

Nutrients, Carbohydrates, Fats, Protein, Vitamins, and Minerals Found in Foods

Carbohydrates are starches, fibers, and sugars in foods from which the body manufactures glucose, the blood sugar used to fuel all activities of the body, including the brain. Glucose is accessed immediately, or stored in the muscles and liver for future use. Fats are necessary to normal bodily functions, but only 20-35 percent of the human diet need be from fats (preferably oils, not solid fats; and unsaturated, not saturated or trans fats). USDA recommends eating below 10 percent of calories in saturated fats, and avoiding trans fats. (Trans fats are found in partially hydrogenated oils.) Humans need to eat foods with protein, which our bodies digest into amino acids to rebuild all organs, tissues, and cells, which contain protein that is continually broken down and replaced. Animal proteins are complete protein sources with all essential amino acids. Vegetable proteins are incomplete sources, but combining complementary vegetable proteins can yield complete protein. Ten to 35 percent of daily calories should be protein; Americans typically eat enough. Vitamins and minerals are important to health; for instance, vitamin D aids calcium absorption, and calcium promotes strong bone formation. Supplements are available, but eating a nutritious and varied diet usually provides enough vitamins and minerals.

FAT

Fats are divided into two main categories: saturated and unsaturated. Saturated fats are mostly found in meat, lard, butter, coconut, and palm oil. Doctors consider these fats to be the most hazardous to health because they increase the risk of heart disease and certain kinds of cancer. Unsaturated fats include sunflower oil, corn oil, olive oil, and canola oil. The last two oils are called monounsaturated fats and are particularly good for the body because they lower cholesterol. Recent research has concluded that the most harmful kinds of fats are trans fats, which are formed when liquid vegetable oil is processed to make table spreads and cooking fats. Trans fats have been consistently shown to create buildup in arteries, a process which can impair heart health.

Many fats can increase cholesterol, a substance in the body which has consistently been linked with heart disease. Cholesterol has many positive uses in the body, like helping the liver operate and helping to form many hormones, but if cholesterol becomes too abundant, it can build up in the arteries and impede the flow of blood. Research has shown that saturated fats cause a more significant buildup of cholesterol than unsaturated fats or other foods that contain cholesterol. In order to minimize cholesterol in the diet, individuals should cut back on fats altogether, but especially limit their intake of saturated fats. Monounsaturated fats, like canola and olive oil, are a good, low-cholesterol source of fat.

FIBER

Whole grains, fruits, and vegetables are all excellent sources of fiber. Fiber can be either insoluble or soluble. Insoluble fibers (cellulose and lignin, for example) speed digestion and can reduce the risk of colon cancer and heart disease. Wheat and corn bran, leafy vegetables, and fruit and vegetable skins are all great sources of insoluble fiber. Soluble fibers (pectins and gums, for example) lower cholesterol levels and help manage the level of blood sugar. They can be found in the pulp of fruits and in vegetables, oats, beans, and barley. Doctors warn that most Americans do not eat nearly enough fiber. However, increasing fiber in your diet should be done gradually, as a sudden increase in fiber can result in bloating, cramps, and diarrhea.

WATER

A person should drink 7 to 10 average sized glasses of water daily. Water is probably the most important substance a person can consume. Water carries nutrients throughout the body and regulates body temperature. Water lubricates joints, aids digestion, and helps speed waste matter out of the body. Losing even 5% of the body's water causes immediate physical symptoms, like dizziness, fatigue, and headache; losing 15% of the body's water can be fatal. The normal daily loss is between 64 and 80 ounces of water a day, which is equal to about 9 large glasses of water. Many fruits and vegetables contain helpful water, but people should still consume the recommended amount of water each day. People who are active, live at a high altitude, or travel a great deal should be sure to drink even more water.

Water Supply

Even though Americans have generally been able to rely on the water supply, in recent years some concerns have been raised about the prevalence of potentially dangerous chemicals in water. Fluoride, which has greatly improved dental health by strengthening teeth since it was added to the water supply, may be damaging to bone strength if it is consumed in great volume. Chlorine, which is often added to water to kill bacteria, may increase the risk of bladder cancer. One of the most dangerous chemicals that can affect water is lead, which is known to leach from pipes and enter the drinking supply. High amounts of lead in the body can cause serious damage to the brain and heart.

Review Video: Dehydration
Visit mometrix.com/academy and enter code: 283651

Alcoholism

The National Council on Alcoholism and Drug Dependence considers alcoholism as a disease that is influenced by social, environmental, and genetic factors. The common features of alcoholism are the inability to control consumption, continued drinking despite negative consequences, and distorted thinking patterns (like irrational denial). It is important to note that alcoholism is not simply the result of a weak will but is a physiological state that requires medical treatment so that it can be controlled. Many individuals may have a problem with alcoholism but not realize it if they are still functioning well overall and only drink in social situations. Alcoholics tend to be those who, even when they aren't drinking, place an undue amount of psychological emphasis on alcohol.

Drug Abuse

A drug is any chemical substance that changes the way a person acts or feels. Drugs may affect a person's mental, physical, or emotional state. Though many drugs are taken to improve the condition of the body or to remedy personal problems, drugs can also undermine health by distorting a person's mind and weakening a person's body. According to the World Health Organization, drug abuse is any excessive drug use that is not approved by the medical profession. The use of some drugs in any quantity is considered abuse; other drugs must be taken in large quantities before they are considered to have been abused. There are health risks involved with the use of any drug, legal or illegal, insofar as they introduce a foreign substance into the balanced system of physical health.

Appetite

The feeling of hunger can be caused by up to 12 different hormones and areas of the brain. There is even some speculation that the size of an individual's fat cells may cause them to feel hungry. The appetite is the physiological desire to eat, and though it is thought to be the body's means of avoiding failure, it can also be stimulated when the body does not really need food. Humans tend to stop eating when they reach the point of satiety, in which they are no longer hungry and feel full. Scientists have advanced the set-point theory of appetite, which contends that each individual has an internal system that is geared to regulate hunger and satiety so as to keep body fat at a certain rate.

Health and Wellness

Health

Quite simply, **health** is the state of being sound in mind, body, and spirit. According to the World Health Organization, health is not only the absence of disease, but the presence of physical, mental, and social well-being. When assessing an individual's health, a professional is likely to examine

amount the body is using. Jogging, walking, or riding a bike are all examples of aerobic activity. The heart also gets an excellent workout during anaerobic exercise, in which the body takes in less oxygen than it needs to maintain the activity. Sprinting or swimming fast can be anaerobic exercises, if they leave the person breathless. Nonaerobic exercise, like bowling or golf, does not challenge the heart and lungs and therefore will not improve cardiovascular fitness.

Muscle Strength and Endurance

Developing healthy muscles is not simply a matter of lifting the heaviest possible object. The ability to use your muscles over and over without getting tired is also an important part of physical fitness. Developing **muscular strength** and **endurance** will help make body tissue firmer and more resilient. Well-maintained muscles tend to work more efficiently, and they can withstand more strain. Furthermore, muscular development aids in circulation, with the result that the whole body absorbs and makes use of nutrients in the blood more quickly. Strength and endurance training has also been shown to be one of the most effective ways to lose weight, as developed muscles burn more calories than fat does.

Exercising Muscles

Muscles are in a constant state of change. If muscles are not used, they will atrophy and weaken; on the other hand, if they are regularly exercised, they will grow stronger and possibly larger. Muscles are best exercised when they are overloaded or asked to do more than they usually do. When you are training your muscles, you will need to gradually increase the amount of the weight or the number of repetitions to ensure that your muscles are always receiving a challenge. Many fitness professionals contend that a good muscular workout will be somewhat painful because muscles can only be developed by exceeding their normal requirements. However, not every kind of pain is profitable for a muscular workout, and individuals should be careful to distinguish muscular fatigue from injury, particularly when they are lifting heavy loads.

Recreation Safety Laws

A number of laws exist to ensure the safety of individuals when they engage in recreational activities. For instance, an individual may not be intoxicated while driving a boat. There are also laws governing the use of jet-skis, wave runners, snowmobiles, and all-terrain vehicles. The federal government cites these last vehicles as particularly dangerous for young, untrained, or intoxicated drivers. A set of regulations governs the use of the national and state parks (e.g., restrictions on the building of fires). The federal government recently reported a sharp increase in the number of in-line skating and skateboarding injuries; skaters are now denied access to many recreational areas because of the increased safety risk.

Vital Signs

Every individual should be able to identify the **vital signs** and know how to measure them. The four common measures considered to be vital signs are body temperature, blood pressure, pulse rate, and respiration rate. Body temperature can be taken with a thermometer and should register between 96º and 99.9º Fahrenheit, depending on the time of day and sex (women tend to have slightly higher temperatures). Measuring blood pressure requires some equipment; a normal blood pressure is between 120/70 and 140/90, depending on age and sex. A normal pulse rate is ranges between 60 and 80 beats per minute. For athletes, resting pulse rates can be even lower. A normal respiration rate is between 15 to 20 breaths a minute.

Warming Up and Cooling Down

There are important reasons for warming up before and cooling down after exercise. For one thing, performance is always enhanced by warming up. Muscles tend to work more effectively when their temperature has been slightly raised; they are also more resistant to strains and tears at a higher temperature. Warming up directs the blood to working muscles and gives the heart time to adjust to the increased demands of the muscles. Warming up also stimulates the secretion of synovial fluid into the joints, which makes them less likely to suffer wear and tear. Warming up should include slow stretching and low-impact cardiovascular exercise. Cooling down is important for easing the body's transition to a normal resting condition. By stretching and slowly decreasing cardiovascular workload, the heart is aided in its readjustment.

Recreation Versus Competition

One of the perennial issues facing physical educators is whether activities should be promoted as forms of recreation or as competition. If competition is to be the dominant feature, then activities must have explicit rules, a formal way of keeping score, and identifiable winners and losers. When students are taught activities for competition, the emphasis will be on practicing specific skills and avoiding mistakes as much as possible. When sports are taught as recreation, participation is the most important factor for students. Each student should get an equal amount of experience and performance time, regardless of their skill level. Although score is typically not kept in strictly recreational activities, students may receive certificates for good sportsmanship or diligent participation.

Health Education as a Community Service

In a general sense, all of health education can be seen as community service. By teaching positive health behaviors and promoting good health to students, health teachers are improving the quality of life for everyone in the community. More specifically, though, the Center for Disease Control has recommended that health educators use their special training to improve health through work outside of school. Many health educators participate in fundraising for health charities, give speeches on health-related topics, or work in the community to generate enthusiasm for exercise and nutrition. According to the Code of Ethics for health educators, it is imperative for those with knowledge and skills to advance positive health behaviors whenever possible and, thus, help their community.

Aerobic Fitness

A minimum of **aerobic fitness** has been achieved when you are able to exercise three times a week at 65% of your maximum heart rate. The easiest means of achieving this level of fitness is by running for 30 minutes three or four times a week. Moderate aerobic fitness is achieved by exercising four or more times a week for at least 30 minutes at a heart rate that is 75% or more of maximum. This level of aerobic fitness is appropriate for athletes who are seeking to play vigorous sports like football or tennis. Maximum aerobic fitness can only be achieved by working close to maximum heart rate several times a week and by exercising vigorously almost every day. In order to achieve this level of fitness, you must consistently work beyond your anaerobic threshold. A good way to do this is having interval training or brief, high-intensity workouts.

Skeletal System

The **skeletal system** is composed of about 200 bones which, along with the attached ligaments and tendons, create a protective and supportive network for the body's muscles and soft tissues. There are two main components of the skeletal system: the axial skeleton and the appendicular skeleton. The axial skeleton includes the skull, spine, ribs, and sternum; the appendicular skeleton includes the pelvis, shoulders, and the various arm and leg bones attached to these. The skeleton does not move, but it is pulled in various directions by the muscles. There are few differences between the male and female skeleton: the bones of a male tend to be a bit larger and heavier than those of the female, who will have a wider pelvic cavity.

Review Video: Skeletal System
Visit mometrix.com/academy and enter code: 256447

Lymphatic System

The **lymphatic system** is connected to the cardiovascular system through a network of capillaries. The lymphatic system filters out organisms that cause disease, controls the production of disease-fighting antibodies, and produces white blood cells. The lymphatic system also prevents body tissues from swelling by draining fluids from them. Two of the most important areas in this system are the right lymphatic duct and the thoracic duct. The right lymphatic duct moves the immunity-bolstering lymph fluid through the top half of the body, while the thoracic duct moves lymph throughout the lower half. The spleen, thymus, and lymph nodes all generate and store the chemicals which form lymph and which are essential to protecting the body from disease.

Nervous System

The **nervous system** collects information for the body and indicates what the body should do to survive in the present conditions. For instance, it is the nervous system that administers a bad feeling when the body is cold, and then sends a more positive message when a person warms up. These important messages are sent by the nerves, which vary in size and cover the entire body. The central nervous system is composed of the brain and spinal cord, and the peripheral nervous system is composed of the rest of the body, including those organs which a person does not voluntarily control. The peripheral nervous system is divided into sympathetic and parasympathetic systems, which counterbalance one another to allow for smooth function.

Review Video: Function of the Nervous System
Visit mometrix.com/academy and enter code: 708428

Digestive System

The **digestive system** is composed of organs that convert food into energy. This process begins with the teeth, which grind food into small particles that are easy to digest. Food is then carried through the pharynx (throat) and esophagus to the stomach. In the stomach, it is partially digested by strong acids and enzymes. From there, food passes through the small and large intestines, the rectum, and out through the anus. On this journey, it will be mixed with numerous chemicals so that it can be absorbed into the blood and lymph system. Some food will be converted into immediate energy, and some will be stored for future use; whatever cannot be used by the body will be expelled as waste.

Review Video: Gastrointestinal System
Visit mometrix.com/academy and enter code: 378740

Muscular System

The muscles of the body are attached to the skeleton by tendons and other connective tissues. Muscles exert force and move the bones of the body by converting chemical energy into contractions. Every muscular act is the result of some muscle growing shorter. The muscles themselves are composed of millions of tiny proteins. Muscles are stimulated by nerves that link them to the brain and spinal cord. There are three types of muscles: cardiac muscles are found only in the heart and pump the blood through the body; smooth muscles surround or are part of internal organs; skeletal muscles are those a person controls voluntarily. Skeletal muscles are the most common tissue in the body, accounting for between 25 and 40% of body weight.

Review Video: Muscular System
Visit mometrix.com/academy and enter code: 967216

ENDOCRINE SYSTEM

The **endocrine system** creates and secretes the hormones that accomplish a wide variety of tasks in the body. The endocrine system is made up of glands. These glands produce chemicals that regulate metabolism, growth, and sexual development. Glands release hormones directly into the bloodstream, where they are then directed to the various organs and tissues of the body. The endocrine system is generally considered to include the pituitary, thyroid, parathyroid, and adrenal glands, as well as the pancreas, ovaries, and testes. The endocrine system regulates its level of hormone production by monitoring the activity of hormones; when it senses that a certain hormone is active, it reduces or stops production of that hormone.

Review Video: Endocrine System
Visit mometrix.com/academy and enter code: 678939

CIRCULATORY SYSTEM

The **circulatory system** is composed of the heart, the blood vessels, and the blood. This system circulates the blood throughout the body, giving nutrients and other essential materials to the body cells and removing waste products. Arteries carry blood away from the heart, and veins carry blood back to the heart. Within body tissues, tiny capillaries distribute blood to the various body cells. The heart takes oxygenated blood from the lungs and distributes it to the body; when blood comes back bearing carbon dioxide, the heart sends it to the lungs to be expelled. Other organs not always considered to be a part of this system (for instance, the kidneys and spleen) help to remove some impurities from the blood.

Review Video: Cardiovascular System
Visit mometrix.com/academy and enter code: 376581

HORMONE SYSTEM

Hormones are the chemicals that motivate the body to do certain things. They are produced in the organs that make up the endocrine system. With the exception of the sex organs, males and females have identical endocrine systems. The actions of the hormones are determined by the hypothalamus, an area of the brain about the size of a pea. The hypothalamus sends messages to the pituitary gland, which is directly beneath it. The pituitary gland turns on and off the various glands that produce hormones. Hormones, once released, are carried to their targets by the bloodstream, at which point they motivate cells and organs to action. Hormones can influence the way a person looks, feels, behaves, or matures.

CARTILAGE

The areas of bones that are close to joints are covered in a shiny connective tissue known as **cartilage**. Cartilage supports the joint structure and protects the fragile bone tissue underneath. Cartilage is susceptible to injury because it is subject to gravitational pressure as well as pressure born of joint movement itself. Long-term stress to cartilage can result in rheumatoid arthritis and osteoarthritis. There are no blood vessels in cartilage; nutrients are delivered by the synovial fluid, and from nearby blood vessels. Cartilage contains a huge number of spongy fibers because it needs to absorb a great deal of shock. Especially resilient cartilage, known as fibrocartilage, is found between the vertebrae and in the knees, among other places.

LIGAMENTS

Ligaments are dense bundles of fibers running parallel to one another from one bone in a joint to another. Ligaments are a part of the joint capsule, although they may also connect to other nearby bones that are not part of the joint. Ligaments are not like muscles; they cannot contract. Instead, ligaments passively strengthen and support the joints by absorbing some of the tension of movement. Ligaments do contain nerve cells which are sensitive to position and speed of movement, and so ligaments can hurt. One function of this pain is to alert the person to an unnatural or dangerous movement of the joint. Ligaments may also be strained or rupture if they are placed under unnecessary or violent stress.

MUSCLE TISSUE

Muscle tissue is made up of bundles of fibers which are held in position and separated by various partitions. These partitions range from large (deep fascia, epimysium) to small (perimysium, endomysium), and often extend beyond the length of the muscle and form tendons connecting to a bone. Each muscle cell is extremely long and has a large amount of nuclei. Every muscle cell contains a number of smaller units called sarcomeres; these contain thick filaments of the protein myosin and thin filaments of the protein actin. Muscle tissue contracts when a nerve stimulates the muscle and the thin filaments compress within the sarcomere, causing a general muscle contraction.

SETTING GOALS

Individuals who are most likely to make positive permanent changes in their health set realistic goals along the way. When setting goals, individuals should identify what resources (time, money, and effort) are available to achieve them. Individuals should also identify the potential barriers to success and consider ways to minimize or remove these problems. It is always better to set a number of small, attainable goals rather than goals that may be difficult to achieve.

PHYSICAL FITNESS

Physical fitness is the body's ability to perform all of its tasks and still have some reserve energy in case of an emergency. People who are physically fit can meet all of their daily physical needs, have a realistic and positive image of themselves, and are working to protect themselves against future health problems. Physical fitness has three main components: flexibility, cardiovascular fitness, and muscular strength or endurance. Some other factors, like agility and balance, are also often considered when assessing physical fitness. The benefits of pursuing physical fitness throughout life are not only physical but mental and emotional; regular exercise is proven to reduce the risk of disease and increase life expectancy.

FLEXIBILITY

A person's **flexibility** is their range of motion around particular joints. An individual's flexibility will vary according to age, gender, and posture. Some individuals may be less flexible because of bone spurs, and some individuals may be less flexible because they are overweight. Typically, an individual's flexibility will increase through childhood until adolescence, at which point joint mobility slows and diminishes for the rest of the individual's life. Muscles and the connective tissue around them (tendons and ligaments) will contract and become tighter if they are not used to their potential. Lack of flexibility can lead to a buildup of tension in the muscles and can increase the risk of injury during exercise.

Heart, Lungs, and Bones

Maintaining physical fitness has a number of advantages besides improving personal appearance. It has been shown time and again that habitual exercise is the best way to prevent coronary death. In fact, individuals who don't exercise are twice as likely as active individuals to die of a heart attack. Exercise makes the lungs more efficient, as they are able to take in more oxygen and make better use of it. This provides the body with more available energy. Exercise also benefits the bones. Individuals who do not exercise are more likely to have weak or brittle bones, and they are more prone to osteoporosis, in which bones lose their mineral density and become dangerously soft.

Mood, Disease Prevention, and Body Weight

The benefits of regular exercise are both physical and mental. It is well documented that frequent exercise improves a person's mood, increases energy, focus, and alertness, and reduces anxiety. In fact, long workouts cause the release of mood-elevating chemicals called endorphins into the brain. Exercise also reduces the risk of disease. By aiding in the proper digestion, exercise reduces the risk of colon and rectal cancers. Studies have also indicated that women who exercise are less likely to develop breast cancer. Finally, exercise is beneficial because it helps people lose weight and keep it off. The body's metabolism remains elevated for a prolonged period after exercise, which means food is processed more quickly and efficiently. In addition, regular exercise helps suppress the appetite.

Nutrition and Exercise

For most people, the balanced diet depicted in realfood.gov's New Pyramid will supply all the healthy nutrients the body needs to maintain a program of physical fitness. However, individuals who are seriously testing their endurance by exercising for periods of more than an hour at a time will need to increase their intake of complex carbohydrates, which keep the level of blood sugar stable and increase the amount of available glycogen. Contrary to popular thought, heavy workouts do not require a diet high in protein, and in fact, consuming too much protein can put a severe strain on the kidneys and liver. Similarly, most health experts discourage the use of dietary supplements and body-building foods unless under supervision because these products can easily result in nutritional imbalances.

Water and Exercise

Water is the most important thing for a person to consume before, during, and after exercise. On hot days, active people can sweat up to a quart of water. If you become dehydrated, your heart will have a difficult time providing oxygen and nutrients to muscles. Even sports drinks cannot provide the hydrating effect of cool water because the sodium, sugar, and potassium in them delay their absorption into the body. Salt tablets should be avoided as well; they are potentially dangerous and unnecessary. Although people do lose a bit of sodium when they sweat, this is more than offset by the huge amount of salt in the average American diet.

Review Video: Dehydration
Visit mometrix.com/academy and enter code: 283651

First-Aid Tips and Supplies

Since it is necessary to act fast when an emergency happens, it is a good idea to think ahead and have a plan in place. If you are in a public place, you may want to begin by shouting for help to see if a doctor is available. Someone should immediately dial 911. Do not attempt any resuscitation techniques unless you are trained. If you have a car and it is appropriate, you should immediately take the victim to the nearest hospital. Furthermore, every home should have some basic first-aid supplies. A good first-aid kit will include bandages, sterile gauze pads, scissors, adhesive tape, calamine lotion, cotton balls, thermometer, ipecac syrup (to induce vomiting), a sharp needle, and safety pins.

Smog

Smog is the informal name given to the combination of smoke, gases, and fog that accumulates in major industrial or metropolitan areas. Most smog is created by motor vehicles, wood-burning stoves, industrial factories, and electric utilities plants. Gray smog, which is mainly sulfur dioxide, is common in the eastern United States because of the high concentration of industry. This kind of smog acts like cigarette smoke on the lungs, impairing the ability of the cilia to expel particulates. Brown smog comes from automobiles and is mainly composed of nitrogen dioxide. Ozone, one of the other components of brown smog, can impair the immune system. Automobiles are also known to produce carbon monoxide, which diminishes the ability of the red blood cells to carry oxygen.

Pollution

Many people do not consider **pollution** a personal health issue, but polluted air and water can affect every aspect of a person's life. Scientists define pollution as any change in the air, soil, or water that impairs its ability to host life. Most pollution is the byproduct of human acts. Some of the common health problems associated with pollution are nasal discharge, eye irritation, constricted air passages, birth defects, nausea, coughing, and cancer. Environmental agents that change the DNA of living cells are called mutagens, and they can lead to the development of cancer. Pollutants that can pass through the placenta of a woman and cause damage to an unborn child are called teratogens.

Exercise and Weight Loss

Despite the appeal of quick solutions to obesity, exercise remains the best way to reduce weight and maintain weight loss. Many people think that increasing exercise will make them want to eat more; in actuality, frequent exercise tends to reduce the appetite, and since it raises the rate of metabolism, it also helps keep weight off. There are numerous other advantages to exercise in regard to weight; exercise burns off fat reserves and increases muscle mass. Since muscle tends to use calories more quickly than fat, this means it will be more difficult for the individual to put on pounds of fat. In study after study, individuals who exercise regularly are shown to be more likely to lose weight and keep it off.

Communicable and Non-Communicable Diseases

Communicable diseases are those that are caused by microorganisms and can be transferred from one infected person or animal to a previously uninfected person or animal. Although some diseases are passed on by direct contact with an infected individual, many can be spread through close proximity: airborne bacteria or viruses account for most communication of disease. Some examples of communicable disease are measles, smallpox, influenza, and scarlet fever. Some communicable diseases require specific circumstances for transmission; for instance, tetanus requires the presence of infected soil or dirt. Any disease that cannot be transferred from one person or animal to another is considered non-communicable.

Infectious and Non-Infectious Diseases

Infectious diseases are those that are caused by a virus, bacterium, or parasite. Infectious diseases are distinguished from **non-infectious diseases** in that they stem from biological causes, rather than from physical or chemical causes (as in the case of burns or poisoning). An infectious disease will always have an agent (something that has the disease and spreads it to others) and a vector (a way of transmitting the disease). In the case of malaria, for instance, a parasite contains the disease,

allergic reaction, but the most common are sneezing, hives, eye irritation, vomiting, and nasal congestion. In some extreme cases, the person may collapse and even die. Allergic triggers, or allergens, can be anything from peanuts to pollen, or from insect bites to mold. Although there is no way to reverse or eliminate a personal allergy, science has made progress in treating the allergic reaction. These days, it is possible to be treated for an allergic reaction without becoming drowsy or sluggish.

Immunizations

Despite the overwhelming evidence supporting the use of **immunization** in preventing potentially life-threatening diseases, many Americans still neglect to get the basic immunizations. At present, the American Academy of Pediatrics recommends that every child be immunized against measles, mumps, smallpox, rubella, diphtheria, tetanus, and hepatitis B. Some vaccinations will need to be repeated on a certain schedule. Basically, a vaccination is the intentional introduction of a small amount of an antigen into the body. This stimulates the immune system to learn how to fight that particular antigen. There are certain vaccinations that a pregnant woman should not get, and a person should never be vaccinated if they are currently sick.

Common Cold

The **common cold** is one of the most pesky and irritating of viruses, though it is rarely a great risk to long-term health. One reason the cold is so difficult to fight is that there are over 200 varieties of the virus, so the body is never able to develop a comprehensive immunity. The cold virus is typically spread through the air or through contact. There is no completely effective medical treatment, either. Indeed, doctors warn that taking aspirin and acetaminophen may actually suppress the antibodies that the body needs to fight the infection and may therefore contribute to some symptoms. There is also no conclusive evidence to support taking vitamin C in large doses. Antihistamines, which many people credit with relieving the symptoms of the common cold, may make the user drowsy.

Cancer

Cancer is the uncontrolled growth and spread of abnormal cells throughout the body. Cancer cells, unlike the regular cells of the body, do not follow the instructions encoded in the body's DNA. Instead, these cells reproduce themselves quickly, creating neoplasms, or tumors. A tumor may be either benign, when it is not considered dangerous, or malignant (cancerous). Unless they are stopped, cancer cells continue to grow, crowding out normal cells in a process called infiltration. Cancer cells can also metastasize, or spread to the other parts of the body by entering the bloodstream or lymphatic system. The gradual overtaking of the body by these cancer cells will eventually make it impossible to sustain human life.

Every cancer has some characteristics in common with other cancers, but it may be more or less treatable depending on its particular nature. The most common forms of cancer are carcinoma, sarcoma, leukemia, and lymphoma. **Carcinoma** is the most common kind of cancer; it originates in the cells that line the internal organs and the outside of the body. **Sarcomas** are those cancers that develop in the connective and supportive tissues of the body, namely bones, muscles, and blood vessels. **Leukemias** are cancers that originate in the blood-creating parts of the body: the spleen, bone marrow, and the lymph nodes. **Lymphomas** are cancers that originate in the cells of the lymph system where impurities are filtered out.

By now, most Americans are aware that the risk of developing cancer is increased more by cigarette smoking than by any other single behavior. Not only do cigarettes lead to lung cancer, but they also lead to cancer of the mouth, pharynx, larynx, esophagus, pancreas, and bladder. The risk of developing cancer is not limited to cigarettes: pipes, smokeless tobacco, and cigars all put a person at risk. Second-hand smoke has a similar effect; scientists have shown that individuals who are exposed to environmental smoke for more than three hours a day are three times more likely to develop cancer than those not exposed. In addition to tobacco, other acknowledged carcinogens are asbestos, dark hair dye, nickel, and vinyl chloride. Individuals should always try to make certain their living and working spaces are well ventilated to reduce the harmful substances in the air.

Hygiene

Besides helping you maintain an attractive appearance, **hygiene** is essential for keeping you healthy and free of disease. The body is usually covered with a certain amount of bacteria, but if this number is allowed to grow too high, you may place yourself at risk for disease. Individuals who fail to regularly wash their hair are more likely to have head lice, and those who fail to properly clean their genitals are more susceptible to urinary tract infections. Good hygiene also reduces an individual's contagiousness when sick. Hygiene is especially important when dealing with food: failing to wash everything involved in the preparation of a meal can result in the spread of bacterial infections like *E. coli* and hepatitis A.

To stay clean and reduce the risk of disease, students should practice daily basic hygiene. Everyone should wash hair and body daily and should wash the hands more frequently than that. Teeth should be brushed between one and three times daily. Always wash hands before eating, avoid spitting or nose-picking, and cover your mouth when sneezing. Try to avoid coming into contact with any bodily fluids, and keep clothes and living space clean. Finally, avoid putting your fingers in your mouth, and try not to touch any animals before eating.

Similarities Between Basic Food Preparation and Medical Hygiene

There are a few basic hygiene habits that every individual should practice when preparing food or performing basic medical procedures. Always clean off the areas where food will be prepared, and wash your hands after touching any uncooked foods. Do not use the same tools to prepare different foods. Always refrigerate foods before and after they are used. Label stored food to indicate when it was produced. Dispose of any uneaten food that cannot be stored. When performing basic medical procedures, always use sterile bandages and any necessary protective clothing, like masks, gloves, or eyewear. Always make sure any medical waste, like used bandages, is disposed of securely.

Meal Planning and Service

Ethics and Religion

An individual's religious, moral, and ethical beliefs can play a large part in their diet because some beliefs can result in the exclusion of certain foods. Ethical and moral considerations such as preventing animal cruelty or showing disapproval of environmentally unfriendly practices that are used to acquire certain types of food can lead some individuals to reduce the amount of certain foods they consume, such as meat, or eliminate those foods from their diet completely. For example, many individuals who choose to be vegetarians or vegans usually do so to avoid supporting what they believe to be the unethical slaughter of animals.

Religious considerations can also determine a person's dietary choices, as some religions dictate what sort of foods the members of that faith can consume. People practicing Judaism, for instance, can consume only kosher foods, which are foods that conform to a series of guidelines established by the faith and include restrictions related to how an animal was slaughtered, what type of animal was slaughtered, and who prepared the food.

Ethnicity

Ethnicity can play an important role in the foods that an individual includes in their diet because people living in different regions of the world usually have different customs regarding the foods that they eat and have different foods commonly available as well. For example, individuals who live in a country like Japan that has easy access to waters suitable for fishing will typically consume significantly more fish than individuals who live in countries that are landlocked. Individuals who have immigrated to another country or who are descendents of immigrants will often include foods in their diet that are traditional foods from their native countries even if those foods are not easily accessible in their new home. People in some regions may include foods in their diets that are not based on how readily available a particular food is, but rather this diet is based on the culture's customary eating habits, as seen in France, where the population as a whole tends to eat foods that are fresh rather than processed.

Planning a Nutritional Meal for a Group

Individuals or organizations that are planning to prepare a nutritionally-sound meal for a group of people should first decide what the nutritional goals of the meal are. Once these goals have been established, the individual or organization should continue the planning process by researching which foods will best satisfy these goals without exceeding the available time and resource limits. The individual or organization should then put together a written meal plan that details what foods will be included, the average time it takes to prepare and cook each of these meals, and the cost associated with the preparation of these meals. Next, they should determine the best method of preparing food for these meals, including which foods should be prepared first, the best way to handle or prepare the foods to minimize the risk of illness, and what techniques can be used to reduce the cooking time. Finally, the individual or organization can prepare the meal according to the plan that has been established.

Reducing the Amount of Time It Takes to Prepare a Meal

Some of the methods that can be used to reduce the time it takes to prepare a meal include keeping the cooking equipment clean and organized, making sure frozen products are thawed ahead of time, preparing foods that have long cooking times beforehand, and preparing foods in order of their cooking times. By keeping the cooking area and the necessary cooking utensils clean, organized, and easily accessible, individuals or organizations can make sure that the time they have available is used as efficiently as possible. Along with thawing products beforehand, preparing foods that require long cooking times ahead of time and re-heating them later can also greatly reduce the amount of time it takes to prepare a meal. As noted, preparing foods according to their cooking times, from longest to shortest, also greatly reduces preparation time because the individual or organization can prepare other foods for the meal while the foods that take the longest are already cooking.

Choosing What Foods to Prepare for a Meal

An individual or organization that is deciding what foods to prepare for a meal should consider the food's nutritional value, the time it takes to prepare each food, the number of people to be served, and the cost of preparing each food. Each food has its own cooking time and offers different nutrients, so it is important that whoever is preparing the meal chooses to serve foods that satisfy people's nutritional goals without exceeding their time constraints for cooking the meal. Since most individuals or organizations will likely have a budgeted amount of money for the meal, they must consider the number of people to be served and how much each food item costs to prepare. If the cost is too high, some meals may not be practical to serve to large groups of people. For example, if an individual or organization is interested in serving a good source of protein for a meal, steak might be a good option for a small group of people but it would probably be too expensive for a larger group.

Food Found in the Marketplace

The two primary types of food that an individual can find in the marketplace are animal products and plant products. Animal products include any foods that either are a part of an animal or are produced by an animal, such as meat, milk, assorted dairy products, and honey. Plant products are foods that originate from some sort of vegetation and include foods such as fruits, vegetables, syrup, nuts, oils, and rice. It is important for an individual to understand the difference between these two groups, as animal products often have more foodborne illnesses associated with them than do plant products; because of this, there are different methods of storing and preparing each type of food to minimize these risks. Animal products and plant products provide different nutrients that the body requires to continue functioning normally, so a healthy individual would need to eat both animal and plant products to receive enough of the various nutrients they require without the use of vitamin supplements.

Factors That Might Influence What Foods are Available to a Consumer

Some of the major factors that might influence what foods are available to a consumer include the region in which a consumer lives, current weather conditions and seasonal effects, and the demand for a particular type of food. Each geographical region has its own natural resources and different climate, which results in foods unique to the area being more readily available. For example, certain crops that react poorly to sudden temperature drops, such as oranges, are easier to grow in areas that have relatively stable and warm climates. Weather and seasonal conditions also play a large role in whether a particular food is available, as certain foods may not be produced during some times of the year in a particular region, while other crops may die as a result of hurricanes, droughts, and other natural disasters. The demand for a particular type of food can also have a significant impact on the availability of that food. Oftentimes, if the demand for a particular type of food increases suddenly, suppliers may have difficulty in meeting that demand.

Food Distributors

Some of the major types of food distributors a consumer might find in the marketplace are chain supermarkets, independent grocery stores, farmers' markets, restaurants, and local farm stands. Each type of food distribution center has its own distinct advantages and disadvantages. Chain supermarkets are large stores that usually offer a large selection of foods at mid-range prices, but foods are often not as fresh and are sometimes of poorer quality than a consumer might find at a farm stand or farmers' market. Independent grocery stores and farmers markets usually have a smaller selection and are slightly more expensive than chain supermarkets, but many consumers prefer to purchase food from these sources because the food is usually locally grown and more fresh. Even though restaurants are the most expensive distributors of food because the food is prepared by the restaurant for the consumer, many people appreciate the quality and variety of foods a restaurant can offer. Local farm stands frequently offer the freshest, highest-quality foods for the lowest prices, but they are open for shorter periods of time, often seasonally.

Food Consumption Patterns

Food consumption patterns are comprised of an analysis of the eating habits of people belonging to a particular region or area. Food consumption patterns indicate that individuals from different regions have very different diets and therefore receive more or less of certain nutrients than individuals from other regions. Knowing the kinds of nutrients that are common in the everyday diets of individuals native to a particular area is important, as that information allows health care professionals to understand what types of diseases and deficiencies pose a threat to the people who live there. By understanding which diseases and vitamin deficiencies the individuals of a particular area may be suffering from, resources can be allocated for treating and preventing those diseases. Monitoring food consumption patterns also allows researchers to study these patterns and determine the ingredients in certain foods that may be contributing to illnesses that are prevalent in a particular region.

Eating Preferences

Just as every individual is fond of certain foods, her or she will dislike or refuse to eat other types of foods as well. In general, people will find the foods that offer them the most physiological and psychological satisfaction to be the most appealing. Most people who refuse to eat a particular food usually do so for one of four basic reasons: they dislike the taste of the food, the condition of the food, how the food was produced, or they fear that the food will make them ill. Such preferences are important because they serve as one of the body's safety mechanisms to prevent individuals from eating foods that may be contaminated or foods that the individual may be allergic to. In some circumstances, however, an individual's eating preferences may need to be ignored so that the body can receive all of the nutrients it needs.

Protecting Consumers from Unsafe Foods and Help Keeping Food Affordable

A number of government regulations in the United States are intended to protect consumers from unsafe foods and help keep food affordable. Some of the most important regulations include food subsidies, price ceilings, tariffs and import quotas, quality-control inspections, and sanitation regulations. Government regulations can manage and even reduce the price of food in several ways: imposing tariffs on imported foods, limiting the amounts of particular foods that can enter the country, offering American farmers and food production facilities financial incentives, and then limiting how much those subsidized producers can charge for their goods. Ideally, these techniques help keep local food sources in business so that the United States does not have to rely too much on outside food sources. Government regulated quality-control inspections and sanitation regulations, such as the Egg Products Inspection Act, the Federal Meat Inspection Act, and the Sanitary Food Transportation Act, all have the purpose of ensuring that foods are produced, stored, and transported following certain guidelines to prevent the food from being contaminated.

Reducing the Risk of Foodborne Illnesses

Some of the methods that can be used to reduce the risk of foodborne illnesses include thawing food appropriately to prevent the growth of bacteria, storing perishable foods in airtight containers and keeping them refrigerated or frozen, and sanitizing all cooking areas and cooking equipment. Individuals should wash their hands before and after handling raw foods, and they should also wash fruits and vegetables before eating to remove any bacteria and pesticides. In addition, foods, especially meats, should be cooked at a high enough temperature for a long enough period of time to avoid the growth of bacteria that lead to foodborne illnesses. Perishable foods should be stored in temperatures of 40 degrees Fahrenheit or below to prevent the growth of bacteria, and frozen foods should be thawed in the refrigerator rather than left out to thaw at room temperature because bacteria grows most quickly at temperatures between 41 degrees and 140 degrees Fahrenheit.

Cooking Equipment

Some of the factors that a consumer should consider when choosing what cooking equipment to purchase include the materials used in the construction of the cooking equipment, the heat conductivity of the materials used, the energy conservation of the cooking equipment, and how easy the equipment is to use and clean. The materials that are used in the construction of a piece of cooking equipment are important because certain materials conduct heat more effectively, retain heat more efficiently, are more durable, and are less expensive than other materials. Certain materials are also rustproof, stick-resistant, and easy to clean, do not flake, and do not react to certain foods. Both the heat conductivity and energy conservation of a piece of cooking equipment are also relevant because the greater the conductivity of the material used in the construction of the equipment and the more efficiently it can retain the heat, the faster the food will cook. Consumers should also know that some types of cooking equipment are easier to use and clean than others, which can make food preparation and clean-up faster.

Food Irradiation and Shelf Life

Food irradiation is the process of using radioactive materials to disinfect and preserve certain types of food. Shelf life refers to how long food will last before it spoils and begins to breed bacteria. Irradiation is important for food manufacturers because it allows a facility to use varying amounts of radiation to kill any bacteria that might be growing on a particular food item, thereby reducing a consumer's risk of foodborne illness. As a result of irradiation techniques that prevent or slow the growth of bacteria, the shelf life of many foods is significantly lengthened. Studies have shown that irradiation methods are completely safe and pose no noticeable risk to individuals who eat foods that have been treated. Foods that are commonly irradiated include apples, bananas, fish, onions, poultry, potatoes, red meats, and strawberries.

Food Preparation

Menu planning is the first step. The menu is the food service's marketing and planning tool to ensure profit. It takes advantage of seasonal and locally-grown food to keep costs under control, and it includes items that can be prepared without excessive time and labor.

Menu specifications are developed so that exactly the right amount of food is purchased. Standardized recipes result in food items being consistent. They also help managers develop work schedules for the kitchen staff.

Foods are cooked by dry heat, moist heat, frying, or a combination of the three. Foods can also be simmered in broth or stock. Appropriate seasonings are used. Foods are portioned using a portion scale. If the food is not served immediately, then proper holding or storage techniques are used to avoid bacteria growth.

The Hazard Analysis Critical Control Point (HACCP) is followed to prevent food-borne illness. It follows the food from purchasing through serving.

Food Presentation

Customers like to see attractively prepared meals served to them; they become eager to eat. The plate needs to be balanced, nicely garnished, and well arranged. Color, shape, arrangement, and texture all influence a diner's appetite. Potatoes, pasta, and rice can be livened up with herbs sprinkled on top. Many food service facilities opt for neutral dishware so that a diner's attention is drawn to the food, not to designs on the rim of the plate.

When plating the food, grease and moisture must be wiped away. An appetizing picture should be created, with garnishes added to provide contrast. Garnishes can also be added to beverages. Food items should be separated from each other. When foods are served on hot plates, customers should be warned by the server so they will not touch the plate and get burned.

Etiquette and Table Service Appropriate for Specific Occasions

Different food operations have different styles. A fast food chain will offer its customers a different atmosphere than an expensive, up-scale operation will offer, and the servers will perform differently. The concept follows from what the food service operation is and what it has on its menu, and this includes types of food, service, price, target market, and theme.

Types of service in the food service industry include counter service, where a customer orders and receives food at a counter before taking it to a table. At a cafeteria service, customers select food from a serving line with a number of choices and serve themselves. A buffet service is similar, except customers receive assistance at an omelet or meat carving station.

Seated service operations seat customers at a table where orders are taken and served by servers. At a banquet service a set menu is served to a group of customers, similar to a catered event.

USDA Dietary Guidelines

The United States Department of Agriculture began issuing nutrition guidelines in 1894, and in 1943 the department began promoting the Basic 7 food groups. In 1956, Basic 7 was replaced with the Basic Four food groups. These were fruits and vegetables, cereals and breads, milk, and meat. Basic Four lasted until 1992, when it was replaced with the Food Pyramid, which divided food into six groups:

1. Bread, cereal, rice, pasta
2. Fruit
3. Vegetables
4. Meat, poultry, fish, dry beans, eggs, nuts
5. Milk, yogurt, cheese
6. Fats, oils, sweets

The Food Pyramid also provided recommendations for the number of daily servings from each group. In 2011, the Food Pyramid was replaced with MyPlate, which used a plate image to show fruits, vegetables, grains, protein, and dairy in a simpler visual format.

The New Pyramid

For the 2025-2030 Dietary Guidelines for Americans, the federal nutrition guide shifted from MyPlate to realfood.gov's New Pyramid. Unlike MyPlate, which focused mainly on dividing a meal into food groups, The New Pyramid emphasizes eating real, whole foods dense in nutrients. Its central message is to get people to choose foods in their natural or minimally processed forms, including protein, dairy, vegetables, fruits, healthy fats, and whole grains. The New Pyramid also encourages people to limit their intake of highly processed foods, refined carbohydrates, added sugars, sugary drinks, excess sodium, unhealthy fats, and artificial ingredients. This change reflects a much broader approach to healthy eating. Instead of focusing solely on portions or food groups, The New Pyramid focuses on the quality of food and encourages people to build meals around foods that provide essential nutrients and support lifelong health.

Chapter Quiz

Ready to see how well you retained what you just read? Scan the QR code to go directly to the chapter quiz interface for this study guide. If you're using a computer, simply visit the online resources page at **mometrix.com/resources719/aafcsfamconsc** and click the Chapter Quizzes link.

Customers demand sanitary conditions in lodging and food service businesses. Employees in food service and lodging establishments must follow all sanitation procedures and should practice good personal hygiene. The businesses must comply with all health codes. Training should be given on how to make surfaces sanitary and prevent mildew from forming in bathrooms. Testing pH levels in swimming pools is important. Kitchen equipment must be cleaned frequently.

Maintenance workers need to be taught landscaping with plants that do not require much water and to use products that will not adversely affect the environment. In lodging facilities, housekeepers can notify guests that they now have the option of hanging towels to dry instead of having towels laundered after each use. Food service workers should be trained on using the **Hazard Analysis Control Point Model** (HACCP) to prevent food-borne illnesses.

Management

The general manager supervises the entire food service or lodging operation and its employees and regularly checks with the managers they oversee to make sure everything is running smoothly. When employee performance excels, then customer satisfaction is high. Competition is stiff in the hospitality industry. Guests can simply go to another hotel or restaurant. Thus, management must continually strive to improve the operation.

There are four styles of management: bureaucratic, democratic, autocratic, and spectator. A good general manager knows when and with whom to use which style. To be a successful manager, a person needs good communication and leadership skills, as well as the ability to perform many tasks at the same time. A manager needs to be good at training and up to date with technology.

Employment Opportunities

The hospitality industry offers service as its product. It is one of the few employment opportunities where one can start at the bottom and work one's way up to a management level position with hard work. Department of Labor projections indicate a strong growth in the coming years.

Employees fall into two categories: **front of the house** and **back of the house** (i.e., those who are seen by guests and those who are not). Entry-level positions are many and include server, bell person, housekeeper, dishwasher, and front desk clerk. Positions requiring some training would include bell captain, concierge, and executive steward. Those requiring a college education include executive chef, front office manager, executive housekeeper, and public relations director. These are just a few of the many opportunities available. Training is possible through job site training, continuing education, and independent study. Grants and scholarships are also available.

Entrepreneurial Opportunities

An **entrepreneur** is someone who starts a new business and who organizes and manages it, taking the associated risks. In the hospitality industry, there are many small businesses that employ fewer than 10 employees, including small boutique hotels, inns, bed and breakfasts, and restaurants.

Entrepreneurs might choose the name, logo, advertising, décor and menus, do the cooking, and clean the rooms. Alternatively, they might decide to buy a franchise, which offers more support and management guidelines with a name that is already known. Most entrepreneurs start small, but those planning to start on a larger scale will most likely need a bank loan and should gain experience in the industry by working for a large facility to start.

Public Policy

The **Occupational Safety and Health Act of 1970** (OSHA) requires businesses to make the workplace free of hazards that could cause injury or death to employees. Compliance officers inspect businesses to ensure that there are no hazards; if any are found, they must be corrected.

OSHA also requires businesses to keep records of all employment-related accidents and to display posters listing the legal rights of employees. If chemicals are handled, Safety Data Sheets (SDS) must be kept on file with details on safe handling and first aid.

The Americans with Disabilities Act of 1992 (ADA) requires that public facilities be accessible fully to disabled persons. Wheelchair ramps, automatic doors, restrooms that are wheelchair accessible, and Braille signs on doors are just a few of the requirements.

Review Video: Intro to OSHA
Visit mometrix.com/academy and enter code: 913559

Industry Standards

A standard of ethics and professionalism is important because hospitality's main purpose is to serve others. A code of ethics should provide specific guidelines for behavior that is morally right and that requires truthfulness in specific situations and applies to all employees. Management should make sure the employees are fully aware of the code and the consequences of not following it.

Theft is a problem in the industry. Employees are exposed to large amounts of cash and credit cards; they see jewelry in people's rooms, place valuables in a safe, and work around large supplies of food and other materials. Kickbacks might be offered by suppliers. Employees might be given bribes to keep an incident secret. Management can minimize these problems by careful screening and background checks of prospective employees and by taking strict action when employees display a breach in the ethics code.

Review Video: Ethical and Professional Standards
Visit mometrix.com/academy and enter code: 391843

Job-Related Competencies and Skills

Having good communication skills is of utmost importance. Getting your ideas across and understanding others is basic. Much communication is by telephone. Proper tone of voice on the phone, pronunciation, and pitch are important to give the right impression. Listening is important, both to guests and co-workers. It involves understanding what is heard and not just daydreaming while someone is talking. Hospitality employees need to have patience, tact, and diplomacy when interacting with guests. They should always be positive and cheerful. They should be attentive to the needs of customers and remember that "the customer is always right." Employees should maintain a professional appearance and be well groomed. Being a team player with fellow workers is important, and computer skills are essential in today's world.

Marketing Techniques

Hospitality businesses have to set goals through marketing plans so they can attract customers, remain profitable, and compete in the market. Even if the business is part of a large corporation or a franchise, it needs to reach its own local market. The plan is a road map used to help sell its services or products and is vital to the success of any business.

A marketing plan helps businesses attract customers away from the competition by persuading them they can better meet the needs and desires of those customers. Using a technique known as positioning, the business paints a specific image of itself that distinguishes it from other businesses. Then the business needs to decide how it is going to communicate its product (through TV, magazines, Internet, etc.).

In the end, marketing helps the business define who its customers are, what they need, and what they want.

Public Relations Techniques

Public relations is communication for the purpose of influencing opinion and promoting a specific image. Publicity received from a public relations release that is presented in the media will benefit the business by drawing attention to it in the community.

There are many occasions where publicity can benefit the business. Rooms or meals could be donated to a local nonprofit agency or charity. The generosity would impress the public when people hear about it. Similarly, a special event that people are interested in might occur at the business.

Public relations keeps the business name familiar to people in the area. Community members are reminded that the business is interested in them. A publicity release about sponsoring a local Little League, for example, provides great public relations and may give more benefit than an advertising campaign.

Quality Assurance

Guestology is the study of guests' needs, wants, and expectations. Lodging guests expect a clean, sanitary room with a comfortable bed. They expect recognition as a guest and want access to laundry and dry cleaning services. They look for information about the city and special events. They like having transportation available.

Guests like having toiletries available if they forgot to pack something. Employees can help by making guests feel important and welcome, and by offering assistance.

Food service customers are hungry and thirsty. They want attention. They want a comfortable and sanitary environment. Customers want the server to remain attentive to their needs. Increasingly, they want healthy food, low in fat and sodium. In general, customers also want variety.

Health Procedures

If a customer is seen to be in need of medical attention, then the front desk manager in a lodging facility or the general manager in a food service operation should be called immediately, and that person should call 911. The manager should then remain with the customer, reassuring the customer that help is on the way and keeping other people away.

If the victim is taken to the hospital by emergency personnel, their belongings will have to be put away for safe keeping until further instructions are received. Asking how to be of assistance in an emergency is consoling and can make the situation easier.

Choking can occur unexpectedly. There should be someone on staff who can administer the Heimlich maneuver to dislodge the object. First aid kits should be maintained to deal with burns, cuts and bruises, and shock, and employees should be trained in how to use them.

Safety Procedures

Employees in the hospitality industry should observe all safety procedures, report any hazards that could cause an accident, and know the location of emergency equipment and how to use it. If the building needs to be evacuated, employees need to know all evacuation routes.

Fire emergencies happen very quickly. A lodging or food service business should have emergency plans that detail employee responsibilities. Emergency meeting locations should be established, and regular emergency drills should be conducted. Employees should be trained in using the proper fire extinguishers and other equipment based on the type of fire. Some employees should be trained to perform CPR.

Procedures should also be established for dealing with weather emergencies and power outages. Employees should be trained on how to avoid panic if any kind of emergency happens.

Supplies

Food service establishments use universal food products (common, everyday products that are always readily available), as well as seasonal and hard-to-get products. Convenience foods that are partially prepared are increasingly being used. Spices, herbs, extracts, and condiments are kept in supply. Baked products are either purchased or made on premises.

Vegetables and fruits are ordered in smaller quantities to avoid spoilage. Various types of salad greens are needed. Gelatin is used to make salads and as a thickener. Many kinds of beverages are supplied, including water, coffee, tea, sodas, and alcoholic beverages.

Lodging and food service operations also both need to have cleaning and sterilizing supplies in stock, as well as bathroom supplies.

Tools

Tools are hand-held equipment used in preparing and serving food. Measuring cups and spoons are used to ensure quality through careful measuring. Thermometers are used to measure the temperature of food; thermometer varieties are confectionary, meat, deep-fat, and dial thermometers. Cutting tools include knives of many different shapes and uses, blade peelers, graters, utility shears, and specialized cutters used for cheese, eggs, tomatoes, and potatoes.

Food handling tools are used for stirring, turning or blending food while cooking, and for serving food in portions. Ladles, scoops, tongs, slotted, perforated and solid spoons, and whisks are commonly used. Rotary beaters are used for beating eggs or batter.

Many kinds of brushes are found in food service facilities. They are used for cleaning food and utensils. Bakers use sifters, pastry blenders, rolling pins, and dough scrapers. Pastry bags are used to apply icing, and pastry wheels cut the dough.

Equipment

Equipment is used for weighing ingredients, opening cans, cutting foods, mixing foods, and cooking food, as well as making and dispensing beverages. Scales are used to save time and ensure accuracy of recipes. Commercial can openers can open everything from small to large institutional size cans. Cutting equipment includes food slicers, vertical food cutters, food processors, and vegetable peelers.

Mixers can be mounted on the floor, bench, or table, with attachments for different ingredients. Blenders are used for high-speed mixing and blending. Beverage equipment includes equipment for making and dispensing coffee, tea, and cold beverages, as well as ice-making machines.

Foods are cooked on grills, range tops, broilers and ovens, and deep-fat fryers. Foods can be steamed or microwaved. New impingement ovens are faster and can make for tastier food. Employees always need to use safety precautions when using and cleaning equipment.

Other Technologies

Up until recently it was very time consuming to make reservations for lodging or dining out. Now, consumers can easily and quickly make online reservations. Computers are used for faster check in and check out.

Key cards are used for entry to rooms, providing extra security. Room TV access was once limited to a few local channels. Now satellite TV provides hundreds of channels and 24-hour news. Rooms are equipped with data ports or wireless access for laptop computers. Employees use computers to clock in and out, and payrolls are electronically deposited. Managers use software programs to make their work more efficient.

Food service servers enter orders using touch screens. Recipes are stored on in-house databases. New advances like cook-chill equipment (which allows food to be prepared in advance and then stored at temperatures just above freezing) provide better-tasting food than frozen food, and they also save time.

Lodging Liability

If a lodging facility refuses to accept or serve a guest, in so doing it might commit a criminal offense that can be punishable by fine or jail term. However, a lodging facility can refuse service to a guest if there are no rooms available, or if the guest is a difficult character, or if the guest is not able to pay for a room. The facility must be able to prove that the guest was exhibiting unfit behavior.

Because a facility is responsible for protecting all its guests, it can refuse to serve a guest who might pose a danger. When a guest becomes ill, the lodging facility is responsible for contacting a physician. If a guest stays longer than originally stated, the guest can be notified that the room has been taken by another guest and arrange for another room. The facility is responsible for guests' personal property up to a posted limit of liability.

Court-Mandated Lodging Security Standards

Court cases have established security standards for the lodging industry. A lodging facility must install adequate locking systems that provide ample security. Public areas and hallways should be well lit. Access to guest areas must be controlled to exclude unauthorized persons.

Monitoring and surveillance systems must be installed in public areas. Adequately trained security personnel should regularly patrol the facility. Staff should be trained in procedures for reporting and dealing with any criminal activity that might occur.

A system for informing guests of any activity around the facility or in the immediate neighborhood that might endanger them should be established and all staff should be trained in advising guests of such activity, as well as when to call for police intervention.

Safety Codes and Laws

There are laws and codes to ensure the safety of guests in lodging facilities. These include laws and codes that mandate fire escape stairways or ladders, fire extinguishers, and smoke detectors in rooms and hallways. There should be well-marked, lighted signs indicating exits and escape routes. Signs in banquet rooms or public spaces must indicate the maximum number of people allowed, so as to allow for safe and orderly evacuation in case of fire.

Fire escape routes must be posted conspicuously in each room. Security bars on windows must have a safety release so they can be opened from the inside to allow egress in an emergency. Laws also mandate a safe water supply for the facility and properly functioning plumbing and ice machines that are available for guests. Elevators are to be inspected regularly. Rooms must be cleaned and sanitized after each use, with clean linens provided.

Accommodating Guests with Disabilities

The Americans with Disabilities Act (ADA) was passed in 1992. It requires that public facilities be open and accessible to people with disabilities. Existing buildings were to be modified by adding wheelchair ramps, automatic doors, wheelchair lifts, wheelchair accessible restrooms, wider parking spaces, guest room numbers in Braille, closed captioning on televisions, and numerous other accommodations.

Facilities that do not comply can be sued for damages by disabled persons. The resulting bad publicity that could result from a lawsuit is usually enough to encourage compliance. Newer facilities use the principles of Universal Design, such as putting entrances at ground level instead of using ramps, having all entrance doors be automatic opening and wide, installing cabinets with shelves that pull out, and light switches with flat panels rather than toggle switches, to name a few.

Legal Expectations of Customers

Facilities that serve customers should admit the customers, operate the business in such a way that customers are not harmed, and sell products and services in an appropriate manner. They have a legal right to refuse admittance to or remove any customer who is a threat or annoyance to other customers or employees.

Food and beverage facilities give customers an implied warranty or unspoken guarantee that their products are fit for human consumption. If unfit, the facility is liable and the customer can be compensated for any damages they might have suffered. The customer must prove that an illness or injury was caused by the food or beverage.

The laws concerning alcohol consumption vary from state to state; however, some activities (such as selling alcohol to a minor or to someone already intoxicated) carry stiff penalties.

Liability in the Food and Lodging Industry with Respect to Employees

Owners, managers, and employees are all responsible for ensuring that their facility complies with all local, state, and federal laws. Workers' Compensation is insurance that pays employees who are injured or become disabled while on the job. Employees cannot sue their employer if injured. A written record must be kept on file of all accidents.

The Occupational Safety and Health Administration (OSHA) is a government agency that enforces safety regulations in the workplace. Employees should report conditions that are unsafe to their supervisors. OSHA compliance officers visit businesses on a regular basis to make sure there are no violations.

Safety Data Sheets must be maintained when chemicals are used. They provide information on health hazards and precautions for safe handling, as well as first aid procedures. The Fair Labor Standards Act regulates minimum wages and rules regarding child labor.

New and Emerging Careers

As in all businesses, there are always changing trends in the food and lodging industry. Because of the ever-increasing presence of computers and smart devices in all businesses, those seeking a career in the hospitality industry should be sure to have a solid knowledge of computer systems and software programs used in the industry.

Certain trends in the hospitality industry indicate that career seekers should get additional training in the fields of adventure tourism and ecotourism, since these types of traveling are extremely popular now. Understanding Universal Design principles would be useful as well. Grounds maintenance is becoming "greener" or more environmentally friendly, and recipes are becoming more health-conscious, so more employees with a background in these fields will be needed.

Retired people are traveling more than in previous generations. Those with a background in geriatric care will be much in demand. Another trend is toward having front desk people who are skilled in more than one language.

Communication Skills

It is important to be able to communicate successfully with guests and co-workers, both verbally and in writing, as well as via electronic communication. Successful employees will project sincerity and courtesy when dealing with guests. One needs to learn when to listen and when to speak, and when to refer a guest to another employee.

Telephone communication skills are very important in the industry, as much of the day-to-day business may be done over the phone. Proper tone of voice and correct pronunciation are necessary. Callers expect to receive information promptly. Having pen and paper at hand allows the employee to write down names, messages, and other important information, and many places have a computer system for data entry. Try to avoid putting callers on hold.

Learning to be a good listener improves relationships with guests and other workers. It is important not to daydream while in conversation. Concentrate on the person who is speaking. When speaking with guests personally, learn to use appropriate body language.

Work Ethics

Set good work habits. Be on time to work. Show a supervisor a willingness to learn new tasks, and complete tasks on time or ahead of time. Personal neatness and cleanliness are important. Be a team player; cooperate with and help other employees. Exhibit a positive attitude in the workplace.

It is important to be able to follow directions. Pay attention to details. If unsure, it is better to ask questions than to perform a task incorrectly. Realize that criticism from a supervisor is meant to improve performance, not to hurt feelings. When an assigned task is complete, look for another job, or offer to help a co-worker.

There will always be situations when problems occur. Learn to remain calm and pause before doing or saying anything to guests, other workers, or supervisors.

An institutional food service can be self-operated or contract-managed. An example of a self-operated unit is a hospital in which the responsibility of management and operation of the food service lie with the hospital. Alternatively, the hospital could choose to hire another company to manage and operate the food service operation.

Customer and Guest Service Practices

Service includes tasks performed for the benefit of guests and customers, as well as types of contact the guest has with the business. Service includes things like checking a guest into a hotel or serving a meal; it also includes greeting a guest with a big smile or helping a customer on the phone.

Guests expect to be served in a short time. They like attention and to feel that they are special during lodging or dining experience. The paramount rule in guest and customer service is "the customer is always right." All employees, from the manager to the housekeeper and from the bus person to the chef, are responsible for providing service.

Lodging guests want a clean, comfortable room. Food service customers want an enjoyable meal in a comfortable setting. Making customers happy results in good word-of-mouth advertising for the operation.

Development and Design

As a result of the Americans with Disabilities Act of 1992 (ADA), hospitality facilities are developed and designed with Universal Design features, which provide the same means of use for everybody, no matter what the disability, without the need for specialization.

Universal Design follows the principle that if it works better for those with disabilities, it works better for everyone. Additionally, with a growing population of elderly persons, it makes life easier for the "baby boomers." It includes wider doors that open automatically, flat entrances, and pull-down cabinets. Universally Designed systems are simple to use and easy to understand and require less physical effort. They have been adopted by hospitality facilities to increase market share and include customers who might have been otherwise passed over.

Maintenance

Maintenance and care of hospitality property means guests are satisfied, business operates smoothly, and equipment investment is protected. Kitchen equipment is exposed to grease, dirt, and food build-up, and must be cleaned frequently for fire prevention and sanitation safety.

Plumbing problems result from normal wear and tear, and systems need to be maintained to prevent damage and waste from leaking pipes. Mildew can be a hazard in bathrooms and kitchens and needs to be treated to prevent disease. All areas need to be treated for insects on a regular basis.

Heating, air conditioning, and refrigeration equipment need to be regularly checked for refrigerant levels and motor condition. They must also be kept clean to run efficiently. Swimming pools need to be checked daily for chlorine and pH levels. Regular landscaping ensures a neat appearance to attract customers. Parking lots have to be repaired of cracks caused by weather and traffic.

Responsibilities of Employees, Supervisors, and Managers

Everyone in the hospitality industry focuses on meeting guests' needs. Customers expect professional service. General managers need to make sure the facility conforms to all health and safety regulations; they develop a management plan and delegate authority to the managers below them to oversee their departments. They also strive to hire employees who project a professional appearance even though they might be young and new to the business.

Supervisors directly oversee employees. They assign tasks according to the area they supervise. If a task is not completed correctly, they must show the employee what was wrong and how to correct it. Both supervisors and employees can be front of the house (meaning they interact with guests, as a front office clerk does), or back of the house (meaning they are usually not seen by guests, as an engineering supervisor is not).

Chapter Quiz

Ready to see how well you retained what you just read? Scan the QR code to go directly to the chapter quiz interface for this study guide. If you're using a computer, simply visit the online resources page at **mometrix.com/resources719/aafcsfamconsc** and click the Chapter Quizzes link.

Consumer and Resource Management

Economics and Resource Management

Time Management

Time management is the process of using skills, tools, principles, and practices together in order to most efficiently use the time that an individual or group of individuals has available. The efficiency goal of time management is similar to that of work simplification; however, time management itself does not actually make work easier. Instead, time management involves finding the most proficient way to complete a task at hand in the amount of time given. Time management is extremely important in both business and personal settings, as people often do not have enough time available to complete every single task they face. Planning how to complete the greatest number of tasks possible in the amount of time available allows people to get more value out of their time and efforts.

Some of the techniques used in time management include setting goals, planning the best way to complete a task, scheduling, prioritizing tasks, making sure that all necessary equipment is easily accessible, and effectively managing one's workload. A successful time management system successfully incorporates all of these techniques, with the end result of accomplishing the most tasks possible in the amount of time given.

Prioritizing Activities

One of the most important techniques in time management is **prioritizing** so that the most important tasks are completed first and the least important tasks are left for when a person has more time available to do them. This system allows people to rank both goals and the corresponding tasks that lead to achieving those goals based on how essential they are to the individual or the organization. An individual, family, or organization can determine what activities are important by first identifying and defining which goals need to be achieved in order to continue functioning. Once this step is completed, people can then decide what tasks are the most crucial for the individual or organization to continue functioning. These tasks should be given the highest priority, while less important tasks, although they also can lead to goal accomplishment, should be considered a lower priority.

Goals and Decision-Making

A **goal** is a particular purpose that an individual or organization wants to achieve in the near or distant future. In a corporate setting, for example, a goal could be achieving a specified amount of sales in the next quarter. In a family setting, a goal might be for a teenager in the family to find a job before the end of summer.

Decision-making is the process by which an individual or group attempts to determine what would be the best managerial selection from a set of possible options and/or actions from a managerial perspective, usually for the purpose of achieving a particular goal. For instance, a couple looking for a new home may make a list of the advantages and disadvantages of living in each house they view to help them decide which house to buy.

Goal-Setting

Goal-setting is important because it allows individuals and organizations to identify what objectives they seek to accomplish and then determine the best means by which to accomplish those objectives. Goal-setting typically involves establishing a particular timeframe in which a goal should be achieved. Knowing what goals an individual is striving to reach in a set amount of time helps motivate each individual involved to reach that ultimate objective. Goals also provide the individual or organization with a means of measuring the amount of effort—or lack thereof—that each individual puts into obtaining those goals. In a corporate setting, goals can also provide the organization with a way to measure its overall success, which in turn can be important in determining what future actions the organization will need to take.

CREATING A VALID GOAL

Individuals or organizations attempting to set a goal can follow the common management mnemonic SMART (specific, measurable, achievable, relevant, and time-bound) to make sure that their goal is well-defined, valid, and useful when both measuring success and motivating the organization as a whole. Established goals should be specific and well-defined, have some way of being measured accurately, be attainable, and be relevant to the tasks that need to be completed for the success of the organization. The individual or organization should also decide upon an appropriate amount of time that should be spent in achieving a goal not only to make sure that it is achieved in a timely fashion, but also to ensure that the goal can be compared to other goals in the future.

EXPLAIN WHETHER OR NOT THE FOLLOWING GOALS ARE VALID:

A manager of a local retailer that usually does somewhere between $35,000 and $60,000 a week decided that their store should set a goal of making $50,000 in sales a week each week for the month of November.

This example demonstrates a completely valid goal, as the goal is well-defined and quantifiable. It can be measured by examining the amount of sales the store has done each week. In addition, the goal is relevant to the store's ability to make a profit within a set amount of time, as the goal of $50,000 is achievable because it falls within the store's normal sales range.

The mother of a child who is having difficulty in school has decided that her son needs to improve his grades.

While the goal of improving the child's grades may be achievable and relevant to the success of the child, it is not well-defined enough to be considered a valid, effective goal. Although the goal could be measured, no criteria are established as to what would be satisfactory improvement in the boy's grades. Furthermore, the mother has not established a timeframe in which her son is expected to reach the goal.

WORKING TOWARD ACHIEVING ANY GIVEN GOAL

Both internal and external factors can have an effect on whether an individual or organization will be able to achieve a particular goal. However, the most important thing for an individual or organization to keep in mind, regardless of the situation, is that everyone involved in the goal-attaining process should remain focused on the goal and the appropriate ways by which they can achieve that goal. This means that if a goal is important to the success of the individual or the organization as a whole, people must be committed to giving their best efforts to achieve that particular goal. Any less important or irrelevant goals should not be allowed to distract the individual or the organization from the primary goal; any secondary goals should ultimately be put aside until the primary goal is achieved. From a managerial standpoint, it is important that people are regularly encouraged to recognize that the goal is achievable so that no one gives up before the goal is actually reached.

IMPORTANCE OF THE DECISION-MAKING PROCESS

The **decision-making process** is important because understanding the process can aid in the discovery of a future rational and reasonable course of action when an individual or organization is presented with a decision that must be made. Each individual has a unique perspective and therefore a different way of reaching a particular decision; through a combination of intuition, knowledge, and an understanding of how the decision-making process works, a decision that is best suited for the goals of a particular individual or group can be made. Effective, knowledgeable decision making consists of identifying the decision that needs to be made, recognizing the benefits of each choice related to that decision, analyzing the potential drawbacks of making each choice related to that decision, and, finally, actually making the choice. By understanding what specifically needs to be identified at each stage of the decision-making process, an individual or organization will be able to make a more informed decision that will be more likely to lead to the achievement of a particular goal.

ATTEMPTING TO MAKE A WELL-INFORMED DECISION

Some of the most common pitfalls that should be avoided when attempting to make a well-informed decision are entering the decision-making process with a preconceived notion that the individual is unwilling to abandon, allowing peer pressure to influence a decision, and overgeneralizing. Entering the decision-making process with a prejudice either for or against a particular idea, source of information, or choice will often lead to the selection of an option that is not the best or most logical choice available. While outside sources should be used to gather information about the decision that needs to be made, no single source should be allowed to pressure the individuals involved in making the decision into making a particular choice. Furthermore, the overuse of generalizations, stereotypes, and attempts to attribute effects to causes that may not have any logical link will often lead to a decision being made that may not be the best course of action.

TECHNIQUES THAT AN INDIVIDUAL MIGHT USE WHEN MAKING A DECISION

An individual who is in the process of making a well-informed decision will often gather as much information as possible from as many reliable sources as possible, list the advantages and disadvantages of each choice, and then compare each choice with the others. In an employment or business situation where finances are involved, an individual may use a mathematical approach and calculate exactly how much money each job choice or each potential new product could offer, as well as how much a poor decision might cost if something goes wrong. A mathematical approach can be an extremely solid strategy when one is attempting to determine what choice might be the most appropriate. Unfortunately, many individuals do not make well-informed decisions because they oftentimes employ faulty methods of decision making. In such instances, they may rely too heavily on the opinions of their peers, gather information from such unreliable sources as erroneous web sites, or even leave the decision to random chance by flipping a coin.

STRATEGIES THAT A BUSINESS OR ORGANIZATION MIGHT USE WHEN MAKING A DECISION

Some of the most common decision-making strategies that a business or other organization might use when making a decision include the **Pareto analysis system**, a cost/benefit analysis strategy, a force field analysis strategy, a grid analysis strategy, and a scenario analysis strategy. Because each strategy has its own advantages and disadvantages, businesses and other types of organizations will usually select the system or systems that best support the kind of decision they are attempting to make. For example, a Pareto analysis system can be useful when an organization is attempting to handle a group of related problems that are discovered to be responsible for most of the difficulties that the organization is experiencing. However, a Pareto analysis system will not work as well, if at all, in situations where a large number of problems are caused by completely unrelated factors.

Pareto Analysis System

Fundamentally, the Pareto analysis system is a decision-making model which assumes that approximately 80% of the benefits that an organization receives from a particular task are the result of 20% of the effort that the various individuals within the organization put into the task. It also assumes that 80% of the problems that the organization faces are produced by approximately 20% of the factors that may be causing them. The first step necessary in using the Pareto analysis system is to list all of the problems that need to be addressed and/or the choices that are available. Next, each of those problems or choices should be grouped so that the choices offering similar benefits and/or the factors leading to larger related problems are grouped together. Each group is then given a score based on how much that particular group affects the overall organization.

Cost/Benefit Analysis

Cost/benefit analysis is a decision-making strategy that examines the total estimated cost of each option that is available alongside the total estimated benefit of each option available and then compares the cost with the benefit to determine if the benefits of the option outweigh their cost. Usually, a cost/benefit analysis refers to the financial cost and benefit of a particular decision, but it can actually be used in any situation in which resources are used. For example, if a corporation has two clothing materials, A and B, that both cost the same but material B can be used more efficiently, the cost/benefit analysis would show that material B is the best choice because it allows for the production of the most clothing using the most efficient amount of material. While cost/benefit analysis can be useful when deciding which option will have the greatest benefit, this method also relies heavily on estimation, which makes it not as accurate as other methods of analysis.

Force Field Analysis

Force field analysis is a decision-making strategy that examines all of the factors that affect a particular situation and identifies those factors as either aiding the organization in achieving a goal or ultimately causing the organization to fall short of reaching its specified goal. Basically, force field analysis is the process of identifying and listing which factors involved with each option help the organization and which ones hurt the organization. The first step of the force field analysis strategy is to make a list of all of the factors involved with a particular option and then identify them as either aiding the organization in moving towards a particular goal or hindering the organization's movement towards that goal. Each factor is then given a rating on a scale of 1 to 5, with 1 being the weakest and 5 being the strongest. If the forces that hinder the organization have higher numbers, indicating that they are stronger overall, the option is probably not a sensible or realistic choice; however, if the forces that aid the organization have higher numbers, then the option may prove beneficial.

Grid Analysis

Grid analysis is a decision-making strategy that takes all of the factors involved with each option, rates each factor, and then gives each factor a weight based on its importance to the overall decision. For example, if the owner of a clothing factory had three materials and was attempting to decide which material would provide the most profit, they might consider each material's cost, how much material is wasted during the manufacturing process, and how difficult each material is to use. The owner would then rate the materials using a set scale—from 0 to 5, for example, with 0 being the most expensive or most wasteful and 5 being the least expensive or least wasteful—and rate the cost, waste, and difficulty of each material's use. By using a similar scale, the owner would then assign the factors of each option a weight based on the effect each factor has on the company's overall profit. Finally, the owner would determine a weighted score by multiplying the weight by the rating that each factor received. The material with the highest total score would be the best option.

Scenario Analysis

The **scenario analysis** strategy is a decision-making method whereby individuals or organizations use their experience, knowledge, and intuition to predict what kind of situations may arise from each option if that particular option were chosen. In other words, decision-makers employing the scenario analysis strategy attempt to determine all of the possible outcomes of a particular choice and what effect each outcome might have on the organization or individual as a whole if that choice were made. In some instances, especially in a business setting, each of these potential outcomes might be assigned a score based on how likely it is that each scenario will actually occur. However, anticipating every possible outcome that might occur and accurately predicting what events are most likely to occur in the future can be impossible.

Resources

A **resource** is anything that can be used to aid either in the daily functioning of an individual or organization or in the achievement of a particular goal. The four primary types of resources are land and natural resources, labor resources, capital and capital goods, and information resources. Land and natural resources consist of anything that comes from the environment, such as water, oil, soil, timber, and the land itself. Labor resources are comprised of the actual effort that various people involved in an organization put into a project, leading to the completion of that project. Capital and capital goods consist of any financial and human-made resources, including money, tools, equipment, buildings, and houses. Information resources are any resources that allow an individual or organization to find, compile, and put to use knowledge that might help in achieving a particular goal.

Determining If Enough Resources Are Available for a Particular Project

An individual or organization that is assessing whether enough resources are available for a particular project might begin by determining exactly what the project consists of and researching how much the project will cost in both time and money. After the individual or organization has developed a basic outline of what the project needs, an inventory of what resources are readily available should be conducted. This inventory would include noting such resources as how much money has been allocated for the project, how many people are available to work on the project, what kind of equipment is available to complete the project, and whether workers have sufficient information regarding the project to complete it. If the individual or organization determines that the resources necessary to complete the project are not available, then the project is not realistic.

Steps a Family Should Take Towards Financial Resources

A family that is attempting to determine the status of its financial resources should first collect all of the financial records from each source of income that it has. The family should then list all of its assets, including the amount of cash the family has on hand, along with the amount of cash it could get by selling any stocks, bonds, mutual funds, and property that it owns. After adding up all of its on-hand assets and potential assets, the family should then list all of its loans, unpaid bills, balances due, and financial obligations and add the numbers together to get the family's total amount of liability. By subtracting its estimated total liability from its estimated total assets, the family will get a good estimation of its financial standing in terms of its net worth.

the board of directors could encourage some of their employees to seek further education and, as a result, be better trained for their professional lives.

The **normative-reeducative strategy** of managing change revolves around the idea that peer pressure can bring about the changes that an organization needs. Basically, this strategy assumes that people rely heavily on social interaction and therefore will normally behave according to the expectations of society. Using this assumption, an organization can institute new changes by gradually redefining aspects of the organization's structure and purpose, and each individual will then begin to accept the changes as part of the social norm. For example, if the board of directors of a retail chain realizes that customer service is becoming a more important marketing point than the price of goods, the directors may institute a normative-reeducative strategy. The company would then begin emphasizing desirable customer-service behaviors by placing posters in the break room that affirm the importance of customer service, holding meetings and performance reviews that address appropriate ways to treat customers, and training new employees in good customer service tactics.

The **power-coercive strategy** of managing change assumes that people will listen to authority figures and do as they are told. This strategy works by preventing the members of the organization from choosing any option other than the path that the manager wants them to follow, mainly because no other acceptable options are available to them. If an individual still refuses to accept the changes that the organization has implemented, the organization might punish the individual for not complying. In a family situation, if a child is doing poorly in school because he is spending too much time watching television, the child's mother might simply turn off the TV and tell her child to do his homework. If the child turns the television back on, his mother might send him to his room and ground him for a week as punishment for disobeying.

While the empirical-rational strategy is very effective when an organization has enough available resources to offer the incentives necessary to make changes more acceptable to employees, it is much less successful if the organization does not offer substantial enough incentives. The changes must have an obvious benefit to the individuals who need to implement them. The normative-reeducative strategy is useful when the managers and workers of an organization have a strong relationship and have time available to implement the changes; however, this strategy is much less effective when relations between management and staff are strained or when employees feel rushed or obligated to execute the changes. The power-coercive strategy is particularly useful in situations where time is limited and the threat to the organization or individual is more crucial; however, this method can often promote unrest among members of the organization, especially in organizations that have traditionally allowed more individual freedoms.

Review Video: Organizational Change Management Theories
Visit mometrix.com/academy and enter code: 404217

Identifying and Defining Goals and Values Before Making a Decision

Identifying and defining the goals that a family or other organization needs to meet and the moral values that the family or organization wants to uphold can be extremely important prior to any decision because doing so offers a guide as to what options are unacceptable before an incorrect choice is made. By setting well-defined goals to achieve in the future, a family or organization can choose the options that appear to move it closer to the achievement of those goals. Defining the family's or organization's moral values is also important, because doing so aids in eliminating choices that might be directly opposed to what the family or organization considers correct and ethical.

Consumerism

Consumer Rights

Consumer rights consist of a series of protections, usually guaranteed by law, that seek to prevent individuals from being taken advantage of in the marketplace. In the United States, a large number of state and federal laws establish the rights of consumers, rights that are enforced by the consumer affairs departments of each state, the Federal Trade Commission, and the US Justice Department. Some examples of federal acts that have guaranteed certain rights to consumers are the Equal Credit Opportunity Act, the Fair Housing Act, the Fair Credit Reporting Act, the Fair Debt Collection Practices Act, and the Truth in Lending Act. All of these federal acts and state laws are important because they help prevent consumers from wasting their finances by spending them on products or services that have been sold to them under unethical business practices. Some of these acts also protect the consumer from being discriminated against by businesses or individuals that harbor prejudices against certain consumers.

Consumer Laws

The overall purpose of any **consumer law** is to uphold the three basic rights of the consumer: the right to safety, the right to be informed, and the right to be treated fairly. Federal and state laws are usually designed to protect one or more of these basic rights and may include protections against products that are dangerous, protections against deceptive business tactics or hidden costs associated with a transaction, and protections against certain types of discrimination within the marketplace. Ultimately, state and federal laws are designed to hold businesses accountable for the manner in which they conduct business.

Equal Credit Opportunity Act and the Fair Housing Act

The **Equal Credit Opportunity Act** is a federal act prohibiting any organization that issues credit to consumers from denying credit to a consumer on the basis of race, gender, age, marital status, religion, national origin, or skin color. The act also prohibits any organization from denying a consumer credit if they are receiving federal aid or has previously used another of their rights granted to them by federal law so long as the consumer used that right without malicious intent. If a consumer is denied credit, the organization issuing the denial is required by law to send the denial notification in writing, and the consumer has 60 days to request that the organization send the reason for the denial in writing as well. The **Fair Housing Act** is a similar anti-discrimination act that prevents a seller or organization offering financing for a home from denying to sell, rent, or finance a residence on the basis of race, color, gender, religion, family status, national origin, or handicap.

Fair Credit Reporting Act

The **Fair Credit Reporting Act** states that consumer reporting agencies collecting and distributing information about consumers for credit related purposes are required to follow certain guidelines concerning how they maintain and distribute the information they collect. This act requires consumer reporting agencies to follow a series of procedures established by law to verify and correct mistakes on a consumer's credit record if a consumer makes a dispute. Agencies must also keep records confidential and release them only to businesses that have a legitimate need for those records. This act also sets limits as to how long certain negative information can be kept on record. If a negative piece of information is removed through a dispute, that information cannot be added back onto a consumer's record without notifying the consumer in writing within 5 days. This act was later amended by the Fair and Accurate Credit Transactions Act to allow consumers the right to receive one free copy of their credit report from the three major credit reporting agencies each year.

Fair Debt Collection Practices Act

The **Fair Debt Collection Practices Act** requires third party debt collectors, businesses that collect debts due to other individuals or businesses, to refrain from using abusive or deceptive practices in collecting those debts. This act prohibits such actions as calling consumers at times other than in between 8:00 a.m. and 9:00 p.m. local time and contacting consumers in any manner other than filing a lawsuit if the consumer has given written notice not to contact them regarding the debt. This act also prohibits the collection agency from such practices as adding extra fees and charges to the original balance unless allowed by law, threatening consumers with arrest or legal action that is not permissible by law, or reporting false information on a consumer's credit report. This act also requires third party debt collectors to identify themselves as a debt collector to the consumer during every communication and to provide verification of the original debt upon request.

Truth in Lending Act

The **Truth in Lending Act** requires any organization issuing credit to a consumer to disclose the full terms of the lending arrangement and all costs associated with entering into that arrangement. This means that when a consumer applies for a loan, a credit card, or any other type of credit, the organization must provide the consumer with information about finance charges, the annual percentage rate for the loan, and any additional fees associated with the loan arrangement. In addition to establishing a procedure for the fair and timely handling of billing disputes related to credit transactions, this act also limits the amount of money that a consumer can be held liable for if their credit card is lost or stolen.

Responsible Consumer

Even though state and federal laws exist to protect consumers from unethical business practices, preventing every type of unscrupulous tactic from being used is not always possible. As a result, consumers should be knowledgeable about their rights to avoid being scammed. Many businesses adhere to legal standards and do not employ tactics that are illegal or unethical. However, some businesses choose to do what is profitable instead of what is ethical. Finding loopholes in regulations allows businesses to make claims about their products that may not necessarily be true without being held legally accountable for those claims. Unethical business operations have also been known to find ways to charge consumers additional fees that are legally questionable. In spite of the number of laws designed to protect them, consumers must be diligent about protecting their own interests when making transactions in the marketplace.

Caveat Emptor

The Latin phrase "**caveat emptor**" literally means "let the buyer beware." This expression suggests that a buyer needs to be aware of products or services that may be defective or not as useful as the buyer had anticipated. Some states have laws that create implied warranties regarding certain types of products and allow a consumer the right to return a product that is defective. However, even implied warranties do not protect the consumer from some kinds of defects and only protect the consumer from defects that show up before a certain period of time has passed. If the product or service is not what the consumer thought it was and the seller made no deliberate attempt to defraud the consumer, the seller is typically under no legal obligation to take the product back or refund the consumer's money.

Before making any purchase that requires a financial investment, a consumer should gather as much information as possible about the product or service they plan to buy. By finding out what other consumers have said about a particular product and discovering exactly what the product is and is not capable of doing, a consumer can avoid purchasing a product that is defective or less useful than they originally thought. A consumer can also research what kinds of problems other consumers have had with the product, as well as compare the quality and uses of similar products to determine if another product would be a better match for the needs of the consumer. Because the seller of a product is not under any obligation to inform a consumer about how the product compares in price, quality, and usefulness to other products, it is the consumer's responsibility to know exactly what they are getting when making a purchase.

Wants and Needs

A **want** is anything that a person desires, as opposed to a **need**, which is anything that a person practically requires to continue functioning. A plasma television, for example, is a want because it is not necessary to an individual's continued survival. Examples of needs would be food and water, both of which are things an individual requires to survive. Because people are inherently different, wants or desires vary from person to person. Typically, a person has more desires than their resources can fulfill. Needs, however, are universal because they are everyone's basic necessities and must be met for the continued existence of an individual.

Decision Process Before Making a Purchase

The first step of the decision process for any consumer planning to make a purchase is recognizing that they have a need or a want that to fulfill. After the want or need has been identified, the consumer typically gathers information about products or services that will fulfill that need or want by searching online, examining magazines and advertisements, or talking with other people. The consumer will then evaluate the various product or service choices available in order to determine which option will best fulfill that need or want. Once the consumer's options have been evaluated, they will decide which product or service is most suitable and make the purchase. After the purchase is complete, the consumer will be able to decide whether or not that product or service really satisfied the initial want or need that initiated the decision process.

Marketing

Marketing is the process of informing as many consumers as possible of the existence of a particular product or service and the benefits of that product or service with the intent of influencing a consumer's purchasing decisions. In other words, marketing is a combination of tactics that includes many different forms of advertising that individuals, businesses, and other organizations use to convince consumers to purchase their products. Marketing is important because it allows businesses to play a large role in what consumers actually purchase. Consumers may have the final say in what products or services they decide to use, but marketing can make certain products seem more appealing, more useful, and of a higher quality than a consumer would normally believe them to be. Marketing can also be useful to consumers by informing them about various useful products within the marketplace; however, consumers should be wary of claims and advertisements made by marketers that may make products seem better than they really are.

Marketing can influence almost every stage of the consumer's purchase decision process, but it has the greatest direct effect on the first three stages of the process. The first stage of the process can be triggered by an advertisement marketing a certain product that the consumer may not normally consider to be necessary. By creating the idea or image that a product or service is something necessary to the consumer, a business can make something the consumer wants appear to be something the consumer needs. Marketing can also directly influence the second and third stages of the decision making process by making a product appear better or more useful than competing products, even though this may not necessarily be true. The marketing process can even indirectly influence the fourth stage of the process; consumers actually purchase the product because they have the preconceived notion that a certain product is better than other similar products as a result of the advertisements that they have seen.

State of the Economy

The economy plays an important role in how careful consumers are when using their resources and what they perceive as needs as opposed to what they perceive as wants. When the economy is doing well, unemployment figures are low, which means that people can easily attain their basic necessities. As a result, consumers are typically more willing to spend their financial resources. Consumers will also be more willing to spend their resources on products and services that are not necessary to their survival, but are instead products and services that consumers enjoy having and which they believe increase their quality of life. On the other hand, when the economy is in a slump, consumers are much more likely to cut back on their spending because they perceive a significantly higher risk of being unable to acquire basic necessities due to a lack of financial resources.

Environmental Concerns and Economics

The condition of the natural environment can affect the purchasing decisions of a consumer in two major ways. First, most consumers, due to their own personal values or the values of society, are at least somewhat concerned with the state of the environment around them and will seek ways to prevent unnecessary waste. However, while hardly anyone will always purchase the most environmentally-friendly products and conserve every possible resource at all times, consumers will often choose to purchase products that are recycled or less wasteful when they are available and are not significantly more expensive. The second major way that the environment can affect the purchasing decisions of a consumer involves a consumer's decision to conserve a particular resource because that resource has become more expensive due to a shortage. For example, if an area experiences a gas shortage, the price of gas will rise significantly, and many consumers will conserve gas to avoid paying higher prices.

The primary way that consumers can help protect the environment is by conserving natural resources and finding ways to reduce pollution and waste. Some of the best ways to conserve natural resources include reducing the amount of resources that a household uses, finding ways to use resources more efficiently, and reusing resources rather than simply discarding them. For example, a household can reduce the amount of resources it consumes by turning off the lights and heat in an unused room or by insulating exterior walls to reduce the amount of heat that is lost in a room. A household can also participate in area recycling programs or find new uses for items, such as using a cardboard box that a product came in to store other household items.

Environmental economics is the study of the effects that businesses and consumers have on the overall financial and natural environment when operating within set limits. In other words, environmental economics studies how much pollution actually costs based on the costs or savings associated with pollution and the amount of environmental resources lost due to that pollution. Environmental economics is important to businesses and consumers because it can influence what products are available and what processes can be used in manufacturing those products. Based on environmental economists' recommendations, laws and regulations are made to reduce the amount of damage done to financial and natural environments due to the inefficient or irresponsible use of natural resources. Even though these laws ultimately exist to prevent the waste of natural resources, these regulations can result in the elimination of some products from the marketplace, as well as significant cost increases to the end consumer as businesses incur additional costs in meeting those regulations.

Agencies That Provide Resources to Consumers

A number of agencies can provide resources that help inform consumers about the advantages, disadvantages, and even the scams related to certain products and services. Some of the government agencies that are most useful to consumers are the Federal Citizen Information Center, the Consumer Product Safety Commission, the National Fraud Information Center, and the Federal Trade Commission. All of these agencies have extensive information online, as well as available print publications, about many of the most common products on the market. Probably the most well known and useful private agencies to consumers are Consumer's Research, a non-profit agency that publishes the magazine *Consumer Reports*, and the Better Business Bureau, which is a private organization made up of businesses that sets certain standards which its members are expected to follow.

Health Care Services

A consumer who is in need of health care services might take into account the number of complaints and the types of complaints a facility or specific doctor has received and consequently examine the qualifications of the doctors who practice in that facility. These factors are important to a consumer for any health care decision ranging from choosing a physician for a simple examination to choosing a surgeon for a complex surgery because they give the consumer a basic idea of the overall quality and reliability of the doctor and medical facility. By determining the number of complaints and the types of complaints that have been filed against a doctor or facility, a consumer can get an idea of what kinds of mistakes and how many of those mistakes the facility or doctor has made. Experience can also be a key factor in how reliable a particular doctor is and can greatly affect the quality of their care, as more experienced doctors are usually better capable of providing more effective care.

Unfortunately, determining the exact number of complaints a physician or facility has had filed against them is difficult because no federal law exists that requires physicians or facilities to disclose or even report information regarding complaints. However, the United States Department of Health and Human Services does keep track of reported complaints and is a good source of information for a consumer who is attempting to gather information about a specific doctor or medical facility. Despite the lack of federal laws, many states have laws that do require medical facilities to report the number of complaints and the nature of those complaints filed against the facility or doctors employed by the facility. In many of these states, the facility may not be obligated to disclose the information directly to patients, but a consumer can oftentimes access a state database containing the information online.

Child Care Services

A consumer should consider many factors when attempting to decide which child care service to use. Some of the most important factors include how many children each teacher or caregiver is in charge of, what qualifications the teachers or caregivers have, how long those teachers or caregivers have worked for the facility, and whether the facility is appropriately licensed and accredited. The number of children that a particular caregiver has to care for at one time, as well as how many children attend the facility, can play a large role in how much time a caregiver can devote to each child. A consumer should also verify that not only the facility itself is accredited and licensed, but also that the teachers or caregivers who work with the children have the knowledge and experience necessary to care for them.

While many resources are available to consumers seeking the best child care in their area, one of the best sources of information is usually the child care facility itself. By visiting child care centers under consideration, consumers can gather firsthand information about a particular facility, such as how many children the facility cares for at one time. A consumer should also be able to find information regarding the facility's accreditation while visiting the facility and should later verify that the accreditation is valid and accurate. The Child Care Bureau, which is a subdivision of the United States Department of Health and Human Services, has services available to help consumers locate local organizations that will verify the accreditation of a child care provider. The Child Care Bureau can also be a useful resource for finding additional information about local child care programs.

Elder Care Services

The first step in the process of choosing a facility to care for an elderly family member is recognizing that the family member needs assistance beyond what the rest of the family can provide. After the family realizes that the elderly individual needs additional care, the second step is to evaluate the needs of the family member. By having a basic idea of what the elderly individual is capable of doing on their own and what they need assistance with, a consumer will more effectively be able to research facilities that best suit the needs of the individual. The next step of the process involves gathering information regarding the appropriate facilities for the elderly individual. Finally, the family should choose the facility that appears to best suit those needs.

Knowing the needs of an elderly person and finding the elder care facility that best suits those needs is important because each elderly individual functions at a different level of ability. Because each person has different capabilities and needs, facilities should be evaluated on how well they are able to take care of the individual needs of each patient. Ensuring that an elderly individual is placed in the environment that allows them the most freedom and is the closest match to their current level of functioning is of utmost importance. For example, a woman who is capable of performing normal activities such as bathing, eating, and moving around on her own should not be placed in a nursing home. However, for a woman who has a high risk of injuring herself due to medical conditions, an assisted living home may be an appropriate option.

A consumer who is examining possibilities for the appropriate care facility for an elderly family member has the option of using resources available from the Administration on Aging. The Administration on Aging is a division of the Department of Health and Human Services that offers resources for locating local care facilities, as well as state and local agencies that monitor those facilities, so that consumers can get a better idea of what kinds of facilities are available. Another source of information available to consumers is the National Institute of Aging, an organization that provides a variety of publications regarding the needs and health concerns of elderly individuals. The National Institute of Aging also offers an online database for finding local care facilities. Regardless of which resources a consumer chooses to use, the best resource is often visiting the facilities in person to see what kind of services they can offer and what problems the facility might have.

Consumer Advocacy Groups

Consumer advocacy groups such as the Consumer's Union, the Better Business Bureau, the Consumer Federation of America, the Center for Science in the Public Interest, and Consumer Alert are organizations that attempt to protect consumers from unsafe products and unethical business practices. Many of these groups conduct research on various products and offer the results of their research to consumers either online or through their publications to warn consumers about safety concerns, defects, or scams associated with certain products or services. Consumer advocacy groups also organize and fund campaigns targeted at encouraging support for legislation that promotes fair and ethical business practices, as well as any other legislation that protects the rights of consumers. In addition to being a good source of information for consumers who are making a decision about which product to buy or service to use, consumer advocacy groups are an important ally for consumers who have had problems with a particular product or service and need assistance.

costs per unit of measurement. For example, a 12-pack of 12-ounce cans of soda might be $2.99 and have a unit price of $0.70 per liter. Unit pricing is important because it can be extremely useful to a consumer who is attempting to compare two similar products of different sizes to determine which product is less expensive. If, for instance, an individual is trying to decide between a 12-pack of soda that costs $2.99 and has a unit price of $0.70 per liter and a six pack that retails for $1.99 and has a unit price of $0.93 per liter, they can easily compare unit prices and see that the 12 pack is the better buy.

Creating a Family Budget

A family that has decided to create a family budget should begin by calculating how much money it has on hand and how much income it receives annually. Next, the family should determine what expenses it has and which of those expenses are necessary and which expenses are for luxuries. Necessary expenses might include housing, food, utilities, child care, health care, transportation, and insurance. Luxuries might include eating at restaurants, going to movies, buying a bigger television, getting manicures and pedicures, and other activities that are not necessary to the family's continued survival. The third step that a family should take when making a budget is to add up all of the total necessary expenses and subtract that total from the family's total annual income to find its total annual income after necessary expenses. Next, the family should decide how much of that money will be applied to the family's savings. Any remaining money can then be budgeted for less necessary items and activities.

Checking and Savings Account

An individual or family should consider several factors when opening a checking or savings account: What are the bank's minimum balance requirements? What bank fees are associated with the account? What is the annual interest rate earned by the account? Most banks charge account holders if they have less than the minimum balance required in their accounts. Banks often charge additional fees for ATM use and other services, which makes an account more expensive to the consumer than they might realize. If the financial institution charges a significant number of fees without offering a high interest rate on the account, the consumer should move their accounts to a different institution. Consumers should also verify that a bank, credit union, or any other institution that they are considering placing money into is insured by the federal government through the Federal Deposit Insurance Corporation (FDIC) or the National Credit Union Administration (NCUA).

Credit

Credit is the ability of a consumer to purchase a product or service now and pay for that product or service at a later time, either in full or in installments. Credit can be useful for consumers because it offers a way for handling larger expenses, such as the purchase of a house, when the consumer does not have enough funds on hand to make the purchase. However, consumers must keep track of the purchases they make with credit and realize that purchases made using credit must be paid back, usually with interest. If consumers either fail to pay back the bank or credit card company that loaned them the money for the purchase or do not pay installments on time, their ability to get credit in the future is greatly impacted. Some tactics that might be useful to consumers attempting to manage their credit more carefully include paying bills on time, avoiding purchases that they cannot afford, keeping track of how much debt they have, and carefully monitoring their credit report. Consumers who avoid making purchases that they cannot afford are much more likely to pay off loans on time, which will help keep their credit in good standing. If consumers make too many purchases, accumulate too much debt, or take out too many loans, they will most likely struggle to pay their bills. Extensive debt often indicates that consumers cannot manage their credit, which adversely affects their credit scores. Consumers should also closely monitor their credit reports so that they can dispute erroneous claims that may be negatively influencing their credit.

Investments and Retirement Plans

Investment is the process of committing an individual's resources to a particular activity, organization, or fund, with the hope that such a commitment will provide additional resources for the individual in the future. Investing and saving for retirement early in a person's career is important because careful investment planning guarantees that a person's financial resources have the opportunity to grow as much as possible. Individuals may have difficulty maintaining their customary standard of living if they do not have sufficient funds set aside when they retire because living expenses increase yearly. Individuals who do not have enough money set aside when they retire could become a burden to family members, as they may depend on the family to help support them financially. Some of the factors that an individual or family might consider when making investment decisions are the risk of the investment, the potential return of the investment, and how difficult liquidating the investment would be. The risk of the investment refers to how likely it is that the individual or family will gain some benefit from the investment versus how likely it is that the individual or family will lose the resources that they have invested. The potential return of the investment refers to how much money the individual or family will likely earn from the investment's interest or dividends over time. The degree of difficulty in cashing-out the investment is an important factor because different investments have different costs, taxes, lengths of time to mature, and stipulations regarding the access of the money associated with them.

Investments such as checking accounts, savings accounts, and CDs are considered low-risk investments, but they also offer a lower return. These investments are safe because banks guarantee that these accounts will earn a certain amount of interest; however, the rate of interest is relatively low in comparison with other investment options. While CDs are low-risk investments, they do offer slightly higher rates of interest than checking or savings accounts, but they can be liquidated only after they mature over a length of time set by the bank. Mutual funds are usually considered low-risk to mid-risk investments, as people have the potential to lose the resources they have invested, but mutual funds also have the potential to yield a higher return, though not as high as high-risk stocks. Because the stock portfolios that mutual funds are invested in are typically diversified and comprised of many different types of relatively stable stocks, mutual funds are generally regarded to be safe investments.

Families that are selecting what investments to make should first consider the financial goals that they hope to attain from these investments. For example, if a family is investing its resources to send family members to college, the family should estimate how much money it will need to pay for each member who plans to attend college. After a family has determined its financial needs, it can begin researching investment options to determine which investments are likely to yield enough financial gain for the family to reach its goals. Once it has identified which investments best suit its goals, the family should assess how much risk each investment carries. Finally, the family should choose the investment that has the lowest risk but will still meet its financial goals.

An individual who is attempting to determine how much money they need to invest or put aside for retirement should start by adding up how much they spend monthly on living expenses, with the exception of health care. That number is then multiplied by 1.5, while health care expenses are multiplied by 2 to determine how much those expenses will increase in ten years. The individual then needs to estimate how many years they have before retirement; for every ten years, the health care expenses once again are multiplied by 2 and all other expenses by 1.5. Finally, adding these two results together equals an individual's estimated monthly expenses adjusted for inflation, thereby providing the individual with a rough estimate of how much they will need each month to maintain the same standard of living after retirement.

Insurance

The major types of insurance that an individual or family might need are health insurance, homeowner's insurance, life insurance, and car insurance. The reason that a family should have these types of insurance is that a problem in any of these areas can have a significant impact on the financial security of the individual or the family as a whole. Health problems, fire, theft, the death of a provider in the family, or a car accident can all lead to additional expenses that the individual or family may not be able to afford. Having these types of insurance offers individuals and families a way to pay unexpected expenses without significantly impacting their overall financial well-being.

Chapter Quiz

Ready to see how well you retained what you just read? Scan the QR code to go directly to the chapter quiz interface for this study guide. If you're using a computer, simply visit the online resources page at **mometrix.com/resources719/aafcsfamconsc** and click the Chapter Quizzes link.

Textiles and Apparel

Planning a Wardrobe

Individuals who are planning their wardrobes should begin by considering what kinds of clothing they usually wear, along with what kinds of clothing they will need for casual, business, and other environments. For example, the president of a major corporation would need to focus more on purchasing formalwear and business attire as opposed to buying a closetful of very casual clothing. After an individual has established a basic outline of the kinds of clothing that best suit their wardrobe needs, the individual should take an inventory of what clothing they already have in order to determine what types of clothing need to be purchased. Once the individual has determined which types of clothing are lacking in their wardrobe, the individual will decide what factors are important for each of those types of clothing. For example, an individual searching for formalwear that will be worn only occasionally might be more concerned with corresponding colors and overall appearance than how durable the materials are. Finally, the individual can select clothing to build a wardrobe based on the factors they have established.

Selecting Clothing

The four major factors that an individual should consider when selecting a piece of clothing for their wardrobe include:

1. Price
2. Purpose
3. Quality
4. Overall Appearance

A consumer who is considering purchasing a garment would be wise to compare the price of the item that they are interested in to other similar items to see what price range is reasonable. A consumer should also consider the purpose of the piece of clothing to be purchased. Under what conditions the clothing will be worn can dictate not only the style of clothing desired, but also the materials that the clothing is made of. Someone camping at a remote wilderness site, for instance, would not purchase a satin robe to wear during the trip. The quality of the clothing is also significant because garments that are better constructed can be worn for a longer period of time; paying more for a high-quality piece of clothing instead of choosing a less expensive garment may cost the consumer less money on additional clothing or mending later on. The overall appearance of a garment may be the most influential factor in a person's choice because a primary purpose of any garment is to help the individual convey a certain image.

Reducing the Cost of Purchasing Additional Clothing

An individual whose goal is to purchase clothing as cheaply as possible might begin their search by making a detailed list of exactly what types of clothing they are looking for. This list should include what types of clothing are desired, an estimate of how many garments of each type the individual wants, and an estimate of how much they are willing to spend overall. The individual should then decide how much time they are willing to devote to comparing various garments and their prices, where to go shopping, and when to go. Because many retailers offer deep discounts on certain days of the year, such as the day after Thanksgiving, consumers who want to reduce the cost of adding to their wardrobes might consider shopping during those special sales. Consumers also have the option of shopping online and looking through store circulars to get a general idea of how much an item costs, to compare prices, and to see what sales are currently being offered. Less expensive alternatives to department stores are second-hand and discount stores, along with online auction sites that give consumers the opportunity to purchase inexpensive pieces of clothing that might never have been worn.

Well-Constructed Clothing

The three universal concepts that an individual should consider when determining the quality of any garment include:

- **Functional**: A garment is regarded as functional if an individual can wear it and move in it without difficulty or fear of damaging the garment.
- **Inconspicuous**: An inconspicuous garment has no apparent flaws and the garment was assembled using techniques that hide any stitching or seams that should not be showing.
- **Durable**: A garment is considered durable if the garment was constructed with strong, high-quality materials, with the seams and stitching done correctly so that they hold together.

Several factors indicate that a ready-made garment is well-constructed: the fabric is used and seamed together correctly; the various colors, fibers, and designs appear to go well together; the garment appears to have a smooth appearance; and the garment does not have any obvious loose threads. Well-constructed garments should also have a seam allowance of half an inch or more, which means that there should be at least half an inch between the edge of one material and the seam stitching that material to the second material. If any of these basic elements is lacking, it usually indicates that the garment may have been assembled incorrectly. If the garment does not have a smooth appearance or the seam allowances are less than half an inch, it is likely that the garment will not hold together very well.

A consumer can conclude that a belt is well-constructed if it has well-finished ends, is made out of a durable material, and has a buckle that is securely attached and closes the belt appropriately.

A consumer can conclude that a collar is well-constructed if both the right and left sides of the collar appear to be the same shape and size, if the collar has both smoothly curved edges and sharp flat points, and if the collar lays flat or sits correctly on the shirt. It is also important that the underside of the collar does not show and that there does not appear to be an excessive number of seams across the collar.

A consumer can conclude that a neckline is well-constructed if the design is the same on both sides and matches correctly, the neckline does not appear to be bulky, and the facings are flat and have stitches that are not obvious.

A consumer can conclude that the darts of a garment have been constructed correctly if the darts are smooth and pressed in the appropriate direction, if the darts have been placed in positions that improve the overall appearance of the garment, and if each dart is the same size and shape. The direction that a dart should be pressed depends on the type of dart: Vertical darts should be pressed toward the center of the garment, while horizontal darts should be pressed toward the bottom of the garment.

A consumer can conclude that the hem of a garment has been constructed correctly if it is flat, hangs smoothly, and sits parallel to the floor. A properly constructed hem has an edge that is smooth and secure with no loose threads. The seams of the hem should be pressed and concealed as much as possible. It is also important that the hem is the appropriate width and length for the individual who will be wearing the garment.

A consumer can conclude that the buttons of a garment have been well-constructed if the buttons are attached securely to the garment, the buttons are the appropriate size to fit through the button holes and stay secure once they are through the hole, and the buttons are spaced evenly on the garment. The buttons should also be stitched starting and ending at the bottom of the button. A consumer can conclude that the buttonholes of a garment have been well-constructed if the buttonholes are spaced evenly and are even in width and length, the buttonholes are the correct size, they lay flat, and they are stitched correctly. The edge of the garment that has the buttonholes should also be interfaced, and buttonholes should be cut on grain at points of strain and carefully stitched to make sure that the buttonhole is secure and does not cause the garment to come apart.

Importance of Correctly Fitting Garments

Several factors indicate that a garment fits correctly: the garment is comfortable and allows the wearer free movement; each part of the garment, such as the neckline or the waistline, sits properly; and the garment lies flat. Garments that fit correctly should also have a smooth appearance with no wrinkles and should not have any folds in the fabric other than the folds that are part of the garment's design. If a garment does not fit an individual properly, that person will not be able to function normally because they will have difficulty moving in the garment. Additionally, the garment should fulfill its intended purpose while also giving the wearer a certain look. A garment that does not fit correctly and does not look right will usually not convey the image the individual intends to project.

Elements of Design

There are four basic elements that make up the design of any garment:

- **Color**: Color refers not only to the shade of the garment, such as red or blue, but also to how bright or dull and how light or dark that shade is.
- **Shape or form**: Shape refers to whether a garment or detail is flat, such as a T-shirt lying on a table, while form refers to the depth that garments have, such as the tiers of a wedding gown.
- **Texture**: Texture refers to how the garment feels to the touch, like rough, soft, coarse, or silky.
- **Line**: Line refers to the seams and lines that are inherent to the fabric, as well as any designs present in the garment.

The principal elements of design are important because they allow an individual who is attempting to construct or select a garment to understand the basic parts that comprise a design. An individual who understands these parts can determine what makes a garment look good and what sorts of garments will complement each other. The elements of design may not be of utmost importance if an individual is not concerned with the garment's overall appearance and is simply looking for a comfortable or useful piece of clothing; however, understanding these elements can be extremely useful when an individual is looking to improve their overall appearance. Ultimately, the elements of design are the basic components that make up a garment and determine whether or not that garment will give the right look when used with other garments.

Principles of Design

Five basic principles are used in the construction of a garment to make its design more appealing:

- **Balance**
- **Emphasis**
- **Proportion**

- **Rhythm**
- **Unity**

Balance

A garment is **balanced** if both the left and right sides of the garment appear to be similar and correspond to each other. The emphasis of a garment refers to any designs, details, or parts of the garment that draw people's attention. A design is correctly proportioned if the attributes of the various parts of a garment, such as size, shape, and color, appear to be appropriate when compared to each other or to the garment as a whole. The rhythm of a design refers to any feature that causes an individual to look at one part of a design and then leads that individual's attention to another part of the garment's design. Finally, the unity of a design is when a garment looks appropriate after all of the parts of the design have been put together.

Balance refers to the fact that both the left and ride sides of the garment appear to be similar to each other, but this does not necessarily mean that both sides of the garment are actually the same. In fact, there are two different types of balance that an individual might find when examining a garment: symmetrical balance and asymmetrical balance. In order for a garment to have symmetrical balance, both sides of the garment must sit at the same level, and the parts on one side of the centerline of the garment have to be an exact mirror of the other side. A garment is considered to have asymmetrical balance if both sides are not physically identical, but are similar enough that an individual believes that both sides seem to have the same "weight." Balance is extremely important to the overall appearance of a garment because a garment that is comprised of unevenly positioned parts will not look as appealing.

Emphasis

The **emphasis** of a design is important because the overall purpose of using the principles and elements of design is to create a certain appearance that will draw people's attention to the garment. The emphasis of a design can refer to virtually anything from the use of a bright color to a subtle form to an intricate image, such as an ivy vine that wraps its way around the garment. However, any individual attempting to decide whether the emphasis of a garment is effective should consider whether it fulfills its purpose of drawing attention without being too flashy or gaudy because garments that place too much emphasis on flashy and gaudy features are usually unappealing.

Proportion

In some ways, **proportion** is similar to the design principle of balance in that it pertains to how the various images, lines, and other attributes of the garment appear to fit the garment when they are compared with one another. For example, a garment that has several roses of a similar size evenly spaced on it would probably be considered correctly proportioned. However, a garment with one enormous rose on it and several other roses of differing sizes placed randomly across the garment would probably be considered incorrectly proportioned, as the roses do not appear to complement or relate to one another. Proportion also refers to how those individual design elements relate to the garment as a whole, so an enormous image of a rose covering an entire garment would not be proportional to the garment and would therefore appear to be overwhelming and unappealing.

Rhythm

If a garment employs the important design technique of **rhythm**, its overall design will have a basic sense of flow, which means that one part of the design seems to ease into the next, leading the viewer's attention to the next detail in the design. The most common way of accomplishing a sense of rhythm in a design is to repeat patterns or colors within a garment, which draws the viewer's

Washing Techniques with Specific Care Instructions

- **Machine-washable** garments can be cleaned using any normal method of washing and drying unless the care instructions state otherwise. This means that machine-washable garments can be hand-washed, cleaned using a washing machine and dryer, dry-cleaned, washed by commercial laundry services, or cleaned using any other reasonable method.
- **Home launder** means that a garment can be cleaned by any method except for commercial laundry services because the procedures frequently used in commercial laundering may damage the garment.
- **No chlorine bleach** means that an individual can use only bleaches that do not contain chlorine, such as oxygen bleaches, because chlorine bleaches may damage or discolor the garment. A label indicating no bleach is a warning that no bleach of any kind should be used when washing the garment to avoid permanently damaging it.
- **Cold wash/cold rinse** means that the garment should be washed in cold water; warm wash/warm rinse means that the garment should be washed in warm water; and hot wash means that the garment should be washed in hot water.
- Applicable only if a garment is being machine-washed, **no spin** means that the garment should be removed from the washer before the final spin cycle to avoid being damaged.
- When the care instructions indicate **delicate cycle**, **gentle cycle**, **durable cycle**, or **permanent press cycle**, the instructions refer to the cycle setting on a washing machine that should be used to minimize the chances of damage to a garment. Stronger materials can usually be washed on any setting, but some materials are more delicate and therefore more likely to be damaged during longer cycles when the inside of the washing machine is spinning faster for a longer period of time.
- **Wash separately** means that a garment should not be washed with other garments or should be washed with garments of a similar color. Garments usually need to be washed separately because either the colors in the garment are likely to run and damage other clothing or the garment is likely to pick up colors from other clothing.
- The instruction **hand-wash** means that a garment should not be machine-washed, but it can be washed by hand- or dry-cleaned. Garments that should be hand-washed have this particular instruction because they are extremely delicate. Hand-wash garments are usually washed in warm water. Hand-wash only means that the garment should be washed by hand; any other method, including dry cleaning, should not be used.
- **Dry-clean only** means that the garment should not be cleaned by a commercial laundry service or a home method and should only be dry-cleaned. **No dry clean** means that the garment should never be dry-cleaned.
- **Tumble-dry** means that a garment can be dried using a normal dryer. The temperature that should be used—low, medium, or high—usually follows the "tumble-dry" instruction. Drip-dry means that a garment should be hung while it is still wet and let dry.

Yarn, Thread, and Fabric

Yarn is a collection of long lengths of tightly spun fibers that are used in the production of fabric. **Thread** is a collection of long, thin lengths of fibers that are used for sewing two fabrics together. **Fabric** is a material constructed by weaving or knitting yarn together to form a type of cloth. Fabrics come in a wide range of colors, textures, and appearances and can be used to make such items as carpeting, towels, and clothing.

Fibers

Fibers are thin filaments similar to thread found in plants or the fur of animals. Fibers can also be man-made filaments that are formed by forcing chemicals through small holes called spinnerets.

These fibers are used for a number of different purposes including the construction of yarn, thread, and fabric.

The two major types of fibers that are used in the construction of fabrics are **natural fibers** and **synthetic fibers**. Natural fibers are any fibers that are taken from plants or animals, while synthetic fibers are fibers that are created in a laboratory or factory through the use of chemicals. Each type of fiber has its own distinct characteristics. Synthetic fibers are often similar to natural fibers, as researchers will base a synthetic fiber on a natural fiber in order to create a stronger or less expensive alternative to the natural fiber. However, synthetic fibers often react very differently under certain conditions such as heat and sunlight than natural fibers, so it can be extremely important for an individual to understand the differences between synthetic and natural fibers.

Acetate fibers are synthetic fibers that are soft, smooth, dry, weak, lustrous, and heat-sensitive. Acetate fibers are used in the construction of a variety of fabrics including satins and taffetas primarily to give the fabric a shiny, lustrous look. Fabrics that are made from acetate fibers are more resistant to shrinking.

Acrylic fibers are synthetic fibers that are soft, lightweight, resilient, and heat-sensitive. Acrylic fibers are often used as a replacement for wool because they have a very similar feel to wool. Acrylic fibers are also often used as a less expensive alternative to cashmere, as they have a very similar appearance.

Cotton fibers are natural fibers that are relatively strong, soft, light sensitive, elastic, and breathable and can absorb and retain water effectively. Many fabrics are made completely out of cotton fibers or a blend of cotton fibers with other fibers such as rayon and polyester. Fabrics made from cotton fibers are usually both soft and heat resistant, but they are also more likely to shrink.

Similar to cotton fibers, **flax** fibers are natural fibers that are soft, breathable, and strong but lack the elasticity that cotton fibers have. High-quality flax fibers are used primarily in the construction of linens. Fabrics made from flax fibers are usually similar to fabrics made from cotton in that they are soft and heat resistant and also likely to shrink.

Nylon fibers are synthetic fibers that are smooth, strong, lightweight, elastic, and lustrous. Nylon fibers are primarily used as a cheaper alternative to silk. Fabrics made from nylon are relatively inexpensive, strong, and extremely light, but they will melt at high temperatures.

Polyester fibers are synthetic fibers that are strong, lightweight, resilient, and resistant to many of the conditions that might normally harm a fabric, such as mildew, sunlight, harsh weather, and moths. Polyester fibers are used in a wide range of fabrics and are often used in a cotton/polyester blend.

Ramie fibers are natural fibers that are very strong, have a lustrous appearance similar to silk, and absorb and retain water effectively. However, they are also brittle, stiff, inelastic, and not very resilient. Ramie fibers are very difficult work with and are usually used in blends with other fibers, such as wool, to make a variety of fabrics.

Rayon fibers are synthetic fibers that have many of the same characteristics as cotton fibers and are relatively soft, smooth, strong, resistant to moths, and can retain water extremely well. Rayon fibers are used in a variety of fabrics, including most types of velvet because it can imitate the feel of silk and other natural fibers.

Silk fibers are natural fibers that are extremely soft, strong, and lustrous, and they absorb and retain water effectively. They are also resistant to mildew, molds, and moths. Silk fibers are used in the construction of a wide range of fabrics such as chiffon, organza, and taffeta because of the silk's soft feel and extremely lustrous appearance.

Wool fibers are natural fibers that normally are coarse, relatively strong, resilient, and can retain water well. Wool fibers are commonly used in a variety of fabrics such as flannel because wool fibers improve a fabric's ability to retain heat. Most types of wool have a scratchy feel to them, which is why many individuals believe that they are allergic to wool, but some wool fabrics such as cashmere are very soft.

Making Fabric from Natural or Synthetic Fibers

A textile manufacturer making a particular type of fabric would begin by acquiring the appropriate fibers from plant or animal sources or by manufacturing the fibers that are needed for that particular fabric. The manufacturer would then convert those fibers into yarn through a process called spinning in which the fibers are spun together tightly. The yarn is then knitted or woven together by hand or by machine to form fabric. The fabric is then put through a series of treatments that are designed to improve the overall appearance and feel of the fabric. These treatments include processes such as dyeing, which the manufacturer would use to achieve the desired color for the fabric unless the natural or plant fibers are already that color. Finally, the fabric is treated with a series of finishing chemicals that are designed to improve the fabric's resistance to harmful conditions such as weather or sunlight.

Chapter Quiz

Ready to see how well you retained what you just read? Scan the QR code to go directly to the chapter quiz interface for this study guide. If you're using a computer, simply visit the online resources page at **mometrix.com/resources719/aafcsfamconsc** and click the Chapter Quizzes link.

Environmental Design

Functions That Any Type of Housing Is Expected to Serve

The primary purpose for any type of housing is to act as a form of shelter. Shelter in this context refers to protecting people from harsh weather such as rain, snow, extreme heat, extreme cold, harsh winds, extensive exposure to UV radiation, or any other conditions that might prove unpleasant or harmful. Housing can also provide security by making it more difficult for harmful things, such as animals, insects, unscrupulous individuals, and other threats, to reach people and their belongings. In some cases, housing can also serve as a location for individuals to conduct business in and work from. A house itself can even act as a symbol of the wealth and status of its owners.

Security can be extremely important for an individual or group that is wealthy, well known, or at high risk of being victimized by thieves or other criminals. A home can provide a wide range of protective measures to help keep people and their belongings safe, including locks on doors and windows, burglar alarms, fire alarms, and carbon monoxide detectors to warn residents of potentially deadly threats. Some homes, especially homes that are located in high-risk areas, might contain a large percentage of an individual's personal assets. Residences that house individuals who are at high risk due to their social status, political rank, or amount of personal assets may also have additional security measures in place. These measures include security features such as strong deadbolts locks, security cameras, gates, and checkpoints.

Running a Business from Home

A number of advantages are associated with running a business from home. Probably the most important advantage of running a home-based business is the fact that individuals or groups can usually significantly reduce their overhead costs as a result of using a single building for both business and living purposes. Overhead costs that can be reduced include expenses related to transportation, business clothing, office furnishings, maintenance, mortgage or rent, taxes, and other similar expenses. Some of the other major advantages of running a home based business are that the individual has a more flexible schedule, has the ability to put most of the time that would be spent commuting to better use, and the individual is readily available to fulfill their domestic responsibilities as well.

For an individual to be able to work successfully and efficiently in any environment, they need a private place to work that is relatively free from distractions. Regardless of what type of business an individual is attempting to run, they will also require time to accomplish the work that needs to be done, as well as the necessary equipment to perform that work. Equipment an individual may require for their home business includes a computer, printer, telephone, and computer programs appropriate to the type of work to be performed. It is also important that an individual is able to find the storage space necessary for any equipment that the business will need to function appropriately.

The single largest problem that home business owners face is the conflict that arises between the needs of the family and the needs of the business. Every business has its own distinct needs, and those needs can be difficult to achieve in a home setting because a conflict can arise between the space, time, and equipment that the family needs and the space, time, and equipment that the business needs. For example, a child or other family member may need the business space or equipment, such as a computer, to complete their homework at the same time that the business owner is using the space and equipment for conducting business. On the other hand, the children of the family may not need to use the computer but instead need the business owner to step into a parental role and feed them or help them with their homework, which distracts the business owner from the tasks they need to complete for work purposes. Any equipment purchased specifically for the business will also require space that the family might have already had dedicated for another purpose in the household.

Using a Home as a Status Symbol

The major reason that a home can act as a status symbol for an individual or family that lives in it is because of the financial investment necessary to own a home. Regardless of the type of home that someone lives in, purchasing or leasing a home is usually the single largest expense that anyone will take on during their life. As a result, the home of an individual or family can act as an indication of the overall wealth that a person or family possesses. This means that the larger and more luxurious a house is, the greater the amount of wealth a person must have had in order to acquire that home. Since wealth is commonly associated with the status of an individual, the type of home that an individual or family lives in acts as an indicator of their social status as well.

Major Types of Housing

The two major types of residential housing are **houses** and **apartments**. A house is an entire building with the primary purpose of providing shelter for the individual or family that resides in it. An apartment is a section of a building that has been split into units, and the primary purpose of each unit is to provide shelter for the individual or family that lives in that unit. The major difference between and house and an apartment is that a building is considered a house if the whole building belongs to one individual, family, or several families that share the entire building or at least most of the building. On the other hand, a building is considered to be an apartment building or a complex if it contains separate sections or units that each allow individuals or families to live on their own without sharing a large portion of the building.

Common Types of Houses and Apartments Found in the United States

The three major types of houses found in the United States are **single-family houses**, **townhouses**, and **semi-detached houses**. A single-family house is any building used for residential use that is separate from any other buildings. Townhouses are a series of buildings where all of the houses share both of their side walls with the buildings on each side (with the exception of the houses at the ends of the row, which only share one wall). Semi-detached houses are two residential buildings that share only a single wall and have empty space on their other side. These three major types of houses are typically determined by how much space is available around the building. A single-family home has space available on all sides, a semi-detached house has space clear in front, behind, and on one side of the house, and townhouses, with the exception of the townhouses on the end of the row, have space only in front and behind the house.

The most common types of apartments found in the United States are single or multi-bedroom apartments and studio apartments. Single or multi-bedroom apartments are units that have one or more bedrooms that are separated from the rest of the unit. Studio apartments are units that have one large room that makes up the entire unit and performs the functions of a bedroom, living room, dining room, kitchen, and any other rooms that the unit's tenants require. Studio apartments and single or multi-bedroom apartments are usually rented, but it is becoming more and more common for apartments to be sold to consumers as condominiums or cooperatives. Regardless of whether an apartment is rented or purchased, studio apartments are almost always less expensive than multi-bedroom apartments; the price of an apartment increases significantly with each extra separate bedroom attached to the unit.

Condominiums vs. Cooperatives

A **condominium**, also known as a condo, is a type of apartment ownership where an individual has purchased complete ownership of a single unit but has shared ownership of certain common areas and common expenses associated with the building. This means that the individual living in the condo owns the unit and is responsible for all of the expenses associated with that unit, but shares the rest of the building and the expenses associated with the remainder of the building. A **cooperative**, also known as a co-op, is a type of apartment where an individual owns shares in a corporation, and that corporation owns the entire building. An individual who owns the shares has the right to live in a particular unit within the building as long as they follow certain guidelines set by the corporation.

The major difference between a condominium and a cooperative is that a condo owner actually owns the unit and shares ownership of the building, while a co-op owner does not actually own the unit or the building, but rather owns part of a corporation and that corporation owns the building. By paying condo fees and discussing problems with a condo association, the owner of a condo directly shares the decisions and expenses related to the upkeep of the building. A co-op owner, on the other hand, is not directly responsible for the expenses and problems of the building as the corporation that owns the building handles these expenses and problems using the funds they gather from the rent or subscription fees paid by the co-op owners.

Methods Used to Pay for Housing

The two most common methods that an individual or family might use to pay for their housing are *renting*, also known as leasing, and *purchasing* a house through a mortgage or loan. When an individual does not actually own a house or apartment where they reside, but rather pays a monthly fee for the use of that house or apartment, that is renting. People who rent a place to live will never own the property no matter how long they live there and how much rent they pay. In contrast, an individual who takes out a mortgage on a particular type of housing actually purchases that house or apartment using the money acquired from a loan.

Advantages and Disadvantages of Renting

The two major advantages of renting housing rather than purchasing housing are that renting is usually less expensive in terms of monthly payments and the individual or family does not have to be concerned with costs associated with maintaining the building. A renter or lessee does not actually own the building and has to pay only whatever fee the building's owner has set for that particular unit, regardless of any work that the rest of the building may require. For example, if a building's plumbing needs work, any cost that is associated with repairing the plumbing is the responsibility of the building's owner and not the responsibility of the apartment's tenant. The major disadvantage of renting is that the individual does not gain any long-term financial advantage from paying rent because each rent payment is not being applied toward the purchase of the property.

Advantages and Disadvantages of Purchasing

The major advantage of purchasing housing rather than renting is that an individual or family actually owns the property and can sell that property or rent it to someone else at a later time. Each monthly payment is applied to paying off the loan that was used to purchase the property, and the owner can decide to sell that property at any point and keep any money that is left over after paying the bank back. An individual who is renting can leave their apartment at any time but will receive no benefit from the money they paid while living there. The two major disadvantages of purchasing a home are that the monthly payments are usually higher than what an individual renting a home pays and the individual or family is responsible for all of the expenses associated with maintaining the home.

An individual who is attempting to decide whether to rent or purchase a particular type of housing should base the decision on their income. This can be done by estimating how much buying a piece of property will cost the individual per month, how much renting will cost the individual per month, and the individual's monthly income. If the individual has the money available to put a reasonable down payment on a piece of property that they are interested in and has the income to make the monthly mortgage payments on the property without a problem, buying is usually the better option. This is because purchasing a piece of property, unlike renting, is an investment that can be sold, usually with some degree of financial return for the original property owner. On the other hand, if an individual cannot afford to make a reasonable down payment and the monthly mortgage payments are more than they can comfortably afford, renting is a better option.

An individual or family that is considering what building to rent or purchase should first determine how much they can afford to spend on housing. Purchasing or even renting housing is always a major investment, and the individual or family needs to make sure that they have the funds available to make the required payments. Next, the individual or family should determine what attributes of the house or apartment are important to them, such as the location of the building, how big the building should be, how much storage space the building has, and how much yard space it includes. Once the individual or family has made a list of all of the things that should be ideal features of their house or apartment, they can then begin looking for houses or apartments that are within their price range and best suit their needs.

Housing Considerations

Factors to Consider When Looking for Housing

Some of the most important factors regarding the surrounding area of a house or apartment that an individual should consider include the current state of the neighborhood, how accessible the property is, and the quality of local schools and recreational facilities. The current state of the neighborhood refers to factors such as the overall appearance of the neighborhood, the traffic and noise level, the number of children living in the area, parking, zoning or other regulations and restrictions, and the level of police, fire, and community involvement in the neighborhood. The accessibility of a piece of property refers to its location and how close it is to stores, public transportation, major roads and highways, and a resident's place of employment. For active individuals and families with children, quality local schools and recreational facilities are important. Considerations include how close the house or apartment is to these facilities, how well maintained these facilities are, and the reputations of these facilities.

Importance of Floor Plans

Some of the factors that an individual should consider when designing a floor plan include what size the room or rooms should be, what furniture needs to go in each room, how often each room will be used, and the traffic patterns of the room. The individual should also consider how many electrical outlets the room requires, how much lighting is necessary for the room, and whether the room has any special requirements to fulfill the purpose of the room. For example, a photographer who is planning to add on a darkroom to his house would need a room that is located away from outside walls and windows, lets in virtually no light, has access to a number of electrical outlets and countertops for developing equipment, and has sinks for rinsing film. A well-designed floor plan is important because it allows an individual to eliminate problems such as not enough storage space, not enough room for furniture and equipment, not enough light, and not enough access to power outlets.

AAFCS Practice Test

Want to take this practice test in an online interactive format?
Check out the online resources page, which includes interactive practice questions and much more: **mometrix.com/resources719/aafcsfamconsc**

1. In Piaget's theory of cognitive development, the thinking of children in the later part of the preoperational stage becomes less egocentric but still remains animistic and magical. This corresponds to which tasks in Duvall's family development theory?

a. Encouraging the child to fit into the community and to participate in education
b. Helping the child find career interests and a balance of freedom and responsibility
c. Adapting to the child's needs and coping with a lack of privacy and depleted energy
d. Stages of establishment, expectancy, and integration of a child into the family

2. Which of the following is true regarding critical evaluation of information in modern media?

a. Methods used to capture attention are irrelevant.
b. Various potential interpretations must be considered.
c. The receiver must consider the medium over the message.
d. Who is sending the message matters more than the content.

3. In his developmental theory, what did Erikson identify as the positive outcome of successfully resolving the nuclear conflict of basic trust versus mistrust?

a. Will
b. Hope
c. Purpose
d. Competence

4. Which of the following solutions would be most practical in housing for a family?

a. Moving to other houses as children grow older
b. Remodeling the house as family needs change
c. A house having some rooms with flexible uses
d. Parents get an apartment when children move

5. What do interpreter-sensitive values mean relative to communication styles?

a. The listener is expected to be responsible for interpreting meaning.
b. The speaker is expected to be responsible for giving a clear message.
c. The speaker avoids nonverbal behaviors for the listener to interpret.
d. The listener need not "read between the lines" as decoding is simple.

6. According to wardrobe experts, which is most accurate about women's wearing habits?

a. Most women tend to wear fewer than half the clothes they buy.
b. Most women tend to wear nearly all of the clothes that they buy.
c. Most women tend to wear over half of the clothes that they buy.
d. Most women tend to wear varying amounts of clothes they buy.

7. Which of the following is true about purchasing a home, according to realtors?

a. To buy a house with a mortgage, some down payment is always necessary.
b. When purchasing a home, the buyer is always responsible for closing costs.
c. Some loan programs give buyers no money down and few/no closing costs.
d. Monthly mortgage payments are not affected by the down payment made.

8. Which of the following most accurately represents a concept of Bandura's social learning theory?

a. Children can learn vicariously by observing and imitating others' behaviors.
b. Children only learn directly through consequences of their own behaviors.
c. Children observe and imitate others' behaviors but do not learn from this.
d. Children learn better from positive reinforcement than from punishment.

9. Which is *most* accurate today about type 1 and type 2 diabetes?

a. Type 1 diabetes is only inherited, while type 2 diabetes only develops from lifestyle.
b. Type 1 diabetes is from lack of insulin; type 2 diabetes is from insensitivity to insulin.
c. Type 1 diabetes occurs in childhood, while type 2 diabetes occurs during adulthood.
d. Type 1 diabetes accounts for half of all cases and type 2 diabetes for the other half.

10. For conflict prevention, which of these is an example of cooperative communication rather than of person-centered expressions that put others on the defensive?

a. "You always interrupt me before I can finish what I'm saying."
b. "It will help me to finish, and then I want to hear your ideas."
c. "Can't you just be quiet and listen to me for even a minute?"
d. "You don't know what you're talking about. You don't listen."

11. Which of these is generally accepted as indicating possible delayed or atypical intellectual development in a child?

a. A standardized IQ test score that is 10 points lower than the mean
b. A standardized IQ test score one standard deviation below the mean
c. A standardized IQ test score two standard deviations below the mean
d. A standardized IQ test score is never used for indicating this possibility

12. In the process of finishing knitted or woven fabrics, which of these reflects the correct sequence?

a. Bleaching or brightening, cleaning, dyeing, singeing, Mercerizing
b. Cleaning, singeing, bleaching or brightening, Mercerizing, dyeing
c. Mercerizing, bleaching or brightening, dyeing, cleaning, singeing
d. Bleaching or brightening, dyeing, Mercerizing, singeing, cleaning

13. Some color analysts divide people's coloring into the four seasons for choosing the most flattering palette of colors in clothing and makeup. Others use adjectives associated with temperaments for the same purpose. Which of the following correctly equates these two systems?

a. "Passionate" = "Autumn"; "Dramatic" = "Winter"; "Vibrant" = "Spring"; "Romantic" = "Summer"
b. "Dramatic" = "Autumn"; "Passionate" = "Winter"; "Romantic" = "Spring"; "Vibrant" = "Summer"
c. "Vibrant" = "Autumn"; "Romantic" = "Winter"; "Passionate" = "Spring"; "Dramatic" = "Summer"
d. "Romantic" = "Autumn"; "Vibrant" = "Winter"; "Dramatic" = "Spring"; "Passionate" = "Summer"

14. When reorganizing clothes, which is the best advice regarding clothing colors?

a. We should discard any items in colors outside of our most flattering color group.
b. We should keep a few items in colors outside our color group to ensure variety.
c. We should coordinate all the colors of our clothes for maximum outfit potential.
d. We should do (A) and (C) to look our best and make creating outfits the easiest.

15. Which of the following events decreases metabolism?

a. Rapid weight loss
b. Increase in muscle mass
c. Slow weight gain
d. Moderate workout

16. Baumrind and other researchers applying her theory of parenting styles have found that children who function at low levels in multiple areas of life are most likely to have had parents with which style?

a. Permissive
b. Authoritarian
c. Uninvolved
d. Authoritative

17. The certified dietary manager (CDM) is monitoring foodservice and notices a delay in plating. When randomly checking the temperature of pieces of the roast chicken that is to be served, the CDM finds that temperatures range from 108-120 °F. The most appropriate intervention is to:

a. authorize serving the chicken as is.
b. reheat the meat to 165 °F.
c. reheat the meat to 140 °F.
d. pull the meat and discard it.

18. Which of these correctly relates a stage in each theory to the same age periods and family stages?

a. Freud's Oedipal; Erikson's Autonomy vs. Shame, Doubt; Piaget's Preoperational; Duvall's School Age
b. Freud's Anal; Erikson's Industry vs. Inferiority; Piaget's Formal Operations; Duvall's with Preschoolers
c. Freud's Genital; Erikson's Initiative vs. Guilt; Piaget's Concrete Operations; Duvall's with Teenagers
d. Freud's Oral; Erikson's Basic Trust vs. Mistrust; Piaget's Sensorimotor; Duvall's Families with Infants

19. Obesity is correlated with which of the following health conditions?

a. Diabetes
b. Hypertension
c. All these and more
d. Hypercholesterolemia

20. When a consumer hires a realtor to help in buying a home, which information that the realtor gives the consumer is important for the realtor to update regularly?

a. The current market conditions
b. Various options for financing
c. All these should be updated
d. Specific negotiating methods

21. According to Robert Havighurst, in which stage do an individual's developmental tasks include learning to relate the self emotionally to other members of the family?

a. Middle Childhood (6 to 12 years)
b. Pre-Adolescence, Adolescence (12 to 18 years)
c. Infancy and Early Childhood (0 to 6 years)
d. Early Adulthood (18 to 35 years)

22. Piaget's first stage of cognitive development involves:

a. Operations with concrete things
b. Sensory input and motor output
c. Operations with abstract things
d. Intuition and animism, not logic

23. Some home construction companies design houses based on certain principles. Which of the following is a general rule reflecting good design principles?

a. There should be only one way to get to the kitchen.
b. It is better to have several ways to get to any room.
c. Traffic patterns going through rooms disrupt activity.
d. Separate rooms look bigger than in open floor plans.

24. Which of the following statements is true?

I. Bulimia can lead to tooth decay.
II. Anorexics tend to have a distorted self-image.
III. Bulimia does not always include purging.

a. I only
b. II only
c. I and II only
d. I, II, and III

25. Which of the following forms of insurance related to owning a home covers a loss of personal property in the home?

a. Title insurance
b. Flood insurance
c. Home warranties
d. Mortgage insurance

26. Of the following, which best reflects a principle of effective classroom management?

a. Classroom management should address only disruptive, not off-task behaviors.
b. Classroom management should teach students to manage their own behaviors.
c. Classroom management is only needed if things take on-task students off-task.
d. Classroom management need not teach students to be on-task, which is normal.

27. A consumer reads an over-the-counter, FDA-approved medication's instructions, dosage, and side effects before taking it. He takes the proper dosage over the correct time frame to control his symptoms. This consumer is demonstrating a high level of:

a. Organizational health literacy
b. Personal health literacy
c. Critical thinking skills
d. Evaluating claims

28. Which provision of the Wall Street Reform and Consumer Protection Act of 2010 fulfills the consumer rights and responsibilities regarding redress?

a. Brokers have advisors' duties if advising on investments.
b. Advisors must give information to the SEC upon request.
c. Whistleblowers who disclose wrongdoing are protected.
d. Consumers can sue credit rating agencies for negligence.

29. Which of these cooking methods is the most energy-efficient?

a. A microwave oven
b. A convection oven
c. An electric oven
d. A gas oven

30. Which program focuses on understanding the economic behavior of groups of families, individuals, or consumers?

a. Consumer economics
b. Home economics
c. Consumer science
d. Financial science

31. Which of these do experts identify as the most accurate definition of a conflict?

a. The most accurate definition of a conflict is simply a disagreement.
b. A conflict is a situation wherein somebody perceives some threat.
c. One central characteristic of a conflict is that it undermines growth.
d. A conflict is a temporary upset that will go away if people ignore it.

32. Research finds that a large proportion of adults working in demanding fields like politics and medicine most often had which birth order in their families?

a. Middle children
b. Second children
c. Youngest children
d. First-born children

33. Which of the following food additives is the safest to consume?

a. Sodium pectinate
b. Sodium bisulfite
c. Sodium nitrate
d. Sodium nitrite

34. An ovo-lacto-vegetarian is a person who eats

a. Fruits, vegetables, and grains
b. Fruits, vegetables, grains, and poultry
c. Fruits, vegetables, grains, and dairy products
d. Fruits, vegetables, grains, dairy products, and eggs

35. Regarding the essential human right to housing, which of the following UN events took place the earliest in history?

a. The United Nations appointed its first Special Rapporteur on Adequate Housing.
b. The United Nations held its first Conference on Human Settlements, or Habitat I.
c. The United Nations declared the International Year of Shelter for the Homeless.
d. The United Nations' Universal Declaration of Human Rights, including the right to housing.

36. Which of the following best represents good advice to job applicants for successful interviews?

a. It is good to back up statements about oneself with specific examples.
b. Direct eye contact with interviewers is intimidating and to be avoided.
c. Be prepared to answer interviewer questions but not to ask questions.
d. If one does not understand a question, do not let on and seem foolish.

37. When children become part of a remarried family with other children, which of these is true?

a. A child's rank in the family never changes.
b. The role each child plays remains distinct.
c. All these factors are most likely to change.
d. Incest taboos are the same for all siblings.

38. Business experts find the most important element of total quality management to be effective communication. Which of the following is true about business communication?

a. A company can maintain high quality without good communication.
b. Poor communication leads to interpersonal mistrust in a company.
c. Communication quality has no effect on productivity in a company.
d. Poor communication causes misunderstandings rather than anger.

39. The nutritional requirements of women:

a. Differ from the requirements of men.
b. Are accurately described by all these.
c. Vary throughout the menstrual cycle.
d. Differ when pregnant and/or nursing.

40. For which pair would sibling rivalry likely be greatest?

a. Sister and brother, ages 5 and 10, respectively
b. Sisters, ages 10 and 5
c. Brother and sister, ages 8 and 10, respectively
d. Brothers, ages 3 and 5

41. Vegetarians can get non-heme iron from plant sources, but to improve absorption of these, they should eat them together with foods rich in:

a. Calcium
b. Vitamin C
c. Vitamin D
d. Vitamin E

42. Freud's Latency stage corresponds most closely in terms of child development with

a. Kohlberg's Pre-Conventional stage in his theory of moral development.
b. Erikson's psychosocial development stage, Identity vs. Role Confusion.
c. Piaget's stage of cognitive development he termed Formal Operations.
d. Duvall's Families with School-Age Children stage of family development.

43. In management decision-making, managers may evaluate alternatives according to four parameters. Determining whether a company's performance goals can maintain a particular alternative or not reflects which of these parameters?

a. Whether an alternative is legal
b. Whether an alternative is ethical
c. Whether an alternative is practical
d. Whether an alternative is economical

44. How does effective marketing benefit society?

a. It utilizes technology, increases product support, and decreases consumer waste.
b. It eliminates unnecessary products, increases customer value, and improves sales.
c. It decreases competition, improves customer service, and increases product choices.
d. It increases competition, improves product quality, and lowers product prices.

51. What does the word *carded* mean when it appears on a clothing label?

a. The garment has been evaluated by a licensed inspector.
b. The garment is made of short and thick cotton fibers.
c. The garment was not created in a sweat shop.
d. The garment only contains one type of fiber.

52. What is the name for wool that has been spun into a fine yarn from parallel fibers?

a. Worsted
b. Cashmere
c. Angora
d. Polyester

53. Of the following fabrics, which one is NOT made using a plain weave?

a. Denim
b. Muslin
c. Percale
d. Burlap

54. Which of these is true about the %DV (Daily Value) the FDA requires on food labels?

a. They make it easier for consumers to know in numbers how much they need of nutrients in a day.
b. They make it easier for consumers to know how much of a day's allowance of nutrients a food has.
c. They make it harder for consumers to comparison shop for foods by the relative nutrient amounts.
d. They were developed by the FDA to replace RDAs because the RDAs were found to be inaccurate.

55. A family wherein the parents, children, and grandmother live together is defined as a(n):

a. Nuclear family
b. Extended family
c. Blended family
d. Combined family

56. As adults get older, they are *most* likely to have poorer absorption from foods of which vitamin?

a. Vitamin A
b. Vitamin C
c. Vitamin B_{12}
d. Vitamin E

57. Where can a consumer obtain a mortgage loan to buy a house?

a. Consumers can acquire loans from any and all of these sources today.
b. Mortgage bankers or mortgage brokers are the only sources for loans.
c. Savings and loans, credit unions, various banks, or insurance companies
d. Increasing numbers of realtors can arrange mortgage financing today.

58. Of the following, which is generally a good practice for writing a résumé?

a. Students with no work history should place education first on a résumé.
b. Education or work history should start with the earliest and go forward.
c. As employers may request personal references, put them in a résumé.
d. In work histories, it is best to include reasons for leaving previous jobs.

59. Which of the following fabric care symbols indicates a garment should be line-dried or hang-dried?

a. A square with diagonal lines in a corner
b. A square with one horizontal line inside
c. A square with three vertical lines inside
d. A square with a horizontal arc at the top

60. Lenders generally require mortgages to be guaranteed by outside third parties (e.g., the FHA, VA, or a private mortgage insurer [PMI]) if the home buyer makes a less than __ down payment.

a. 20%
b. 10%
c. 50%
d. 30%

61. Lightweight, usually unlined fabric panels hung directly on a rod that cover either the extreme ends of the window or the entire window are called

a. Curtain
b. Louvered shutter
c. Drapery
d. Grille

62. When a consumer purchases a new mattress from a manufacturer, they have the right to expect that mattress to meet safety standards and to have the features the manufacturer claims it has, even if there is no explicit promise made. What form of contract does this represent?

a. Express contract
b. Written contract
c. Implied in-law contract
d. Implied in-fact contract

63. According to Bowen Family Systems Theory, when both partners in a couple project their expanding anxieties into their relationship, this reflects which basic relationship pattern identified by Bowen?

a. Emotional distance
b. Impairment of child(ren)
c. Dysfunction in one spouse
d. The pattern of marital conflict

64. The FTC does NOT mandate that clothing labels include

a. The country of origin
b. Whether the garment contains mink or rabbit
c. An indication of whether wool is new or recycled
d. The Registered Identification Number or name of the manufacturer

65. Which of the following is NOT a typical developmental task for children at the earliest school ages?

a. Conservation of amount
b. Identifying with sex roles
c. Developing early morality
d. Learning to play in groups

66. Which of the following is NOT a way the FDA protects consumers?

a. Regulating drugs for humans
b. Recalling products that have been deemed unsafe
c. Regulating tobacco products
d. Approving dietary supplements

67. The law requires food labels to include which of the following information?

a. Product name, form, amount, name and address, and ingredients are required.
b. The product name is required, but not the address of the company.
c. The weight or fluid ounces must be listed, but not every ingredient.
d. All ingredients must be listed, but weight or fluid ounces is optional.

68. Which of the following represents a discretionary expense?

a. Textbook
b. Rent
c. DVD
d. Groceries

69. Banks require ____ when issuing a secured loan, giving them more security should the customer default on their payments.

a. Insurance
b. Cosigners
c. Tier 1 credit scores
d. Collateral

70. If a menu item is a green salad made with lettuce, tomatoes, cucumbers, green onions, and carrots, the most appealing name to list it on the menu as is:

a. GREEN SALAD (in all-capital letters).
b. tossed green salad.
c. green salad with mixed vegetables.
d. fresh garden salad.

71. Regarding financial institutions, what is the current status of careers for FCS graduates?

a. Brokerage firms want to hire qualified financial planners, who are among FCS graduates.
b. Banks and savings and loan associations are not presently accepting many job applicants.
c. FCS financial planners are needed at insurance companies but not at counseling agencies.
d. FCS financial planners need not understand community relationship and family dynamics.

72. Among the following developmental theories, what is correct relative to development across the life span?

a. Freud and Piaget had theories that cover the life span, while Erikson's ends at adolescence.
b. Piaget's and Erikson's theories do not go beyond adolescence while Freud's goes to death.
c. Freud's and Erikson's theories cover all of life while Piaget's theory stops with adolescence.
d. Erikson's theory has developmental stages across the life span; Freud's and Piaget's do not.

73. In which of Erikson's stages of development do individuals measure their success by what they contribute to their families and society?

a. Intimacy vs. Isolation
b. Ego Integrity vs. Despair
c. Identity vs. Role Confusion
d. Generativity vs. Stagnation

74. Which of the following represents good dining etiquette at a dinner party?

a. Only giving an RSVP if one will not attend
b. Asking what food the hosts will be serving
c. Asking what dress is suitable when invited
d. Beginning to eat as soon as one sits down

75. According to Urie Bronfenbrenner's ecological systems theory of development, in which system does an individual experience the most interpersonal and social interactions?

a. The exosystem
b. The mesosystem
c. The macrosystem
d. The microsystem

76. Which of the following best reflects a disadvantage of stage theories regarding families?

a. The stages express typical family concerns in each stage.
b. Stage theories provide convenient common vocabularies.
c. They focus on dynamics the same as systems theories do.
d. Many diverse families today are not normative or typical.

77. Which of the following is NOT true of supply and demand?

a. At higher prices, demand is lower and sellers will supply more goods.
b. Supply and demand is a theory explaining the interaction between the amount buyers are willing to pay for an item (demand) and the amount sellers will provide of the commodity at a given price (supply).
c. If supply is constant and demand decreases, the equilibrium price will be lower.
d. The relationship between supply and demand is shown on a graph with a positive slope.

78. In making good consumer decisions, which of the following should smart shoppers do first?

a. Narrow the field of potential choices
b. List attributes of products or services
c. Determine their needs as consumers
d. The order of doing these is irrelevant.

79. According to personal finance experts, what is the maximum percentage of income a family should spend on housing?

a. 5%
b. 50%
c. 10%
d. 25%

80. When describing moral development, Piaget characterized two types of adult-child relationships. In the ______ relationship, the adult ______ the child; in the ______ relation, the adult _____ the child.

a. Heteronomous, "co-operates" with; autonomous, coerces
b. Unilateral, respects; reciprocal, controls and/or constrains
c. Interactive, constrains; heteronomous, "co-operates" with
d. Heteronomous, coerces; autonomous, "co-operates" with

81. Among these food preparation factors, which can cause food poisoning or other food-borne illness?

a. All of these can cause it with certain foods.
b. Cooking foods for the wrong lengths of time
c. Cooking foods at the wrong temperatures
d. Not pasteurizing or refrigerating some foods

82. Which of the following websites is best for finding free information about energy-efficient home appliances?

a. www.sears.com/
b. http://products.construction.com/
c. www.consumerreports.org
d. www.energystar.gov

83. Of the following, which food preparation(s) is/are LEAST likely to cause food-borne illnesses?

a. Improperly canned or fermented foods
b. Uncooked meats/produce
c. Undercooked meats/produce
d. Overcooked meats and produce

84. According to Evelyn Duvall's description of the Family Life Cycle, which of these accurately represents two consecutive stages of the family life cycle?

a. Married couples without children; families with children from birth to 6 years
b. Families with children from birth to 6 years; families with children 6-13 years
c. Families with children 0–30 months; families with children 24 months-6 years
d. Families with teen children 13–20 years; "empty nest" couples to retirement

85. We are advised to analyze our financial goals and values periodically to aid in financial planning. What is true about this process?

a. The reason for analysis is to distinguish between wants and needs.
b. Analyzing our financial values excludes how we feel about money.
c. Financial values analysis omits why we feel as we do about money.
d. Once we analyze financial goals and values, results will not change.

86. Which of these reflects proper dining etiquette when attending dinner parties?

a. When passing dishes, they should be passed clockwise.
b. Bread only should be passed around the table clockwise.
c. Guests should pass all dishes including bread to the right.
d. It is up to the individual guests whether to cut up foods.

87. Which of the following is the most important component of preparation for inspection?

a. Review of compliance issues.
b. Network with similar organizations.
c. Prepare an alert system so the staff is aware of an upcoming inspection.
d. Conduct self-inspections.

88. The research of Walter Toman (also adopted by Murray Bowen) found that in the workplace, which relationship is likely to be most harmonious and effective?

a. A boss and an assistant who were both oldest children working together
b. A boss and assistant who were both youngest children working together
c. A boss who was an oldest child and an assistant who was a youngest child
d. A boss who was a youngest child and an assistant who was an oldest child

89. For many first-time homebuyer programs in the U.S., a 'first-time buyer' is someone who:

a. Anyone who has never owned real estate property before.
b. Anyone who has only owned property less than six months.
c. Anyone who has not owned a principal residence in the last three years.
d. Anyone who did own, but does not currently own, property.

90. What is the best way to sort clothing in the closet to find items and assemble outfits easily?

a. Sort first by colors, and then sort second by category.
b. Sort by placing favorite items to be most within reach.
c. Sort by placing item types wherever they will fit best.
d. Sort first by category, and then sort second by colors.

91. A person who is 41 to 100% heavier than his or her ideal weight is likely

a. Mildly obese
b. Hypertensive
c. Diabetic
d. Moderately obese

92. One factor that influences consumers to consider prices more in their health care decisions is:

a. When they have no prior ideas of provider quality.
b. When their health insurance plan is a PPO or HMO.
c. When they have a severe and/or urgent condition.
d. When they have health care practitioners they like.

93. Many researchers have found that family recreation affords benefits both to the family unit and to its individual members. Of the following, which reflects a benefit to individual family members?

a. Improvements in communication
b. The development of lifelong skills
c. Increased stability and interaction
d. The maintenance of relationships

94. Which of the following foods is a source of the most vitamin A?

a. Sweet potatoes
b. Animal livers
c. Carrots
d. Kale

95. Of the following types of fat, which is considered the *most* harmful?

a. Partially hydrogenated vegetable oil
b. Fully hydrogenated vegetable oil
c. Monounsaturated fat
d. Polyunsaturated fat

96. Erikson's theory of development focuses MOST on a person's relationship with which other aspect?

a. Self
b. Society
c. Other individuals
d. The parents exclusively

97. What is true about information on clothing labels, tags, or packages?

a. Manufacturers are required by law to provide only registered ID number identification.
b. Manufacturers are required by law to show ID, fiber content, country, and care instructions.
c. Manufacturers are required by law to indicate only the country where clothing is made.
d. Manufacturers are required by law to show fiber content; care instructions are optional.

Answer Key and Explanations

1. C: Piaget's later preoperational stage, when children's thinking becomes less egocentric but still retains animistic and magical characteristics, takes place during the preschool years (ages 3–6 years). This corresponds to Duvall's family development stage of families with preschoolers, when parents must adapt to the needs of young children and cope with young children's intrusions on their privacy and the energy drain of caring for and supervising preschoolers. Duvall described answer choice A as tasks of her families with school-age children stage, answer choice B as tasks of her families with teenagers stage, and answer choice D as tasks of her families with infants stage.

2. B: For critically evaluating information delivered through any of today's many media channels, consumers should consider who is delivering the message, what the message is, why they are delivering it, what medium they are using, how they are getting attention, and how the message can be interpreted. All of these are crucial elements, so no particular one can be ignored, and no particular one is more important than the others. Choices A, C, and D all state that one or more of these elements is on a different level of importance.

3. B: Erikson described the nuclear conflict of infancy as basic trust versus mistrust: A baby whose needs are met fully and consistently develops trust in the world, while a baby whose needs are inadequately or inconsistently met develops mistrust. Erikson found that the positive outcome of this stage was hope. He identified will (A) as the positive outcome of successfully resolving the nuclear conflict of autonomy versus shame and self-doubt during toddlerhood. He identified purpose (C) as the positive outcome of successfully resolving the nuclear conflict of initiative versus guilt during the preschool years. He identified competence (D) as the positive outcome of successfully resolving the nuclear conflict of industry versus Inferiority during the elementary school years.

4. C: A design innovation in home construction is including some rooms planned to be flexible in purposes and use. This allows homes to change along with families. For example, when a grown child moves out, that child's bedroom can easily be converted to a home gym, home theater, craft room, etc. As children grow older, a playroom can become a game room, homework/studying room, hobby room, etc. This eliminates having to move to another house (A), which takes more money, time, and effort; and many families love their homes, neighbors, and neighborhoods and do not want to move. Remodeling (B) also costs more, and ongoing work inconveniences home living. While some "empty nest" parents may want to downsize to an apartment (D), this again involves more effort and time, even if proceeds from a house sale cover moving expenses. And some parents want to keep the space in their house for grown children, their families, and others to visit.

5. A: Interpreter-sensitive values are characteristic of high-context communication and mean the listener is expected to take responsibility for interpreting the speaker's meaning, much of which is conveyed through nonverbal behaviors, which the listener must "read between the lines" to infer. Conversely, low-context communication uses sender-oriented values, meaning the speaker is expected to take responsibility for communicating clearly and directly to make decoding simple for the listener.

6. A: Experts find that the majority of women tend to wear a small proportion of all the clothes they buy, from less than half to 20% or less. Few women wear over half of the clothes they buy (C), and fewer still wear most of them (B). It is also less accurate to say that women vary that much in how many of the clothes they buy that they actually wear (D), because the most common pattern among women is to buy far more clothes than they end up wearing.

7. C: When people buy houses with mortgages, they usually make a down payment in some amount; however, this is not always required (A). Buyers usually pay closing costs as well, but also not always (B). There are some loan programs that allow the borrower to purchase a house with no money down; and in buyers' markets (i.e., when the housing market affords the buyer more leverage), realtors can help buyers negotiate deals wherein the owner or seller pays some or all of the closing costs. While some buyers want to eliminate a down payment, many others prefer to pay something down to lower their subsequent monthly mortgage payment: paying nothing down results in higher monthly payments (D).

8. A: Bandura proposed that children learn not only through the direct consequences of their own behaviors but also through observational learning, in which they watch others' behaviors and the consequences that follow and then imitate those behaviors. Therefore, children can learn vicariously by observing and imitating others (A). Choice B is incorrect because Bandura argued that learning can occur without direct experience. Choice C contradicts the theory by claiming that observation does not lead to learning. Choice D describes a behaviorist principle regarding reinforcement but does not represent the central concept of Bandura's social learning theory.

9. B: In type 1 diabetes, the pancreas does not produce insulin, the hormone necessary for converting sugars, starches, and other foods to energy. In type 2 diabetes, the pancreas does produce insulin, but the body does not respond properly to the hormone; this is called insulin insensitivity. Both type 1 and type 2 diabetes have both genetic and environmental factors, so (A) is incorrect. While type 1 diabetes was previously called "juvenile diabetes," it can develop in childhood or young adulthood; and though most cases of type 2 diabetes used to develop during adulthood, in recent years many children and adolescents have been diagnosed with type 2 diabetes, so (C) is incorrect. Type 1 accounts for only 5% of diabetes cases while the rest are type 2, so (D) is incorrect.

10. B: Cooperative communication prevents conflict by suggesting rather than accusing; using "I" rather than "You"; and focusing on solving problems rather than focusing on the other person. Example (B) has these qualities. The examples in (A), (C), and (D) are all person-centered by accusing the other person of interrupting, not listening, etc.; whereas the example in (B) expresses the speaker's desires both to finish, and also to hear what the other person has to say. It also positively communicates the speaker's need to complete a message instead of negatively accusing the other person of not listening to it.

11. C: Scoring two or more standard deviations (SDs) below the mean (average) on a standardized IQ test is commonly accepted as indicating that a child may have delayed or atypical intellectual development. Thus, one SD (B) is incorrect. On standardized IQ tests used today, one SD = 15 points. Hence, 10 points (A) would not be enough to indicate potentially delayed or atypical intellectual development. Since IQ scores are used to indicate possible delayed/atypical intellectual development, answer choice D is incorrect.

12. B: Unfinished knitted or woven fabrics are first cleaned to remove dirt and oils from the manufacturing process. Then many fabrics are singed with heat to eliminate protruding ends of fibers, which prevents fabric pilling and promotes even color when dyeing. After singeing, cotton and other natural fibers are bleached to get rid of impurities in their natural color and appearance. Fabrics that do not respond well to bleach are treated with optical brighteners instead. Manufacturers may then Mercerize certain linens, cottons, and rayons using alkali to make them stronger, softer, more lustrous, and more readily accepting of dyes. Fabrics are then dyed to add colors.

13. A: Many color analysts use the four seasons to identify the main categories of color palettes. In this system, "Autumn" represents warm, yellow-based earth tones; "Winter" represents cool, blue-based colors and black and white; "Spring" represents bright, vibrant, clear colors; and "Summer" represents soft, light pastel tints of colors. Some color analysts use adjectives descriptive of temperaments for the same groups. In this system, "Passionate" equals "Autumn"; "Dramatic" equals "Winter"; "Vibrant" equals "Spring"; and "Romantic" equals "Summer."

14. D: Everybody has a certain group or family of colors that are most flattering to them based on their skin tones. Experts advise getting rid of clothes with colors outside of this group as we do not look as well in them (A). They also advise keeping and buying clothes whose colors all coordinate, which makes it easiest to mix and match items into different outfits and gives the largest potential numbers of outfits we can create (C). Keeping items in less flattering colors just to provide more variety or a change of pace (B) is not the best advice, because we do not look as well wearing colors we may like to see but which don't flatter us. Others respond less favorably to us when we wear unflattering colors, and we are likely not to feel as good about how we look when we wear them.

15. A: Losing weight quickly decreases the metabolism. Nutritionists and doctors believe that this is a form of self-defense by the body, which senses that food is not as available and therefore tries to limit its use of calories. All of the other answer choices are events that increase metabolism. Weight gain always raises metabolism, in part because there is more muscle or fat to provide with nutrients. Muscle burns more calories than fat, so an increase in muscle mass will correlate with an increase in metabolism. Any exercise tends to increase metabolism.

16. C: Baumrind and others after her who have applied her theory in research have found that children who function at low levels in multiple life areas are likely to have had parents with the uninvolved parenting style. This style makes few demands of children, but also demonstrates little warmth toward them. The children thus lack both parental discipline and parental nurturing. Extremes of the uninvolved parenting style include rejecting and/or neglecting children. As a result, children perform poorly and are likely to be depressed and engage in delinquent behaviors. Children with permissive (A) parents are likely to be impulsive, seem immature, lack self-control, and avoid responsibility. Children with authoritarian (B) parents are likely to behave well, but also to lack initiative, avoid leadership roles, and are moody. Children with authoritative (D) parents are likely to be well-behaved but also independent; self-reliant, socially responsible, achievement-oriented, and to have high self-esteem.

17. B: If the CDM is monitoring foodservice and notices a delay in plating and, when randomly checking the temperature of pieces of the roast chicken that is to be served, the CDM finds that temperatures range from 108-120 °F, the most appropriate intervention is to reheat the meat to 165 °F. Food should be served at a temperature of at least 140 °F, but safe reheating of food requires that it be heated to 165 °F to ensure that all pathogens are destroyed.

18. D: Freud's Oral stage corresponds to infancy, as does Erikson's stage of Basic Trust vs. Mistrust, Piaget's Sensorimotor stage, and Duvall's Families with Infants stage. Freud's Oedipal stage (A) corresponds to preschool; Erikson's corresponding stage is Initiative vs. Guilt, rather than toddlerhood's Autonomy vs. Shame and Doubt; Piaget's Preoperational stage corresponds to preschool, but Duvall's corresponding stage is Families with Preschoolers, not School Age. Freud's Anal (B) stage corresponds to toddlerhood; Erikson's corresponding stage is Autonomy vs. Shame and Doubt, not Industry vs. Inferiority, which occurs during elementary/middle school ages.

Piaget's stages corresponding to toddlerhood are the end of the sensorimotor and beginning of the preoperational, not formal operations, which develops around adolescence. Toddlerhood falls between Duvall's Families with Infants and Families with Preschoolers stages. Freud's Genital (C) stage is in adolescence, as is Duvall's Families with Teenagers, but Erikson's corresponding stage is Identity vs. Role Confusion, not early childhood's Initiative vs. Guilt; Piaget's is Formal Operations, not the elementary/middle school years' Concrete Operations.

19. C: Obesity is correlated with diabetes (A); hypertension, i.e., high blood pressure (B); hypercholesterolemia, i.e., high cholesterol (D); heart disease; certain cancers; autoimmune diseases and other inflammatory conditions; and many other diseases and adverse health conditions.

20. C: The realtor should regularly update information s/he gives the consumer about the conditions of the current housing market (A), available financing options (B), and situation-specific negotiating tactics (D). Buying a house can be a lengthy process, and market conditions can change, sometimes in relatively short periods. So can financing options: lenders can offer new products; different providers may offer different financing plans, and the realtor can recommend some that may suit the consumer better than others; the buyer's credit score and financial resources can change, etc. Also, some strategies for negotiating can apply to some situations, but not others wherein different techniques are more appropriate.

21. C: Havighurst identified developmental tasks in each of the six stages of human development he defined. Learning to walk, talk, eat, eliminate, stabilize physically; sexual modesty and differences, simple physical and social reality concepts, right and wrong; developing a conscience, and learning to relate the self emotionally to parents and siblings are tasks he assigned to the period from birth to 6 years. He identified learning of basic reading, writing, and calculation skills; physical game skills, getting along with peers; developing everyday living concepts, morality, values, personal independence, a gender role, healthy self-attitudes, and attitudes toward institutions and social groups as tasks for those from 6 to 12 years (A). He included tasks in pre-teen and teen years (B) as maturing peer relations; emotional independence from adults; career preparation and choice; marriage and family life preparation; developing concepts and skills for civic competence; developing social responsibility and an ethical system, etc. Among tasks of early adulthood (D), Havighurst included marrying, starting families, raising children, managing households, beginning careers, etc.

22. B: Piaget's first stage of cognitive development is called the sensorimotor stage. This is when infants respond to input they receive from the environment through their sensory organs by engaging in motor actions. When they realize how the environment then reacts to some of these actions, they respond again; this is called circular reactions. Operations with concrete things (A) occur during Piaget's third stage called concrete operations. Operations with abstract things (C) occur during Piaget's fourth and final stage called formal operations. Intuition and animism but not logic (D) occur during Piaget's second stage, which he called preoperational.

23. B: One design principle embraced by some home design companies is that there should be at least three ways to get to a room in the house. They find this especially important for getting to the kitchen, which is the center of many family homes; therefore (A) is incorrect. Floor plans can be designed wherein people get to different rooms by passing through other rooms in an open floor plan, yet without disrupting how the rooms are used (C). These open floor plans make the rooms look bigger than if they were separated by halls (D) used as passageways. They also make it easier to see farther through the house.

24. D: All three of the statements are true. Bulimia can lead to tooth decay, due to a combination of malnutrition and corrosion by stomach acid. Anorexics often have a distorted self-image, believing themselves to be much heavier than they actually are. Finally, bulimia does not always include purging (induced vomiting or defecation). Bulimics may obsessively exercise or abstain from food in order to lose weight. Both anorexia and bulimia are extremely dangerous and should be discussed at great length in the nutrition component of a family and consumer science class. These conditions are especially common among middle-school and high-school students.

25. B: Flood insurance not only covers the loss of or damage to a home due to flooding; it also covers the loss or damage of personal property and other contents within the home. Title insurance (A) only protects home buyers up to the real estate's mortgage value ("lenders" policies), and protects homeowners up to the purchase price of the home including down payments ("owners" policies) if the title to the real property turns out to be invalid. It does not cover personal property. Home warranties (C) cover workmanship in a new home, usually for the first year; wiring, plumbing, and other mechanical issues for the first two years; and structural defects in the building for up to ten years. They do not cover personal property in the home. Mortgage insurance (D) covers the lender, not homeowner property.

26. B: One basic principle of effective classroom management is that it should teach the students to manage their own behaviors, rather than teachers' always having to manage them. Though classroom management techniques are necessary to address disruptive behaviors when they do happen, they should also focus on reducing nondisruptive behaviors that are nevertheless off-task (A). Students who are off-task are not engaged in learning. Students are not simply taken off-task by external distractions (C); it is more normal and natural for them to be off-task than on-task (D). Therefore, classroom management is necessary for students to learn to be engaged in learning activities and stay on-task.

27. B: Personal health literacy is the ability to find, understand, and use information to make health-related decisions. By reading the information associated with the medication and adhering to the manufacturer instructions, the consumer is demonstrating the ability to make safe health-related decisions. Organizational health literacy is how organizations enable individuals to make health-related decisions. This scenario does not require critical thinking skills, as there is no need for the consumer to make inferences, problem solve, or communicate in order to properly take the medication. There is also no need to evaluate claims in this scenario, as they are taking an over-the-counter, FDA-approved medication that has been approved for the usage on the label.

28. D: The provision of the Dodd–Frank Wall Street Reform and Consumer Protection Act of 2010 (Dodd–Frank Act) that consumers can sue credit rating agencies for negligence fulfills the consumer's right and responsibility concerning redress, i.e., to request compensation for an agency's wrongdoing. This law's provision that brokers have the same fiduciary duties as advisors if they give investment advice to consumers (A) fulfills the consumer's right to be given accurate product information, and the responsibility to analyze and apply that information judiciously; and the consumer's right to have access to education to inform purchasing decisions, and the responsibility to pursue consumer education about products as they change. The law's provision that advisors disclose information to the SEC on request (B) also fulfills the consumer's right and responsibility to information [see (A) above]. The law's provision of protection to whistleblowers (C) fulfills the consumer right and responsibility to be heard.

29. A: Microwave ovens are by far more energy-efficient than the others named, in both cooking efficiency and energy factor. (Cooking efficiency equals the fraction of the total energy the oven consumes that is used to cook food. Energy factor equals the ratio of energy used for cooking food to the total energy consumed.) Convection ovens (B) powered by electricity are more energy-efficient than convection ovens powered by gas; and both kinds of convection ovens are more energy-efficient than traditional electric or gas ovens; but a microwave oven is about 7–8 times more energy-efficient than electric or gas convection ovens, about 5 times more efficient than standard electric ovens, and about 9–10 times more efficient than regular gas ovens. (Note that efficiency refers to energy used, not cost in money, which depends on local rates charged for electricity and gas.)

30. A: Traditional branches of economics focus on governments and businesses. Consumer economics looks at the buying choices made in a particular economy. These economists observe consumer patterns and then identify the needs and problems of the observed groups.

31. B: Whether it is real or imagined, when a threat is perceived by one or more persons in a situation, this defines a conflict, which experts say is not simply a disagreement (A). An additional characteristic of conflicts is that they present opportunities for the people involved to experience growth (C) when they succeed in resolving them, which builds interpersonal trust. Conflicts will not go away if people ignore them (D) because the threats people perceive that trigger conflicts are to their survival and welfare. Hence, they will continue indefinitely or until the relationship ends if they are ignored.

32. D: First-born children tend to grow up being responsible and dependable because parents tend to have higher expectations of their first child and give them more responsibility. Middle children (A) tend to feel less pressure as their parents have more experience with parenting by the time they have them. Parents also tend to expect less from their second children (B) and focus less attention on them than first-borns. Parents often give more special attention to their youngest children (C), but also expect less of them.

33. A: Sodium pectinate, like pectin, is a safe carbohydrate added to fruits and vegetables to strengthen their cell walls and to jellies, jams, sauces, yogurts, and frostings as a gelling agent to thicken them. Sodium bisulfite (B) is safe for some people, though it destroys vitamin B_1; however, for individuals sensitive to it, particularly those with asthma, it can cause severe reactions and even death. It is used to prevent bacteria from growing in wines and keep certain foods from discoloring. Sodium nitrate (C) and sodium nitrite (D) are used in cured meats like hot dogs and bacon to preserve color and flavor and prevent bacterial growth, but are also associated with formation of carcinogenic (cancer-causing) nitrosamines. (Today, the addition of ascorbic acid or erythorbic acid mitigates nitrosamine formation, but nitrates and nitrites still present some risk. Also, these additives are mainly used in fatty, salty foods—another reason to avoid them.)

34. D: An ovo-lacto-vegetarian is a person who eats fruits, vegetables, grains, dairy products, and eggs. These people get most of their proteins from beans, eggs, milk, and cheese. There are several other kinds of vegetarians. A vegan is someone who eats only plant foods; vegans do not eat any meat or animal product, including honey. Lacto-vegetarians eat fruits, vegetables, grains, and dairy products.

35. D: The UN issued its Universal Declaration of Human Rights, which includes the right to adequate housing, in 1948. The UN appointed its first Special Rapporteur on Adequate Housing (A) in 2000. The UN held its first Conference on Human Settlements, which it called Habitat I (B), in 1976. (Its second Conference on Human Settlements, called Habitat II, was held in 1996.) The UN declared the International Year of Shelter for the Homeless (C) in 1987.

36. A: One good piece of advice for job applicants preparing for interviews is, when answering questions about themselves, they should support statements they make about their own positive attributes with specific examples to illustrate those qualities whenever they can. However, it is not good advice to avoid eye contact with interviewers (B). This would be good advice in Japan, where direct eye contact is found intimidating and is avoided; but in America, it is a sign that one is paying attention and is interested in the conversation. Avoiding eye contact in job interviews in America can be interpreted as a lack of confidence or as dishonesty. Applicants should be prepared not only to answer questions, but also to ask them (C). Applicants can ask interviewers what the company is looking for in an employee; interviewers' responses can provide applicants with opportunities to explain how they meet those needs. If applicants do not understand an interviewer question, they should not try to hide it (D), but rather should request clarification.

37. C: When adults remarry and their respective children are blended into one family, a child's family rank often changes (A); e.g., a child who was formerly the eldest may now be the second or third oldest child, and is treated as such by the rest of family. The roles played by each child also often become less clear (B) than they were in their previous family's dynamics. Additionally, because stepsiblings have no blood relation, incest taboos among them are also less clear-cut (D) than among biological siblings.

38. B: Research into business communication has demonstrated that poor communication in an organization eventually causes individual employees to mistrust each other. This research also shows that poor communication undermines quality (A); compromises productivity (C); and causes not only lack of understanding among members, but also development of anger (D) among individuals in an organization. Researchers conclude that the difference between profit and loss, or between success and failure, in a business is most often attributable to whether communication in the business is effective or not. These findings are the reasons that experts identify effective communication as the most important element of total quality management in business.

39. B: All of these accurately describe women's nutritional requirements. Women differ from men not only in numbers of calories, but also in the amounts they need of various nutrients. And even among women alone, their nutritional needs will not be the same at different times during their menstrual cycles (C), or when they are pregnant, or when they are breastfeeding (nursing) infants (D).

40. D: Of the given pairs, sibling rivalry would likely be greatest for brothers aged 3 and 5. In general, sibling rivalry is most pronounced in same-sex siblings within three years of age. In the first 8–10 years of life, siblings tend to alternate between cooperation and competition. As they grow older, they will often spend little time together for a few years, but during adolescence will gradually develop empathy for one another. Most research suggests that adult relationships between siblings simply exaggerate the tone of the relationship of youth; that is, good relationships get better and bad relationships get worse.

41. B: Vitamin C increases the body's absorption of non-heme iron, i.e., iron from plant sources such as beans, tomatoes, prunes, spinach, pumpkins, etc. Calcium (A), vitamin D (C), and vitamin E (D) are not known to increase the absorption of iron from vegetables, fruits, and legumes.

42. D: Freud's Latency stage of psychosexual development corresponds with the elementary and middle school years; therefore, Duvall's family development stage of Families with School-Age Children corresponds most closely to it. Kohlberg's Pre-Conventional stage of moral development (A) corresponds to the preschool years; Freud's corresponding stage is the Oedipal. Erikson's psychosocial stage of Identity vs. Role Confusion (B) and Piaget's Formal Operations stage of

rebellious children; parents with high hostility and low control reject and/or neglect children, producing dysfunctional individuals.

50. D: In infancy (A), Maslow's first level of Physiological needs takes priority. In adulthood (B), people may give priority to any or several of his needs levels depending on their circumstances. However, assuming they have met all the other levels, adults are most likely to be concerned with the highest level of self-actualization needs. In adolescence (C), the third and fourth levels of need, for love and belonging and for esteem, are likely to have priority. Whereas infants are most concerned with meeting their needs to sleep, be fed, changed, etc., in toddlerhood the emphasis typically progresses to meeting their needs for safety and security.

51. B: When it appears on a clothing label, the word *carded* means that the garment was made from short, thick cotton fibers. The use of carded fibers creates a fabric that is soft and strong. When the garment is made of long, straight cotton fibers, it is said to be combed. A combed fabric is smoother and shinier. A garment that only contains one type of fiber is considered pure. Garments that have been subjected to a permanent or durable press are more resistant to wrinkles. These processes are usually noted on the clothing label. A label may also indicate whether a garment has been inspected and that it was not created in a sweat shop.

52. A: Wool that has been combed into parallel fibers and then spun into a fine yarn is known as worsted. Renowned for its strength, worsted is used in dresses and suits. Cashmere is an especially soft form of wool that is created from the hair of a specific Indian goat. Angora is another high-end wool made from the long hairs of the angora rabbit. Polyester is a synthetic textile, meaning that it is not made from plants or animals. It is often blended with other fibers in the creation of clothing. Spandex is a synthetic textile with amazing flexibility.

53. A: Denim, as we commonly see in blue jeans, is made using a twill weave rather than a plain weave. Plain-weave fabrics are made by weaving one yarn over the other and then under it. Twill-weave fabrics are made by weaving the weft yarn under and then over two or more warp yarns at regular intervals. This creates a diagonal "twill line" or wale. Muslin (B), percale (C), and burlap (D) are woven using the plain weave, which is the least expensive and simplest way to weave fabrics.

54. B: %Daily Values (DVs) make it easier for consumers to know how much of what they need daily of certain nutrients a food supplies. For example, if a food contains 33%DV of sodium, the consumer knows that food supplies 1/3 of the sodium s/he should consume in a day. This means consumers need *not* know the actual numerical amounts of nutrients they need daily, or how much of that number a food has (A); e.g., they need not know that 2300 mg of sodium is how much to consume daily, or that a food with 800 mg of sodium has 1/3 the DV. %DVs also make it *easier* for consumers to comparison shop, not harder (C); e.g., they can pick a food lower in sodium instead of one with too much. The FDA developed DVs *not* because the RDAs were inaccurate (D), but because certain nutrients the FDA wanted to require on labels did not have RDAs established for them.

55. B: This is defined as an extended family because a grandparent lives with the parents and children. A nuclear family (A) is defined as only the parents and children living together. A blended family (C) is formed when two parents, each with children from a previous marriage or relationship, marry each other and both sets of children live together with both parents. A combined family (D) is not a commonly used term, but may be used as a synonym for a blended family.

56. C: Aging adults are most likely to have poorer absorption of vitamin B_{12} from foods because their stomachs produce less acid with aging, and stomach acid extracts this vitamin from the proteins containing it. Vitamins A (A), C (B), and E (D) are important to older adults for their antioxidant properties (antioxidants can prevent heart disease and some cancers, and may mitigate the process of aging), but aging is not known to decrease absorption of these from foods as it does with vitamin B_{12}.

57. A: Today, consumers can apply for mortgage loans from diverse sources including mortgage bankers and mortgage brokers, but these are not the only sources (B); they can also apply at savings and loan associations, credit unions, commercial banks, mutual savings banks, and insurance companies (C); and today, more and more realtors are also able to arrange mortgage financing for home buyers (D).

58. A: In résumés, those with work experience should place their job histories first; students with no work history should place their educational histories first. When writing one's educational or work history in a résumé, one should list the most recent item first and work backward, not vice versa (B). One should NOT include personal references in the résumé itself (C). For employers who request personal references, the applicant should submit these separately. Résumés should NOT include reasons for leaving previous jobs (D). If these are relevant, they can be addressed during job interviews.

59. D: The fabric care symbol of a square with an arc resembling a slightly slack clothesline hanging from between the top corners indicates the garment should be line-dried or hang-dried. The symbol of a square with diagonal lines in the upper left corner (A) indicates the garment should be dried in the shade (i.e., not in direct sunlight if this would fade the dye). A square with one horizontal line inside (B) indicates the garment should be dried flat. A square with three vertical lines inside (C) indicates the garment should be drip-dried.

60. A: In general, mortgage lenders require an outside third-party guarantee when the home buyer pays less than 20% of a house's list price as a down payment. Thus, they would definitely require such a guarantee if the down payment were less than 10% (B), but between 10% and 20% would still not be enough to waive this requirement. They do not require as much as 50% (C) down, which would more than double the millions of guaranteed mortgages generated annually. Lenders do not usually require 30% (D) down to waive the guarantee either.

61. A: Curtains are lightweight, usually unlined fabric panels hung directly on a rod. They may cover the entire window or be positioned at the sides of the window for decoration, privacy, or light control. Draperies (c) are typically heavier and often lined, while louvered shutters (b) are rigid window coverings with adjustable slates. A grille (d) is a decorative or structural framework within a window rather than a fabric window treatment.

62. D: An implied in-fact contract exists when two parties enter into an agreement, such as the purchase of a mattress. The manufacturer is expected to produce a mattress that meets all advertised expectations, and the customer is expected to pay the price the manufacturer charges and to use the product as it is intended to be used.

63. D: As described by Dr. Murray Bowen in his Family Systems Theory, the pattern of marital conflict involves both spouses' projecting their escalating anxieties into their relationship. The pattern of emotional distance (A) occurs in relation to the other three patterns, as family members seek to escape the excessive intensity of their interactions by withdrawing emotionally. The pattern of impairment of one or more children (B) involves the parents' projecting their anxiety by focusing excessively on the child(ren). Dysfunction in one spouse (C) results when one partner becomes overly dominant and hyperfunctional, while the other becomes overly dependent and dysfunctional in response. This pattern can emerge out of the pattern of marital conflict.

64. B: The FTC does not require clothing labels to indicate whether a garment contains mink or rabbit. FTC textile labeling rules require disclosure of fiber content using approved generic names, generally listing fibers that make up a significant portion of the product. In addition, imported textile and wool products must disclose country of origin, and labels must include the manufacturer's name or Registered Identification Number (RN). Wool products must also specify whether the wool is new or recycled. Fur content is regulated under separate fur-labeling rules, but mink or rabbit content is not a standard required disclosure on general clothing labels.

65. A: Conservation of amount, i.e., understanding that there is the same amount of a substance even if its shape, appearance, or arrangement is changed, is a typical developmental task for children at the middle school ages, e.g., around six to twelve years old. This is a part of what Piaget called Concrete Operations, which children at the earliest school ages (e.g., around four to six years old) typically cannot yet understand or perform. Typical developmental tasks for these younger school-age children include identifying with female or male sex roles (B), developing early concepts of basic morality (C), and learning to play in groups (D) with other children.

66. D: The FDA does not approve dietary supplements or their claims. However, the FDA does regulate supplements and will pull ones they've found to be dangerous off the market. The FDA also regulates drugs for humans and animals; recalls unsafe medical products, foods, and drugs; and regulates the growth, manufacturing, and consumption of tobacco.

67. A: By law, food labels must include the name of the product (e.g. "chunk light tuna"), its form (e.g., "in spring water" or "in soybean oil"); its amount (i.e., weight in ounces, fluid ounces, pounds, grams, milligrams, etc.); the name and address of the company that packed, processed, or distributed the food; and every ingredient included in the food.

68. C: Purchasing a DVD is a discretionary expense because it is based on personal desire rather than need. In other words, it is an expense made at the discretion of the consumer. Discretionary expenses are those over which a consumer has the most control. A comprehensive budget must include discretionary expenses as well as fixed and variable expenses. Fixed expenses, such as rent, are the same every month. Variable expenses, including groceries, school supplies, and heating oil, are always present but vary in amount over the course of a year.

69. D: Collateral is an asset owned by the borrower. With a secured loan, the borrower uses this asset to guarantee the lender that they will repay the loan. If they default on the loan, the bank will sell the asset to recover funds that were not repaid. For example, with a mortgage loan, the property is the collateral.

70. D: If a menu item is a green salad made with lettuce, tomatoes, cucumbers, green onions, and carrots, the most appealing name to list it on the menu as is "fresh garden salad." Menu items should not be listed in all caps, and "tossed green salad" may be interpreted as only being made with greens. "Green salad with mixed vegetables" is not the usual way that salads are described, so consumers may be unclear about what to expect. "Fresh garden salad" suggests a salad made with different fresh ingredients from the garden, an appealing option.

71. A: Within the wide range of specialties available in FCS education, financial planning is one that is currently needed by many financial institutions. This includes brokerage firms; banks and savings and loan associations, which *are* also looking for financial planners (B); and insurance companies and also counseling agencies (C). Qualified financial planners *do* need to understand both the relationship with community and also family dynamics (D) to serve these financial institutions best.

72. D: Erikson's theory of psychosocial development is the only one of the three named whose stages include young adulthood, middle adulthood, and old age until death. Freud's and Piaget's theories both have stages that end with adolescence. Freud's theory of psychosexual development ended with adolescence because he believed the personality was essentially formed by then. Piaget's theory of cognitive development ended at adolescence because he believed his highest stage of Formal Operations was attained by then. (Later researchers have found some adults never attain it.) Of the three, only Erikson included adult developmental stages in his theory.

73. D: Erikson's stage of Generativity vs. Stagnation occurs in middle adulthood, when people focus on creating legacies to leave for future generations, such as children, businesses, homes, inventions, and other contributions to family and society. Those who stagnate rather than generate become self-absorbed. Intimacy vs. Isolation (A) occurs in young adulthood, when people either succeed at forming intimate relationships with others or become isolated from them. Ego Integrity vs. Despair (B) occurs during old age, when people review their lives and feel either satisfaction or regret. Identity vs. Role Confusion (C) occurs during adolescence, when teens either succeed at establishing a personal identity or become confused about what their role is.

74. C: It is considered acceptable etiquette when invited to a dinner party to ask what kind of dress is expected; however, it is *not* acceptable to ask what will be served (B). If a guest has dietary restrictions, it is politer to inform the hosts of these rather than to ask what foods they are serving. Etiquette dictates that it is courteous to RSVP to dinner party invitations both if attending and if not attending (A). Another rule of dining etiquette is that guests only begin to eat when the host/ess raises his/her fork (or spoon if the first course is soup), not as soon as they sit down to the table (D).

75. D: What Bronfenbrenner termed the microsystem is a person's immediate environment, including family, friends, communities, religious groups, and others with which the person has direct contact and regular interactions. The exosystem (A) is what Bronfenbrenner named the wider social system of events and experiences that directly affect a person's microsystems but which the person does not construct; e.g., getting or losing a job. The mesosystem (B) is what Bronfenbrenner called the system of connections between elements of the microsystem; e.g., between a child's parents and teachers, friends and relatives, etc. The macrosystem (C) is what Bronfenbrenner labeled the cultural beliefs, values, laws, rules, and customs that influence the person. (Bronfenbrenner also identified the chronosystem, a system in the dimension of time that includes processes and events like biological maturation, parental death, etc.)

76. D: American society today encompasses so many and diverse versions of families that stage theories of family development often do not describe them adequately. Families may progress through stages out of their usual sequence, or revisit some stages when parents remarry after being widowed or divorced. This reflects a disadvantage of stage theories, which describe more normative or typical family stages. That the stages in these theories express the typical foremost concerns of families in each stage (A) is an advantage of stage theories (for example, referring to an "infant and toddler" stage evokes health, safety, and protection concerns; labeling periods of childhood as "teething", "toileting," or "tantrum" stages instantly evokes the family's main focus). Providing convenient common vocabularies (B) for describing family development is another advantage of stage theories. Systems theories focus more on less linear family dynamics and family responses to change, whereas stage theories focus more on sequential phases of family development (C).

77. D: Supply and demand theory is a microeconomic theory of the relationship between the amount of a good that is available (supply) to the need or want for that good (demand). A good example of this is in natural disasters, such as a hurricane, where the demand for a product is exaggerated as consumers try to purchase extra supplies. As a result, the supply is heavily reduced temporarily, which may drive prices up until the demand is fully met, at which point the supply, demand, and price will equalize. Equilibrium price is where the supply of goods matches its demand. If price is above equilibrium, the quantity supplied is greater than the quantity demanded, which results in a surplus of the good.

78. C: The first thing consumers should do before making a decision to purchase goods or services is to identify their needs, e.g., how they will use the product or service; how often; where to store a product; the location of a service. The second thing they should do is to list the attributes they want the product or service to have (B); e.g., a doctor has long enough office visits, short enough patient waiting times, and a good bedside manner; an insurance company settles claims promptly, subrogates effectively, and charges low premiums; or a power tool is cordless, lightweight, and powerful. The third thing is to narrow the field of possible choices (A); e.g., which are available for sale, which have prices the consumer can afford; product support and customer service; and how each choice meets the consumer's needs and desired product/service attributes. Thus the order of these steps is not irrelevant (D).

79. D: It is generally agreed that a family should spend no more than 25% of its income on housing. Although some lenders are willing to give money to homebuyers who will spend up to 40% of their income on housing, this is considered a risky loan. Such a loan is especially risky when the borrower has other long-term debt besides housing costs. As a consumer, it is wise to create a detailed budget before committing any funds to housing costs.

80. D: Piaget named a unilateral (B) or one-sided morality and adult-child relationship, based on obedience to authority, heteronomous or controlled by others. He named an independently constructed morality, and an adult-child relationship of reciprocal respect, as autonomous. The adult coerces the child in the heteronomous relation and "co-operates" (i.e., collaborates in performing operations) with the child in the autonomous one, rather than vice versa (A). The autonomous relation involves interaction and cooperation whereas the heteronomous involves adult constraint of the child rather than the opposite (C).

81. A: Food poisoning can be caused by all of these preparation factors with certain foods. For example, cooking for the wrong lengths of time (B) and/or at the wrong temperatures (C) can cause perfringens food poisoning due to the microorganism *Clostridium perfringens*. Unpasteurized milk can cause illness from the bacteria *Campylobacter jejuni; Salmonella; E. coli* 0157:H7 infection; and/or *Listeria monocytogenes;* and unrefrigerated (or improperly refrigerated) meats, cream pastries, and egg or potato salads with mayonnaise can cause food poisoning from the bacterium *Staphylococcus aureus.*

82. D: The government's Energy Star website is best because it lists all home appliances and equipment with Energy Star labels for meeting EPA-set energy-efficiency specifications, and includes annual electricity and water use, free of charge. The Sears website (A) is good for consumers who want to buy specifically from Sears, because they can type "Energy Star appliances" into the site's search bar and it will display all Energy Star products sold by Sears. The Energy Star.gov site is better by not restricting customers to one store for buying energy-efficient home equipment and appliances. McGraw-Hill's Sweets Network site at http://products.construction.com/ (B) is an excellent site, but only for locating construction materials and building products and equipment; it does not sell appliances. The Consumer Reports website (C) rates and recommends appliances among many other products, including whether they meet Energy Star requirements; however, while consumers can read its buying guides online, they must pay for a subscription to view the actual ratings, including Energy Star information.

83. D: Cooking food for too long can burn it or dry it out, causing poor taste and texture and removing nutrients. Grilling or charring foods at high temperatures can produce carcinogens (cancer-promoting agents), as can burning them; however, cancers are not food-borne illnesses, which are caused by viral, bacterial, or parasitic microorganisms present in foods. Improperly canned and fermented foods can cause botulism, which can be deadly. Uncooked or raw meats and produce can cause many food-borne illnesses, as can undercooked foods.

84. C: Duvall's Stages of the Family Life Cycle are: (1) childless married couples; (2) families with children 0–30 months; (3) families with children 24 months to 6 years; (4) families with children 6–13 years; (5) families with children 13–20 years; (6) families launching children, from the first child gone to the last child leaving home; (7) middle-age parents, from "empty nest" to retirement; and (8) aging family members, from retirement to both spouses' deaths. The second stage in (A) incorrectly ends with 6 years instead of 30 months. The first stage in (B) incorrectly represents Stage 2 as ending at 6 years instead of 30 months, and the second stage named in (B) represents Stage 4, skipping Stage 3; thus these two are not consecutive. The two stages in (D) are also not consecutive: they name Stages 5, families with teenagers; and Stage 7, middle-age couples from "empty nest" to retirement, skipping Stage 6, families launching children.

85. A: The reason we are advised to analyze our values regarding money and our financial goals is to help us distinguish between what we want and what we need. We may not have enough income for everything we want, but giving our true needs priority before spending on wants can help us stay within our budgets. Analyzing our financial values does include defining how we feel about money (B) and also why we feel as we do about it (C). The reason we are advised to conduct such analyses periodically is because once we have analyzed our financial goals and values, the results we get can still change over time (D): not only can the things we need and want change, but also the ways we feel about money and the reasons for those feelings can change with changes in life circumstances, personal maturity, etc.

86. B: Dining etiquette dictates that hosts and guests should pass dishes to the right, not clockwise (A) with the exception of bread, which should be passed clockwise; hence (C) is incorrect. The rule for cutting up one's food is to cut up anything larger than one's thumb at the second joint rather than being up to the individual guest (D).

87. D: The most important component of preparation for inspection is to conduct self-inspections using the checklists that the inspectors will use so that any noncompliance can be identified before the inspection and can be corrected. Self-inspections should be carried out periodically even if an inspection is not anticipated. All regulations should be reviewed before the inspection. Networking with similar organizations can help identify resources and strategies. The training and education of staff should be ongoing.

88. C: Toman found in his research that an oldest child, who grew up being comfortable with a leadership role, will have a complementary relationship in the workplace with an assistant who grew up as a youngest child and is more comfortable with following the leader. (Youngest children may also like being in charge, but usually have different leadership styles than oldest children.) A boss and assistant who both grew up with the same sibling positions {(A), (B)} are more likely to clash by being too similar in motivations and styles. A boss who was a youngest child and an assistant who was an oldest child (D) are less likely to have as effective a relationship, as their respective roles are more likely mismatched to their interactional personality characteristics. (Murray Bowen also incorporated Toman's findings into his Family Systems Theory, making it better known.)

89. C: One might assume that the term "first-time buyer" refers to anybody who has never owned any real property before (A), but this is not true in the realty market. In most US states, this term actually refers to anyone who has not owned property within the past three years. It does not refer to anyone who has owned real estate property for less than six months (B), or to anyone who once owned but does not currently own any real estate property (D).

90. D: The most efficient way to sort clothes in the closet for finding them easily is first to sort them by category; e.g., pants all together, shirts all together; for women, skirts all together and dresses all together, etc. After sorting clothing categories, within each category they should then be sorted by color. It is not as easy to find individual items and assemble outfits by sorting into colors first and categories second (A). Putting favorite items most within reach (B) sounds like a good idea if we wear these the most, but is not as efficient as (D). Simply putting things wherever they can most easily be fitted in the closet (C) may be quicker and easier when putting clothes away, but will take longer and be harder to find things when getting dressed.

91. D: A person who is 41 to 100% heavier than his or her ideal weight is moderately obese. Health professionals have divided obesity into three degrees: mild, moderate, and severe. A person who is less than 20% heavier than his or her ideal weight is considered merely overweight, but a person who is 20 to 40% heavier is considered mildly obese. People who are more than 100% heavier than their ideal weight are severely obese. Moderately obese people are much more likely to have diabetes or hypertension, but do not necessarily suffer from these conditions.

92. A: According to research, consumers who have not formed any preconceived notions about the quality of various providers through what they have heard from others, through advertising they have seen/heard, or through their own prior experiences are more interested in price as a factor in their health care decisions. However, when consumers have traditional health insurance plans, like PPOs or HMOs (B), they are less interested in price because these plans control price. When consumers have severe or urgent medical conditions (C), they are less interested in price because obtaining immediate care takes priority. And when they like their current health care practitioners (D), they are less interested in price because they want to see the same provider(s) regardless.

93. B: One benefit of family recreation to individual family members is that it can teach members of the family skills that they will use for the rest of their lives. Benefits of family recreation to the family unit as a whole include improved communication within the family (A), increased stability of the family and of interactions within it (C), and a way to maintain family relationships (D), among others.

94. B: Liver from any animal has the highest amount of vitamin A, with turkey liver being the highest, providing over 1500% of the Daily Value (DV) per 100-gram (3½-ounce) serving. Sweet potatoes (A) provide 384% of the DV of vitamin A per 100-gram serving. Carrots contain beta-carotene, which is converted by the body to vitamin A, providing 334% of the DV for a 100-gram (raw) serving. Kale (D) provides 308% of the DV of vitamin A per 100-gram serving via its carotenes.

95. A: Partially hydrogenated vegetable oil is produced through a chemical reaction of liquid vegetable oil with hydrogen to make it semi-solid, as in shortening. This process creates trans fats, which contribute to heart disease. The US FDA has found trans fats even more dangerous than

How to Overcome Test Anxiety

Just the thought of taking a test is enough to make most people a little nervous. A test is an important event that can have a long-term impact on your future, so it's important to take it seriously and it's natural to feel anxious about performing well. But just because anxiety is normal, that doesn't mean that it's helpful in test taking, or that you should simply accept it as part of your life. Anxiety can have a variety of effects. These effects can be mild, like making you feel slightly nervous, or severe, like blocking your ability to focus or remember even a simple detail.

If you experience test anxiety—whether severe or mild—it's important to know how to beat it. To discover this, first you need to understand what causes test anxiety.

Causes of Test Anxiety

While we often think of anxiety as an uncontrollable emotional state, it can actually be caused by simple, practical things. One of the most common causes of test anxiety is that a person does not feel adequately prepared for their test. This feeling can be the result of many different issues such as poor study habits or lack of organization, but the most common culprit is time management. Starting to study too late, failing to organize your study time to cover all of the material, or being distracted while you study will mean that you're not well prepared for the test. This may lead to cramming the night before, which will cause you to be physically and mentally exhausted for the test. Poor time management also contributes to feelings of stress, fear, and hopelessness as you realize you are not well prepared but don't know what to do about it.

Other times, test anxiety is not related to your preparation for the test but comes from unresolved fear. This may be a past failure on a test, or poor performance on tests in general. It may come from comparing yourself to others who seem to be performing better or from the stress of living up to expectations. Anxiety may be driven by fears of the future—how failure on this test would affect your educational and career goals. These fears are often completely irrational, but they can still negatively impact your test performance.

Elements of Test Anxiety

As mentioned earlier, test anxiety is considered to be an emotional state, but it has physical and mental components as well. Sometimes you may not even realize that you are suffering from test anxiety until you notice the physical symptoms. These can include trembling hands, rapid heartbeat, sweating, nausea, and tense muscles. Extreme anxiety may lead to fainting or vomiting. Obviously, any of these symptoms can have a negative impact on testing. It is important to recognize them as soon as they begin to occur so that you can address the problem before it damages your performance.

The mental components of test anxiety include trouble focusing and inability to remember learned information. During a test, your mind is on high alert, which can help you recall information and stay focused for an extended period of time. However, anxiety interferes with your mind's natural processes, causing you to blank out, even on the questions you know well. The strain of testing during anxiety makes it difficult to stay focused, especially on a test that may take several hours. Extreme anxiety can take a huge mental toll, making it difficult not only to recall test information but even to understand the test questions or pull your thoughts together.

Effects of Test Anxiety

Test anxiety is like a disease—if left untreated, it will get progressively worse. Anxiety leads to poor performance, and this reinforces the feelings of fear and failure, which in turn lead to poor performances on subsequent tests. It can grow from a mild nervousness to a crippling condition. If allowed to progress, test anxiety can have a big impact on your schooling, and consequently on your future.

Test anxiety can spread to other parts of your life. Anxiety on tests can become anxiety in any stressful situation, and blanking on a test can turn into panicking in a job situation. But fortunately, you don't have to let anxiety rule your testing and determine your grades. There are a number of relatively simple steps you can take to move past anxiety and function normally on a test and in the rest of life.

Physical Steps for Beating Test Anxiety

While test anxiety is a serious problem, the good news is that it can be overcome. It doesn't have to control your ability to think and remember information. While it may take time, you can begin taking steps today to beat anxiety.

Just as your first hint that you may be struggling with anxiety comes from the physical symptoms, the first step to treating it is also physical. Rest is crucial for having a clear, strong mind. If you are tired, it is much easier to give in to anxiety. But if you establish good sleep habits, your body and mind will be ready to perform optimally, without the strain of exhaustion. Additionally, sleeping well helps you to retain information better, so you're more likely to recall the answers when you see the test questions.

Getting good sleep means more than going to bed on time. It's important to allow your brain time to relax. Take study breaks from time to time so it doesn't get overworked, and don't study right before bed. Take time to rest your mind before trying to rest your body, or you may find it difficult to fall asleep.

Along with sleep, other aspects of physical health are important in preparing for a test. Good nutrition is vital for good brain function. Sugary foods and drinks may give a burst of energy but this burst is followed by a crash, both physically and emotionally. Instead, fuel your body with protein and vitamin-rich foods.

Also, drink plenty of water. Dehydration can lead to headaches and exhaustion, especially if your brain is already under stress from the rigors of the test. Particularly if your test is a long one, drink water during the breaks. And if possible, take an energy-boosting snack to eat between sections.

Along with sleep and diet, a third important part of physical health is exercise. Maintaining a steady workout schedule is helpful, but even taking 5-minute study breaks to walk can help get your blood pumping faster and clear your head. Exercise also releases endorphins, which contribute to a positive feeling and can help combat test anxiety.

When you nurture your physical health, you are also contributing to your mental health. If your body is healthy, your mind is much more likely to be healthy as well. So take time to rest, nourish your body with healthy food and water, and get moving as much as possible. Taking these physical steps will make you stronger and more able to take the mental steps necessary to overcome test anxiety.

Mental Steps for Beating Test Anxiety

Working on the mental side of test anxiety can be more challenging, but as with the physical side, there are clear steps you can take to overcome it. As mentioned earlier, test anxiety often stems from lack of preparation, so the obvious solution is to prepare for the test. Effective studying may be the most important weapon you have for beating test anxiety, but you can and should employ several other mental tools to combat fear.

First, boost your confidence by reminding yourself of past success—tests or projects that you aced. If you're putting as much effort into preparing for this test as you did for those, there's no reason you should expect to fail here. Work hard to prepare; then trust your preparation.

Second, surround yourself with encouraging people. It can be helpful to find a study group, but be sure that the people you're around will encourage a positive attitude. If you spend time with others who are anxious or cynical, this will only contribute to your own anxiety. Look for others who are motivated to study hard from a desire to succeed, not from a fear of failure.

Third, reward yourself. A test is physically and mentally tiring, even without anxiety, and it can be helpful to have something to look forward to. Plan an activity following the test, regardless of the outcome, such as going to a movie or getting ice cream.

When you are taking the test, if you find yourself beginning to feel anxious, remind yourself that you know the material. Visualize successfully completing the test. Then take a few deep, relaxing breaths and return to it. Work through the questions carefully but with confidence, knowing that you are capable of succeeding.

Developing a healthy mental approach to test taking will also aid in other areas of life. Test anxiety affects more than just the actual test—it can be damaging to your mental health and even contribute to depression. It's important to beat test anxiety before it becomes a problem for more than testing.

Study Strategy

Being prepared for the test is necessary to combat anxiety, but what does being prepared look like? You may study for hours on end and still not feel prepared. What you need is a strategy for test prep. The next few pages outline our recommended steps to help you plan out and conquer the challenge of preparation.

Step 1: Scope Out the Test

Learn everything you can about the format (multiple choice, essay, etc.) and what will be on the test. Gather any study materials, course outlines, or sample exams that may be available. Not only will this help you to prepare, but knowing what to expect can help to alleviate test anxiety.

Step 2: Map Out the Material

Look through the textbook or study guide and make note of how many chapters or sections it has. Then divide these over the time you have. For example, if a book has 15 chapters and you have five days to study, you need to cover three chapters each day. Even better, if you have the time, leave an extra day at the end for overall review after you have gone through the material in depth.

If time is limited, you may need to prioritize the material. Look through it and make note of which sections you think you already have a good grasp on, and which need review. While you are studying, skim quickly through the familiar sections and take more time on the challenging parts.

Write out your plan so you don't get lost as you go. Having a written plan also helps you feel more in control of the study, so anxiety is less likely to arise from feeling overwhelmed at the amount to cover.

Step 3: Gather Your Tools

Decide what study method works best for you. Do you prefer to highlight in the book as you study and then go back over the highlighted portions? Or do you type out notes of the important information? Or is it helpful to make flashcards that you can carry with you? Assemble the pens, index cards, highlighters, post-it notes, and any other materials you may need so you won't be distracted by getting up to find things while you study.

If you're having a hard time retaining the information or organizing your notes, experiment with different methods. For example, try color-coding by subject with colored pens, highlighters, or post-it notes. If you learn better by hearing, try recording yourself reading your notes so you can listen while in the car, working out, or simply sitting at your desk. Ask a friend to quiz you from your flashcards, or try teaching someone the material to solidify it in your mind.

Step 4: Create Your Environment

It's important to avoid distractions while you study. This includes both the obvious distractions like visitors and the subtle distractions like an uncomfortable chair (or a too-comfortable couch that makes you want to fall asleep). Set up the best study environment possible: good lighting and a comfortable work area. If background music helps you focus, you may want to turn it on, but otherwise keep the room quiet. If you are using a computer to take notes, be sure you don't have any other windows open, especially applications like social media, games, or anything else that could distract you. Silence your phone and turn off notifications. Be sure to keep water close by so you stay hydrated while you study (but avoid unhealthy drinks and snacks).

Also, take into account the best time of day to study. Are you freshest first thing in the morning? Try to set aside some time then to work through the material. Is your mind clearer in the afternoon or evening? Schedule your study session then. Another method is to study at the same time of day that you will take the test, so that your brain gets used to working on the material at that time and will be ready to focus at test time.

Step 5: Study!

Once you have done all the study preparation, it's time to settle into the actual studying. Sit down, take a few moments to settle your mind so you can focus, and begin to follow your study plan. Don't give in to distractions or let yourself procrastinate. This is your time to prepare so you'll be ready to fearlessly approach the test. Make the most of the time and stay focused.

Of course, you don't want to burn out. If you study too long you may find that you're not retaining the information very well. Take regular study breaks. For example, taking five minutes out of every hour to walk briskly, breathing deeply and swinging your arms, can help your mind stay fresh.

As you get to the end of each chapter or section, it's a good idea to do a quick review. Remind yourself of what you learned and work on any difficult parts. When you feel that you've mastered the material, move on to the next part. At the end of your study session, briefly skim through your notes again.

But while review is helpful, cramming last minute is NOT. If at all possible, work ahead so that you won't need to fit all your study into the last day. Cramming overloads your brain with more information than it can process and retain, and your tired mind may struggle to recall even

previously learned information when it is overwhelmed with last-minute study. Also, the urgent nature of cramming and the stress placed on your brain contribute to anxiety. You'll be more likely to go to the test feeling unprepared and having trouble thinking clearly.

So don't cram, and don't stay up late before the test, even just to review your notes at a leisurely pace. Your brain needs rest more than it needs to go over the information again. In fact, plan to finish your studies by noon or early afternoon the day before the test. Give your brain the rest of the day to relax or focus on other things, and get a good night's sleep. Then you will be fresh for the test and better able to recall what you've studied.

Step 6: Take a Practice Test

Many courses offer sample tests, either online or in the study materials. This is an excellent resource to check whether you have mastered the material, as well as to prepare for the test format and environment.

Check the test format ahead of time: the number of questions, the type (multiple choice, free response, etc.), and the time limit. Then create a plan for working through them. For example, if you have 30 minutes to take a 60-question test, your limit is 30 seconds per question. Spend less time on the questions you know well so that you can take more time on the difficult ones.

If you have time to take several practice tests, take the first one open book, with no time limit. Work through the questions at your own pace and make sure you fully understand them. Gradually work up to taking a test under test conditions: sit at a desk with all study materials put away and set a timer. Pace yourself to make sure you finish the test with time to spare and go back to check your answers if you have time.

After each test, check your answers. On the questions you missed, be sure you understand why you missed them. Did you misread the question (tests can use tricky wording)? Did you forget the information? Or was it something you hadn't learned? Go back and study any shaky areas that the practice tests reveal.

Taking these tests not only helps with your grade, but also aids in combating test anxiety. If you're already used to the test conditions, you're less likely to worry about it, and working through tests until you're scoring well gives you a confidence boost. Go through the practice tests until you feel comfortable, and then you can go into the test knowing that you're ready for it.

Test Tips

On test day, you should be confident, knowing that you've prepared well and are ready to answer the questions. But aside from preparation, there are several test day strategies you can employ to maximize your performance.

First, as stated before, get a good night's sleep the night before the test (and for several nights before that, if possible). Go into the test with a fresh, alert mind rather than staying up late to study.

Try not to change too much about your normal routine on the day of the test. It's important to eat a nutritious breakfast, but if you normally don't eat breakfast at all, consider eating just a protein bar. If you're a coffee drinker, go ahead and have your normal coffee. Just make sure you time it so that the caffeine doesn't wear off right in the middle of your test. Avoid sugary beverages, and drink enough water to stay hydrated but not so much that you need a restroom break 10 minutes into the

test. If your test isn't first thing in the morning, consider going for a walk or doing a light workout before the test to get your blood flowing.

Allow yourself enough time to get ready, and leave for the test with plenty of time to spare so you won't have the anxiety of scrambling to arrive in time. Another reason to be early is to select a good seat. It's helpful to sit away from doors and windows, which can be distracting. Find a good seat, get out your supplies, and settle your mind before the test begins.

When the test begins, start by going over the instructions carefully, even if you already know what to expect. Make sure you avoid any careless mistakes by following the directions.

Then begin working through the questions, pacing yourself as you've practiced. If you're not sure on an answer, don't spend too much time on it, and don't let it shake your confidence. Either skip it and come back later, or eliminate as many wrong answers as possible and guess among the remaining ones. Don't dwell on these questions as you continue—put them out of your mind and focus on what lies ahead.

Be sure to read all of the answer choices, even if you're sure the first one is the right answer. Sometimes you'll find a better one if you keep reading. But don't second-guess yourself if you do immediately know the answer. Your gut instinct is usually right. Don't let test anxiety rob you of the information you know.

If you have time at the end of the test (and if the test format allows), go back and review your answers. Be cautious about changing any, since your first instinct tends to be correct, but make sure you didn't misread any of the questions or accidentally mark the wrong answer choice. Look over any you skipped and make an educated guess.

At the end, leave the test feeling confident. You've done your best, so don't waste time worrying about your performance or wishing you could change anything. Instead, celebrate the successful completion of this test. And finally, use this test to learn how to deal with anxiety even better next time.

Review Video: Test Anxiety
Visit mometrix.com/academy and enter code: 100340

Important Qualification

Not all anxiety is created equal. If your test anxiety is causing major issues in your life beyond the classroom or testing center, or if you are experiencing troubling physical symptoms related to your anxiety, it may be a sign of a serious physiological or psychological condition. If this sounds like your situation, we strongly encourage you to seek professional help.

Online Resources

Due to our efforts to try to keep this book to a manageable length, we've created a link that will give you access to all of your online resources:

mometrix.com/resources719/aafcsfamconsc

It's Your Moment, Let's Celebrate It!

Share your story @mometrixtestpreparation

DEAR FUTURE EXAM SUCCESS STORY:

First of all, **THANK YOU** for purchasing Mometrix study materials!

Second, congratulations! You are one of the few determined test-takers who are committed to doing whatever it takes to excel on your exam. **You have come to the right place.** We developed these study materials with one goal in mind: to deliver you the information you need in a format that's concise and easy to use.

In addition to optimizing your guide for the content of the test, we've outlined our recommended steps for breaking down the preparation process into small, attainable goals so you can make sure you stay on track.

We've also analyzed the entire test-taking process, identifying the most common pitfalls and showing how you can overcome them and be ready for any curveball the test throws you.

Standardized testing is one of the biggest obstacles on your road to success, which only increases the importance of doing well in the high-pressure, high-stakes environment of test day. Your results on this test could have a significant impact on your future, and this guide provides the information and practical advice to help you achieve your full potential on test day.

Your success is our success

We would love to hear from you! If you would like to share the story of your exam success or if you have any questions or comments in regard to our products, please contact us at **800-673-8175** or **support@mometrix.com**.

Thanks again for your business and we wish you continued success!

Sincerely,
The Mometrix Test Preparation Team

Need more help? Check out our flashcards at:
http://MometrixFlashcards.com/CCSS

Written and edited by the Mometrix Exam Secrets Test Prep Team
Printed in the United States of America

TABLE OF CONTENTS

Introduction

Thank you for purchasing this resource! You have made the choice to prepare yourself for a test that could have a huge impact on your future, and this guide is designed to help you be fully ready for test day. Obviously, it's important to have a solid understanding of the test material, but you also need to be prepared for the unique environment and stressors of the test, so that you can perform to the best of your abilities.

For this purpose, the first section that appears in this guide is the **Secret Keys**. We've devoted countless hours to meticulously researching what works and what doesn't, and we've boiled down our findings to the five most impactful steps you can take to improve your performance on the test. We start at the beginning with study planning and move through the preparation process, all the way to the testing strategies that will help you get the most out of what you know when you're finally sitting in front of the test.

We recommend that you start preparing for your test as far in advance as possible. However, if you've bought this guide as a last-minute study resource and only have a few days before your test, we recommend that you skip over the first two Secret Keys since they address a long-term study plan.

If you struggle with **test anxiety**, we strongly encourage you to check out our recommendations for how you can overcome it. Test anxiety is a formidable foe, but it can be beaten, and we want to make sure you have the tools you need to defeat it.

Secret Key #1 – Plan Big, Study Small

There's a lot riding on your performance. If you want to ace this test, you're going to need to keep your skills sharp and the material fresh in your mind. You need a plan that lets you review everything you need to know while still fitting in your schedule. We'll break this strategy down into three categories.

Information Organization

Start with the information you already have: the official test outline. From this, you can make a complete list of all the concepts you need to cover before the test. Organize these concepts into groups that can be studied together, and create a list of any related vocabulary you need to learn so you can brush up on any difficult terms. You'll want to keep this vocabulary list handy once you actually start studying since you may need to add to it along the way.

Time Management

Once you have your set of study concepts, decide how to spread them out over the time you have left before the test. Break your study plan into small, clear goals so you have a manageable task for each day and know exactly what you're doing. Then just focus on one small step at a time. When you manage your time this way, you don't need to spend hours at a time studying. Studying a small block of content for a short period each day helps you retain information better and avoid stressing over how much you have left to do. You can relax knowing that you have a plan to cover everything in time. In order for this strategy to be effective though, you have to start studying early and stick to your schedule. Avoid the exhaustion and futility that comes from last-minute cramming!

Study Environment

The environment you study in has a big impact on your learning. Studying in a coffee shop, while probably more enjoyable, is not likely to be as fruitful as studying in a quiet room. It's important to keep distractions to a minimum. You're only planning to study for a short block of time, so make the most of it. Don't pause to check your phone or get up to find a snack. It's also important to **avoid multitasking**. Research has consistently shown that multitasking will make your studying dramatically less effective. Your study area should also be comfortable and well-lit so you don't have the distraction of straining your eyes or sitting on an uncomfortable chair.

The time of day you study is also important. You want to be rested and alert. Don't wait until just before bedtime. Study when you'll be most likely to comprehend and remember. Even better, if you know what time of day your test will be, set that time aside for study. That way your brain will be used to working on that subject at that specific time and you'll have a better chance of recalling information.

Finally, it can be helpful to team up with others who are studying for the same test. Your actual studying should be done in as isolated an environment as possible, but the work of organizing the information and setting up the study plan can be divided up. In between study sessions, you can discuss with your teammates the concepts that you're all studying and quiz each other on the details. Just be sure that your teammates are as serious about the test as you are. If you find that your study time is being replaced with social time, you might need to find a new team.

Secret Key #3 – Practice the Right Way

Your success on test day depends not only on how many hours you put into preparing, but also on whether you prepared the right way. It's good to check along the way to see if your studying is paying off. One of the most effective ways to do this is by taking practice tests to evaluate your progress. Practice tests are useful because they show exactly where you need to improve. Every time you take a practice test, pay special attention to these three groups of questions:

- The questions you got wrong
- The questions you had to guess on, even if you guessed right
- The questions you found difficult or slow to work through

This will show you exactly what your weak areas are, and where you need to devote more study time. Ask yourself why each of these questions gave you trouble. Was it because you didn't understand the material? Was it because you didn't remember the vocabulary? Do you need more repetitions on this type of question to build speed and confidence? Dig into those questions and figure out how you can strengthen your weak areas as you go back to review the material.

Additionally, many practice tests have a section explaining the answer choices. It can be tempting to read the explanation and think that you now have a good understanding of the concept. However, an explanation likely only covers part of the question's broader context. Even if the explanation makes perfect sense, **go back and investigate** every concept related to the question until you're positive you have a thorough understanding.

As you go along, keep in mind that the practice test is just that: practice. Memorizing these questions and answers will not be very helpful on the actual test because it is unlikely to have any of the same exact questions. If you only know the right answers to the sample questions, you won't be prepared for the real thing. **Study the concepts** until you understand them fully, and then you'll be able to answer any question that shows up on the test.

It's important to wait on the practice tests until you're ready. If you take a test on your first day of study, you may be overwhelmed by the amount of material covered and how much you need to learn. Work up to it gradually.

On test day, you'll need to be prepared for answering questions, managing your time, and using the test-taking strategies you've learned. It's a lot to balance, like a mental marathon that will have a big impact on your future. Like training for a marathon, you'll need to start slowly and work your way up. When test day arrives, you'll be ready.

Start with the strategies you've read in the first two Secret Keys—plan your course and study in the way that works best for you. If you have time, consider using multiple study resources to get different approaches to the same concepts. It can be helpful to see difficult concepts from more than one angle. Then find a good source for practice tests. Many times, the test website will suggest potential study resources or provide sample tests.

Practice Test Strategy

If you're able to find at least three practice tests, we recommend this strategy:

UNTIMED AND OPEN-BOOK PRACTICE

Take the first test with no time constraints and with your notes and study guide handy. Take your time and focus on applying the strategies you've learned.

TIMED AND OPEN-BOOK PRACTICE

Take the second practice test open-book as well, but set a timer and practice pacing yourself to finish in time.

TIMED AND CLOSED-BOOK PRACTICE

Take any other practice tests as if it were test day. Set a timer and put away your study materials. Sit at a table or desk in a quiet room, imagine yourself at the testing center, and answer questions as quickly and accurately as possible.

Keep repeating timed and closed-book tests on a regular basis until you run out of practice tests or it's time for the actual test. Your mind will be ready for the schedule and stress of test day, and you'll be able to focus on recalling the material you've learned.

Secret Key #4 – Pace Yourself

Once you're fully prepared for the material on the test, your biggest challenge on test day will be managing your time. Just knowing that the clock is ticking can make you panic even if you have plenty of time left. Work on pacing yourself so you can build confidence against the time constraints of the exam. Pacing is a difficult skill to master, especially in a high-pressure environment, so **practice is vital**.

Set time expectations for your pace based on how much time is available. For example, if a section has 60 questions and the time limit is 30 minutes, you know you have to average 30 seconds or less per question in order to answer them all. Although 30 seconds is the hard limit, set 25 seconds per question as your goal, so you reserve extra time to spend on harder questions. When you budget extra time for the harder questions, you no longer have any reason to stress when those questions take longer to answer.

Don't let this time expectation distract you from working through the test at a calm, steady pace, but keep it in mind so you don't spend too much time on any one question. Recognize that taking extra time on one question you don't understand may keep you from answering two that you do understand later in the test. If your time limit for a question is up and you're still not sure of the answer, mark it and move on, and come back to it later if the time and the test format allow. If the testing format doesn't allow you to return to earlier questions, just make an educated guess; then put it out of your mind and move on.

On the easier questions, be careful not to rush. It may seem wise to hurry through them so you have more time for the challenging ones, but it's not worth missing one if you know the concept and just didn't take the time to read the question fully. Work efficiently but make sure you understand the question and have looked at all of the answer choices, since more than one may seem right at first.

Even if you're paying attention to the time, you may find yourself a little behind at some point. You should speed up to get back on track, but do so wisely. Don't panic; just take a few seconds less on each question until you're caught up. Don't guess without thinking, but do look through the answer choices and eliminate any you know are wrong. If you can get down to two choices, it is often worthwhile to guess from those. Once you've chosen an answer, move on and don't dwell on any that you skipped or had to hurry through. If a question was taking too long, chances are it was one of the harder ones, so you weren't as likely to get it right anyway.

On the other hand, if you find yourself getting ahead of schedule, it may be beneficial to slow down a little. The more quickly you work, the more likely you are to make a careless mistake that will affect your score. You've budgeted time for each question, so don't be afraid to spend that time. Practice an efficient but careful pace to get the most out of the time you have.

Secret Key #5 – Have a Plan for Guessing

When you're taking the test, you may find yourself stuck on a question. Some of the answer choices seem better than others, but you don't see the one answer choice that is obviously correct. What do you do?

The scenario described above is very common, yet most test takers have not effectively prepared for it. Developing and practicing a plan for guessing may be one of the single most effective uses of your time as you get ready for the exam.

In developing your plan for guessing, there are three questions to address:

- When should you start the guessing process?
- How should you narrow down the choices?
- Which answer should you choose?

When to Start the Guessing Process

Unless your plan for guessing is to select C every time (which, despite its merits, is not what we recommend), you need to leave yourself enough time to apply your answer elimination strategies. Since you have a limited amount of time for each question, that means that if you're going to give yourself the best shot at guessing correctly, you have to decide quickly whether or not you will guess.

Of course, the best-case scenario is that you don't have to guess at all, so first, see if you can answer the question based on your knowledge of the subject and basic reasoning skills. Focus on the key words in the question and try to jog your memory of related topics. Give yourself a chance to bring the knowledge to mind, but once you realize that you don't have (or you can't access) the knowledge you need to answer the question, it's time to start the guessing process.

It's almost always better to start the guessing process too early than too late. It only takes a few seconds to remember something and answer the question from knowledge. Carefully eliminating wrong answer choices takes longer. Plus, going through the process of eliminating answer choices can actually help jog your memory.

Summary: Start the guessing process as soon as you decide that you can't answer the question based on your knowledge.

How to Narrow Down the Choices

The next chapter in this book (**Test-Taking Strategies**) includes a wide range of strategies for how to approach questions and how to look for answer choices to eliminate. You will definitely want to read those carefully, practice them, and figure out which ones work best for you. Here though, we're going to address a mindset rather than a particular strategy.

Your odds of guessing an answer correctly depend on how many options you are choosing from.

Number of options left	5	4	3	2	1
Odds of guessing correctly	20%	25%	33%	50%	100%

You can see from this chart just how valuable it is to be able to eliminate incorrect answers and make an educated guess, but there are two things that many test takers do that cause them to miss out on the benefits of guessing:

- Accidentally eliminating the correct answer
- Selecting an answer based on an impression

We'll look at the first one here, and the second one in the next section.

To avoid accidentally eliminating the correct answer, we recommend a thought exercise called **the $5 challenge**. In this challenge, you only eliminate an answer choice from contention if you are willing to bet $5 on it being wrong. Why $5? Five dollars is a small but not insignificant amount of money. It's an amount you could afford to lose but wouldn't want to throw away. And while losing $5 once might not hurt too much, doing it twenty times will set you back $100. In the same way, each small decision you make—eliminating a choice here, guessing on a question there—won't by itself impact your score very much, but when you put them all together, they can make a big difference. By holding each answer choice elimination decision to a higher standard, you can reduce the risk of accidentally eliminating the correct answer.

The $5 challenge can also be applied in a positive sense: If you are willing to bet $5 that an answer choice *is* correct, go ahead and mark it as correct.

Summary: Only eliminate an answer choice if you are willing to bet $5 that it is wrong.

Which Answer to Choose

You're taking the test. You've run into a hard question and decided you'll have to guess. You've eliminated all the answer choices you're willing to bet $5 on. Now you have to pick an answer. Why do we even need to talk about this? Why can't you just pick whichever one you feel like when the time comes?

The answer to these questions is that if you don't come into the test with a plan, you'll rely on your impression to select an answer choice, and if you do that, you risk falling into a trap. The test writers know that everyone who takes their test will be guessing on some of the questions, so they intentionally write wrong answer choices to seem plausible. You still have to pick an answer though, and if the wrong answer choices are designed to look right, how can you ever be sure that you're not falling for their trap? The best solution we've found to this dilemma is to take the decision out of your hands entirely. Here is the process we recommend:

Once you've eliminated any choices that you are confident (willing to bet $5) are wrong, select the first remaining choice as your answer.

Whether you choose to select the first remaining choice, the second, or the last, the important thing is that you use some preselected standard. Using this approach guarantees that you will not be enticed into selecting an answer choice that looks right, because you are not basing your decision on how the answer choices look.

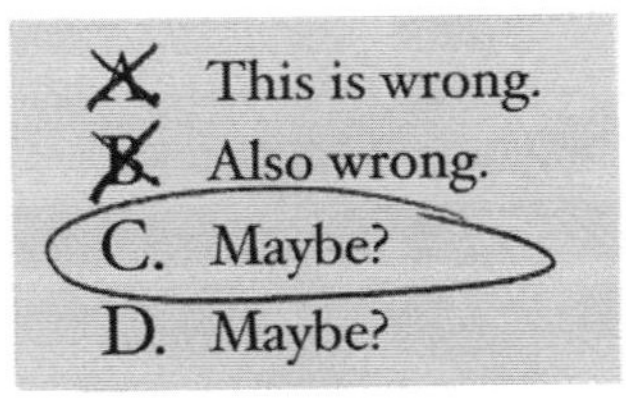

This is not meant to make you question your knowledge. Instead, it is to help you recognize the difference between your knowledge and your impressions. There's a huge difference between thinking an answer is right because of what you know, and thinking an answer is right because it looks or sounds like it should be right.

Summary: To ensure that your selection is appropriately random, make a predetermined selection from among all answer choices you have not eliminated.

Test-Taking Strategies

This section contains a list of test-taking strategies that you may find helpful as you work through the test. By taking what you know and applying logical thought, you can maximize your chances of answering any question correctly!

It is very important to realize that every question is different and every person is different: no single strategy will work on every question, and no single strategy will work for every person. That's why we've included all of them here, so you can try them out and determine which ones work best for different types of questions and which ones work best for you.

Question Strategies

☑ Read Carefully

Read the question and the answer choices carefully. Don't miss the question because you misread the terms. You have plenty of time to read each question thoroughly and make sure you understand what is being asked. Yet a happy medium must be attained, so don't waste too much time. You must read carefully and efficiently.

☑ Contextual Clues

Look for contextual clues. If the question includes a word you are not familiar with, look at the immediate context for some indication of what the word might mean. Contextual clues can often give you all the information you need to decipher the meaning of an unfamiliar word. Even if you can't determine the meaning, you may be able to narrow down the possibilities enough to make a solid guess at the answer to the question.

☑ Prefixes

If you're having trouble with a word in the question or answer choices, try dissecting it. Take advantage of every clue that the word might include. Prefixes can be a huge help. Usually, they allow you to determine a basic meaning. *Pre-* means before, *post-* means after, *pro-* is positive, *de-* is negative. From prefixes, you can get an idea of the general meaning of the word and try to put it into context.

☑ Hedge Words

Watch out for critical hedge words, such as *likely, may, can, often, almost, mostly, usually, generally, rarely,* and *sometimes.* Question writers insert these hedge phrases to cover every possibility. Often an answer choice will be wrong simply because it leaves no room for exception. Be on guard for answer choices that have definitive words such as *exactly* and *always.*

☑ Switchback Words

Stay alert for *switchbacks.* These are the words and phrases frequently used to alert you to shifts in thought. The most common switchback words are *but, although,* and *however.* Others include *nevertheless, on the other hand, even though, while, in spite of, despite,* and *regardless of.* Switchback words are important to catch because they can change the direction of the question or an answer choice.

⊘ BENCHMARK

As you read through the answer choices and you come across one that seems to answer the question well, mentally select that answer choice. This is not your final answer, but it's the one that will help you evaluate the other answer choices. The one that you selected is your benchmark or standard for judging each of the other answer choices. Every other answer choice must be compared to your benchmark. That choice is correct until proven otherwise by another answer choice beating it. If you find a better answer, then that one becomes your new benchmark. Once you've decided that no other choice answers the question as well as your benchmark, you have your final answer.

⊘ PREDICT THE ANSWER

Before you even start looking at the answer choices, it is often best to try to predict the answer. When you come up with the answer on your own, it is easier to avoid distractions and traps because you will know exactly what to look for. The right answer choice is unlikely to be word-for-word what you came up with, but it should be a close match. Even if you are confident that you have the right answer, you should still take the time to read each option before moving on.

General Strategies

⊘ TOUGH QUESTIONS

If you are stumped on a problem or it appears too hard or too difficult, don't waste time. Move on! Remember though, if you can quickly check for obviously incorrect answer choices, your chances of guessing correctly are greatly improved. Before you completely give up, at least try to knock out a couple of possible answers. Eliminate what you can and then guess at the remaining answer choices before moving on.

⊘ CHECK YOUR WORK

Since you will probably not know every term listed and the answer to every question, it is important that you get credit for the ones that you do know. Don't miss any questions through careless mistakes. If at all possible, try to take a second to look back over your answer selection and make sure you've selected the correct answer choice and haven't made a costly careless mistake (such as marking an answer choice that you didn't mean to mark). This quick double check should more than pay for itself in caught mistakes for the time it costs.

⊘ PACE YOURSELF

It's easy to be overwhelmed when you're looking at a page full of questions; your mind is confused and full of random thoughts, and the clock is ticking down faster than you would like. Calm down and maintain the pace that you have set for yourself. Especially as you get down to the last few minutes of the test, don't let the small numbers on the clock make you panic. As long as you are on track by monitoring your pace, you are guaranteed to have time for each question.

⊘ DON'T RUSH

It is very easy to make errors when you are in a hurry. Maintaining a fast pace in answering questions is pointless if it makes you miss questions that you would have gotten right otherwise. Test writers like to include distracting information and wrong answers that seem right. Taking a little extra time to avoid careless mistakes can make all the difference in your test score. Find a pace that allows you to be confident in the answers that you select.

⊘ Keep Moving

Panicking will not help you pass the test, so do your best to stay calm and keep moving. Taking deep breaths and going through the answer elimination steps you practiced can help to break through a stress barrier and keep your pace.

Final Notes

The combination of a solid foundation of content knowledge and the confidence that comes from practicing your plan for applying that knowledge is the key to maximizing your performance on test day. As your foundation of content knowledge is built up and strengthened, you'll find that the strategies included in this chapter become more and more effective in helping you quickly sift through the distractions and traps of the test to isolate the correct answer.

Now that you're preparing to move forward into the test content chapters of this book, be sure to keep your goal in mind. As you read, think about how you will be able to apply this information on the test. If you've already seen sample questions for the test and you have an idea of the question format and style, try to come up with questions of your own that you can answer based on what you're reading. This will give you valuable practice applying your knowledge in the same ways you can expect to on test day.

Good luck and good studying!

Mathematics

Ratio and Proportion

A ratio is a comparison of two numbers by division. The ratio of a to b, where $b \neq 0$, can be written as

$$a \text{ to } b$$
$$a : b$$
$$\frac{a}{b}$$

A proportion is a statement of equality between two ratios. For example, $\frac{a}{b} = \frac{c}{d}$, where $b \neq 0$ and $d \neq 0$, is a proportion equating the ratios $\frac{a}{b}$ and $\frac{c}{d}$.

Integer

The set of integers includes whole numbers and their opposites: {...,-3,-2,-1,0,1,2,3...}.

Rational Number

A rational number is a real number which can be written as a ratio of two integers a and b, where $b \neq 0$ so long as the second is not zero; in other words, any rational number can be expressed in fractional form $\frac{a}{b}$, where $b \neq 0$. Rational numbers include whole numbers, fractions, terminating, and repeating decimals.

Example

Write each rational number as a fraction

- 3
- 0.6

Since dividing any number by one does not change its value, a whole number can be written as a fraction with a denominator of 1. So, $3 = \frac{3}{1}$.

The six in 0.6 is in the tenths place. The number six-tenths can also be written as $\frac{6}{10}$, which reduces to $\frac{3}{5}$.

EXAMPLE

Simply each expression

$$\frac{2}{3}+\frac{1}{2}$$

$$\frac{2}{3}-\frac{1}{2}$$

When combining fractions, it is helpful to write them so that they have the same denominator.

1.

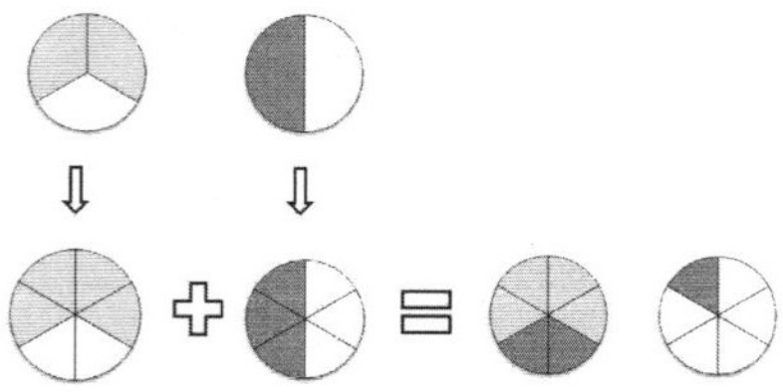

$$\frac{2}{3}=\frac{4}{6} \qquad \frac{1}{2}=\frac{3}{6}$$

$$\frac{4}{6}+\frac{3}{6}=\frac{7}{6}=1\frac{1}{6}$$

2.

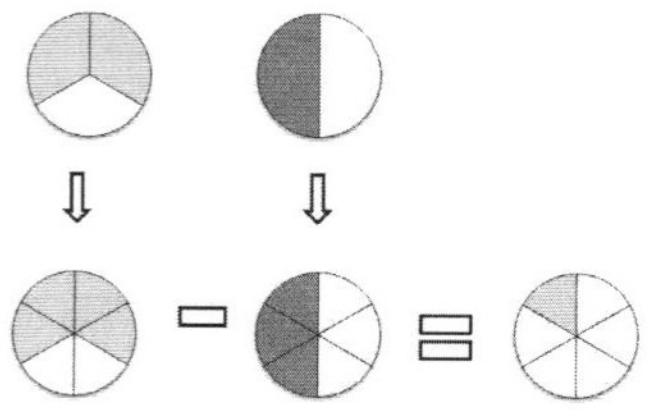

$$\frac{4}{6}-\frac{3}{6}=\frac{1}{6}$$

EXAMPLE

Simplify each expression

$$\frac{1}{2}\times\frac{2}{3}$$

$$\frac{1}{8}\div\frac{1}{2}$$

The numerator of the product of two fractions is the product of their numerators; $1 \times 2 = 2$. Likewise, the denominator of the product of two fractions is the product of their denominators: $2 \times 3 = 6$. Reduce the resulting fraction if necessary.

$$\frac{1}{2}\times\frac{2}{3}=\frac{2}{6}=\frac{1}{3}$$

When dividing fractions, rewrite the expression as the product of the first fraction and the reciprocal (or multiplicative inverse) of the second. Reduce if necessary.

$\frac{1}{8}\div\frac{1}{2}=\frac{1}{8}\times\frac{2}{1}=\frac{2}{8}=\frac{1}{4};$

Example

Convert each fraction to a decimal.

$$\frac{4}{5}$$
$$\frac{5}{6}$$

A fraction is a quotient of two numbers such that the denominator is not zero. So, one way to convert a fraction to a decimal is to divide the denominator into the numerator. The resulting decimal will either terminate or repeat.

$$\frac{4}{5} = 4 \div 5 = 0.8$$

```
   0.8
5 |4.0
   0 ↓
   4 0
   4 0
     0
```

$$\frac{5}{6} = 5 \div 6 = 0.8\overline{3}$$

```
   0.833...
6 |5.000...
   0 ↓
   5 0
   4 8↓
     20
     18↓
      20
      18
       2
```

Percentage

One *Percent* means one part per hundred, so a percentage is the ratio of a number to 100. For example, 42% can be written as the ratio $\frac{42}{100}$ or its reduced equivalent, $\frac{21}{50}$.

Example

Write each percentage as a simplified fraction and as a decimal

- 32%
- 135%

A percentage is a ratio of a number to 100.

$$32\% = \frac{32}{100} = \frac{8}{25} \; or \; \frac{32}{100} = 0.32$$
$$135\% = \frac{135}{100} = 1\frac{35}{100} = 1\frac{7}{20}$$
$$\frac{135}{100} = 1\frac{35}{100} = 1.35$$

EXAMPLE

Write each number as a percentage

$$\frac{4}{5}$$
$$\frac{2}{3}$$
$$0.23 = \frac{23}{100} = 23\%$$
$$\frac{4}{5} = \frac{80}{100} = 80\%$$
$$\frac{2}{3} = 0.\overline{6} = 66.\overline{6}\%$$

EXAMPLE

Express the shaded portion of the circle as a fraction, a decimal, and a percentage.

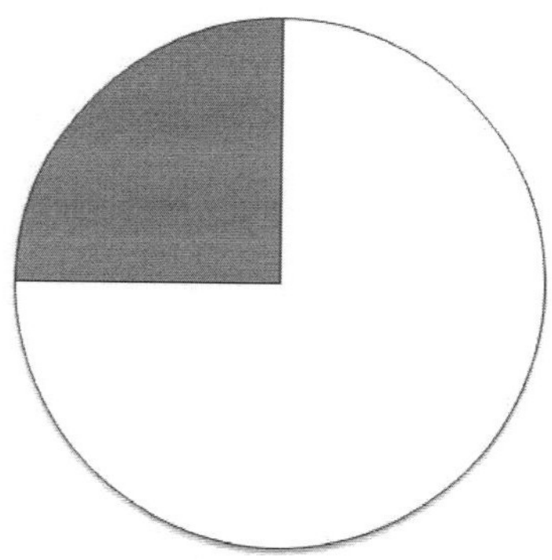

¼ of the circle's area is shaded. $\frac{1}{4} = 0.25 = \frac{25}{100} = 25\%$.

DETERMINING WHETHER OR NOT TWO RATIOS FORM A PROPORTION

Two ratios form a proportion if they are equal. One way to determine if two ratios are equal is to write each ratio as a fraction (as long as neither contains a zero in its denominator) and then cross-multiply: that is, multiply the numerator of the first fraction and the denominator of the second; then, multiply the denominator of the first fraction and the numerator of the second. If these two products are equal, the ratios are equal and therefore form a proportion. $\frac{a}{b} = \frac{c}{d}$ if and only if $a \times d = b \times c$. For example, $\frac{2}{3} = \frac{12}{18}$, and $2 \times 18 = 3 \times 12 = 36$.

Determining Whether Two Quantities Have a Proportional Relationship in a Graph

If two quantities are graphed on the coordinate plane, and the result is a straight line through the origin, then the two quantities are proportional.

Example

For the graphs below, determine whether there exists a proportional relationship between x and y. If a proportional relationship exists, find the constant of proportionality and write an equation for the line.

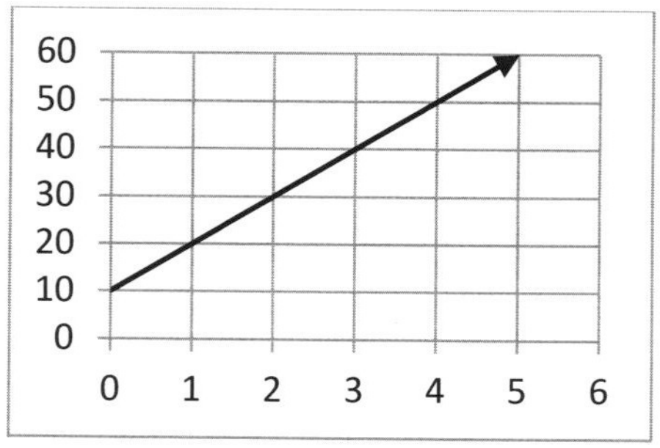

Though the graph of the relationship between x and y is a straight line, it does not pass through the origin. So, though y varies directly as x, the ratio $\frac{y}{x}$ is not constant: for instance, $\frac{20}{1} \neq \frac{30}{2} \neq \frac{40}{3}$..

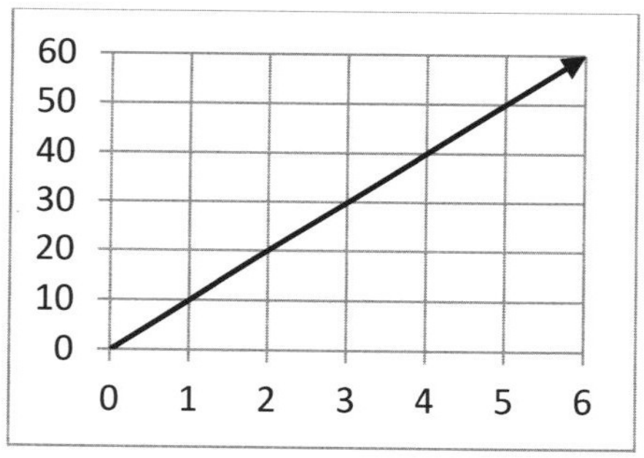

The graph of a proportional relationship is a straight line through the origin.

This graph is a straight line through the origin, so the relationship between x and y is proportional. The constant of proportionality is represented by the ratio $\frac{y}{x}$. This constant is the same as the unit rate. The constant of proportionality is equal to the y-value when $x = 1$. Since the ratio $\frac{y}{x}$ is 10 (see that $\frac{y}{x} = \frac{10}{1} = \frac{20}{2} = \frac{30}{3}$ and so on), or since $y = 10$ when $x = 1$, the constant of proportionality is 10. The relationship between x and y is represented by the equation $y = 10x$.

Determining Whether Two Quantities Have a Proportional Relationship Given a Table of Values

If the ratio of y to x is constant for all values of x and y besides zero, then there is a proportional relationship between the two variables. The value $\frac{y}{x}$ is the constant of proportionality.

Example

Determine whether there exists a proportional relationship between x and y. If a proportional relationship exists, find the constant of proportionality and write an equation to represent the relationship.

If the ratio of y to x is constant for all values of x and y besides zero, then there is a proportional relationship between the two variables. The value $\frac{y}{x}$ is the constant of proportionality.

x	1	2	3	4
y	5	9	13	17

The ratio of y to x is not constant; therefore, the values in the table do not represent a proportional relationship:

x	1	2	3	4
y	5	9	13	17
$\frac{y}{x}$	$\frac{5}{1} = 5$	$\frac{9}{2} = 4.5$	$\frac{13}{3} = 4.\overline{3}$	$\frac{17}{4} = 4.25$

Example

Determine whether there exists a proportional relationship between x and y. If a proportional relationship exists, find the constant of proportionality and write an equation to represent the relationship.

x	1	2	3	4
y	1	4	9	16

If the ratio of y to x is constant for all values of x and y besides zero, then there is a proportional relationship between the two variables. The value $\frac{y}{x}$ is the constant of proportionality.

The ratio of y to x is not constant; therefore, the values in the table do not represent a proportional relationship:

x	1	2	3	4
y	1	4	9	16
$\frac{y}{x}$	$\frac{1}{1} = 1$	$\frac{4}{2} = 2$	$\frac{9}{3} = 3$	$\frac{16}{4} = 4$

Example

Determine whether there exists a proportional relationship between x and y. If a proportional relationship exists, find the constant of proportionality and write an equation to represent the relationship.

x	1	2	3	4
y	2.5	5	7.5	10

If the ratio of y to x is constant for all values of x and y besides zero, then there is a proportional relationship between the two variables. The value $\frac{y}{x}$ is the constant of proportionality.

The ratio of y to x is 2.5; therefore, the values in the table represent the proportional relationship modeled by the equation $y = 2.5x$:

x	1	2	3	4
y	2.5	5	7.5	10
$\frac{y}{x}$	$\frac{2.5}{1} = 2.5$	$\frac{5}{2} = 2.5$	$\frac{7.5}{3} = 2.5$	$\frac{10}{4} = 2.5$

EXAMPLE

Suppose gasoline costs $3 per gallon. Create a table of values for the total cost of gasoline and the gallons of gasoline purchased.

The graph is confined to the first-quadrant of the coordinate plane because neither the amount of gas nor the price of the gas can be negative. Since the cost depends on the number of gallons purchased, plot the number of gallons along the horizontal axis and the cost along the vertical axis.

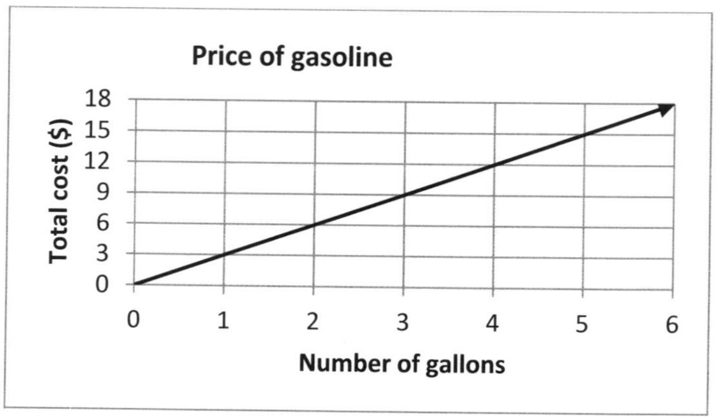

EXAMPLE

Create a graph of the relationship between the total cost and the gallons of gasoline purchased

The graph is confined to the first-quadrant of the coordinate plane because neither the amount of gas nor the price of the gas can be negative. Since the cost depends on the number of gallons purchased, plot the number of gallons along the horizontal axis and the cost along the vertical axis.

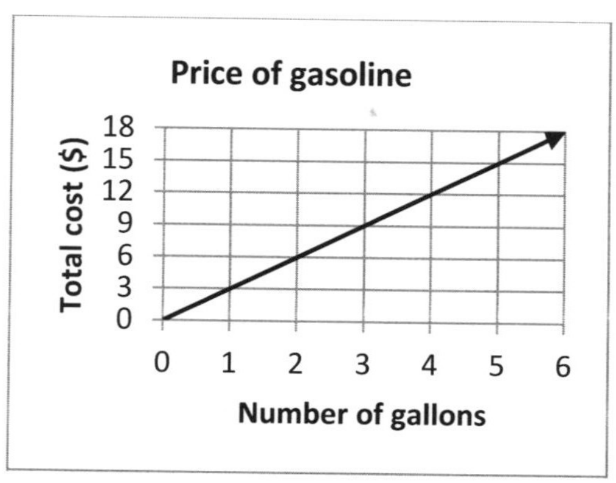

Write an equation which relates total cost to gallons of gasoline purchased.

The equation for the line is y=3x, or total cost=3×number of gallons purchased.

EXAMPLE

Find the rate of travel from a graph representing a proportional relationship between travel time in hours (graphed along the horizontal axis) and distance traveled in miles (graphed along the vertical axis).

The slope of the line represents the rate of travel, or the proportionality constant, in miles per hour. The slope of a line is its vertical change, or rise, divided by its horizontal change, or run. These values can be determined from by counting the vertical and horizontal distances between any two points on the line or by using the equation $\frac{y_2-y_1}{x_2-x_1}$, where (x_1, y_1) and (x_2, y_2) are points on the line.

A point on the line shows the distance traveled at a particular travel time. Since the ratio of distance to time is constant along the graph representing a proportion relationship, any point on the graph can be used to find the rate by simply finding the ratio of y to x. For example, a point (3,90) on the graph indicates that it takes three hours to travel ninety miles, so the rate is $\frac{90\ miles}{3\ hours}$= 30 miles per hour.

Since the unit rate, miles per hour, compares the distance traveled in miles to one hour, the unit rate is the y-coordinate when x=1. For example, if the line passes through (1,30), the rate is 30 miles per hour.

USING PROPORTIONS TO SOLVE PERCENT PROBLEMS

A proportion is a statement of equivalence between two ratios. In percent problems, both ratios compare parts to a whole; in particular, a percentage expresses parts per 100. A proportion which can be used to solved a percent problem is $\frac{part}{whole} = \frac{percent}{100}$.

In the given scenario, 4 is 80% of some number a, so 4 represents part of the unknown number. The proportion, therefore, is $\frac{4}{a} = \frac{80}{100}$. There are many ways to solve proportions. Notice that $\frac{80}{100}$ reduces to $\frac{4}{5}$, so $a = 5$.

EXAMPLE

A family of six dines at a restaurant which charges an automatic gratuity for parties of six or more. A tip of $28 is added to their bill of $80. Determine the percent gratuity charged.

One way to determine the percent gratuity charged is to set up and solve a proportion of the form

$$\frac{part}{whole} = \frac{percent}{100}.$$

$$\frac{28}{80} = \frac{p}{100}$$

There are many ways to solve proportions. Notice that $\frac{28}{80}$ reduces to $\frac{7}{20}$, which can easily be converted to a fraction with a denominator of 100 by multiplying the numerator and denominator by 5.

$$\frac{7 \times 5}{20 \times 5} = \frac{35}{100}$$

So, $p = 35$. The gratuity added is 35%.

Example

The ratio of flour to sugar in a cookie recipe is 3:1. Find the amount of sugar needed for 1 ½ cups of flour

There are many ways to solve this problem using proportional reasoning. One way is to notice that the amount of flour divided by three gives the amount of sugar.

cups of flour: cups of sugar

$$\overset{\div 3}{3:1}$$

So, the amount of sugar needed for 1 ½ cups of flour can be found by dividing $1\frac{1}{2}$ by 3. $1\frac{1}{2} \div 3 = \frac{3}{2} \times \frac{1}{3} = \frac{3}{6} = \frac{1}{2}$.

$$\overset{\div 3}{1\frac{1}{2} : \frac{1}{2}}$$

The amount of sugar needed is ½ cup.

Example

Suppose you purchase a \$7.00 entrée and a \$2.00 drink at your favorite restaurant.

- Determine the amount of a 10% tax on your purchase.
- Determine the amount of a 15% tip on the pre-tax amount.
- Find the total price of the meal, including tax and tip.

The amount of your purchase before tax and tip is \$9.00. There are many way to calculate the amount of tax on the purchase. One method is to set up and solve a proportion:

The amount of the tax will be calculated as a fraction of the purchase price. That fraction comparing the tax amount to the pre-tax price is equal to 10%, or $\frac{10}{100}$. So, $\frac{tax\ amount}{\$9.00} = \frac{10}{100}$. A tax amount of \$0.90 satisfies the proportion.

Another method involves translating the problem into a mathematical expression which represents the tax amount, which is *10% of the purchase.*

A percent is a ratio out of 100, so $10\% = \frac{10}{100} = 0.10$.

The word "of" indicates multiplication.

The purchase price is \$9.

So, *10% of the purchase* translates to $0.10 \times \$9$, which equals \$0.90.

Again, there are many ways to calculate the amount of the tip. 15% of \$9.00 translates to 0.15×\$9, which equals \$1.35.

The total price is the cost of the meal plus the tax plus the tip: \$9.00+\$0.90+\$1.35=\$11.25.

EXAMPLE

A number is decreased by 20%. The resulting number is then increased by 20%. Determine whether the consequent number is greater than, less than, or equal to the original number

If a number is decreased by 20%, and the resulting number is increased by 20%, then the consequent number will be less than the original number.

Consider, for instance, that the original number is 100.

20% of 100 is $0.20 \times 100 = 20$, and $100 - 20 = 80$.

20% of 80 is $0.20 \times 80 = 16$, and $80 + 16 = 96$.

96 is less than 100.

EXAMPLE

A school has 400 students; 220 of these students are girls. If the ratio of boys to girls in a class of twenty is representative of the ratio of boys to girls school-wide, determine how many boys are in the class.

Since the ratio of boys to girls in the class is equal to the ratio of boys to girls in the school, the ratio of boys to students in the class must also equal to the ratio of boys to students in the school.

$$\frac{\text{Number of boys in the school}}{\text{number of students in the school}} = \frac{\text{number of boys in the class}}{\text{number of students in the class}}$$

In a school of 400 students, 220 of which are girls, there are $400 - 220 = 180$ boys. The ratio of boys to total students is 180/400, which reduces to 9/20. So, in the class of twenty students, there must be nine boys.

Check to see that the ratio of boys to girls in the class is indeed equal to ratio school-wide. If there are nine boys in a class of twenty, then there are eleven girls:

$$\frac{9 \text{ boys}}{11 \text{ girls}} = \frac{180 \text{ boys}}{220 \text{ girls}}$$

One way to determine whether or not this statement is true is to cross multiply. If the products are equal, then so are the ratios.

$$9 \times 220 = 1980$$

$$11 \times 180 = 1980$$

EXAMPLE

Use a number line to find the sum of 2.1 and 3.2.

Plot 2.1 on a number line and move three and two tenths spaces to the right.

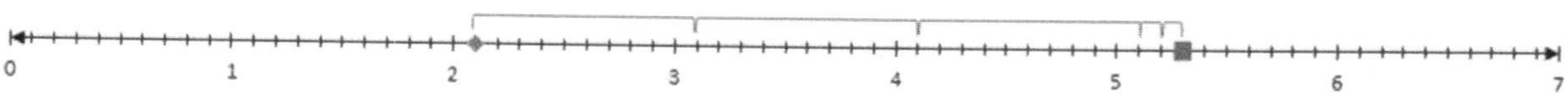

ADDITIVE INVERSE

The sum of a number and its additive inverse, or opposite, is the additive identity, 0.

EXAMPLE

Find the additive inverse of

3
-5
x

The additive inverse of 3 is -3 because $3 + (-3) = 0$.

The additive inverse of -5 is 5 because $-5 + 5 = 0$.

The additive inverse of x is $-x$because $x + (-x) = 0$.

MULTIPLICATIVE INVERSE

The product of a number and its multiplicative inverse is the multiplicative identity, 1. The multiplicative inverse of a number is also called its reciprocal. The reciprocal of a non-zero rational number is also rational. Zero does not have a multiplicative because the product of zero and any number is zero and can therefore not equal 1 and because zero can never be in the denominator of a fraction.

EXAMPLE

Find the multiplicative inverse of

5

$-\frac{2}{3}$

x such that $x \neq 0$.

The multiplicative inverse of 5 is $\frac{1}{5}$ because $5\left(\frac{1}{5}\right) = 1$.

The multiplicative inverse of $-\frac{2}{3}$ is $-\frac{3}{2}$ because $-\frac{2}{3}\left(-\frac{3}{2}\right) = 1$.

- A 5°F drop in temperature
- A debit of $1.60
- An extra half-mile

A gain of four yards → $+\mathbf{4}$ yards. The opposite is *a loss of four yards*, or $-\mathbf{4}$ yards.

A deduction of ten points → $-\mathbf{10}$ points. The opposite is *an addition of 10 points*, or $+\mathbf{10}$ points.

A 5°F drop in temperature → $-\mathbf{5}$°F. The opposite is *an increase in temperature of 5 °F*, or $+\mathbf{5}$°F.

A debit of $1.60→ $-\$\mathbf{1.60}$. The opposite is *a credit of $1.60*, or $+\$\mathbf{1.60}$.

An extra half-mile→ +½ mile. The opposite is *a half-mile less*, or −½ mile

ABSOLUTE VALUE

The absolute value of a number is the number's distance from zero on a number line. A measure of distance is always positive, so absolute value is always positive.

EXAMPLE

Show that $|3|=|-3|$.

The absolute value of 3, written as $|3|$, is 3 because the distance between 0 and 3 on a number line is three units. Likewise, the absolute value of -3, written as $|-3|$, is 3 because the distance between 0 and -3 on a number line is three units. So, $|3|=|-3|$.

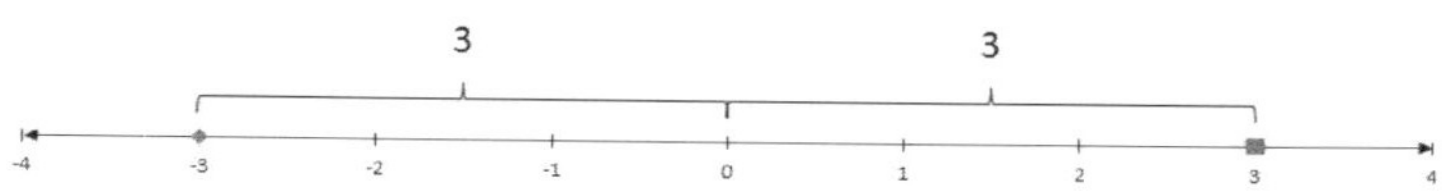

MULTIPLYING AND DIVIDING POSITIVE AND NEGATIVE NUMBERS

The product or quotient of two positive numbers is positive.

$$2 \times 4 = 8$$

$$18 \div 3 = 6$$

The product or quotient of two negative numbers is positive.

$$(-3)(-1) = 3$$

$$\frac{-18}{-9} = 2$$

The product or quotient of a positive and a negative number or a negative and a positive number is negative.

$$4(-2) = -8$$
$$-3 \times 6 = -18$$
$$\frac{20}{-10} = -2$$
$$-15 \div 3 = -5$$

EXAMPLE

For integers p and q, $q \neq 0, -\left(\frac{p}{q}\right) = \frac{-p}{q} = \frac{p}{-q}$. Illustrate this property using an example.

Choose an integer value for p and a non-zero integer value for q to show that $-\left(\frac{p}{q}\right) = \frac{-p}{q} = \frac{p}{-q}$. For instance, when p=10 and q=2,

$$-\left(\frac{p}{q}\right) = -\left(\frac{10}{2}\right) = -5$$
$$\frac{-p}{q} = \frac{-10}{2} = -5$$
$$\frac{p}{-q} = \frac{10}{-2} = -$$

EXAMPLE

A jacket is marked 75% off. Determine which of these methods will give the discounted price of the jacket.

- Find 75% of the jacket's original price and subtract the result from the original price.
- Find 25% of the jacket's original price.
- Divide the original price by four.

All of these methods will give the discounted price of the jacket. If x is the jacket's original price, its new price is $x - 0.75x$. This expression simplifies to $0.25x$. $0.25 = \frac{25}{100} = \frac{1}{4}$, so 0.25x can be rewritten as $\frac{1}{4}x$, which equals $\frac{x}{4}$.

EXAMPLE

A person plans to lose ½ pound each week by following a healthy diet. Write and simplify an expression to show his expected weight loss after six weeks of healthy eating

A loss of ½ pound each week for six weeks translates to $\left(-\frac{1}{2}\right)(6)$, which simplifies to -3. He can expect to lose three pounds in six weeks.

EXAMPLE

A gymnast's routine has a start value of 16 points. During her routine, she incurs three deductions of one-tenth of a point, two deductions of three-tenths of a point, and one deduction of half a point. Determine the score she receives for her performance.

Write an expression to represent the gymnast's score. Each deduction is subtracted from her start value. Write all the deductions as fractions or as decimals. $\frac{1}{10} = 0.1; \frac{3}{10} = 0.3; \frac{1}{2} = 0.5$.

$$16 - 3(0.1) - 2(0.3) - 0.5$$
$$16 - 0.3 - 0.6 - 0.5$$
$$15.7 - 0.6 - 0.5$$
$$15.1 - 0.5 = 14.6$$

The gymnast's score is 14.6.

Example

A service provider charges \$25.75 for phone, \$27.75 for internet, and \$33.50 for cable each month. A one-time credit of \$35.50 is applied towards a customer's bill when the customer opts to prepay for service by quarterly bank draft. After a new customer orders phone, internet, and cable service and signs up for automatic bill pay, she notices a transaction on her bank statement of -\$225.50. Write an expression which justifies this charge made by the service provider.

If the customer pays her bill quarterly, then she pays for three months of service at one time. So, her bill includes three times the total for the phone charge and the internet charge and the cable charge. A credit of \$35.00 is given only once.

$$\begin{aligned}&3[(-\$25.75) + (-\$27.75) + (-\$33.50)] + \$35.00\\&= 3(-\$87.00) + \$35.50\\&= -\$261.00 + \$35.50\\&= -\$225.50\end{aligned}$$

Commutative Property of Addition and Multiplication

The commutative property of addition states that the order in which two numbers are added does not change their sum; the commutative property of multiplication states that the order in which two numbers are multiplied does not change their product.

$$a + b = b + a$$

$$ab = ba$$

Associative Property of Addition and Multiplication

The associate property of addition states that a series of added numbers can be grouped in various ways without affecting the sum; the associative property of multiplication states that a series of multiplied numbers can be grouped in various ways without affecting the product.

$$a + (b + c) = (a + b) + c$$

$$a(bc) = (ab)c$$

Additive Identity and the Multiplicative Identity

The additive identity is the number which can be added to a number without changing its value; that number is zero. The multiplicative identity is the number which can be multiplied by a number without changing its value; that number is one.

Example

Use the distributive property to simplify $-\frac{1}{2}(x-8)$.

The distributive property states that $a(b+c) = ab + ac$ and $a(b-c) = ab - ac$.

$$-\frac{1}{2}(x-8) = \left(-\frac{1}{2}\right)(x) - \left(-\frac{1}{2}\right)(8) = -\frac{1}{2}x + 4.$$

EXAMPLE

Name the property used in each step of simplifying $\frac{3}{4} \cdot \frac{2}{3} + \frac{3}{4} \cdot \frac{1}{3}$.

$$\frac{3}{4} \cdot \frac{2}{3} + \frac{3}{4} \cdot \frac{1}{3} = \frac{3}{4}\left(\frac{2}{3} + \frac{1}{3}\right)$$
$$\frac{3}{4}\left(\frac{2}{3} + \frac{1}{3}\right) = \frac{3}{4}(1)$$
$$\frac{3}{4}(1) = \frac{3}{4}$$

$\frac{3}{4} \cdot \frac{2}{3} + \frac{3}{4} \cdot \frac{1}{3} = \frac{3}{4}\left(\frac{2}{3} + \frac{1}{3}\right)$	Distributive property
$\frac{3}{4}\left(\frac{2}{3} + \frac{1}{3}\right) = \frac{3}{4}(1)$	Substitution property of equality (Since $\frac{2}{3} + \frac{1}{3} = 1$, the number 1 can replace the expression $\frac{2}{3} + \frac{1}{3}$.)
$\frac{3}{4}(1) = \frac{3}{4}$	Multiplicative identity

EXAMPLE

Simplify.

$$-\frac{2}{3}\left(a - \frac{1}{4}\right)$$
$$\left(\frac{3}{4}x + 4\right) + \left(\frac{1}{4}x - 3\right)$$

Use the distributive property to simplify $-\frac{2}{3}\left(a - \frac{1}{4}\right)$.

$$\begin{aligned} -\frac{2}{3}\left(a - \frac{1}{4}\right) &= -\frac{2}{3} \cdot a + \left(-\frac{2}{3}\right)\left(-\frac{1}{4}\right) \\ &= -\frac{2}{3}a + \frac{2}{12} \\ &= -\frac{2}{3}a + \frac{1}{6} \end{aligned}$$

Use the associative and commutative properties of addition to simplify $\left(\frac{3}{4}x + 4\right) + \left(\frac{1}{4}x - 3\right)$.

$$\begin{aligned} \left(\frac{3}{4}x + 4\right) + \left(\frac{1}{4}x - 3\right) &= \frac{3}{4}x + 4 + \frac{1}{4}x - 3 \\ &= \frac{3}{4}x + \frac{1}{4}x + 4 - 3 \\ &= x + 1 \end{aligned}$$

Example

Factor $\frac{1}{2}$ from the expression $\frac{1}{4}x - \frac{1}{2}$.

To factor $\frac{1}{2}$ from the expression $\frac{1}{4}x - \frac{1}{2}$, divide each term in the expression by $\frac{1}{2}$.

$$\frac{1}{4}x \div \frac{1}{2} = \frac{1}{4} \cdot x \cdot \frac{2}{1} = \frac{1}{4} \cdot \frac{2}{1} \cdot x = \frac{2}{4} \cdot x = \frac{1}{2}x$$
$$-\frac{1}{2} \div \frac{1}{2} = -1$$

The factored expression is $\frac{1}{2}\left(\frac{1}{2}x - 1\right)$.

Words and/or Phrases

Addition

Some words and phrases that indicate addition are sum, plus, total, and, increased by, more, together, added to, combined with, gain.

Subtraction

Some words and phrases that indicate subtraction are difference, minus, less, decreased by, take away, fewer than, from, subtracted from, loss.

Multiplication

Some words and phrases that indicate multiplication are product, times, multiplied by, of, twice/double (×2), thrice/triple (×3).

Division

Some words and phrases that indication division are quotient, divided by, into, among, between, over, per, for every, ratio of, out of.

These lists are not exhaustive.

Example

Joshua calculates that the product of 19 $\frac{3}{4}$ and 10$\frac{1}{4}$ is 404 $\frac{7}{8}$. Use mental estimation to determine the reasonableness of his answer.

The product of two numbers is found by multiplying those numbers, so the product of 19 $\frac{3}{4}$ and 10$\frac{1}{4}$is about 200 since 19 $\frac{3}{4}$ is close to 20 and 10$\frac{1}{4}$ is close to 10. An answer of 404 $\frac{7}{8}$ seems unreasonable, so Joshua should check his calculation.

Example

Suppose you wish to center a 3 ¼ ft wide painting over a buffet which is 5 ¾ ft wide. Approximate how far each edge of the painting would be from each edge of the buffet.

The painting is just over 3 ft wide, and the buffet is just under 6 feet wide. So, the difference in the width of the buffet and the painting is about 3 ft.. The painting is centered above the buffet, half of the difference in width will be space to the left of the painting and the other half will be space to the right of the painting. Half of 3 ft is

1 ½ ft. So, each edge of the painting should be about 1 ½ ft from each edge of the buffet.

MATHEMATICAL SYMBOLS

= equals, is equal to, is, was, were, will be, yields, is the same as, amounts to, becomes

>**is** greater than, **is** more than

≥ **is** greater than or equal to, is at least, is no less than

<**is** less than, **is** fewer than

≤ **is** less than or equal to, is at most, is no more than

EXAMPLE

Write three sentences which would translate into the inequality $2(x + 4) \geq 6$.

There are many ways to write $2(x + 4) \geq 6$ as a sentence, including

Two times the sum of a number and four is greater than or equal to six.

Twice the quantity *x* increased by four is at least six.

The product of two and a number to which four has been added must be no less than six.

EXAMPLE

An $80 dress is marked down 25%. Find the price of the dress after the discount.

One way to find the price of the dress after the discount is to calculate the amount of the discount and subtract it from the price of the dress:

The amount of the discount is 25% of $80, or $0.25 \times \$80 = \$20$.

The price of the dress after the discount is $\$80 - \$20 = \$60$.

Another way to find the discounted price is to write, simplify, and evaluate an expression representing the problem:

- If p = original price, then $0.25p$ represents the amount of the discount. So, an expression for the new price of the dress is $p - 0.25p$, which simplifies to $0.75p$. In other words, the discounted price of the dress is 75% of the original price. $0.75p = 0.75(\$80) = \$60$.

EXAMPLE

If *px* represents the price of an item, write an expression can be used to find

- The price of the item with 8% sales tax.
- The pre-tax price of the item during a half-off sale.
- The after-tax price of the item during a half-off sale.

The amount of an 8% sales tax added to an item which costs *px* dollars is $0.08p$. The price of the item with the sales tax is $p + 0.08p$, which simplifies to $\mathbf{1.08p}$.

If the item is half off, the amount of the discount is $\frac{1}{2}p$. The new price of the item is the original price minus the discount, or $-\frac{1}{2}p$, which simplifies to $\mathbf{\frac{1}{2}p}$.

The price of the half-off item with 8% sales tax is $1.08\left(\frac{1}{2}p\right) = 1.08(0.5p) = \mathbf{0.54p}$.

DETERMINING THE ORDER OF OPERATIONS USED TO SIMPLIFY EXPRESSIONS

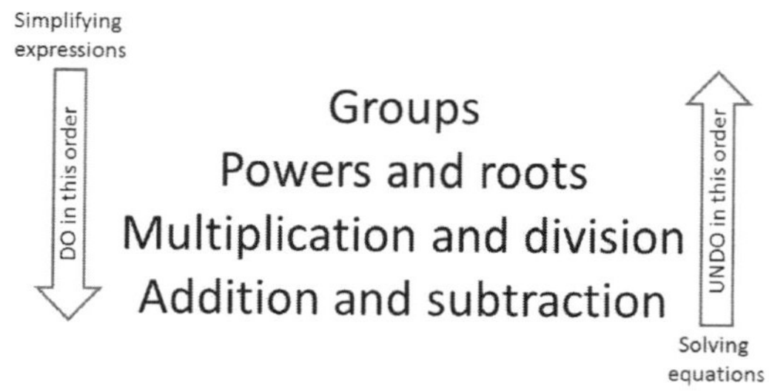

When simplifying an expression, work first within groups, which can be found within grouping symbols such as parentheses but also can be found under radical signs, in numerators or denominators of fractions, as exponents, etc. without such grouping symbols. Next, simplify powers and roots. Then multiply and divide from left to right; finally, add and subtract from left to right.

RELATE THIS TO THE SEQUENCE OF OPERATIONS USED WHEN SOLVING EQUATIONS

When solving an equation, it is often helpful to first use order of operations to simplify the expressions on both sides of the equation, if possible. Then, undo the operations which have been performed on the variable by using inverse operations in reverse order of operations.

EXAMPLE

Alina spent \$20 at the fair. She paid \$2 for admission plus \$1.50 for every ride. Write and solve an equation to determine how many rides she rode.

Let $x =$ the number of rides Alina rode at the fair. The expression $\$1.50x$ represents the amount of money spent riding rides. Alina spent a total of $\$1.50x + \$2$ on rides and admission; this amount equals \$20, so

$$\begin{aligned} \$1.50x + \$2 &= \$20 \\ \underline{\quad -\$2} &\underline{\;-\$2} \\ \$1.50x + \$0 &= \$18 \\ \frac{\$1.50x}{\$1.50} &= \frac{\$18}{\$1.50} \\ x &= 12 \end{aligned}$$

Alina rode twelve rides at the fair.

EXAMPLE

24 feet of fencing was used to enclose a rectangular garden with a width of 8 feet. Write and solve an equation to determine the length of the garden.

The perimeter of a rectangle can be found using the expression $2l + 2w$, where l is the rectangle's length and w is its width. The perimeter of the garden is 24 feet, and the width of the garden is 8 feet, so

$$\frac{150}{2} = x$$
$$75 = x$$

The distance between the cities is 75 miles.

EXAMPLE

A room has dimensions of 12' wide by 15' long. Using a scale of 1/4":1', draw a blueprint of the room.

First, determine the dimensions of the room on the blueprint. Let w represent the width in inches and l represent the length in inches of the room on the scale drawing.

$$\frac{\frac{1}{4}\text{ in}}{1\text{ ft}} = \frac{w}{12\text{ ft}}$$

Using cross-multiplication, $1 \cdot w = \left(\frac{1}{4}\right)(12) = 3$. The width of the room on the blueprint is 3 in.

$$\frac{\frac{1}{4}\text{ in}}{1\text{ ft}} = \frac{l}{15\text{ ft}}$$

Again using cross-multiplication, $1 \cdot l = \left(\frac{1}{4}\right)(15) = \frac{15}{4} = 3\frac{3}{4}$. The length of the room on the blueprint is $3\frac{3}{4}$in.

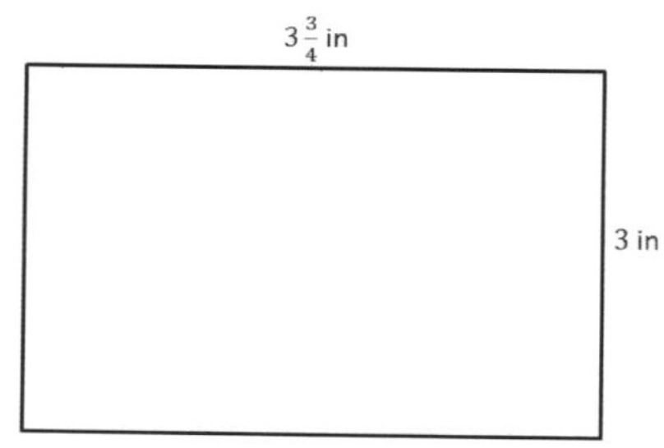

EXAMPLE

Below is a box drawn at 1:4 scale.

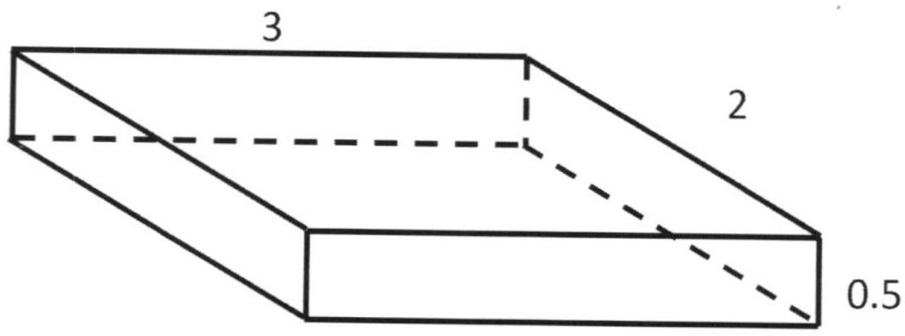

Draw the box at a scale of 1:8.

Determine the actual dimensions of the box.

A scale of 1:4 mean that each side of the box is $\frac{1}{4}$ the size of the actual box.

Redrawing at a scale of 1:8 means that each side of the drawing will be $\frac{1}{8}$ the size of the actual box, or $\frac{1}{2}$ of the size of the given scale drawing since $\frac{1}{2} \times \frac{1}{4} = \frac{1}{8}$. So, draw a box whose dimensions are half of the length, width, and height of the given scale drawing. Since ½=0.5, use either number to calculate the new dimensions.
$0.5 \times 3 \text{ in} = 1.5 \text{ in}$ $0.5 \times 2 \text{ in} = 1 \text{ in}$ $0.5 \times 0.5 \text{ in} = 0.25 \text{ in}$

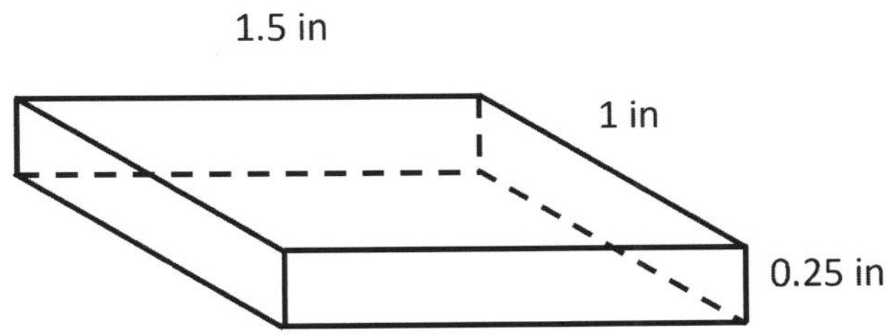

Since each side of the scale drawing is $\frac{1}{4}$ the size of each side of the actual box, the sides of the box are four times the size of the sides in the scale drawing. So, the box's actual dimensions are

- $4 \times 3 \text{ in} = 12 \text{ in}$ $4 \times 2 \text{ in} = 8 \text{ in}$ $4 \times 0.5 \text{ in} = 2 \text{ in}$.

Example

The ratio of the lengths of two squares is 1:5.

- Determine the ratio of the perimeters of the two squares.
- Determine the ratio of the areas of the two squares.

If the ratio of the lengths of two squares is 1:5, the ratio of the perimeters of the two squares is also 1:5. Consider, for instance, a square with a length of 1 cm. Its width would also be 1 cm, so it perimeter would be $2l + 2w = 2(1 \text{ cm}) + 2(1 \text{ cm}) = 2 \text{ cm} + 2 \text{ cm} = 4 \text{ cm}$. A square with a length five times that of the first square would have a perimeter of $2(5 \text{ cm}) + 2(5 \text{ cm}) = 10 \text{ cm} + 10 \text{ cm} = 20 \text{ cm}$. So, the ratio of the perimeters of the squares would be 4:20, which reduces to 1:5.

Considering the same two squares, the area of the first square would be $l \times w = (1 \text{ cm})(1 \text{ cm}) = 1 \text{ cm}^2$. The area of the second would be $(5 \text{ cm})(5 \text{ cm}) = 25 \text{ cm}^2$. So, the ratios of the areas of the squares **would be 1:25.**

Example

Determine whether or not a triangle can be constructed given

- The measures of three angles.
- The lengths of three sides.

The sum of the measures of a triangle's angles is always 180°. So, a triangle can be constructed from given angle measurements only if those measurements add to 180°.

The sum of the lengths of two shorter sides of a triangle must be greater than the length of the third side.

EXAMPLE

Using a ruler and a protractor, draw a right triangle with the two shorter sides measuring 3 in and 4 in. Measure the length of the longest side and the approximate measures of the two non-right angles. Make sure the measurements are consistent with the properties of a triangle.

The length of the longest side is 5 inches. The angle across from the side measuring 4 inches is approximately 53°, and the other angle is approximately 37°. These measures are consistent with the properties of triangles:

- The sum of the angles of the triangle is $53° + 37° + 90° = 180°$. The combined lengths of the two shorter sides exceed the longest side: $3 \text{ in} + 4 \text{ in} > 5\ in$. The smallest angle forms the shortest side, and the largest angle forms the longest side.

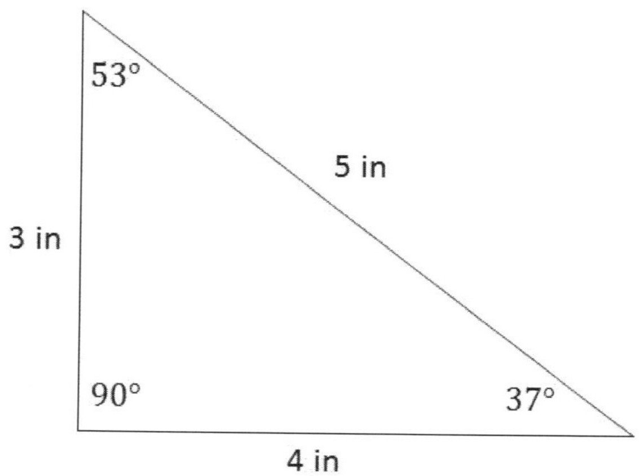

EXAMPLE

An equilateral triangle whose sides each measure two inches is enlarged by a factor of four. Determine the measurements of the enlarged triangle's sides and angles.

A triangle which has three congruent sides also must have three congruent angles. Since the sum of the angles of a triangle is 180°, each angle measures $\frac{180°}{3} = 60°$. When a triangle is dilated, its angles remain the same; the lengths of the sides change but remain proportionally. Since the scale factor is four, the length of each the three sides is $2 \text{ in} \times 4 = 8 \text{ in}$. So, each angle of the enlarged triangle measures 60°, and each side measures 8 in.

EXAMPLE

The area, A, of a parallelogram can be found using the formula $A = bh$, where b is the length of the base and h is the height, which is the distance between the parallelogram's base and its opposite, parallel side. Two congruent triangles are obtained when a parallelogram is cut in half from one of its corners to the opposite corner. From this information, determine the area formula of a triangle.

Two identical triangles make up the parallelogram's area, so the area of one triangle is half the area of the parallelogram. Since the area of a parallelogram is $A = bh$, the area of a triangle is $A = \frac{1}{2}bh$, where b is the base of the triangle and h is the height of the triangle as shown.

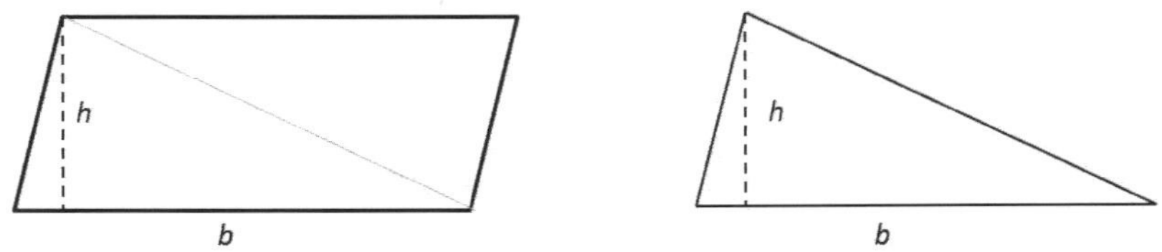

EXAMPLE

Determine if no triangle, one triangle, or more than one triangle can be drawn given the side and/or angle measurements.

- Side lengths 3 cm, 4 cm, and 5 cm.
- Angle measurements 60°, 40°, and 80°.

In order to construct a triangle, the sum of the lengths of two shorter sides of a triangle must be greater than the length of the third side. $3 \text{ cm} + 4 \text{ cm} > 5\ cm$, so these sides can be used to draw one triangle.

The sum of the measures of a triangle's angles is always 180°. Since 60°+ 40°+ 80°=180°, a triangle can be constructed with these angle measurements. In fact, many similar triangles can be constructed with these angle measures.

EXAMPLE

Determine the two-dimensional cross section obtained by slicing a right, triangular prism parallel to its base

A triangular prism contains two congruent triangular bases which lie in parallel planes. The side edges of a right prism are perpendicular to the base.

Taking the cross section of right, triangular prism parallel to its base yields a triangle which is congruent to the prism's base. One such cross section is illustrated.

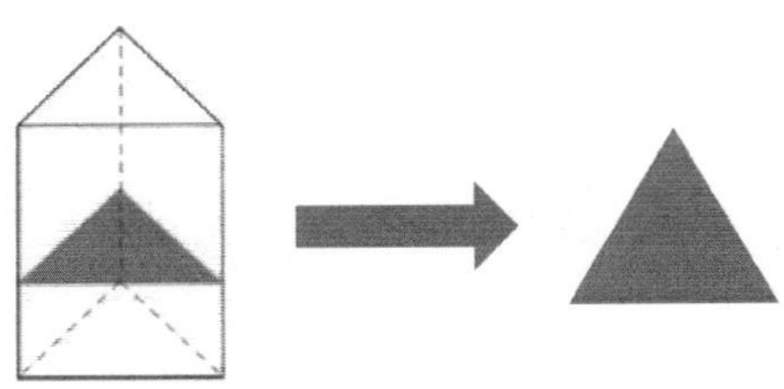

EXAMPLE

Determine the two-dimensional cross section obtained by slicing a right, triangular prism perpendicular to its base

The cross section of right, triangular prism perpendicular to its base yields a rectangle. One such cross section is illustrated.

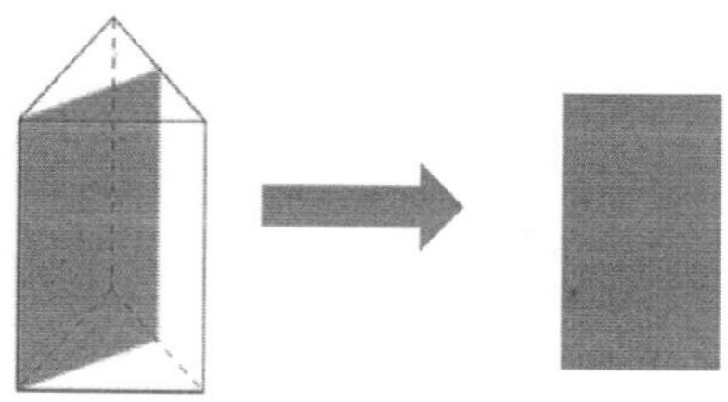

EXAMPLE

Determine the two-dimensional cross section obtained by slicing a right, rectangular pyramid parallel to its base

A rectangular pyramid is one which has one rectangular base and four triangular sides which meet at an apex. The apex of a right pyramid lies directly above the center of the pyramid's base.

The cross section of a right, rectangular pyramid parallel to its base is a rectangle which is smaller than but similar to the rectangular base. One such cross section is illustrated.

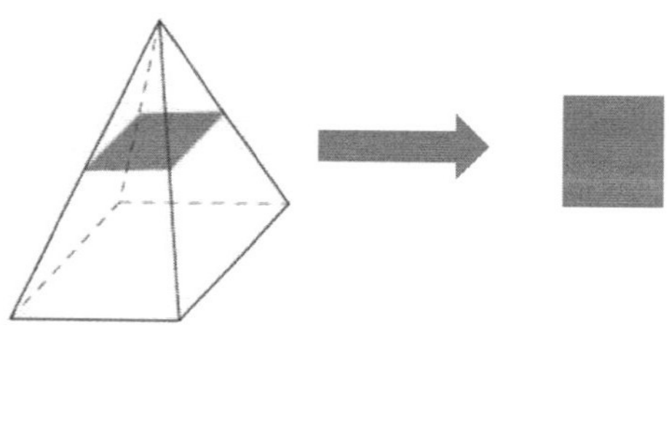

EXAMPLE

Determine the two-dimensional cross section obtained by slicing a right, rectangular pyramid perpendicular to its base and through the apex.

A rectangular pyramid is one which has one rectangular base and four triangular sides which meet at an apex. The apex of a right pyramid lies directly above the center of the pyramid's base.

The cross section of a right, rectangular pyramid perpendicular to its base and through its apex is a triangle whose height is equal to the height of the pyramid. One such cross section is illustrated.

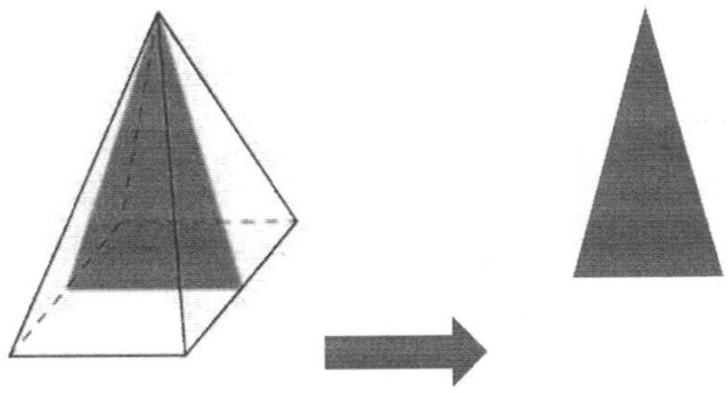

Example

Write the formulas used to find the circumference and area of a circle, respectively.

The formula used to find the circumference, C, of a circle is $C = 2\pi r$ or $C = \pi d$, where r is the radius of the circle and d its diameter. The formula used to find the area, A, of a circle is $A = \pi r^2$, where r is the radius of the circle.

Example

A round table has a diameter of 6 feet. A circular table cloth is cut from a piece of fabric in such a way that it hangs down 6 inches all the way around the table, and a decorative fringe is added along the cut. Using 3.14 as an approximation for π, determine the area of the table cloth and the length of the fringe. Round answers to the nearest tenth.

The area of the table cloth is found using the area formula of a circle, $A = \pi r^2$. First, find the radius of the table cloth.

Since the table cloth hangs down 6", which is equivalent to half a foot, add 0.5 ft to the radius of the table to find the radius of the table cloth. The radius of the table cloth is 3.5 ft, so the area of the table cloth is about 38.5 square feet:

$$A = \pi r^2$$
$$A = (3.14)(3.5 \text{ ft})^2$$
$$A = (3.14)(3.5 \text{ ft})^2$$
$$A = (3.14)(12.25 \text{ ft}^2) = 38.465 \text{ ft}^2$$

The fringe goes around the circular table cloth, so its length can be found using the formula for the circumference of a circle, $C = 2\pi r$. The length of the fringe is about 22.0 ft.

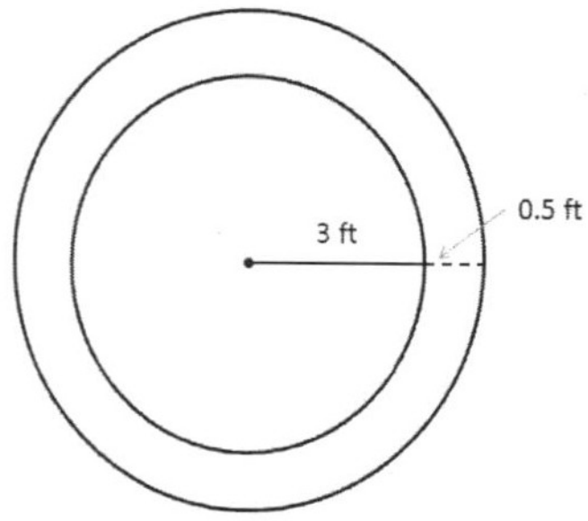

$$C = 2\pi r$$

$$C = 2(3.14)(3.5\text{ ft}) = 21.98\text{ ft}$$

EXAMPLE

Name a pair of:

- Supplementary angles
- Complementary angles
- Vertical angles
- Adjacent angles

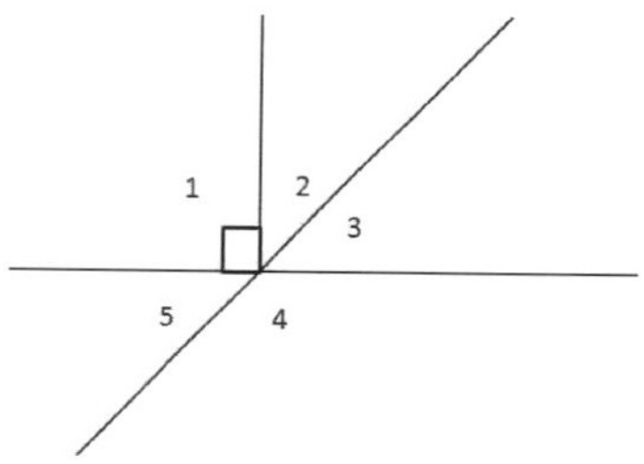

The sum of two supplementary angles is 180°. Angles 4 and 5 are supplementary, as are angles 3 and 4.

The sum of two complementary angles is 90°. Angles 2 and 3 are complimentary.

Vertical angles share a vertex but are not adjacent; rather, vertical angles are congruent angles across from each other in the X made by the intersection of two lines. Angles 3 and 5 are vertical angles.

Adjacent angles share a vertex and a side. Angles 1 and 2, 2 and 3, 3 and 4, 4 and 5, and 1 and 5 are adjacent.

The angles label $(2x + 10)°$ measures $48°$, so its complement measures $90° - 48° = 42°$, and its supplement measures $180° - 48° = 132°$.

EXAMPLE

Match each measurement with the appropriate unit of measurement.

Measurements	Units
Length of a table	Feet
Area of a house	Square feet
Perimeter of a room	Cubic feet
Surface area of a polyhedron	
Volume of a box	
Distance between two houses	
Storage capacity of a refrigerator	

Lengths or distances are measured in units, while area is measured is square units and volume is measured in cubic units. When the unit of measure is feet:

- The length of a table is measured in *feet.*
- The area of a house is measured in *square feet.*
- The perimeter of a room is the distance around the room and is measured in *feet.*
- The surface area of a polyhedron is the sum of the areas of its polygonal faces and is measured in *square feet.*
- The volume of a box is measured in *cubic feet.*
- The distance between two houses is measured in *feet.*
- The storage capacity of a refrigerator is how much space is inside. The refrigerator's volume is measured in *cubic feet.*

EXAMPLE

Find the surface area and volume of the box.

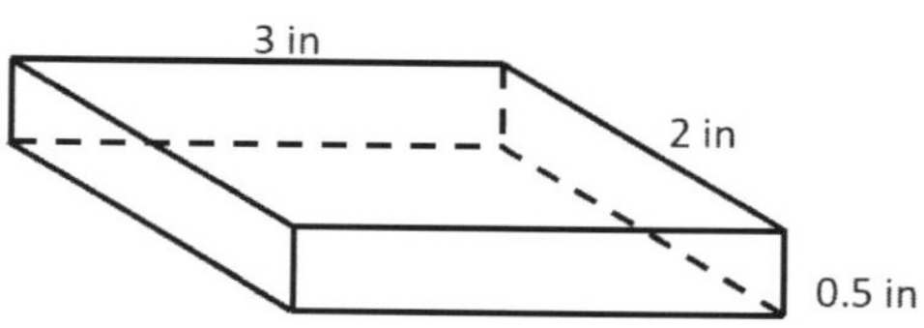

The surface area of a box is the sum of the areas of its six rectangular surfaces.

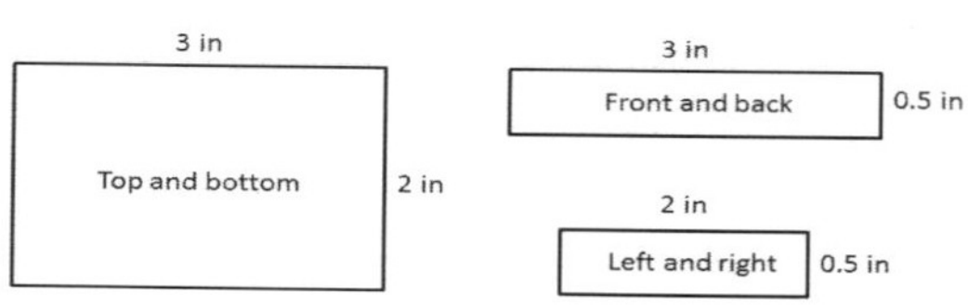

The area of the top rectangle is $A = lw = (3\text{ in})(2\text{ in}) = 6\text{ in}^2$. The area of the top rectangle combined with the bottom rectangle is $6\text{ in}^2 + 6\text{ in}^2 = 12\text{ in}^2$.

The area of the front rectangle is $A = lw = (3\text{ in})(0.5\text{ in}) = 1.5\text{ in}^2$. The area of the front rectangle combined with the back rectangle is $1.5\text{ in}^2 + 1.5\text{ in}^2 = 3\text{ in}^2$.

The area of the left side rectangle is $A = lw = (2 \text{ in})(0.5 \text{ in}) = 1 \text{ in}^2$. The area of the left and right side rectangles together is $1 \text{ in}^2 + 1 \text{ in}^2 = 2 \text{ in}^2$.

The total surface area of the box is $12\ in^2 + 3\ in^2 + 2\ in^2 = 17\ in^2$.

The volume, V, of a box is found by $V = lwh$. o, $V = (3\ in)(2in)(0.5\ in) = 3\ in^3$.

EXAMPLE

An 11'×13' room contains a 3' wide, 7' tall doorway and two 5'x3' windows. The ceiling height is 9'. Determine :

- The price to install baseboards which cost $1.25 per linear foot.
- The price to install flooring which costs $5 per square foot.
- If one gallon of paint which covers 400 square feet of surface is sufficient to paint the walls of the room.

Baseboards run along the edge of the room, but not across doorways. To determine the price for the baseboards, first determine how many feet are needed for the perimeter of the room, excluding the doorway. Use the formula for the perimeter of a rectangle to find the perimeter of the room: $2l + 2w = 2(11 \text{ ft}) + 2(13 \text{ ft}) = 48 \text{ ft}$. After adjusting for the width of the doorway, 45ft of baseboard is needed for the room. The price of baseboards is 45×1.25=$56.25.

To determine the price for flooring, first determine the area of the room. Use the formula for the area of a rectangle to find the area of the room: $lw = (11 \text{ ft})(13 \text{ ft}) = 143 \text{ ft}^2$. The price for the flooring is 143×5=$715.

To determine the amount of paint needed for the walls, first determine the total surface area to be covered. Two walls are 11'×9' and the other two walls are 13'×9'. Disregarding doors and windows, the total area of the walls is $2(11 \text{ ft})(9 \text{ ft}) + 2(13 \text{ ft})(9 \text{ ft}) = 198 \text{ ft}^2 + 234 \text{ ft}^2 = 432 \text{ ft}^2$. Take away the area of the door, which is $(3 \text{ ft})(7 \text{ ft}) = 21 \text{ ft}^2$, and the area of the windows, which is$2(5 \text{ ft})(3 \text{ ft}) = 30 \text{ ft}^2$: $432 \text{ ft}^2 - 21 \text{ ft}^2 - 30 \text{ ft}^2 = 381 \text{ ft}^2$. The amount of surface which needs paint is 381 ft^2. Purchasing one gallon of paint should be sufficient to cover the walls if only one coat of paint is needed.

EXAMPLE

Find the perimeter of the shape below.

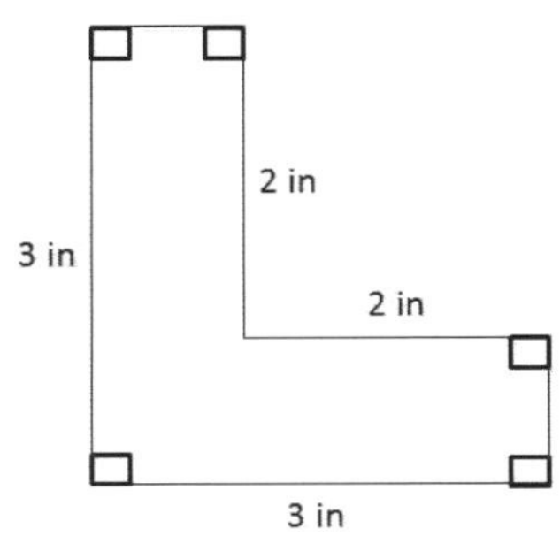

The perimeter of a polygon is the sum of its sides, which is 12 inches.

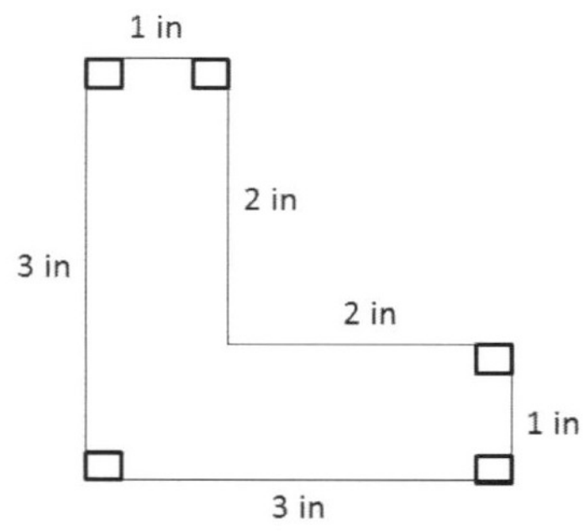

Example

Find the area of the shape below.

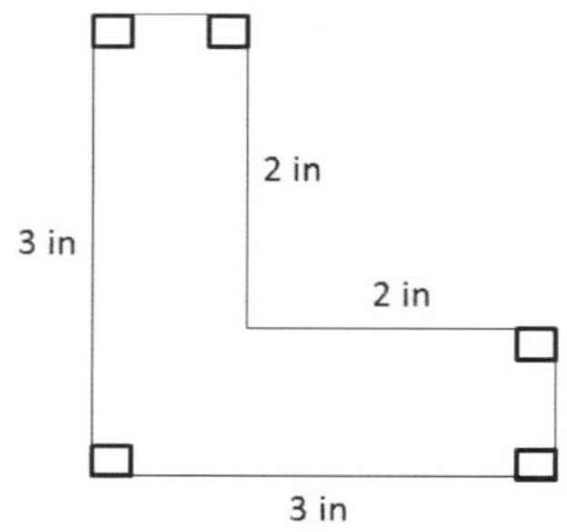

To find the area of an irregular polygon, draw it as a familiar polygon to which another familiar polygon is added or from which a familiar polygon is removed. For example, the shape can be viewed as vertically-oriented rectangle to which a horizontally-oriented rectangle is added, or it can be viewed as a square from which a smaller square is removed.

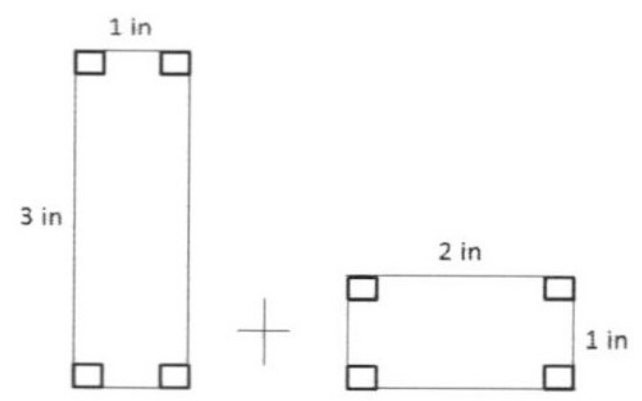

The area of the irregular polygon is the sum of the areas of the two rectangles shown. The area of a rectangle is $A = lw$, so the area of the polygon is $(3 \text{ in})(1 \text{ in}) + (2 \text{ in})(1 \text{ in}) = 3 \text{ in}^2 + 2 \text{ in}^2 = 5 \text{ in}^2$.

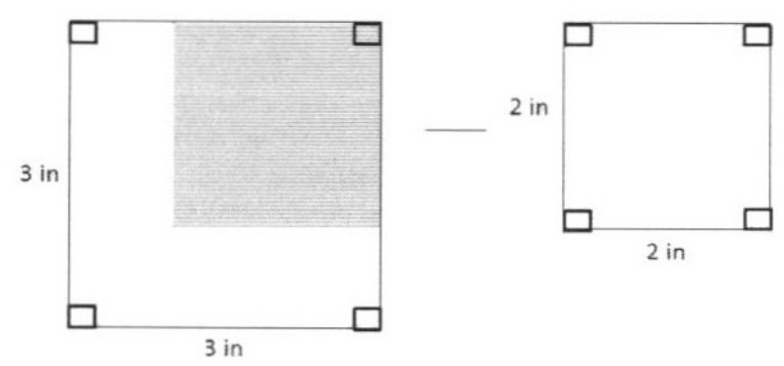

The area of the irregular polygon is the area of the larger square minus the area of the smaller square. The larger square has side lengths of 3 inches, so the area is $(3 \text{ in})(3 \text{ in}) = 9 \text{ in}^2$. The smaller square has an area of 4 in^2. So, the area of the irregular polygon is $9 \text{ in}^2 - 4 \text{ in}^2 = 5 \text{ in}^2$.

Notice that either method of calculating the irregular polygon's area gives the same answer.

MEAN, MEDIAN, AND MODE

Mean, median, and mode are all measures of central tendency. Measures of central tendency summarize a set of data with values which represent the average, middle, or most common value. The mean, or average, of numerical data can be found by dividing the sum of the numbers in a set by how many numbers are in that set. When numerical data are organized from least to greatest, the middle number or average of the two middle numbers is the median of the set. The mode is the value which appears most frequently in a data set; there may be no mode, one mode, or more than one mode in a set.

EXAMPLE

Find the mean, median, and mode of these numbers:

23, 42, 36, 21, 28, 29, 32, 28, 40, 36, 39.

The mean of a set of numbers is the sum of the numbers divided by how many numbers are in the set. So, the mean of this set of numbers is $\frac{23+42+36+21+28+29+32+28+40+36+39}{11} = \frac{354}{11} = 32.\overline{18}$.

To find the median, put the numbers in increasing order and find the number in the middle:

21, 23, 28, 28, 29, **32**, 36, 36, 39, 40, 42. The median is 32.

The mode is the value in the set that occurs most frequently. There may be no mode, one mode, or more than one mode in a set. This set has two values that occur twice while all others occur only once. Thus, the modes are those two values, namely 28 and 36.

RANDOM SAMPLING

A random sample is a collection of members, chosen at random and with equal likelihood, from a group about which information is desired. Rather than collecting information from the entire group, which can be quite difficult when the group is very large, information is collected instead from the sample. If the sample is representative of the group and is sufficiently large, then the information gained from the sample is representative can be used to describe the group as a whole.

EXAMPLE

Determine whether or not each represents a random sample of seventh graders at a middle school:

- A student selects all the seventh graders on her bus.
- A teacher puts the names of her first period students in a hat and draws ten names.
- A principal assigns uses a random number generator to select student ID numbers of seventh graders.

Each seventh grader must be chosen at random and must have an equal chance of being selected. Neither criterion is true in the first scenario. The teacher in the second scenario is selecting a random sample of students in her first period class, but this is not a random sample of the group of seventh graders as a whole. The principle's method of selecting is random and ensures that the likelihood of selecting one student is the same as the likelihood of selecting another.

Example

A random number generator produced the following sets of eight numbers between 1 and 99:

82, 60, 40, 69, 40, 36, 25, 59
98, 28, 15, 91, 51, 74, 11, 36
21, 66, 46, 16, 32, 73, 3, 81
91, 80, 32, 72, 1, 53, 51, 28.

Determine the mean of each set. Describe how well the mean of the each random sample represents the set from which the sample was taken.

The mean, or average, is a measure of central tendency which can be found by dividing the sum of the numbers in a set by how many numbers are in that set. The mean of the numbers 1 through 99 is 50, so the mean of a random sample of numbers taken from 1 to 99 should be approximately 50.

$$\frac{82+60+40+69+40+36+25+59}{8}=\frac{411}{8}=51.375$$
$$\frac{98+28+15+91+51+74+11+36}{8}=\frac{404}{8}=50.5$$
$$\frac{21+66+46+16+32+73+3+81}{9}=\frac{338}{8}=42.25$$
$$\frac{91+80+32+72+1+53+51+28}{8}=\frac{408}{8}=51$$

Samples 1, 2, and 4 seem to be a good representatives of the set from which they were taken. The mean of the values in the third sample varies more from the actual mean than the others. One way to improve the likelihood of a random sample's representation of the actual set is to collect more samples.

Example

Suppose all 567 seventh graders in a school vote on which field trip they would like to take from the following options: science center, art museum, or state capitol. Explain how an administrator might use the surveys to predict the most popular vote without tallying all of the results.

An administrator could compile the results from a random sample of surveys. For instance, if the surveys were shuffled, and 50 randomly chosen surveys showed a strong preference for the science center, then the administrator would likely predict that the science center will be the popular vote. To increase her confidence in the results, the administrator could increase the sample size by examining more surveys.

Example

When determining the reading level of a book, a publisher considers many factors, including the average word length. Explain how a publisher might use random sampling to find the average word length in a book.

To find the average word length in a book, a publisher might randomly select a set of words from the book and find the average length of those words. This average should be representative of the whole book if the words are indeed chosen at random.

Example

The histogram below displays the heights of randomly selected eighteen-year-old boys and girls living in Atlanta, Georgia. Compare the variability within each group and between the groups.

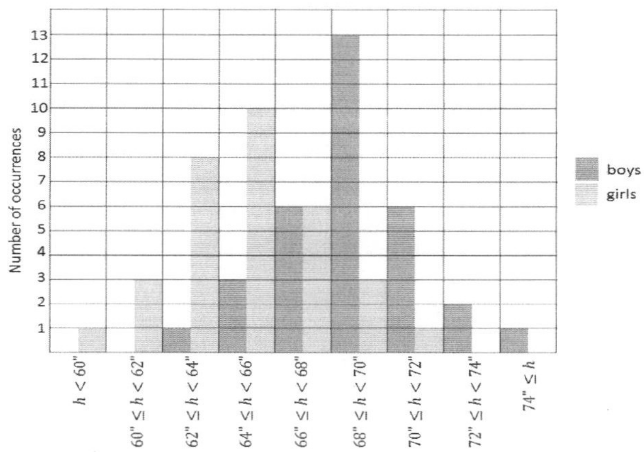

Both of the histograms show a similar, normal distribution of heights, with fewer individuals at the two extremes and the majority clustered at a more central point. The two distributions overlap, which means that some eighteen-year-old boys are the same height as some eighteen-year-old girls. However, there is a noticeable shift to the right in the bell-shaped distributions for boys when compared to girls; this indicates that eighteen-year-old boys are generally taller than girls of the same age. According to the chart, the most common height for eighteen-year-old boys is between 68 and 70 inches, while the most common height among girls is 64 and 66 inches.

Explain what each of the probabilities means in terms of the likelihood of an event:

$$0$$
$$\frac{1}{100{,}000}$$
$$\frac{1}{2}$$
$$99\%$$
$$1$$

The probability of an event occurring ranges from 0 to 1 when expressed as a fraction or decimal and between 0% and 100% when expressed as a percentage. When there are a finite number of outcomes, a probability of 0 means that the occurrence of an event is impossible, while a probability of 1 means the occurrence is certain. The closer a value is to 0, the less likely it is to occur. For example, a probability of $\frac{1}{100{,}000}$ indicates that an event is unlikely to occur, whereas a probability of 99% indicates a likely event. Since ½ is halfway between 0 and 1, a probability of ½, or 50%, means that an event is neither likely nor unlikely.

In Terms of Probability, Define and Give an Example of a

- Sample space
- Simple event
- Compound event

The set of all outcomes of a probability experiment is called the sample space. For example, if a coin is tossed one time, there are two possible outcomes: heads or tails. So, the sample space consists of two elements. If a coin is tossed two times, there are four possible outcomes: heads then tails, heads then heads, tails then heads, tails then tails. So, the sample space consists of four outcomes.

A simple event consists of only one outcome in the sample space. For instance, the event of getting heads in a single coin toss is a simple event.

Compound events consist of more than one outcome in the sample space. For instance, the event of getting heads at least once in two coin tosses is a compound event. The compound event of getting heads is composed of three outcomes: heads then tails, heads then heads, tails then heads.

Determine the Following:

- The probability of winning a single coin toss.
- The probability of a rolling a multiple of 3 on a die.
- The probability of randomly picking a green marble from a bag containing 15 blue marbles and 5 green marbles.
- The probability of randomly picking a red marble from a bag containing 15 blue marbles and 5 green marbles.

Probability is the chance that something will happen. The probability of an event is the ratio of the number of favorable outcomes to the number of possible outcomes when all outcomes are equally likely.

Because there is one favorable outcome of two equally likely outcomes, so the probability of winning a coin toss is ½, or 50%.

Both 3 and 6 are multiples of 3, so there are two favorable outcomes out of six equally likely total outcomes. So, the probability of rolling a multiple of 3 on a die is $\frac{2}{6} = \frac{1}{3} = 33.\bar{3}\%$.

There are five green marbles and fifteen blue marbles in a bag. The probability of picking a green marble is the ratio of green marbles to total marbles, or $\frac{5}{20} = \frac{1}{4} = 25\%$.

Because there are no red marbles in the bag, it is not possible to choose a red marble from the bag. Therefore, the probability of choosing a red marble is 0.

Distinguish Between Theoretical Probability and Experimental Probability.

Theoretical probability is the expected likelihood of an event. Experimental probability is found by conducting trials and comparing the actual occurrence of an event to the number of trials.

For example, the probability of rolling a 2 on a die is 1/6 because there is one favorable outcome, namely rolling a 2, and six equally possible outcomes. So, theoretically, a 2 would appear 100 times if a die is rolled 600 times. Suppose, however, that a die is actually rolled 600 times, and a 2 appears 90 times. The experimental probability is $\frac{90}{600} = \frac{3}{20}$.

expected results based on the experimental probability seen here.

	1	2	3	4	5	6
Number of times rolled	3	4	8	3	2	4
Experimental probability	$\frac{3}{24}=\frac{1}{8}$	$\frac{4}{24}=\frac{1}{6}$	$\frac{8}{24}=\frac{1}{3}$	$\frac{3}{24}=\frac{1}{8}$	$\frac{2}{24}=\frac{1}{12}$	$\frac{4}{24}=\frac{1}{6}$
Theoretical probability	$\frac{1}{6}$	$\frac{1}{6}$	$\frac{1}{6}$	$\frac{1}{6}$	$\frac{1}{6}$	$\frac{1}{6}$
Expected results $\mathbf{\frac{1}{6}\times\frac{24}{1}=\frac{24}{6}=4}$	4	4	4	4	4	4

The results of 24 rolls of a die are tabulated below. Determine the experimental probability from these results and use it to predict the number of times each result will occur if a number cube is rolled 200 times.

	1	2	3	4	5	6
Number of times rolled	3	4	8	3	2	4

The predicted results of 200 rolls based on the experimental probabilities are shown below.

	1	2	3	4	5	6
Number of times rolled	200	200	200	200	200	200
Experimental probability	$\frac{1}{8}$	$\frac{1}{6}$	$\frac{1}{3}$	$\frac{1}{8}$	$\frac{1}{12}$	$\frac{1}{6}$
Predicted results	$\frac{200}{8}=25$	$\frac{200}{6}\approx 33$	$\frac{200}{3}\approx 67$	$\frac{200}{8}=25$	$\frac{200}{12}\approx 17$	$\frac{200}{6}\approx 33$

Note that every predicted result is rounded to the nearest whole number since there cannot exist a fraction of a roll. Check to makesure the sum of the rounded numbers is 200: 25+33+67+25+17+33=200.

If the die is indeed loaded, these would be the predicted results. If the die is fair, the results of 200 rolls should show an even distribution of around 33 or 34 rolls over all outcomes.

Joseph is in a math class with 23 other students, 14 of whom are girls. If a student is selected at random from the class, determine the probability that:

- The student selected is Joseph.
- The student selected is a boy.

Since each student has an equally likely chance of being selected from the class, the probability of selecting Joseph is $\frac{1}{number\ of\ students\ in\ the\ class}$. The number of students in the class, including Joseph, is 24. So, $P(\text{Joseph}) = \frac{1}{24}$.

Since each student has an equally likely chance of being selected from the class, the probability of selecting a boy is $\frac{\textit{number of boys in the class}}{\textit{number of students in the class}}$. Since 14 of the 24 students are girls, there are 10 boys. So, $P(\text{boy}) = \frac{10}{24} = \frac{5}{12}$.

A baby inherited one copy of a beta-globin gene from her mother and one from her father. Both the baby's mother and father are carriers for sickle cell anemia, meaning that each parent contains a normal allele for beta-globin called type A, and a recessive allele which has a single mutation called type S. The Punnet square below shows the possible genotypes (types of genes) and phenotypes (expressions of the genotypes). Determine the probability that the baby has sickle cell anemia. Express the probabilities as a fraction and as a percent.

	A	S
A	AA (no disease)	AS (no disease/carrier)
S	AS (no disease/carrier)	SS (sickle cell anemia)

The probability that the baby will have sickle cell anemia is ¼, or 25%.

	A	S
A	AA (no disease)	AS (no disease/carrier)
a	AS (no disease/carrier)	SS (sickle cell anemia)

DETERMINE THE PROBABILITY THAT THE BABY HAS INHERITED A MUTATED ALLELE FROM AT LEAST ONE OF HER PARENTS. EXPRESS THE PROBABILITIES AS A FRACTION AND AS A PERCENT.

	A	S
A	AA (no disease)	AS (no disease/carrier)
S	AS (no disease/carrier)	SS (sickle cell anemia)

The probability that the baby has inherited a mutated allele from at least one of her parents is ¾, or 75%.

	A	S
A	AA (no disease)	AS (no disease/carrier)
a	AS (no disease/carrier)	SS (sickle cell anemia)

Determine the probability that the baby does not have the disease but carries one copy of the mutated allele. Express the probabilities as a fraction and as a percent.

	A	S
A	AA (no disease)	AS (no disease/carrier)
S	AS (no disease/carrier)	SS (sickle cell anemia)

The probability that the baby does not have the disease but carries one copy of the mutated allele is $\frac{2}{4} = \frac{1}{2}$, or 50%.

	A	S
A	AA (no disease)	AS (no disease/carrier)
a	AS (no disease/carrier)	SS (sickle cell anemia)

Show the sample space for rolling two dice and find the probability of rolling double sixes.

Since there are six outcomes for each die, there are $6 \times 6 = 36$ outcomes for rolling two dice.

1,1	1,2	1,3	1,4	1,5	1,6
2,1	2,2	2,3	2,4	2,5	2,6
3,1	3,2	3,3	3,4	3,5	3,6
4,1	4,2	4,3	4,4	4,5	4,6
5,1	5,2	5,3	5,4	5,5	5,6
6,1	6,2	6,3	6,4	6,5	6,6

For fair dice, each outcome is equally likely. So, the probability can be found $\frac{number\ of\ double\ sixes}{number\ of\ outcomes}$

Only one of the 36 possible outcomes is double sixes, so the probability of getting double sixes is $\frac{1}{36}$.

USE A TREE DIAGRAM TO DETERMINE THE PROBABILITY THAT OF TWO CARDS PULLED FROM TWO DIFFERENT DECKS, AT LEAST ONE IS A HEART.

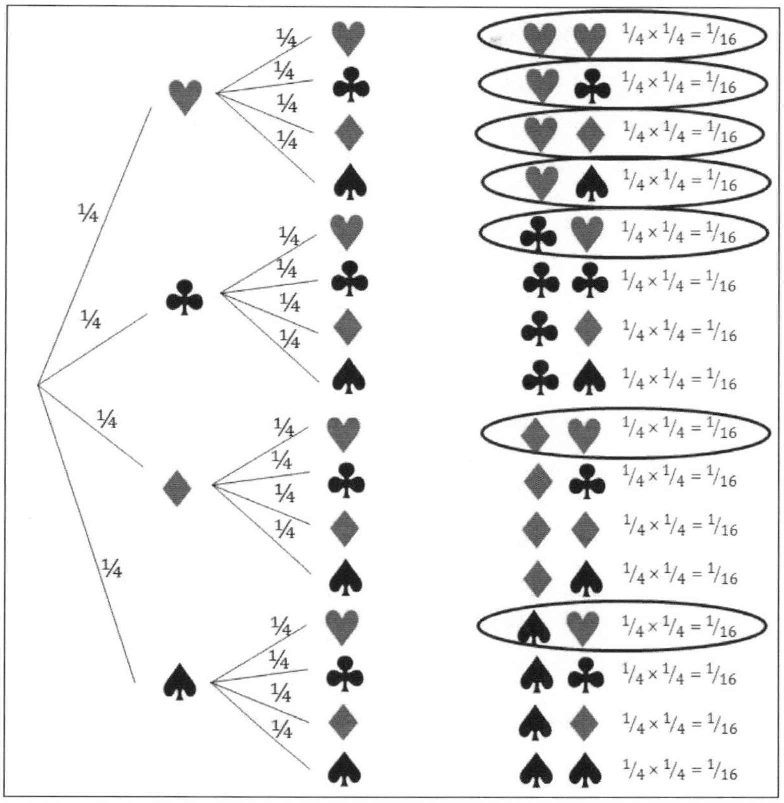

The probability of getting at least one heart includes the outcomes in which a heart is selected from the first deck, in which a heart is selected from the second deck, and in which a heart is selected from both decks. Seven of the sixteen outcomes include at least one heart, so the probability that at least one of the two cards drawn is a heart is $\frac{7}{16}$.

USE A TREE DIAGRAM TO DETERMINE THE PROBABILITY THAT TWO CARDS PULLED FROM TWO DIFFERENT DECKS ARE BOTH SPADES.

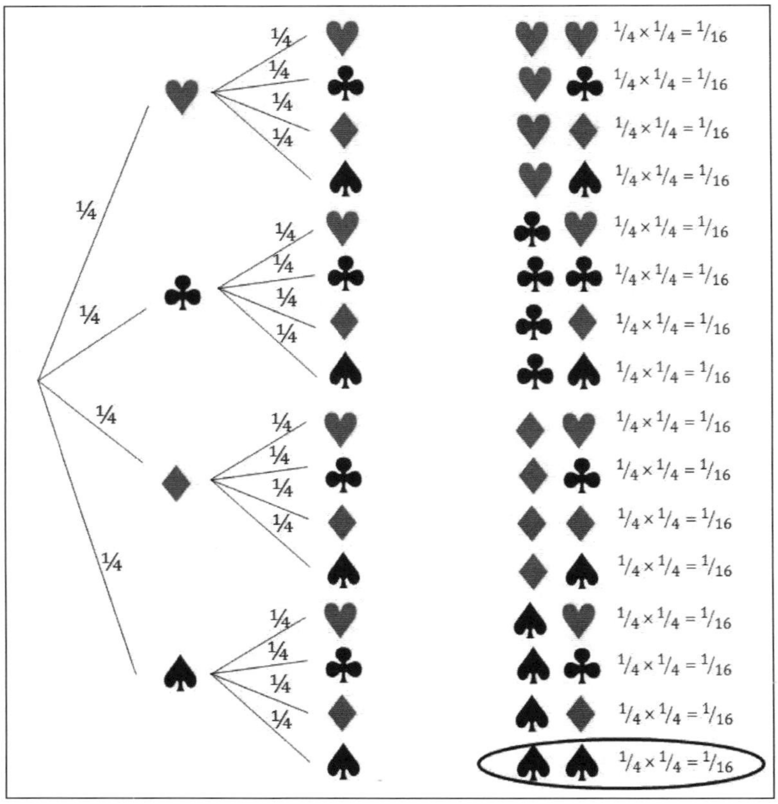

One out of sixteen outcomes is drawing a spade from both decks. So, the probability that both cards are spades is $\frac{1}{16}$.

Notice that the probability of each outcome in the sample space can be found by multiplying the probability of the first event by the probability of the second event.

SUPPOSE A COUPLE HAS A ¼ CHANCE OF HAVING A CHILD WITH SICKLE CELL ANEMIA. EXPLAIN WHETHER OR NOT EACH STATEMENT IS TRUE.

- If the couple's first child has sickle cell anemia, then their second child will not have sickle cell anemia.
- If the couple has eight children, two of them will have sickle cell anemia.
- If the couple has six children, it is unlikely that all of them will have sickle cell anemia.

The statement *If the couple's first child has sickle cell anemia, then their second child will not have sickle cell anemia* is **not necessarily true**. If a couple's first child is affected, it is possible for second child to have the disease, too. Each child conceived by the couple has a ¼ chance of developing the disease. The fact that the first child has the disease has no effect on the second child's chance of having it.

The statement *If a couple has eight children, two of them will have sickle cell anemia* is **not necessarily true**. You can predict based on the probability that one in four of the couple's children will be affected that, of eight children, two will have the disease. However, it is possible that none, all, or any number of the children will have the disease.

The statement *If the couple has six children, it is unlikely that all of them will have sickle cell anemia* is **true**. This statement mentions only the relative likelihood of an occurrence. The probability that all six children would have the disease is $\frac{1}{4} \times \frac{1}{4} \times \frac{1}{4} \times \frac{1}{4} \times \frac{1}{4} \times \frac{1}{4} = \frac{1}{4096}$. A probability of $\frac{1}{4096}$ indicates that is unlikely for all six children to have the disease.

Mathematics Practice Test #1

Practice Questions

1. Ana has completed approximately $\frac{2}{7}$ of her research paper. Which of the following best represents the percentage of the paper she has completed?

a. 24%
b. 26%
c. 27%
d. 29%

2. Simplify the expression: $2n + (3n - 2)^2$

____

3. Elijah has prepared $2\frac{1}{2}$ gallons of lemonade to distribute to guests at a party. If there are 25 guests, how much lemonade is available to each guest, given that each guest receives an equal amount?

a. $\frac{1}{8}$ of a gallon
b. $\frac{1}{6}$ of a gallon
c. $\frac{1}{12}$ of a gallon
d. $\frac{1}{10}$ of a gallon

4. The points M, N, and O are plotted on the number line below. Plot point P based on the equation: $N - M + O = P$

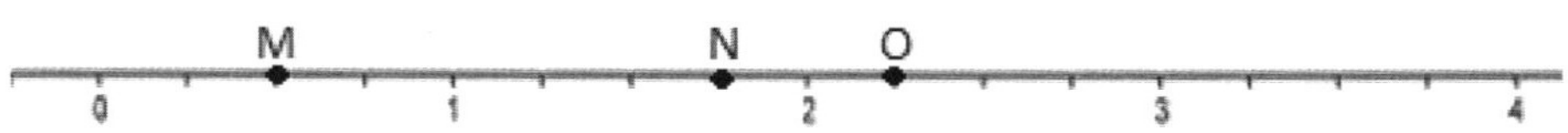

5. Part A: Edward spins the spinner below three times. If the spinner lands on a different number each time what is the highest total he could get?

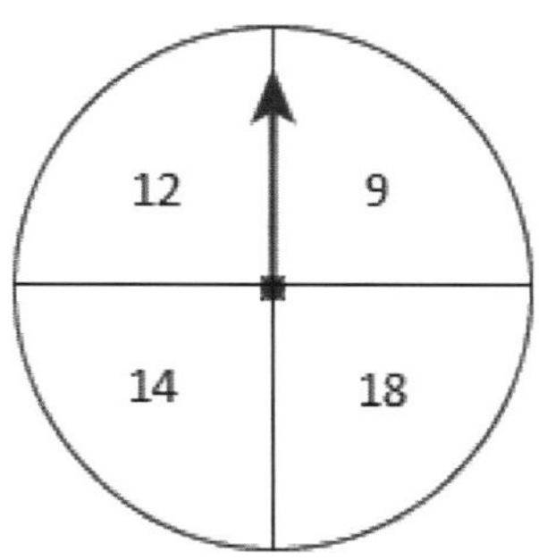

____

10. Martin and his friends are taking a road trip from Houston, TX to Las Vegas, NV. The trip will take two days and they will spend the night in El Paso, TX. The first day they drove for 11 hours at a rate of 69 miles per hour. The next day they drove for 10 hours at a rate of 72 miles per hour. How far is it from Houston to Las Vegas?

____

11. A bottle of lotion contains 20 fluid ounces and costs $3.96. Which of the following best represents the cost per fluid ounce?

a. $0.18
b. $0.20
c. $0.22
d. $0.24

12. Solve the equation for x: $3^2 + 2x = 17$.

____

13. Given the figure below what is the area of the shaded regions? Figure is not to scale.

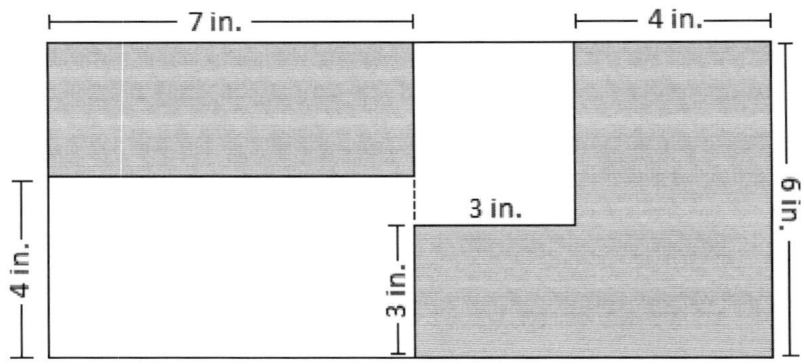

____

14. Given the sequence represented in the table below, where *n* represents the position of the term and a_n represents the value of the term, which of the following describes the relationship between the position number and the value of the term?

n	1	2	3	4	5	6
a_n	5	2	−1	−4	−7	−10

a. Multiply *n* by 2 and subtract 4
b. Multiply *n* by 2 and subtract 3
c. Multiply *n* by −3 and add 8
d. Multiply *n* by −4 and add 1

15. The number 123 is the 11th term in a sequence with a constant rate of change. Which of the following sequences has this number as its 11th term?

a. 5, 17, 29, 41, ...
b. 3, 15, 27, 39, ...
c. −1, 11, 23, 35, ...
d. 1, 13, 25, 37, ...

16. **Kevin pays \$12.95 for a text messaging service plus \$0.07 for each text message he sends. Which of the following equations could be used to represent the total cost, *y*, when *x* represents the number of text messages sent?**

a. $y = \$12.95x + \$0.07$
b. $y = \$13.02x$
c. $y = \frac{\$12.95}{\$0.07}x$
d. $y = \$0.07x + \$12.95$

17. **Hannah draws two supplementary angles. One angle measures 34°. What is the measure of the other angle?**

a. 56°
b. 66°
c. 146°
d. 326°

18. **Part A: Steven's class had a pushup contest and the results are recorded below. What is the median number of pushups the class did?**

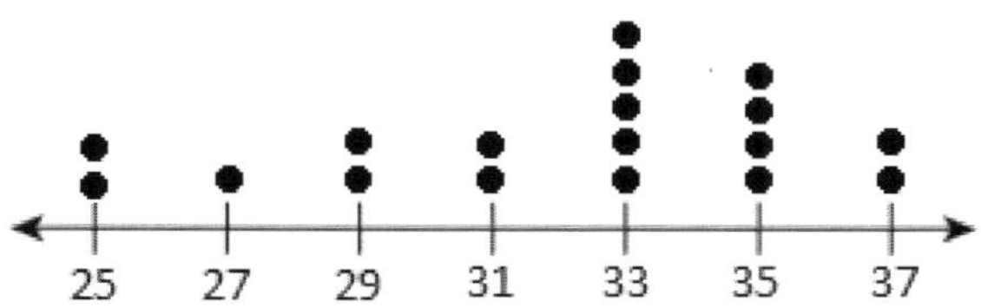

____

Part B: What is the difference between the median and the mean number of pushups?

____

19. **A triangle has the following angle measures: 98°, 47°, and 35°. What type of triangle is it?**

a. Obtuse
b. Right
c. Acute
d. Equiangular

20. **Part A: Jordan has a bag full of red and blue marbles. There are 16 red marbles and 24 blue marbles? What is the probability of him drawing a red marble?**

____

Part B: Jordan has continued to draw marbles and has now taken out 4 red and 6 blue. What is the probability of him drawing a red marble now?

____

21. **Which figure has two circular bases and a lateral face?**

a. Cone
b. Prism
c. Cylinder
d. Sphere

22. **Plot the two ordered pairs on the graph below: (x-3, 5) and (7-x, -2) where x = 2.**

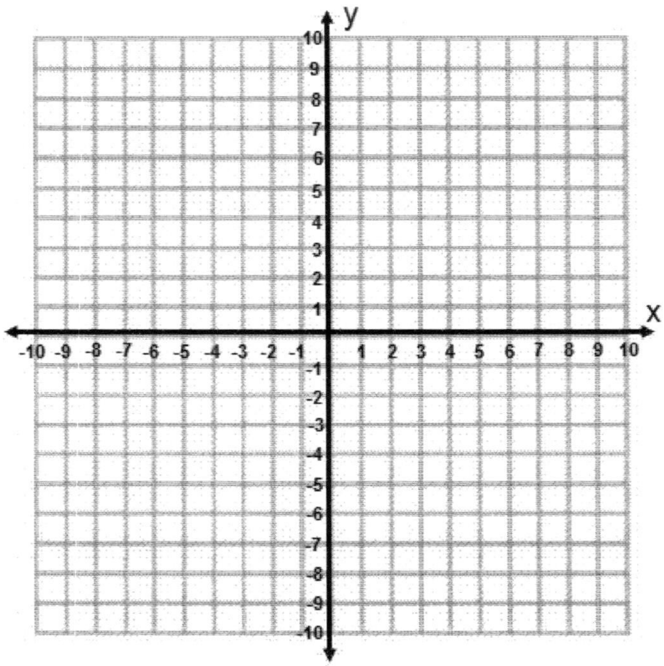

23. **Bookstore A sells a particular book for $15.25, but they have it on sale for 20% off. Bookstore B sells it for $12.45. How much more does it cost at Bookstore B?**

____

24. **A carpenter must fix a broken section of a kitchen cabinet. The intact portion of the cabinet forms a 76 degree angle with the wall. The width of the cabinet is supposed to form a 90 degree angle with the wall. What angle measure should the carpenter use when cutting the piece that will fit next to the 76 degree angle?**

a. 12°
b. 14°
c. 104°
d. 136°

25. Which of the following represents the net of a triangular prism?

a.

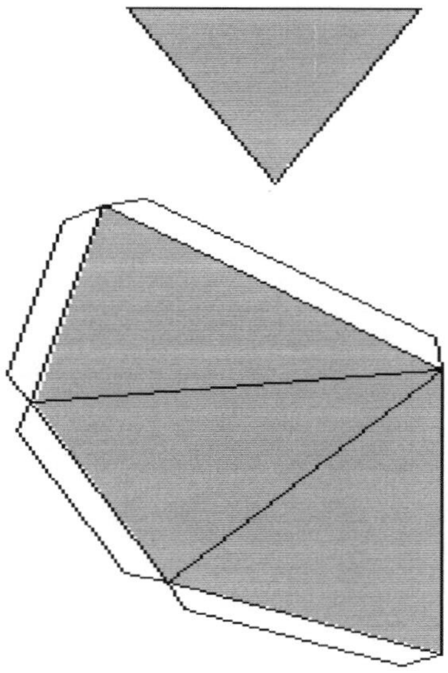

c.

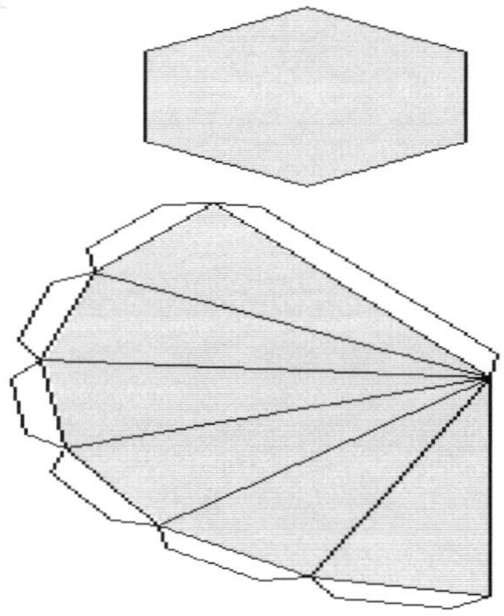

b.

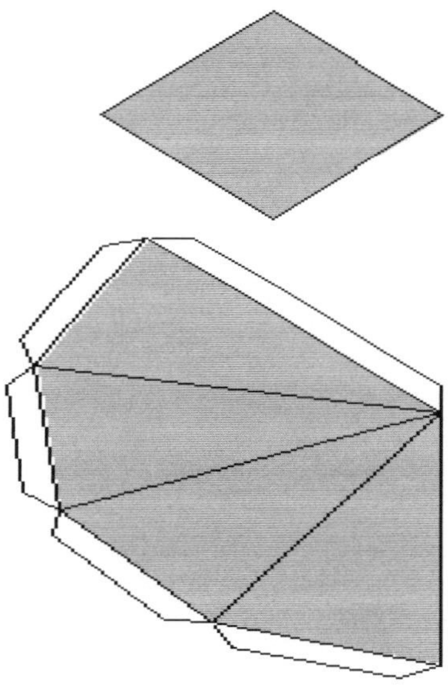

d.

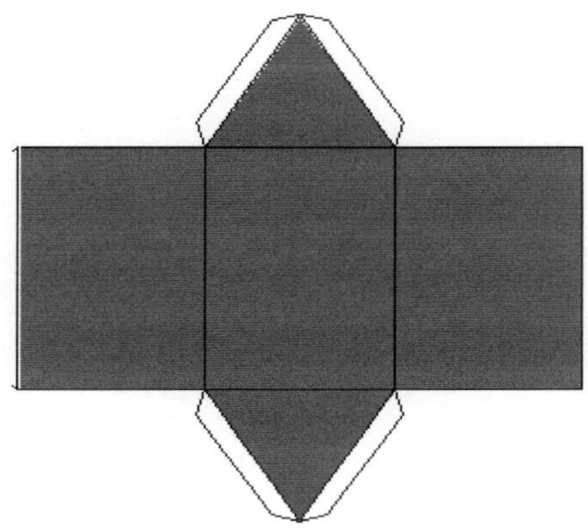

26. A circle has a radius of 23 cm. Which of the following is the best estimate for the circumference of the circle?

a. 71.76 cm
b. 143.52 cm
c. 144.44 cm
d. 72.22 cm

27. Sally is driving to the store. It takes her $\frac{1}{12}$ of an hour to go 3.4 miles. How many miles an hour is Sally driving?

a. 36 mph
b. 17 mph
c. 40.8 mph
d. 42.4 mph

28. Which of the following is also equal to $4n^2 + (3n + 5)^2$? Select all that apply.

1. $2n^2 + 2n^2 + (3n + 5)^2$
2. $4n^2 + (3n^2 + 25)$
3. $4n^2 + (3n + 5)$
4. $13n^2 + 30n + 25$
5. $7n^2 + 25$

29. Ashton draws the parallelogram shown below. How many square units represent the area of the parallelogram?

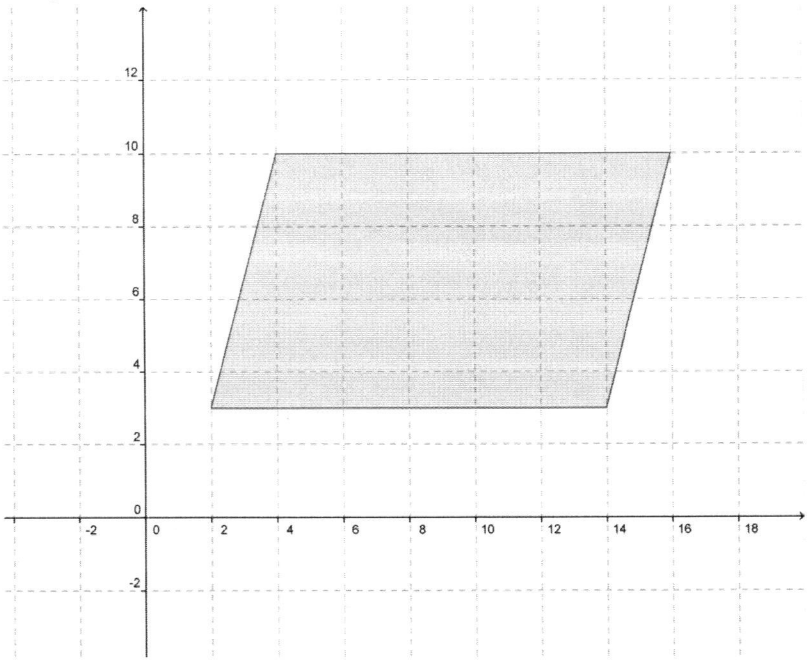

____

30. In the formula for the volume of the figure shown below, written as $V = B \cdot h$, h represents the height of the prism when it rests one of its bases. What does the B represent?

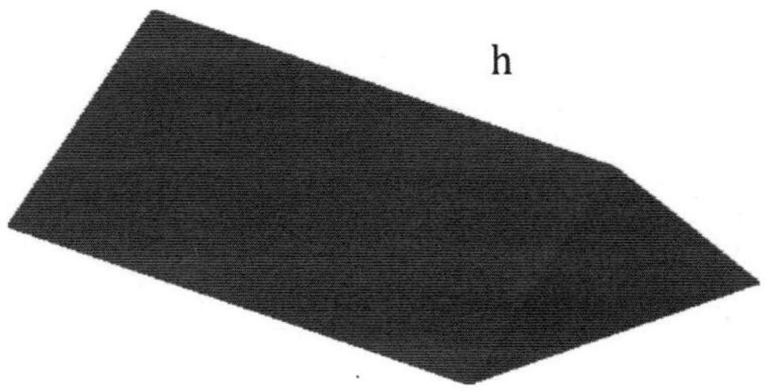

a. $\frac{1}{3}bh$, where b represents the length of the triangle's base and h represents the triangle's height
b. bh, where b represents the length of the triangle's base and h represents the triangle's height
c. $2bh$,where b represents the length of triangle's base and h represents the triangle's height
d. $\frac{1}{2}bh$, where b represents the length of triangle's base and h represents the triangle's height

31. A rectangular prism has a length of 14.3 cm, a width of 8.9 cm, and a height of 11.7 cm. Which of the following is the best estimate for the volume of the rectangular prism?

a. 1,512 cm^3
b. 1,287 cm^3
c. 1,386 cm^3
d. 1,620 cm^3

32. A can has a radius of 3.5 cm and a height of 8 cm. Which of the following best represents the volume of the can?

a. 294.86 cm^3
b. 298.48 cm^3
c. 307.72 cm^3
d. 309.24 cm^3

33. Fred designs a candy box in the shape of a triangular prism. The base of each triangular face measures 4 inches, while the height of the prism is 7 inches. Given that the length of the prism is 11 inches, what is the volume of the candy box?

a. 102 in^3
b. 128 in^3
c. 154 in^3
d. 308 in^3

34. Miranda rolls a standard die and spins a spinner with 4 equal sections. Which of the following represents the sample space?

a. 10
b. 12
c. 24
d. 36

35. A hat contains 6 red dice, 4 green dice, and 2 blue dice. What is the probability that Sarah pulls out a blue die, replaces it, and then pulls out a green die?

a. $\frac{1}{18}$
b. $\frac{1}{3}$
c. $\frac{1}{2}$
d. $\frac{1}{16}$

36. The histogram below represents the overall GRE scores for a sample of college students. Which of the following is a true statement?

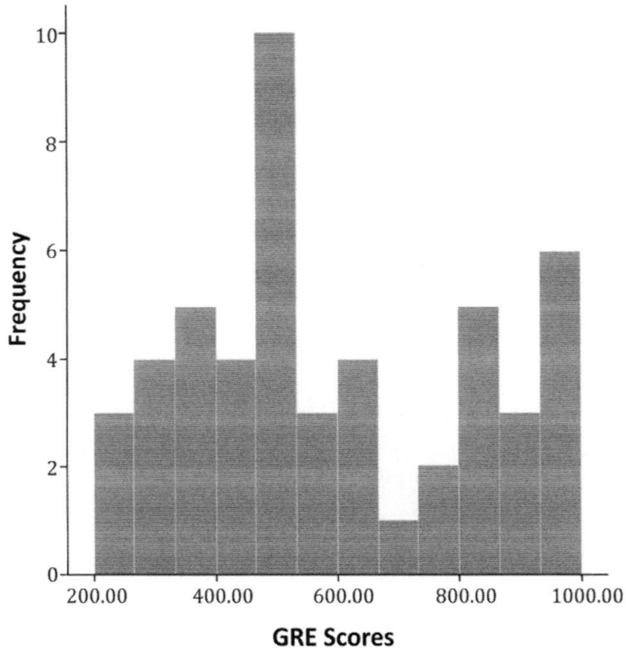

a. The range of GRE scores is approximately 600.
b. The average GRE score is 750.
c. The median GRE score is approximately 500.
d. The fewest number of college students had an approximate score of 800.

37. What is the area of the circle on the graph below? Each square represents 1 inch.

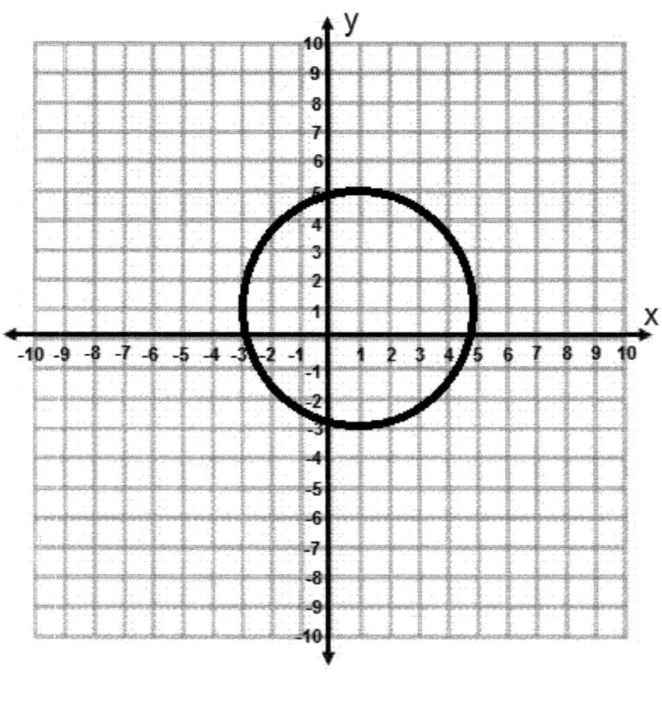

Answers and Explanations

1. D: In order to convert the given fraction to a percentage, divide 2 by 7. Doing so gives a decimal of approximately 0.29. The decimal can be converted to a percentage by multiplying by 100, which moves the decimal point two places to the right and gives 29%.

2. $2n + (3n - 2)^2 = 2n + 9n^2 - 6n - 6n + 4 = 9n^2 - 10n + 4$

3. D: In order to determine the amount available to each guest, the total amount of prepared lemonade should be divided by 25 guests. Thus, the expression $2\frac{1}{2} \div 25$ represents the amount that each guest has available for consumption. The mixed fraction can be rewritten as $\frac{5}{2}$. The expression can be simplified by writing $\frac{5}{2} \div 25 = \frac{5}{2} \times \frac{1}{25}$, which equals $\frac{5}{50}$, or $\frac{1}{10}$.

4. M=$\frac{1}{2}$, N=$1\frac{3}{4}$, and O=$2\frac{1}{4}$, so N-M+O=$3\frac{1}{2}$. The number line is shown below.

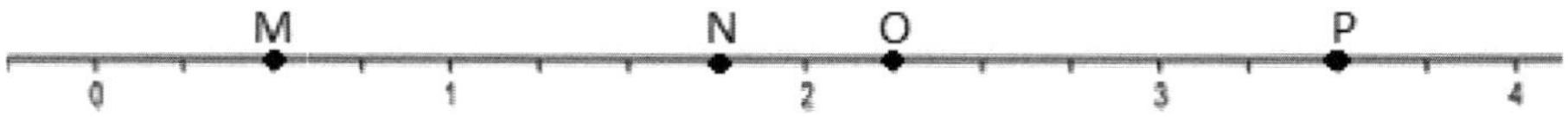

5. Part A: 44: I the spinner must land on a different number each time then the highest three numbers are 12,14, and 18. Added together they equal 44.

Part B: $\frac{3}{4}$: He has already landed on three of the four numbers so the probability is $\frac{3}{4}$.

6. D: The cost per ounce can be calculated by dividing the cost of the bag by the number of ounces the bag contains. Thus, the cost per ounce can be calculated by writing $9.85 ÷ 16, which equals approximately $0.62 per ounce.

7. D: The order of operations states that numbers with exponents must be evaluated first. Thus, the expression can be rewritten as $64 + 12 \div 4 + 64 \times 3$. Next, multiplication and division must be computed as they appear from left to right in the expression. Thus, the expression can be further simplified as $64 + 3 + 192$, which equals 259.

8. Part A: C: Both flour and butter are given in cups so they are easy to compare. First find a like denominator. $\frac{1}{2}$ is equal to $\frac{2}{4}$, and $1\frac{3}{4}$ is equal to $\frac{7}{4}$. So the ratio is 2:7.

Part B: 3:7, 4:7: This question is very similar to Part A. Both milk and flour are given in cups and fourths. It's $\frac{3}{4}$ to $\frac{7}{4}$ or 3:7, and if she puts a whole cup in then its $\frac{4}{4}$ to $\frac{7}{4}$ or 4:7.

9. Part A: B: The discounted price is 25% less than the original price. Therefore, the discounted price can be written as $36.95 - ((0.25)(36.95))$, which equals approximately 27.71. Thus, the discounted price of the jacket is $27.71.

Part B: $29.93: If sales tax is 8% then that can be written as ($27.71)(1.08), which is approximately $29.93.

10. 1479 miles: The first day they went 69 miles per hour for 11 hours, so they went $11 \times 69 = 759\ miles$. The second day they went 72 miles per hour for 10 hours, so they went $10 \times 72 = 720\ miles$. Total they went 1479 miles.

11. B: In order to find the unit rate, the cost of the bottle should be divided by the number of fluid ounces contained in the bottle: $\frac{\$3.96}{20} \approx 0.20$. Thus, the cost per fluid ounce is approximately \$0.20.

12. X=4: $3^2 + 2x = 17, 9 + 2x = 17, 2x = 8, x = 4$

13. 47 square inches: The top left shaded region can be found by first finding the width. Since 6 in. is given as the width of the whole rectangle and 4 in. is given for the width of the non shaded region then the width of the shaded region is the difference of 2 inches. So, the area of that region is $7\ in. \times 2in. = 14\ square\ inches$. The other shaded region can be broken into a 3 in. by 3 in. square and a 4 in. by 6 in. rectangle. So, $3\ in. \times 3\ in. = 9\ ssquare\ inches$ and $4\ in. \times 6\ in. = 24\ square\ inches$. Added together, 14 square inches + 9 square inches + 24 square inches = 47 square inches.

14. C: The equation that represents the relationship between the position number, *n*, and the value of the term, a_n, is $a_n = -3n + 8$. Notice each *n* is multiplied by −3, with 8 added to that value. Substituting position number 1 for *n* gives $a_n = -3(1) + 8$, which equals 5. Substitution of the remaining position numbers does not provide a counterexample to this procedure.

15. B: All given sequences have a constant difference of 12. Subtraction of 12 from the starting term, given for Choice B, gives a *y*-intercept of −9. The equation $123 = 12x - 9$ can thus be written. Solving for *x* gives $x = 11$; therefore, 123 is indeed the 11th term of this sequence. Manual computation of the 11th term by adding the constant difference of 12 also reveals 123 as the value of the 11th term of this sequence.

16. D: The constant amount Kevin pays is \$12.95; this amount represents the *y*-intercept. The variable amount is represented by the expression $\$0.07x$, where *x* represents the number of text messages sent and \$0.07 represents the constant rate of change or slope. Thus, his total cost can be represented by the equation $y = \$0.07x + \$12.95$.

17. C: Supplementary angles add to 180 degrees. Therefore, the other angle is equal to the difference between 180 degrees and 34 degrees: $180 - 34 = 146$. Thus, the other angle measures 146°.

18. Part A: 33: The median number is the middle number out of the group. In this case there are 18 numbers so it is the average of the 9th and 10th number. However, since the 9th and 10th numbers are both 33 it is just 33.

Part B: 1: The mean can be found by adding all of the numbers together and dividing by 18. If you add all of the numbers up and divide by 18 you get 32. The difference in the mean and median is 1.

19. A: A triangle with an obtuse angle (an angle greater than 90°) is called an obtuse triangle.

20. Part A: $\frac{2}{5}$: There are a total of 40 marbles in the bag. The probability of him drawing a red one is $\frac{16}{40}$ or $\frac{2}{5}$.

Part B: $\frac{2}{3}$: If 4 red and 6 blue are missing then there are only 12 red and 18 blue remaining. The probability of drawing a red one then is $\frac{12}{18}$ or $\frac{2}{3}$.

21. C: A cylinder has two circular bases and a rectangular lateral face.

22. The point (x-3, 5) is (2-3, 5) or (-1, 5). The point (7-x, -2) is (7-2, -2) or (5, -2). Both are graphed below.

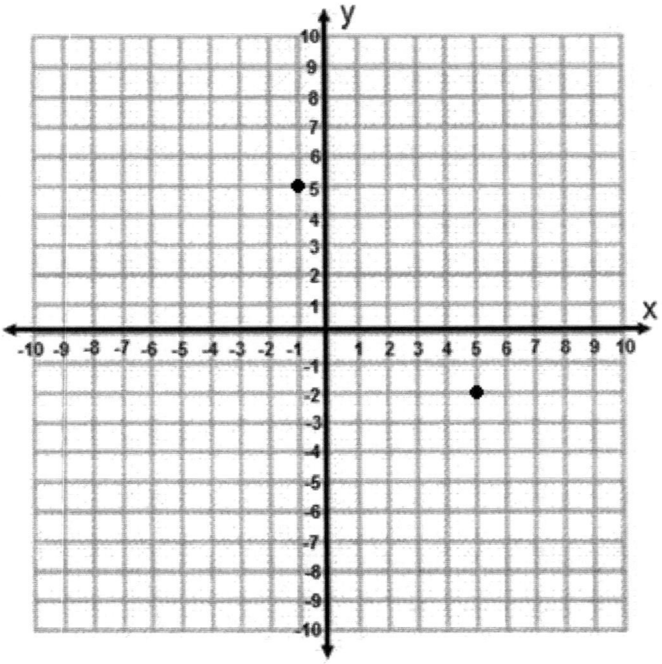

23. $.25: First find the cost of the book at Bookstore A. The price would be $\$15.25(.8) = \$12.20$. The cost at Bookstore B is $12.45 so it is $.25 more.

24. B: Since the intact portion of the cabinet and the missing piece form a 90 degree angle with the wall, the missing piece must have an angle equal to the difference between 90 degrees and 76 degrees. Thus, the newly cut cabinet piece should have an angle measure of 14 degrees.

25. D: A triangular prism has two triangular bases and three rectangular faces.

26. C: The circumference of a circle can be determined by using the formula $C = \pi d$. A radius of 23 cm indicates a diameter of 46 cm, or twice that length. Substitution of 46 cm for d and 3.14 for π gives the following: $C = 3.14 \cdot 46$, which equals 144.44. Thus, the circumference of the circle is approximately 144.44 cm.

27. C: If she goes 3.4 miles in $\frac{1}{12}$ of an hour then just multiply by 12 to see how far she will go in one hour. Then that is her miles per hour.

28. I, IV: Start by simplifying the equation. $4n^2 + (3n + 5)^2 = 4n^2 + 9n^2 + 15n + 15n + 25 = 13n^2 + 30n + 25$. So, you can see that you came up with answer IV right there, and answer I is the same as the original equation except for the $4n^2$ is broken down to $2n^2 + 2n^2$.

29. 84: The area of a parallelogram can be found by using the formula $A = bh$, where b represents the length of the base and h represents the height of the parallelogram. The base and the height of the parallelogram are 12 units and 7 units, respectively. Therefore, the area can be written as $A = 12 \cdot 7$, which equals 84.

30. D: The B in the formula $V = Bh$ represents the area of the triangular base. The formula for the area of a triangle is $\frac{1}{2}bh$, where b represents the length of the triangle's base and h represents the triangle's height.

31. A: The dimensions of the rectangular prism can be rounded to 14 cm, 9 cm, and 12 cm. The volume of a rectangular prism can be determined by finding the product of the length, width, and height. Therefore, the volume is approximately equal to $14 \times 9 \times 12$, or 1,512 cm^3.

32. C: The volume of a cylindrical can be found using the formula $V = \pi r^2 h$, where r represents the radius and h represents the height. Substitution of the given radius and height gives $V = \pi(3.5)^2 \cdot 8$, which is approximately 307.72. Thus, the volume of the can is approximately 307.72 cm^3.

33. C: The volume of a triangular prism can be determined using the formula $V = \frac{1}{2}bhl$, where b represents the length of the base of each triangular face, h represents the height of each triangular face, and l represents the length of the prism. Substitution of the given values into the formula gives $V = \frac{1}{2} \cdot 4 \cdot 7 \cdot 11$, which equals 154. Thus, the volume of the candy box is 154 cubic inches.

34. C: The sample space of independent events is equal to the product of the sample space of each event. The sample space of rolling a die is 6; the sample space of spinning a spinner with four equal sections is 4. Therefore, the overall sample space is equal to 6 × 4, or 24.

35. A: The events are independent since Sarah replaces the first die. The probability of two independent events can be found using the formula $P(A\ and\ B) = P(A) \cdot P(B)$. The probability of pulling out a blue die is $\frac{2}{12}$. The probability of pulling out a green die is $\frac{4}{12}$. The probability of pulling out a blue die and a green die is $\frac{2}{12} \cdot \frac{4}{12}$, which simplifies to $\frac{1}{18}$.

36. C: The median is the value that splits the data in half. Thus, approximately 50% of the scores are above it and 50% are below it. From the histogram, there are a total of 50 data points, so the median would be located between the 25th and 26th scores when the data are ordered from least to greatest. Counting the frequencies from left to right, the halfway point lies in the interval around 500. Therefore, the median GRE score is approximately 500.

37. 16π or 50.27 inches: The area of a circle is πr^2. You can find the radius by counting the number of units across the circle and dividing by 2. So the radius is 4 inches. $\pi 4^2 = 16\pi\ or\ 50.27\ inches$.

38. D: If you were to cut a slice vertically it would produce a triangle. If you were to cut a slice long ways it would produce a rectangle, and if you were to cut a slice that was the front face of the prism it would be an 8 by 8 square.

39. 15%: First find the cost of the first taco shop by dividing $4.14 by 3, which comes out to $1.38 per taco. Then find the price at the second shop by dividing $4.80 by 4 to get $1.20 per taco. The difference is .$18. To get a percentage divide $.18 by $1.20 and multiply by 100. The first shop cost 15% more than the second.

40. D: The probability of getting an even number is $\frac{3}{6}$. The probability of getting heads is $\frac{1}{2}$. The probability of spinning an even number is $\frac{2}{4}$. The probability of all three occurring can be calculated by multiplying the probabilities of the individual events: $\frac{3}{6} \cdot \frac{1}{2} \cdot \frac{2}{4}$ equals $\frac{1}{8}$.

Mathematics Practice Test #2

Practice Questions

1. Which of the following is the largest number?

$$\frac{14}{4}, 3.41, \pi, 3\frac{3}{8}$$

____

2. A plane takes off from Dallas and lands in New York 3 hours and 20 minutes later. The distance from Dallas to New York is 1510 miles. Approximately how fast was the plane traveling?

a. 445 mph
b. 453 mph
c. 456 mph
d. 449 mph

3. According to the order of operations, which of the following steps should be completed immediately following the evaluation of the squared number when evaluating the expression $9 - 18^2 \times 2 + 12 \div 4$?

a. Subtract 18^2 from 9
b. Multiply the squared value by 2
c. Divide 12 by 4
d. Add 2 and 12

4. A parcel of land has 35 mature trees for every 3 acres. How many mature trees can be found on 18 of the acres?

a. 206
b. 212
c. 210
d. 214

5. Which of the following is equivalent to $-8^2 + (17 - 9) \times 4 + 7$?

a. −217
b. 24
c. −64
d. −25

6. Jason chooses a number that is the square root of four less than two times Amy's number. If Amy's number is 20, what is Jason's number?

a. 6
b. 7
c. 8
d. 9

7. Part A: What is the area of the shaded region below?

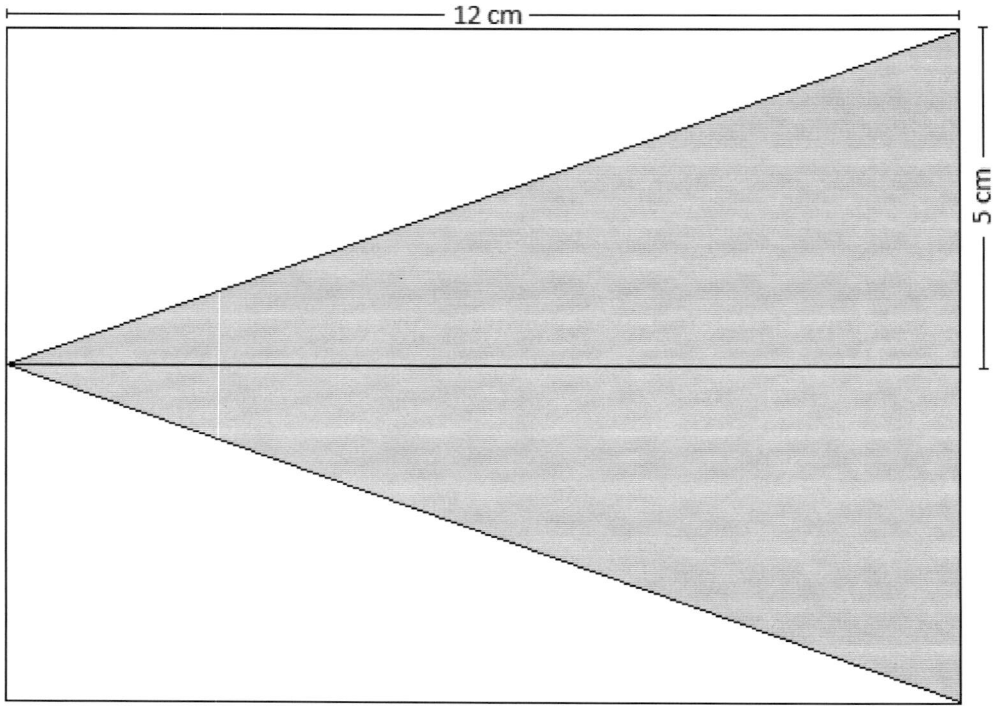

____

Part B: How does the shaded region compare to the non shaded region?

a. The shaded region is bigger than the non shaded region
b. Both the shaded and non shaded region are the same size
c. The non shaded region is bigger than the shaded region
d. The area of the non shaded region cannot be determined

8. Given the table below what would *y* be if *x=5*?

X	-2	0	3	4
y	2	-2	7	14

a. 21
b. 23
c. 19
d. 24

9. Given the following equation what is x equal to if y equals 8? $6x + 4 = 2y - 7$.

____

17. Point S and Point T are shown on the number line below. Which of the following equations produces a point, *R*, not on the number line?

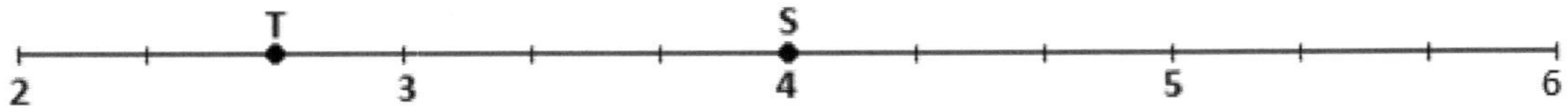

a. 2S-T=R
b. 2T-S=R
c. 3T-S=R
d. 3S-2T=R

18. Part A: The spinner below is spun once. What is the probability of landing on a 12 or greater?

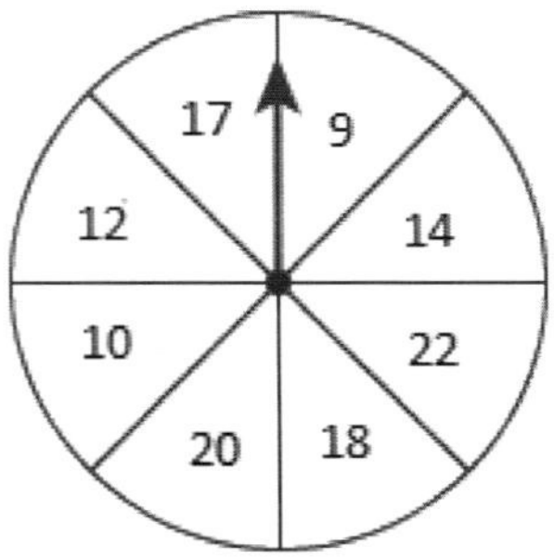

a. $\frac{5}{8}$
b. $\frac{1}{2}$
c. $\frac{3}{4}$
d. $\frac{2}{8}$

Part B: What is the probability of spinning it a second time and getting a 12 or greater both times?

____

19. A toy store owner sells action figures. He buys each one from the manufacture for $4.10. He has labor and other costs of $1.35 per action figure. He wants to make a 32% profit on each one. How much does he need to sell them for?

____

20. Angle A and Angle B are complementary. Angle B measures 28°. What is the measure of Angle A?

a. 62°
b. 92°
c. 72°
d. 152°

21. Which of the following describes *all* requirements of similar polygons?

a. Similar polygons have congruent corresponding angles and proportional corresponding sides
b. Similar polygons have congruent corresponding angles and congruent corresponding sides
c. Similar polygons have proportional corresponding sides
d. Similar polygons have congruent corresponding angles

22. How far apart are the two points on the graph below? Each square represents 3 feet.

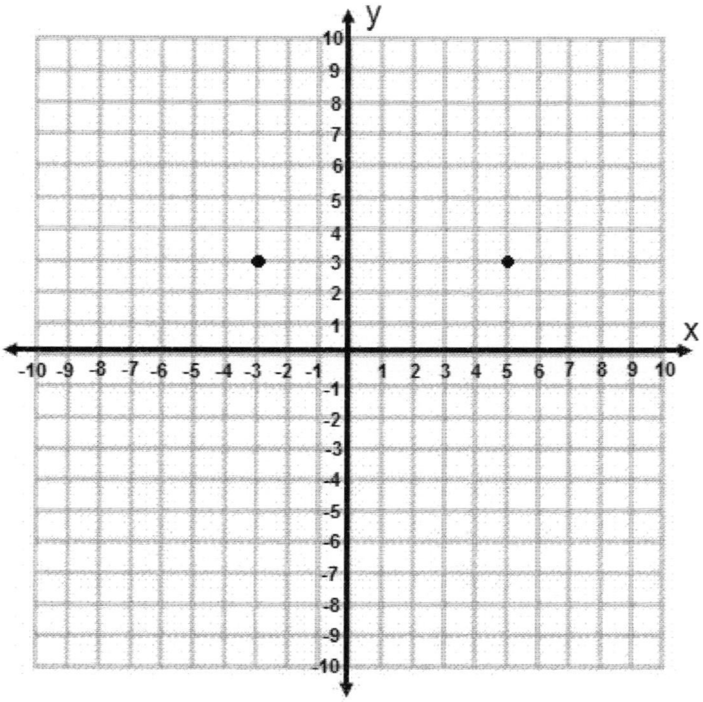

a. 15 feet
b. 22 feet
c. 21 feet
d. 24 feet

23. Eric is able to dribble the soccer ball down $\frac{2}{3}$ of the field in $\frac{2}{5}$ of a minute. How long will it take him to dribble the whole field?

a. 36 seconds
b. 24 seconds
c. 30 seconds
d. 32 seconds

24. Kaleb has a bike rim that is 18 inches in diameter. He puts a tire on it that is 2 inches thick. What is the circumference of the tire?

a. 20π
b. 22π
c. 24π
d. 18π

25. Given the trapezoid shown below, which of the following vertices represent the reflection of the trapezoid across the *y*-axis?

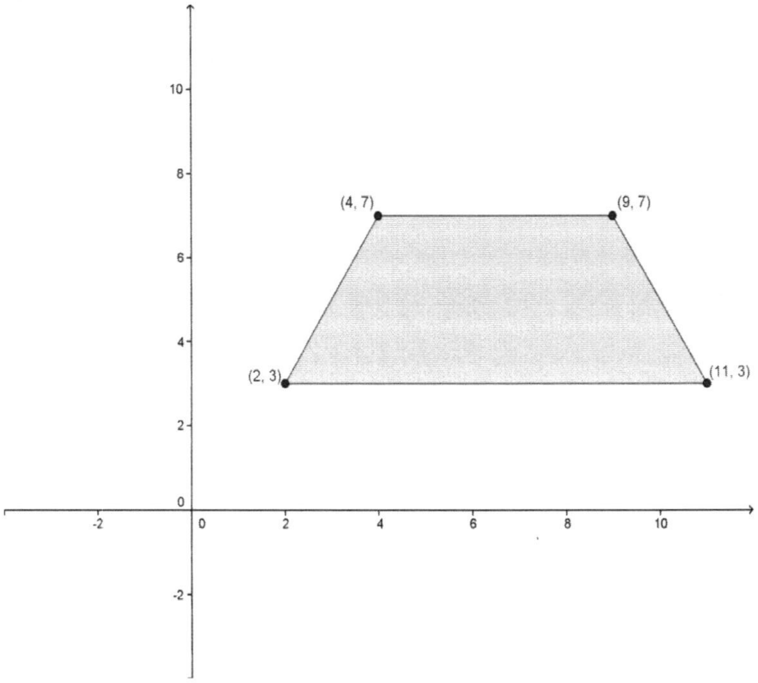

a. (−4, −7), (−9, −7), (−2, −3), (−11, −3)
b. (−4, 7), (−9, 7), (−2, 3), (−11, 3)
c. (7, −4), (7, −9), (3, −2), (3, −11)
d. (4, −7), (9, −7), (2, −3), (11, −3)

26. A regular heptagon has each side length equal to 9.2 cm. Which of the following is the best estimate for the perimeter of the heptagon?

a. 60 cm
b. 63 cm
c. 54 cm
d. 70 cm

27. A parallelogram has two bases, each equal to 18 cm, and a height of 8 cm. What is the area of the parallelogram?

a. 288 cm^2
b. 72 cm^2
c. 144 cm^2
d. 96 cm^2

28. Which of the following represents the area of the triangle shown below?

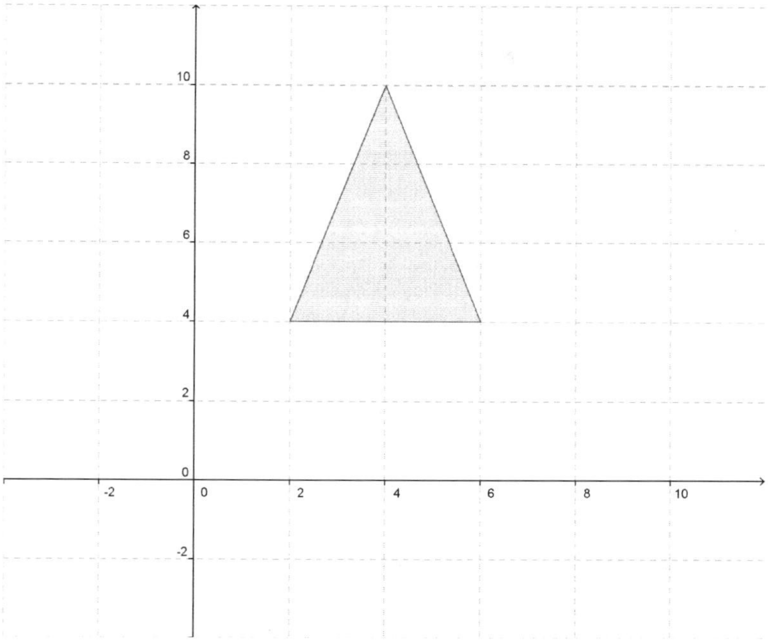

a. 8 square units
b. 9 square units
c. 10 square units
d. 12 square units

29. Judith purchased a box from the U.S. Postal Service with dimensions of 12 inches by 8 inches by 6 inches. How many cubic inches of space inside the box does she have available for use?

____

30. A pothole has a radius of 9 inches. Which of the following best represents the distance around the pothole?

a. 14.13 inches
b. 28.26 inches
c. 42.39 inches
d. 56.52 inches

31. What is the area of a trapezoid with base lengths of 7 cm and 10 cm and a height of 5 cm?

a. 85 cm^2
b. 42.5 cm^2
c. 28 cm^2
d. 8.5 cm^2

32. What is the sample space when rolling two standard dice?

a. 18
b. 6
c. 12
d. 36

33. What is the sample space when flipping a coin 9 times?

a. 256
b. 4,096
c. 512
d. 1,028

34. Kevin spins a spinner with 8 sections labeled 1 through 8. He also flips a coin. What is the probability he will land on a number less than 5 and get tails?

a. $\frac{7}{8}$
b. $\frac{1}{4}$
c. $\frac{5}{16}$
d. $\frac{1}{2}$

35. A box contains 8 yellow marbles, 9 orange marbles, and 1 green marble. What is the probability that Ann pulls out a yellow marble, replaces it, and then pulls a green marble?

a. $\frac{4}{153}$
b. $\frac{1}{2}$
c. $\frac{4}{9}$
d. $\frac{2}{81}$

36. The number of flights a flight attendant made per month is represented by the line graph below.

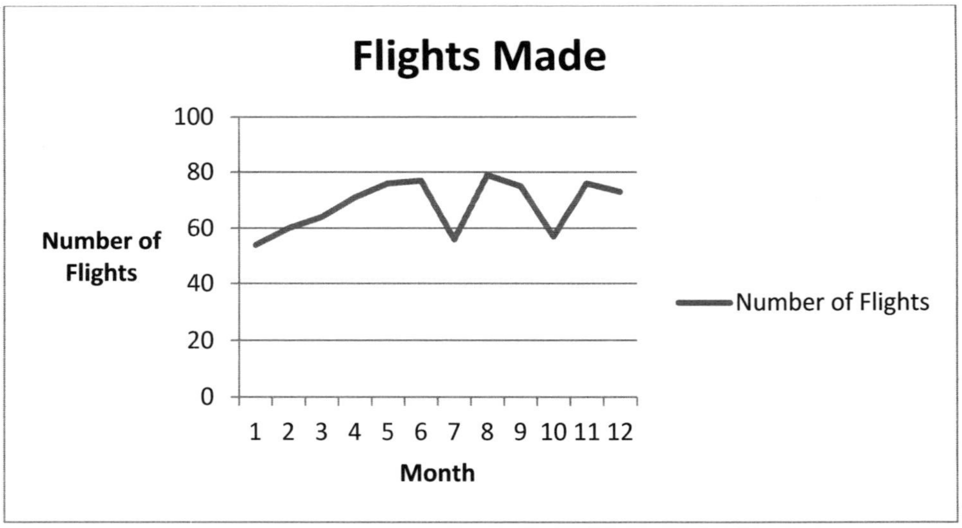

What is the range in the number of flights the flight attendant made?

a. 20
b. 25
c. 29
d. 32

37. Aubrey planted fruit trees on her farm. The number of each type of tree planted is shown in the table below.

Type of Tree	Number of Trees
Apple Tree	8
Peach Tree	18
Fig Tree	12
Pear Tree	3

Which circle graph represents the percentage of each type of tree planted?

a.

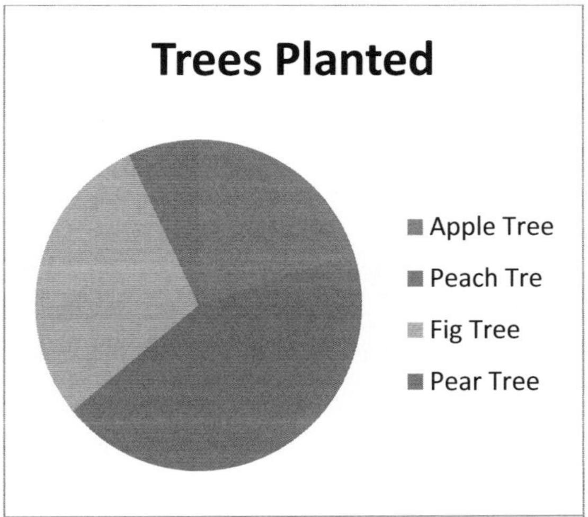

b.

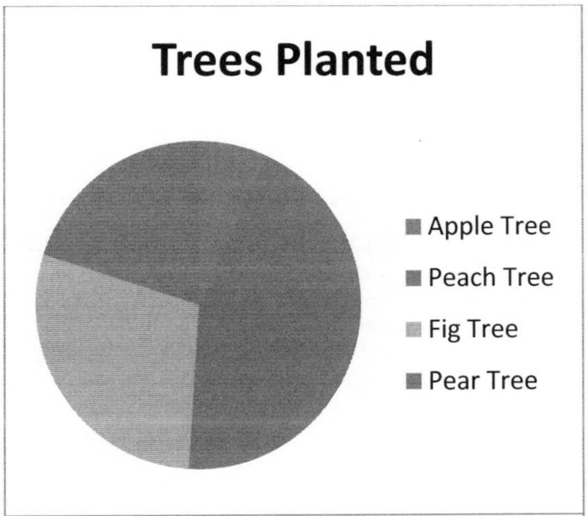

c.

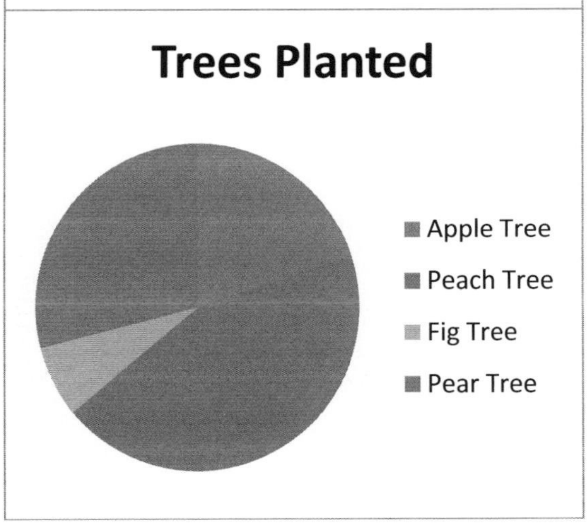

d.

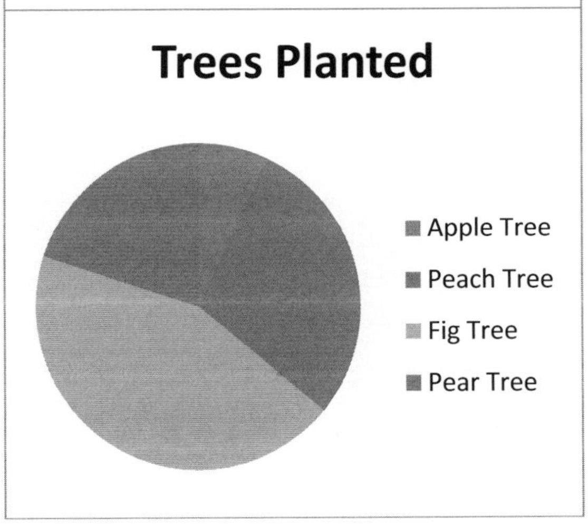

38. Chandler wishes to examine the median house value in his new hometown. Which graphical representation will most clearly indicate the median?

a. Box-and-whisker plot
b. Stem-and-leaf plot
c. Line plot
d. Bar graph

39. Part A: The number of long distance minutes Amanda used per week for business purposes is shown in the table below.

Week	Number of Minutes
1	289
2	255
3	322
4	291
5	306
6	302
7	411
8	418

What is the median number of long distance minutes she used?

____

Part B: How much more is the mean than the median?

____

40. A university reported the number of incoming freshmen from 2002 to 2011. The data is shown in the table below.

Year	Number of Incoming Freshmen
2002	7,046
2003	7,412
2004	6,938
2005	7,017
2006	7,692
2007	8,784
2008	7,929
2009	7,086
2010	8,017
2011	8,225

Based on the 10-year sample of data, which of the following represents the approximate average number of incoming freshmen?

a. 7,618
b. 7,615
c. 7,621
d. 7,624

be used to determine the final charge: $x - 0.20x$, where x represents the original charge. Substitution of 100 for x gives $100 - 0.20(100)$, which equals $100 - 20$, or 80. Thus, the company will charge $80.

11. B: The original price was $290,600 ($278,000 + $12,600). In order to determine the percentage of reduction, the following equation can be written: $\$12{,}600 = \$290{,}600x$, which simplifies to $x \approx 0.04$, or 4%. Thus, the percentage of reduction was approximately 4%.

12. $.47: The total was $5.17 and the 4 apples cost $3.76. So, $5.17-$3.76=$1.41 that was spent on bananas. Since there were 3 bananas, divide $1.41 by 3 to get $.47 per banana.

13. $51.20: First find what Melanie makes in a week. She makes $12 an hour times 40 hours, so she makes $480. Then take off 15% percent for her taxes. $480(.85)=$408. Next find what Lynn makes in one week. She makes $14 an hour times 40 hours, so she makes $560. Then take 18% off for her taxes. $560(.82)=$459.20. Then subtract what Melanie makes from what Lynn makes to find out how much more Lynn makes. $459.20-$408= $51.20.

14. B: The volume of a cone can be determined by using the formula $V = \frac{1}{3}\pi r^2 h$. Substitution of the radius and volume into the formula gives $150.72 = \frac{1}{3}\pi(4)^2 h$, which simplifies to $150.72 = \frac{1}{3}\pi 16h$. Division of each side of the equation by $\frac{1}{3}\pi 16$ gives $h = 9$. Thus, the height of the cone is 9 cm.

15. 3: The range of the numbers is the difference between the largest and smallest numbers in a set of numbers. In this case each tick mark on the number line represents $\frac{1}{2}$. The smallest number plotted is $5\frac{1}{2}$ and the largest number is $8\frac{1}{2}$. The range is 3.

16. $\frac{1}{2}, 1\frac{5}{8}$: $2\frac{1}{2} - \frac{3}{2}$ is already given as$1\frac{1}{8}$. Inside the parentheses convert the $\frac{2}{3}$ to $\frac{4}{6}$. Then you can do $\frac{4}{6} - \frac{1}{6} = \frac{3}{6} = \frac{1}{2}$. The first space is $\frac{1}{2}$. Then $1\frac{1}{8} + \frac{1}{2} = 1\frac{5}{8}$. The second space is $1\frac{5}{8}$.

17. B: Point T is equal to $2\frac{2}{3}$ and S is equal to 4. Perform all of the equations to figure out which one produces a number that is not on the number line. 2T-S= $2(2\frac{2}{3})$-4= $1\frac{1}{3}$ which is not on the number line.

18. Part A: C: There are a total of 8 spaces on the spinner all of equal size. There are 6 spaces that are 12 or greater. $\frac{6}{8} = \frac{3}{4}$.

Part B: $\frac{9}{16}$: The first time the probability was $\frac{3}{4}$ and the second time the probability is also $\frac{3}{4}$. The probability of it happening both times though is $\frac{3}{4} \times \frac{3}{4}$ which is $\frac{9}{16}$.

19. $7.02: If he buys each toy for $4.10 and then has another $1.30 in it, then he has a total of $5.40 in each toy. He marks it up 30% to sell it so his sales price is $5.40(1.30)=$7.02.

20. A: Complementary angles sum to 90 degrees. Since Angle B measures 28°, Angle A measures $90° - 28°$, or 62°.

21. A: Similar polygons must have congruent corresponding angles and proportional corresponding sides. Both requirements must be fulfilled in order to declare similarity in polygons.

22. D: One point is at (-3, 3) and the other is at (5, 3). So, they are at a distance of 8. Since each square is equal to 3 feet they are 24 feet apart.

23. A: If it takes him $\frac{2}{5}$ of a minute to dribble $\frac{2}{3}$ of the field then divide by 2 to get $\frac{1}{5}$ of a minute for $\frac{1}{3}$ of the field. Then you can multiply by 3 to get $\frac{3}{3}$ of the field in $\frac{3}{5}$ of a minute. $\frac{3}{5}$ of a minute equals 36 seconds.

24. C: If he puts a tire on that is 2 inches thick then that adds 4 inches to the overall diameter. Now the radius is 12 inches. The formula for circumference is $2\pi r$. 2 times 12π is 24π.

25. B: A reflection of a figure across the y-axis is achieved by finding the additive inverse of each x-value. The y-values will not change. Therefore, the vertices of the reflected figure are $(-4, 7)$, $(-9, 7)$, $(-2, 3)$, and $(-11, 3)$.

26. B: A regular heptagon has equal side lengths. Thus, an estimate for the perimeter can be computed by rounding the given side length and multiplying by 7 (the number of sides of a heptagon); 9.2 can be rounded to 9, and 9 × 7 =63. Thus, the best estimate for the perimeter of the heptagon is 63 cm.

27. C: The area of a parallelogram can be calculated using the formula $A = bh$. The length of the base of the parallelogram is 18 cm, and the height is 8 cm. Thus, the area is equal to 18 × 8 cm^2, or 144 cm^2.

28. D: The given triangle has a base equal to 4 units and a height equal to 6 units. Thus, the area of the triangle is equal to $\frac{1}{2}(4)(6)$ square units, or 12 square units.

29. The correct answer is **576.** The box is a rectangular prism, and the amount of available space inside the box is synonymous with the volume of the box. The volume of a rectangular prism is calculated by finding the product of the length, width, and height. Thus, the volume of the box is equal to 12 in × 8 in × 6 in, or 576 cubic inches.

30. D: The distance around the pothole indicates the circumference of the pothole. The circumference of a circle can be determined by using the formula $C = \pi d$, where C represents the circumference and d represents the diameter. The diameter of the pothole is 18 inches (9 × 2). Substituting a diameter of 18 inches and 3.14 for the value of pi gives the following: $C = 3.14(18)$, or 56.52. Thus, the distance around the pothole is equal to 56.52 inches.

31. B: The area of a trapezoid can be found by using the formula $A = \frac{1}{2}(b_1 + b_2)h$, where b_1 and b_2 represent the lengths of the bases and h represents the height of the trapezoid. Substituting the given base lengths and height reveals the following: $A = \frac{1}{2}(7 + 10)5$, which equals 42.5. Thus, the area of the trapezoid is 42.5 cm^2.

32. D: The sample space of rolling each die is 6. Thus, the sample space of rolling two dice is equal to the product of the sample spaces. 6 × 6 = 36; therefore, the sample space is equal to 36.

33. C: Flipping a coin one time has a sample space equal to 2, i.e., T or H. Flipping a coin 2 times has a sample space equal to 4, i.e., TT, HH, TH, HT. Flipping a coin 3 times has a sample space of 8, i.e., TTT, HHH, THT, HTH, TTH, HHT, THH, HTT. Notice that 2 is equal to 2^1, 4 is equal to 2^2, and 8 is equal to 2^3. The sample space of flipping a coin 9 times is equal to 2^9, or 512.

34. B: The events are independent since the spin of a spinner does not have an effect on the outcome of the flip of a coin. The probability of two independent events can be found using the formula $P(A \text{ and } B) = P(A) \cdot P(B)$. The probability of landing on a number less than 5 is $\frac{4}{8}$ since there are 4 possible numbers less than 5 (1, 2, 3, and 4). The probability of getting tails is $\frac{1}{2}$. The probability of landing on a number less than 5 and getting tails is $\frac{4}{8} \cdot \frac{1}{2}$, which equals $\frac{4}{16}$, or $\frac{1}{4}$.

35. D: The events are independent since Ann replaces the first marble drawn. The probability of two independent events can be found using the formula $P(A \text{ and } B) = P(A) \cdot P(B)$. The probability of pulling out a yellow marble is $\frac{8}{18}$. The probability of pulling out a green marble after the yellow marble has been replaced is $\frac{1}{18}$. The probability that Ann pulls out a yellow marble and then a green marble is $\frac{8}{18} \cdot \frac{1}{18}$, which equals $\frac{8}{324}$, which reduces to $\frac{2}{81}$.

36. B: The line graph shows the largest number of flights made during a month as 79 with the smallest number of flights made during a month as 54. The range is equal to the difference between the largest number of flights and smallest number of flights, i.e., 79 – 54 = 25. Therefore, the range is equal to 25.

37. A: The percentages of each type of tree are as follows: Apple tree – 20%; Peach tree – 44%; Fig tree – 29%, and Pear tree – 7%. The circle graph for Choice A accurately represents these percentages.

38. A: The median can be determined using any of the given graphical representations. However, a box-and-whiskers plot actually includes a line drawn for the median, thus clearly indicating the value of the median.

39. The correct answer is **304.** The median number of minutes can be determined by listing the number of minutes in order from least to greatest and calculating the average of the two middle values. The number of minutes can be written in ascending order as 255, 289, 291, 302, 306, 322, 411, and 418. The two middle values are 302 and 306. The average of these values can be determined by writing $\frac{302+306}{2}$, which equals 304. Thus, the median number of minutes is 304.

40. B: The average number (or mean) of incoming freshmen can be calculated by summing the numbers of incoming freshmen and dividing by the total number of years (or 10). Thus, the mean can be calculated by evaluating $\frac{76{,}146}{10}$, which equals 7,614.6. Since a fraction of a person cannot occur, the mean can be rounded to 7,615 freshmen.

How to Overcome Test Anxiety

Just the thought of taking a test is enough to make most people a little nervous. A test is an important event that can have a long-term impact on your future, so it's important to take it seriously and it's natural to feel anxious about performing well. But just because anxiety is normal, that doesn't mean that it's helpful in test taking, or that you should simply accept it as part of your life. Anxiety can have a variety of effects. These effects can be mild, like making you feel slightly nervous, or severe, like blocking your ability to focus or remember even a simple detail.

If you experience test anxiety—whether severe or mild—it's important to know how to beat it. To discover this, first you need to understand what causes test anxiety.

Causes of Test Anxiety

While we often think of anxiety as an uncontrollable emotional state, it can actually be caused by simple, practical things. One of the most common causes of test anxiety is that a person does not feel adequately prepared for their test. This feeling can be the result of many different issues such as poor study habits or lack of organization, but the most common culprit is time management. Starting to study too late, failing to organize your study time to cover all of the material, or being distracted while you study will mean that you're not well prepared for the test. This may lead to cramming the night before, which will cause you to be physically and mentally exhausted for the test. Poor time management also contributes to feelings of stress, fear, and hopelessness as you realize you are not well prepared but don't know what to do about it.

Other times, test anxiety is not related to your preparation for the test but comes from unresolved fear. This may be a past failure on a test, or poor performance on tests in general. It may come from comparing yourself to others who seem to be performing better or from the stress of living up to expectations. Anxiety may be driven by fears of the future—how failure on this test would affect your educational and career goals. These fears are often completely irrational, but they can still negatively impact your test performance.

Elements of Test Anxiety

As mentioned earlier, test anxiety is considered to be an emotional state, but it has physical and mental components as well. Sometimes you may not even realize that you are suffering from test anxiety until you notice the physical symptoms. These can include trembling hands, rapid heartbeat, sweating, nausea, and tense muscles. Extreme anxiety may lead to fainting or vomiting. Obviously, any of these symptoms can have a negative impact on testing. It is important to recognize them as soon as they begin to occur so that you can address the problem before it damages your performance.

The mental components of test anxiety include trouble focusing and inability to remember learned information. During a test, your mind is on high alert, which can help you recall information and stay focused for an extended period of time. However, anxiety interferes with your mind's natural processes, causing you to blank out, even on the questions you know well. The strain of testing during anxiety makes it difficult to stay focused, especially on a test that may take several hours. Extreme anxiety can take a huge mental toll, making it difficult not only to recall test information but even to understand the test questions or pull your thoughts together.

Effects of Test Anxiety

Test anxiety is like a disease—if left untreated, it will get progressively worse. Anxiety leads to poor performance, and this reinforces the feelings of fear and failure, which in turn lead to poor performances on subsequent tests. It can grow from a mild nervousness to a crippling condition. If allowed to progress, test anxiety can have a big impact on your schooling, and consequently on your future.

Test anxiety can spread to other parts of your life. Anxiety on tests can become anxiety in any stressful situation, and blanking on a test can turn into panicking in a job situation. But fortunately, you don't have to let anxiety rule your testing and determine your grades. There are a number of relatively simple steps you can take to move past anxiety and function normally on a test and in the rest of life.

Physical Steps for Beating Test Anxiety

While test anxiety is a serious problem, the good news is that it can be overcome. It doesn't have to control your ability to think and remember information. While it may take time, you can begin taking steps today to beat anxiety.

Just as your first hint that you may be struggling with anxiety comes from the physical symptoms, the first step to treating it is also physical. Rest is crucial for having a clear, strong mind. If you are tired, it is much easier to give in to anxiety. But if you establish good sleep habits, your body and mind will be ready to perform optimally, without the strain of exhaustion. Additionally, sleeping well helps you to retain information better, so you're more likely to recall the answers when you see the test questions.

Getting good sleep means more than going to bed on time. It's important to allow your brain time to relax. Take study breaks from time to time so it doesn't get overworked, and don't study right before bed. Take time to rest your mind before trying to rest your body, or you may find it difficult to fall asleep.

Along with sleep, other aspects of physical health are important in preparing for a test. Good nutrition is vital for good brain function. Sugary foods and drinks may give a burst of energy but this burst is followed by a crash, both physically and emotionally. Instead, fuel your body with protein and vitamin-rich foods.

Also, drink plenty of water. Dehydration can lead to headaches and exhaustion, especially if your brain is already under stress from the rigors of the test. Particularly if your test is a long one, drink water during the breaks. And if possible, take an energy-boosting snack to eat between sections.

Along with sleep and diet, a third important part of physical health is exercise. Maintaining a steady workout schedule is helpful, but even taking 5-minute study breaks to walk can help get your blood pumping faster and clear your head. Exercise also releases endorphins, which contribute to a positive feeling and can help combat test anxiety.

When you nurture your physical health, you are also contributing to your mental health. If your body is healthy, your mind is much more likely to be healthy as well. So take time to rest, nourish your body with healthy food and water, and get moving as much as possible. Taking these physical steps will make you stronger and more able to take the mental steps necessary to overcome test anxiety.

Mental Steps for Beating Test Anxiety

Working on the mental side of test anxiety can be more challenging, but as with the physical side, there are clear steps you can take to overcome it. As mentioned earlier, test anxiety often stems from lack of preparation, so the obvious solution is to prepare for the test. Effective studying may be the most important weapon you have for beating test anxiety, but you can and should employ several other mental tools to combat fear.

First, boost your confidence by reminding yourself of past success—tests or projects that you aced. If you're putting as much effort into preparing for this test as you did for those, there's no reason you should expect to fail here. Work hard to prepare; then trust your preparation.

Second, surround yourself with encouraging people. It can be helpful to find a study group, but be sure that the people you're around will encourage a positive attitude. If you spend time with others who are anxious or cynical, this will only contribute to your own anxiety. Look for others who are motivated to study hard from a desire to succeed, not from a fear of failure.

Third, reward yourself. A test is physically and mentally tiring, even without anxiety, and it can be helpful to have something to look forward to. Plan an activity following the test, regardless of the outcome, such as going to a movie or getting ice cream.

When you are taking the test, if you find yourself beginning to feel anxious, remind yourself that you know the material. Visualize successfully completing the test. Then take a few deep, relaxing breaths and return to it. Work through the questions carefully but with confidence, knowing that you are capable of succeeding.

Developing a healthy mental approach to test taking will also aid in other areas of life. Test anxiety affects more than just the actual test—it can be damaging to your mental health and even contribute to depression. It's important to beat test anxiety before it becomes a problem for more than testing.

Study Strategy

Being prepared for the test is necessary to combat anxiety, but what does being prepared look like? You may study for hours on end and still not feel prepared. What you need is a strategy for test prep. The next few pages outline our recommended steps to help you plan out and conquer the challenge of preparation.

Step 1: Scope Out the Test

Learn everything you can about the format (multiple choice, essay, etc.) and what will be on the test. Gather any study materials, course outlines, or sample exams that may be available. Not only will this help you to prepare, but knowing what to expect can help to alleviate test anxiety.

Step 2: Map Out the Material

Look through the textbook or study guide and make note of how many chapters or sections it has. Then divide these over the time you have. For example, if a book has 15 chapters and you have five days to study, you need to cover three chapters each day. Even better, if you have the time, leave an extra day at the end for overall review after you have gone through the material in depth.

If time is limited, you may need to prioritize the material. Look through it and make note of which sections you think you already have a good grasp on, and which need review. While you are studying, skim quickly through the familiar sections and take more time on the challenging parts.

Write out your plan so you don't get lost as you go. Having a written plan also helps you feel more in control of the study, so anxiety is less likely to arise from feeling overwhelmed at the amount to cover.

Step 3: Gather Your Tools

Decide what study method works best for you. Do you prefer to highlight in the book as you study and then go back over the highlighted portions? Or do you type out notes of the important information? Or is it helpful to make flashcards that you can carry with you? Assemble the pens, index cards, highlighters, post-it notes, and any other materials you may need so you won't be distracted by getting up to find things while you study.

If you're having a hard time retaining the information or organizing your notes, experiment with different methods. For example, try color-coding by subject with colored pens, highlighters, or post-it notes. If you learn better by hearing, try recording yourself reading your notes so you can listen while in the car, working out, or simply sitting at your desk. Ask a friend to quiz you from your flashcards, or try teaching someone the material to solidify it in your mind.

Step 4: Create Your Environment

It's important to avoid distractions while you study. This includes both the obvious distractions like visitors and the subtle distractions like an uncomfortable chair (or a too-comfortable couch that makes you want to fall asleep). Set up the best study environment possible: good lighting and a comfortable work area. If background music helps you focus, you may want to turn it on, but otherwise keep the room quiet. If you are using a computer to take notes, be sure you don't have any other windows open, especially applications like social media, games, or anything else that could distract you. Silence your phone and turn off notifications. Be sure to keep water close by so you stay hydrated while you study (but avoid unhealthy drinks and snacks).

Also, take into account the best time of day to study. Are you freshest first thing in the morning? Try to set aside some time then to work through the material. Is your mind clearer in the afternoon or evening? Schedule your study session then. Another method is to study at the same time of day that you will take the test, so that your brain gets used to working on the material at that time and will be ready to focus at test time.

Step 5: Study!

Once you have done all the study preparation, it's time to settle into the actual studying. Sit down, take a few moments to settle your mind so you can focus, and begin to follow your study plan. Don't give in to distractions or let yourself procrastinate. This is your time to prepare so you'll be ready to fearlessly approach the test. Make the most of the time and stay focused.

Of course, you don't want to burn out. If you study too long you may find that you're not retaining the information very well. Take regular study breaks. For example, taking five minutes out of every hour to walk briskly, breathing deeply and swinging your arms, can help your mind stay fresh.

As you get to the end of each chapter or section, it's a good idea to do a quick review. Remind yourself of what you learned and work on any difficult parts. When you feel that you've mastered the material, move on to the next part. At the end of your study session, briefly skim through your notes again.

But while review is helpful, cramming last minute is NOT. If at all possible, work ahead so that you won't need to fit all your study into the last day. Cramming overloads your brain with more information than it can process and retain, and your tired mind may struggle to recall even

test. If your test isn't first thing in the morning, consider going for a walk or doing a light workout before the test to get your blood flowing.

Allow yourself enough time to get ready, and leave for the test with plenty of time to spare so you won't have the anxiety of scrambling to arrive in time. Another reason to be early is to select a good seat. It's helpful to sit away from doors and windows, which can be distracting. Find a good seat, get out your supplies, and settle your mind before the test begins.

When the test begins, start by going over the instructions carefully, even if you already know what to expect. Make sure you avoid any careless mistakes by following the directions.

Then begin working through the questions, pacing yourself as you've practiced. If you're not sure on an answer, don't spend too much time on it, and don't let it shake your confidence. Either skip it and come back later, or eliminate as many wrong answers as possible and guess among the remaining ones. Don't dwell on these questions as you continue—put them out of your mind and focus on what lies ahead.

Be sure to read all of the answer choices, even if you're sure the first one is the right answer. Sometimes you'll find a better one if you keep reading. But don't second-guess yourself if you do immediately know the answer. Your gut instinct is usually right. Don't let test anxiety rob you of the information you know.

If you have time at the end of the test (and if the test format allows), go back and review your answers. Be cautious about changing any, since your first instinct tends to be correct, but make sure you didn't misread any of the questions or accidentally mark the wrong answer choice. Look over any you skipped and make an educated guess.

At the end, leave the test feeling confident. You've done your best, so don't waste time worrying about your performance or wishing you could change anything. Instead, celebrate the successful completion of this test. And finally, use this test to learn how to deal with anxiety even better next time.

Review Video: Test Anxiety
Visit mometrix.com/academy and enter code: 100340

Important Qualification

Not all anxiety is created equal. If your test anxiety is causing major issues in your life beyond the classroom or testing center, or if you are experiencing troubling physical symptoms related to your anxiety, it may be a sign of a serious physiological or psychological condition. If this sounds like your situation, we strongly encourage you to seek professional help.

How to Overcome Your Fear of Math

Not again. You're sitting in math class, look down at your test, and immediately start to panic. Your stomach is in knots, your heart is racing, and you break out in a cold sweat. You're staring at the paper, but everything looks like it's written in a foreign language. Even though you studied, you're blanking out on how to begin solving these problems.

Does this sound familiar? If so, then you're not alone! You may be like millions of other people who experience math anxiety. Anxiety about performing well in math is a common experience for students of all ages. In this article, we'll discuss what math anxiety is, common misconceptions about learning math, and tips and strategies for overcoming math anxiety.

What Is Math Anxiety?

Psychologist Mark H. Ashcraft explains math anxiety as a feeling of tension, apprehension, or fear that interferes with math performance. Having math anxiety negatively impacts people's beliefs about themselves and what they can achieve. It hinders achievement within the math classroom and affects the successful application of mathematics in the real world.

Symptoms and Signs of Math Anxiety

To overcome math anxiety, you must recognize its symptoms. Becoming aware of the signs of math anxiety is the first step in addressing and resolving these fears.

Negative Self-Talk

If you have math anxiety, you've most likely said at least one of these statements to yourself:

- "I hate math."
- "I'm not good at math."
- "I'm not a math person."

The way we speak to ourselves and think about ourselves matters. Our thoughts become our words, our words become our actions, and our actions become our habits. Thinking negatively about math creates a self-fulfilling prophecy. In other words, if you take an idea as a fact, then it will come true because your behaviors will align to match it.

Avoidance

Some people who are fearful or anxious about math will tend to avoid it altogether. Avoidance can manifest in the following ways:

- Lack of engagement with math content
- Not completing homework and other assignments
- Not asking for help when needed
- Skipping class
- Avoiding math-related courses and activities

Avoidance is one of the most harmful impacts of math anxiety. If you steer clear of math at all costs, then you can't set yourself up for the success you deserve.

Lack of Motivation

Students with math anxiety may experience a lack of motivation. They may struggle to find the incentive to get engaged with what they view as a frightening subject. These students are often overwhelmed, making it difficult for them to complete or even start math assignments.

Procrastination

Another symptom of math anxiety is procrastination. Students may voluntarily delay or postpone their classwork and assignments, even if they know there will be a negative consequence for doing so. Additionally, they may choose to wait until the last minute to start projects and homework, even when they know they need more time to put forth their best effort.

Physiological Reactions

Many people with a fear of math experience physiological side effects. These may include an increase in heart rate, sweatiness, shakiness, nausea, and irregular breathing. These symptoms make it difficult to focus on the math content, causing the student even more stress and fear.

Strong Emotional Responses

Math anxiety also affects people on an emotional level. Responding to math content with strong emotions such as panic, anger, or despair can be a sign of math anxiety.

Low Test Scores and Performance

Low achievement can be both a symptom and a cause of math anxiety. When someone does not take the steps needed to perform well on tests and assessments, they are less likely to pass. The more they perform poorly, the more they accept this poor performance as a fact that can't be changed.

Feeling Alone

People who experience math anxiety feel like they are the only ones struggling, even if the math they are working on is challenging to many people. Feeling isolated in what they perceive as failure can trigger tension or nervousness.

Feeling of Permanency

Math anxiety can feel very permanent. You may assume that you are naturally bad at math and always will be. Viewing math as a natural ability rather than a skill that can be learned causes people to believe that nothing will help them improve. They take their current math abilities as fact and assume that they can't be changed. As a result, they give up, stop trying to improve, and avoid engaging with math altogether.

Lack of Confidence

People with low self-confidence in math tend to feel awkward and incompetent when asked to solve a math problem. They don't feel comfortable taking chances or risks when problem-solving because they second-guess themselves and assume they are incorrect. They don't trust in their ability to learn the content and solve problems correctly.

Panic

A general sense of unexplained panic is also a sign of math anxiety. You may feel a sudden sense of fear that triggers physical reactions, even when there is no apparent reason for such a response.

Causes Of Math Anxiety

Math anxiety can start at a young age and may have one or more underlying causes. Common causes of math anxiety include the following:

The Attitude of Parents or Guardians

Parents often put pressure on their children to perform well in school. Although their intentions are usually good, this pressure can lead to anxiety, especially if the student is struggling with a subject or class.

Perhaps your parents or others in your life hold negative predispositions about math based on their own experiences. For instance, if your mother once claimed she was not good at math, then you might have incorrectly interpreted this as a predisposed trait that was passed down to you.

Teacher Influence

Students often pick up on their teachers' attitudes about the content being taught. If a teacher is happy and excited about math, students are more likely to mirror these emotions. However, if a teacher lacks enthusiasm or genuine interest, then students are more inclined to disengage.

Teachers have a responsibility to cultivate a welcoming classroom culture that is accepting of mistakes. When teachers blame students for not understanding a concept, they create a hostile classroom environment where mistakes are not tolerated. This tension increases student stress and anxiety, creating conditions that are not conducive to inquiry and learning. Instead, when teachers normalize mistakes as a natural part of the problem-solving process, they give their students the freedom to explore and grapple with the math content. In such an environment, students feel comfortable taking chances because they are not afraid of being wrong.

Students need teachers that can help when they're having problems understanding difficult concepts. In doing so, educators may need to change how they teach the content. Since different people have unique learning styles, it's the job of the teacher to adapt to the needs of each student. Additionally, teachers should encourage students to explore alternate problem-solving strategies, even if it's not the preferred method of the educator.

Fear of Being Wrong

Embarrassing situations can be traumatic, especially for young children and adolescents. These experiences can stay with people through their adult lives. Those with math anxiety may experience a fear of being wrong, especially in front of a group of peers. This fear can be paralyzing, interfering with the student's concentration and ability to focus on the problem at hand.

Timed Assessments

Timed assessments can help improve math fluency, but they often create unnecessary pressure for students to complete an unrealistic number of problems within a specified timeframe. Many studies have shown that timed assessments often result in increased levels of anxiety, reducing a student's overall competence and ability to problem-solve.

Debunking Math Myths

There are lots of myths about math that are related to the causes and development of math-related anxiety. Although these myths have been proven to be false, many people take them as fact. Let's go over a few of the most common myths about learning math.

Myth: Men Are Better at Math Than Women

Math has a reputation for being a male-dominant subject, but this doesn't mean that men are inherently better at math than women. Many famous mathematical discoveries have been made by women. Katherine Johnson, Dame Mary Lucy Cartwright, and Marjorie Lee Brown are just a few of the many famous women mathematicians. Expecting to be good or bad at math because of your gender sets you up for stress and confusion. Math is a skill that can be learned, just like cooking or riding a bike.

Myth: There Is Only One Good Way to Solve Math Problems

There are many ways to get the correct answer when it comes to math. No two people have the same brain, so everyone takes a slightly different approach to problem-solving. Moreover, there isn't one way of problem-solving that's superior to another. Your way of working through a problem might differ from someone else's, and that is okay. Math can be a highly individualized process, so the best method for you should be the one that makes you feel the most comfortable and makes the most sense to you.

Myth: Math Requires a Good Memory

For many years, mathematics was taught through memorization. However, learning in such a way hinders the development of critical thinking and conceptual understanding. These skill sets are much more valuable than basic memorization. For instance, you might be great at memorizing mathematical formulas, but if you don't understand what they mean, then you can't apply them to different scenarios in the real world. When a student is working from memory, they are limited in the strategies available to them to problem-solve. In other words, they assume there is only one correct way to do the math, which is the method they memorized. Having a variety of problem-solving options can help students figure out which method works best for them. Additionally, it provides students with a better understanding of how and why certain mathematical strategies work. While memorization can be helpful in some instances, it is not an absolute requirement for mathematicians.

Myth: Math Is Not Creative

Math requires imagination and intuition. Contrary to popular belief, it is a highly creative field. Mathematical creativity can help in developing new ways to think about and solve problems. Many people incorrectly assume that all things are either creative or analytical. However, this black-and-white view is limiting because the field of mathematics involves both creativity and logic.

Myth: Math Isn't Supposed To Be Fun

Whoever told you that math isn't supposed to be fun is a liar. There are tons of math-based activities and games that foster friendly competition and engagement. Math is often best learned through play, and lots of mobile apps and computer games exemplify this.

Additionally, math can be an exceptionally collaborative and social experience. Studying or working through problems with a friend often makes the process a lot more fun. The excitement and satisfaction of solving a difficult problem with others is quite rewarding. Math can be fun if you look for ways to make it more collaborative and enjoyable.

Myth: Not Everyone Is Capable of Learning Math

There's no such thing as a "math person." Although many people think that you're either good at math or you're not, this is simply not true. Everyone is capable of learning and applying mathematics. However, not everyone learns the same way. Since each person has a different learning style, the trick is to find the strategies and learning tools that work best for you. Some people learn best through hands-on experiences, and others find success through the use of visual aids. Others are auditory learners and learn best by hearing and listening. When people are overwhelmed or feel that math is too hard, it's often because they haven't found the learning strategy that works best for them.

Myth: Good Mathematicians Work Quickly and Never Make Mistakes

There is no prize for finishing first in math. It's not a race, and speed isn't a measure of your ability. Good mathematicians take their time to ensure their work is accurate. As you gain more experience and practice, you will naturally become faster and more confident.

Additionally, everyone makes mistakes, including good mathematicians. Mistakes are a normal part of the problem-solving process, and they're not a bad thing. The important thing is that we take the time to learn from our mistakes, understand where our misconceptions are, and move forward.

Myth: You Don't Need Math in the Real World

Our day-to-day lives are so infused with mathematical concepts that we often don't even realize when we're using math in the real world. In fact, most people tend to underestimate how much we do math in our everyday lives. It's involved in an enormous variety of daily activities such as shopping, baking, finances, and gardening, as well as in many careers, including architecture, nursing, design, and sales.

Tips and Strategies for Overcoming Math Anxiety

If your anxiety is getting in the way of your level of mathematical engagement, then there are lots of steps you can take. Check out the strategies below to start building confidence in math today.

Focus on Understanding, Not Memorization

Don't drive yourself crazy trying to memorize every single formula or mathematical process. Instead, shift your attention to understanding concepts. Those who prioritize memorization over conceptual understanding tend to have lower achievement levels in math. Students who memorize may be able to complete some math, but they don't understand the process well enough to apply it to different situations. Memorization comes with time and practice, but it won't help alleviate math anxiety. On the other hand, conceptual understanding will give you the building blocks of knowledge you need to build up your confidence.

Replace Negative Self-Talk with Positive Self-Talk

Start to notice how you think about yourself. Whenever you catch yourself thinking something negative, try replacing that thought with a positive affirmation. Instead of continuing the negative thought, pause to reframe the situation. For ideas on how to get started, take a look at the table below:

Instead of thinking	Try thinking
"I can't do this math." "I'm not a math person."	"I'm up for the challenge, and I'm training my brain in math."
"This problem is too hard."	"This problem is hard, so this might take some time and effort. I know I can do this."
"I give up."	"What strategies can help me solve this problem?"
"I made a mistake, so I'm not good at this."	"Everyone makes mistakes. Mistakes help me to grow and understand."
"I'll never be smart enough."	"I can figure this out, and I am smart enough."

Practice Mindfulness

Practicing mindfulness and focusing on your breathing can help alleviate some of the physical symptoms of math anxiety. By taking deep breaths, you can remind your nervous system that you are not in immediate danger. Doing so will reduce your heart rate and help with any irregular breathing or shakiness. Taking the edge off of the physiological effects of anxiety will clear your mind, allowing your brain to focus its energy on problem-solving.

Do Some Math Every Day

Think about learning math as if you were learning a foreign language. If you don't use it, you lose it. If you don't practice your math skills regularly, you'll have a harder time achieving comprehension and fluency. Set some amount of time aside each day, even if it's just for a few minutes, to practice. It might take some discipline to build a habit around this, but doing so will help increase your mathematical self-assurance.

Use All of Your Resources

Everyone has a different learning style, and there are plenty of resources out there to support all learners. When you get stuck on a math problem, think about the tools you have access to, and use them when applicable. Such resources may include flashcards, graphic organizers, study guides, interactive notebooks, and peer study groups. All of these are great tools to accommodate your individual learning style. Finding the tools and resources that work for your learning style will give you the confidence you need to succeed.

Realize That You Aren't Alone

Remind yourself that lots of other people struggle with math anxiety, including teachers, nurses, and even successful mathematicians. You aren't the only one who panics when faced with a new or challenging problem. It's probably much more common than you think. Realizing that you aren't alone in your experience can help put some distance between yourself and the emotions you feel about math. It also helps to normalize the anxiety and shift your perspective.

Thank You

We at Mometrix would like to extend our heartfelt thanks to you, our friend and patron, for allowing us to play a part in your journey. It is a privilege to serve people from all walks of life who are unified in their commitment to building the best future they can for themselves.

The preparation you devote to these important testing milestones may be the most valuable educational opportunity you have for making a real difference in your life. We encourage you to put your heart into it—that feeling of succeeding, overcoming, and yes, conquering will be well worth the hours you've invested.

We want to hear your story, your struggles and your successes, and if you see any opportunities for us to improve our materials so we can help others even more effectively in the future, please share that with us as well. **The team at Mometrix would be absolutely thrilled to hear from you!** So please, send us an email (support@mometrix.com) and let's stay in touch.

If you'd like some additional help, check out these other resources we offer for your exam: http://MometrixFlashcards.com/CCSS

Online Resources

Due to our efforts to try to keep this book to a manageable length, we've created a link that will give you access to all of your online resources:

mometrix.com/resources719/ccg7math

It's Your Moment, Let's Celebrate It!

Share your story @mometrixtestpreparation

Thomas M. Jahns, University of Wisconsin, Madison,
Trevor O. Jones (NAE), Biomec, Inc.,
Fritz R. Kalhammer, Electric Power Research Institute (retired),
Robert J. Nowak, Consultant,
William F. Powers (NAE), Ford Motor Company (retired),
Richard J. Robbins (NAE), The Robbins Group LLC, and
Daniel Sperling, University of California, Davis.

Although the reviewers listed above have provided many constructive comments and suggestions, they were not asked to endorse the conclusions or recommendations, nor did they see the final draft of the report before its release. The review of this report was overseen by William H. Press (NAS) and Maxine L. Savitz (NAE). Appointed by the National Research Council, they were responsible for making certain that an independent examination of this report was carried out in accordance with institutional procedures and that all review comments were carefully considered. Responsibility for the final content of this report rests entirely with the authoring committee and the institution.

tion and control technologies are needed, and the cooperation of energy companies in research programs aimed at these technologies will increase the likelihood of finding solutions using hydrocarbon-based or alternative liquid fuels. Hydrogen might also become a fuel for the ICE if a viable system for its production, distribution, and storage for transportation vehicles is developed and implemented.

Fuel Cells and Hydrogen Storage

Conclusion. The Partnership has an extremely ambitious goal: to develop both vehicle and infrastructure technology that would make it possible for automotive companies to decide in 2015 whether or not to build commercially viable fuel-cell-powered vehicles. The development of commercially viable fuel cells and onboard hydrogen storage is, without question, the most difficult vehicular aspect of this program. Multiple challenges are being addressed: performance, durability, efficiency, and cost, and they are being worked on at all levels: basic technology, the individual components, stacks, and systems. For fuel cells, durability and cost are the most difficult goals, and for hydrogen storage, the most difficult are size, weight, and cost. In most instances, solutions depend on yet-to-be-conceived or -proven component and manufacturing technology rather than incremental improvement. While this makes outcomes difficult to predict, the committee agrees with the strategy and research directions that DOE is taking to address both the fuel cell and hydrogen storage areas; however, some areas need greater effort.

Recommendation. DOE should expand activity and raise priorities on membrane R&D, new catalyst systems, and electrode design (with the BES program). In particular, the Partnership should focus the national laboratories and other appropriate scientific centers on fundamental failure mechanisms, including a better understanding of the chemistry, physics, and materials involved.

Recommendation. In view of the risk posed to the entire hydrogen program by the currently unmet need for a viable hydrogen storage system, the hydrogen storage technical team and the FreedomCAR and Fuel Partnership leadership team should report annually to all program participants, DOE, and Congress on the state of hydrogen storage technology worldwide relative to the goals and targets of the program.

Electrochemical Energy Storage for Electric Vehicles

Conclusion. Since using hydrogen as a transportation fuel would necessitate several significant breakthroughs, other alternatives to achieve the program goals should be explored and additional research supported if such alternatives show

ratories, and private companies. Especially in the case of development activities, projects are often cost shared between the private sector and the federal government (see Chapter 5 for further discussion).

The Partnership plays an important role in the planning, pursuit, and assessment of high-risk R&D for many of the needed vehicle and fuel technologies. Federal funds enable much of this work to move forward. The Partnership also serves as a communication mechanism for the interested players, including government, the private sector, the national laboratories, universities, the public, and others.

GOALS AND TARGETS

The long-term goal of the FreedomCar and Fuel Partnership is to enable the transition to a transportation system "that uses sustainable energy resources and produces minimal criteria or net carbon emissions on a life cycle or well-to-wheel basis" (DOE, 2004b). Starting to reduce the nation's dependence on imported petroleum is central to this goal. The current plan envisions a pathway initially involving more fuel-efficient internal combustion engines, followed by increasing use of hybrid electric vehicles and, ultimately, transition to an infrastructure for supplying hydrogen fuel to fuel-cell-powered vehicles (DOE, 2004b).

To address the technical challenges associated with this envisioned pathway, the FreedomCar and Fuel Partnership has established specific, quantitative 2010 and 2015 technology and cost goals in eight areas:

- ICEs (both petroleum- and hydrogen-fueled),
- Fuel cell power systems,
- Fuel cells,
- Hydrogen storage systems,
- Energy storage systems for hybrid vehicles,
- Electric propulsion systems,
- Materials for lightweight vehicles, and
- Hydrogen production and delivery systems.

These goals and the research related to their attainment will be discussed later in this report. Technical teams, as noted in the next section, "Organization of the Partnership," have also been formed to deal with specific technical areas and other crosscutting needs in the program.

ORGANIZATION OF THE PARTNERSHIP

The FreedomCAR and Fuel Partnership consists of a number of oversight groups and technical teams that have participants from government and industry (see Figure 1-1). The Executive Steering Group, which is responsible for the governance of the Partnership, comprises the DOE assistant secretary for energy